# Gulf Studies

Volume 2

**Series Editor**

Md Mizanur Rahman, Gulf Studies Center, College of Arts and Sciences, Qatar University, Doha, Qatar

This series is dedicated to serving growing academic interests in the dynamic, complex and strategically important Gulf region, offering a publication platform to scholars in the region and globally. The series takes an interdisciplinary approach to documenting the changes taking place in the Gulf societies, and examines the evolving relationship between the Gulf and other regions. The series advances non-Western perspectives in studying the Gulf societies, and their interactions with the rest of the world. The series includes work being done on the member states of the Gulf Cooperation Council (GCC), namely: Saudi Arabia, Oman, the United Arab Emirates, Qatar, Bahrain and Kuwait, in addition to Iran, Iraq, and Yemen.

The series accepts proposals for monographs, edited volumes and handbooks that provide an understanding of the Gulf societies' contemporary social, economic, and political landscapes. The series covers a wide range of topics within four broad themes, as follows:

**Social and Cultural Issues in the Gulf:**

Demography, migration, citizenship, gender, culture and identity, urbanization, education, new media, unemployment, youth, family, women empowerment, leadership, aging, human rights, sports, Islamic ethics, and Islam and society

**Politics and Security in the Gulf:**

Gulf defence and security, the GCC integration, democracy and political reform, Gulf politics and political systems, the GCC's foreign policy

**Energy and Economics in the Gulf:**

The political economy of the Gulf, economic diversification, international investments, international trade, knowledge economy, energy security and geopolitics of energy, renewable energy development, environmental regulations, infrastructure and urban planning, Islamic finance

**The GCC and the World:**

The GCC and Iran, the GCC and Africa, the GCC and South Asia, the GCC and East Asia, the GCC and Southeast Asia, and the GCC and Western Powers

More information about this series at https://www.springer.com/series/16417

Arno Tausch

# The Future of the Gulf Region

## Value Change and Global Cycles

Springer

Arno Tausch
Department of Political Science
University of Innsbruck
Innsbruck, Austria

ISSN 2662-4494 ISSN 2662-4508 (electronic)
Gulf Studies
ISBN 978-3-030-78298-6 ISBN 978-3-030-78299-3 (eBook)
https://doi.org/10.1007/978-3-030-78299-3



This Springer imprint is published by the registered company Springer Nature Switzerland AG
The registered company address is: Gewerbestrasse 11, 6330 Cham, Switzerland

*For my beloved wife Krystyna*

# Preface

It was with great pleasure that the present author received an invitation to submit a work for the new Springer series *Gulf Studies*, edited by Prof. Mizanur Rahman from Qatar University.

As the research project evolved, the countries of the Gulf received an increased international attention by the peace accords between the State of Israel, and Bahrain and the United Arab Emirates, with other Arab nations following. These *Abraham Accords* could be a path for the world: the respectful coexistence and working together of the major world civilizations. As an author who contributed for many years to debates about liberal and democratic Muslim societies, and *Peace, Salam, Shalom* between the major world religions, and who also actively published in the State of Israel for major think tanks of that country, this is a very welcome development, adding yet another motive to proceed with the present book.

At this junction of world history and amidst the unprecedented current economic and social crisis in the wake of the COVID-19 pandemic (World Bank, 2021), the Gulf countries could in future play a major role in the institutions of global governance like the United Nations, the G-20 and, hopefully also one day, the OECD, to stabilize and safeguard the "*Constitution of Liberty*" (Hayek) on a global scale and to work for the tolerance of the global civilizations.

Especially to my audiences in the Gulf another note might be permitted right at the start: this book was written in the tradition of political science as a rigorous quantitative discipline, committed to the idea of testing theories with empirical data and advanced statistical methodologies, initiated by Karl Wolfgang Deutsch (1912–1992) (Taylor & Russett, 2020). Such a political science tradition is especially fruitful for the future of the expanding discipline in the Gulf region: the testing of theories and hypotheses with highly developed statistical methodologies and data, independent from one's own ideological, religious, or whatsoever background. It is all to logical that the present author, given his earlier research (Solomon & Tausch, 2020, 2021; Tausch & Moaddel, 2009; Tausch, Heshmati, & Karoui 2014; Tausch, & Heshmati, 2013; Tausch, & Prager, 1993), again follows these traditions here. In presenting this book, the author also gratefully acknowledges the political science perspective of the European scholars Volker Bornschier, Hartmut Elsenhans, Wilfried Röhrich, and Dieter Senghaas (Bornschier, 1996, 2005, 2016; Bornschier & Chase-Dunn,

1985; Elsenhans, 1983, 1991, 1996, 2011, 2014, 2016, 2021; Elsenhans & Babones, 2020; Röhrich, 1978, 1978, 2004, 2015; Senghaas, 1985, 1988, 2013). For all of them, political science is a discipline informed by theory and empirical data, and they share with Deutsch, 1968 also the perspective of unequal relationships in world society between the centers and the peripheries. With all of them, the author could develop a fruitful exchange of ideas over decades at crucial junctions of his academic life. Their work and support were a continuing inspiration for the present author ever since the days when he defended his doctoral thesis at Salzburg University, 1976 and his "*habilitation*" thesis at Innsbruck University, 1988.

Appearing in a book series, edited at Qatar University, this text is also intended as an input for the political debate in Qatar itself, whose global universities, media and think tanks now play an ever-increasing and very considerable role across the world.[1] This book is published with the hope that the quantitative perspective on the analysis of economic and political cycles and value change, based on freely available international data, will contribute to further analyses by colleagues in the Gulf countries in these research directions.

Besides being linked to the *academia* since the 1970s, the present author also served in the decision-making bureaucracy, dealing with European and international social policy affairs (since 1992, retired 2016), and as such attempts to contribute a work which is not only useful to academic publics, but also for the international decision-makers, the bureaucracies, the analysts in the intelligence agencies, the think tanks, the ministries, the government agencies, the spin doctors of major political parties, the Emirs, the Monarchs and their staff, the armed forces, the trade unions and the NGOs, internationally operating law firms, the international media, etc.

To increase the readability of this study, we present many details of our statistical results intended for the specialists in an openly available electronic appendix, so as to free the text from an overload with statistics, and yet making these data publicly available.[2]

Keeping with the traditions of decision-making bureaucracies in international affairs, I also decided to present concise country information on the Gulf region at the end of this book. These information are based on data from institutions like

- Center for Strategic and International Studies, Washington, D.C.
- European Commission, European Union External Action
- Federal Foreign Office, Federal Republic of Germany
- Freedom House
- IMF World Economic Outlook Database
- Reuters Factbox
- SCIMAGO SIR
- UNDP Global Human Development Indicators
- United States Central Intelligence Agency, The World Factbook

[1] All downloads of this book as of April 21, 2021.

[2] https://www.academia.edu/45321329/Factors_affecting_the_Gulfs_future_Global_cycles_and_value_change_Electronic_Appendix.

- United States Central Intelligence Agency, World Leaders and Cabinet Members of Foreign Governments
- United States Department of State, Country Reports on Human Rights Practices
- World Bank, MENA Economic Update.

With these country files, which also include a summary of our own statistical results, achieved in this study, we hope to describe the situation, the problems, and the evolution for each Gulf country in a concise way.

To achieve a greater clarity of the argument, we also drew a large number of social scientific maps, nowadays called "choropleth maps". For reasons of simplicity, "good" and politically desirable results are always presented in blue color, and "bad" and undesirable results are always presented in red color. For example, a high Human Development Index according to the UNDP Human Development Report will be presented in blue color, and a low Human Development Index will be presented in red color.

In presenting this study, I would like to thank my beloved wife Krystyna for all the support and inspiration which I received from her, ever since we first met in the summer 1977. May this book and my expression of gratitude in love and affection be a source of joy for her and my family.

Innsbruck, Austria — Arno Tausch

## References


Bornschier, V. (1996). *Western society in transition.* New Brunswick, N.J.: Transaction.

Bornschier, V. (2005). *Culture and politics in economic development.* London: Routledge (Routledge frontiers of political economy, 67).

Bornschier, V. (2016). Hegemonic decline, West European unification, and the future structure of the core. *Journal of World-Systems Research, 1*, 69–96. https://doi.org/10.5195/jwsr.1995.42

Bornschier, V. & Chase-Dunn, C. K. (1985). *Transnational corporations and underdevelopment.* New York: Praeger.

Deutsch, K. W. (1968). *The analysis of international relations* (Vol. 12). Englewood Cliffs, NJ: Prentice-Hall.

Elsenhans, H. (1983). *Rising mass incomes as a condition of capitalist growth: Implications for the world economy.* International Organization, 1–39.

Elsenhans, H. (1991). *Development and underdevelopment: The history, economics, and politics of north-south relations.* New Delhi: Sage.

Elsenhans, H. (1996). *State, class, and development.* New Delhi: Radiant.

Elsenhans, H. (2011). *The rise and demise of the capitalist world system.* Leipzig: Leipziger Universitätsverlag.

Elsenhans, H. (2014). *Saving capitalism from the capitalists: World capitalism and global history.* Los Angeles: Sage.

Elsenhans, H. (2016). *Hartmut Elsenhans and a critique of capitalism: Conversations on theory and policy implications.* N. Wilcock & C. Scholz. Houndmills (Eds.), Basingstoke, Hampshire: Palgrave Macmillan. https://doi.org/10.1007/978-1-137-56464-1.

Elsenhans, H. (2016). Overcoming Political Isolation: The Relevance of Keynesian Fundamentals for Moral Economy Approaches in an Increasingly Multipolar International System. 13th Post-Keynesian Conference, University of Missouri, 15 to 18 September 2016, Panel: "Perspectives on Keynesian Economics", 10:30 AM on Thursday, Sept. 15th, 2016. Available at: https://hartmutelsenhans.net/paper/.

Elsenhans, H. (2021). *Development, capitalism, and rent: the political economy of Hartmut Elsenhans*. H. Warnecke-Berger (Ed.). Cham, Switzerland: Palgrave Macmillan. https://doi.org/10.1007/978-3-030-62605-1.

Elsenhans, H. & Babones, S. (2020). *BRICS or bust?* Stanford University Press.

Hayek, F. A. (1960). *The constitution of liberty*. Chicago: University of Chicago Press.

Röhrich W. (2004). *Die Macht der Religionen: Glaubenskonflikte in der Weltpolitik*. Originalausgabe edn. München: Beck (Beck'sche Reihe, 1585).

Röhrich, W. (1978). *Politik als Wissenschaft*. Deutscher Taschenbuch Verlag, Munich.

Röhrich, W. (1978). *Politik und Ökonomie der Weltgesellschaft*. Das Internationale System. (Rowohlts Deutsche Enzyklopädie) (Rowohlt Taschenbuch Verlag GmbH) Reinbek bei Hamburg.

Röhrich, W. (2015). *Die Politisierung des Islam: Islamismus und Dschihadismus*. Springer-Verlag.

Senghaas, D. (1985). *The European experience: a historical critique of development theory*. Leamington Spa, Warwickshire: Berg.

Senghaas, D. (1988). *Konfliktformationen im internationalen System*. 1. Aufl edn. Frankfurt am Main: Suhrkamp (Edition Suhrkamp, 1509 = n.F., Bd. 509).

Senghaas, D. (2013). Dieter Senghaas: pioneer of peace and development research. Berlin: Springer (Springerbriefs on Pioneers in Science and Practice, 6). https://doi.org/10.1007/978-3-642-34114-4.

Solomon, H. & Tausch, A. (2020). *Islamism, crisis and democratization: Implications of the World Values Survey for the Muslim world*. (Perspectives on development in the Middle East and North Africa (Mena region)). Cham: Springer.

Solomon, H. & Tausch, A. (2021). *Arab MENA countries: Vulnerabilities and constraints against democracy on the eve of the global COVID-19 crisis*. (Perspectives on development in the Middle East and North Africa (Mena region)). Cham: Springer.

Tausch A. & Moaddel, M. (2009). *What 1.3 billion Muslims really think. An answer to a recent Gallup study, based on the 'World Values Survey'*. Hauppauge, N. Y.: Nova Science Publishers.

Tausch A., Heshmati A. & Karoui H. (2014). *The political algebra of global value change. General models and implications for the Muslim world*. Hauppauge, N. Y.: Nova Science Publishers.

Tausch, A. (2014). The new global antisemitism: Implications from the recent ADL-100 data. *Middle East Review of International Affairs*, *18*(3) (Fall 2014).

Tausch, A. (2015). The fertile grounds for ISIL terrorism. Telos. *Critical Theory of the Contemporary*, *171*, 54–75 (Summer 2015).

Tausch, A. (2016). Islamism and antisemitism. Preliminary evidence on their relationship from cross-national opinion data. *Social Evolution & History*, *15*(2), 50–99 (Uchitel Publishing House, Moscow) (September 2016).

Tausch, A. (2019). Migration from the Muslim world to the west: Its most recent trends and effects. *Jewish Political Studies Review*, *30*(1–2), 65–225. Available at http://jcpa.org/article/migration-from-the-muslim-world-to-the-west-its-most-recent-trends-and-effects/ (with data definitions and sources. Free data download available from https://www.academia.edu/37568941/Migration_from_the_Muslim_World_to_the_West_Its_Most_Recent_Trends_and_Effects.

Tausch, A. & Heshmati, A. (2013). Globalization, the human condition, and sustainable development in the twenty-first century: Cross-national perspectives and European implications. London, New York and Delhi: Anthem Press.

Tausch, A. & Prager, F. (1993). Towards a socio-liberal theory of world development. Basingstoke and New York: Palgrave Macmillan/St. Martins Press.

Taylor, C. L. & Russett, B. M. (Eds.) (2020). Karl W. Deutsch: Pioneer in the theory of international relations. Springer.

World Bank (2021). Global economic prospects. A World Bank flagship report, January 2021. World Bank, Washington D.C. Available at https://www.worldbank.org/en/publication/global-economic-prospects.

# Praise for *The Future of the Gulf Region: Value Change and Global Cycles*

"As a researcher with a long background in cross-national research (from *The World Handbook of Political and Social Indicators* to my current work leading a global social research company), I was thrilled to read Arno Tausch's latest book. While the very timely focus is on the Gulf countries, the book provides a broad benefit to researchers through its use of multiple data sources, multi-variate statistical analysis, and comparisons of aggregate and survey-based data (such as the *World Values Survey, the Arab Barometer*, and *Pew's Global Attitudes Project*), all of which are available to the public. The author is well aware of how one's personal values and orientation can affect quantitative analysis, as the numbers seldom speak for themselves. The book has abundant (so ramp up your mouse's scroll speed) reference materials, links to data sources, and excellent data presentation with choropleth maps. This book should be on the 'shelf' of every student of the Middle East, development, and public opinion research."

—David A. Jodice, *Founder and CEO, D3 Systems, Inc.*

"Excellent work, well informed about the Arab World, with insightful references to quantitative data and opinion surveys. Few books may rival with this well-documented exploration of the Gulf's future. A must-read."

—Dr. Hichem Karoui, *Director at Large of "The Gulf Futures Center", London*

"This book represents an unprecedented research work relating to a subject that has not received the adequate response from the scientific community. Arno Tausch deserves our admiration for the formidable work done on this subject. No doubt that it is a great contribution to our understanding the role of the different actors referred to in this monumental research book."

—Dr. Jacques Neriah, *Late Prime Minister Yitzhak Rabin's Foreign Policy Adviser and Editor of Jewish Political Studies Review*

"This book is exceptional in scope and right up-to-the-minute in coverage. Dr. Arno Tausch has unusual command of literatures in international relations and comparative politics which he applies to this insightful study of the Gulf States. Tausch presents original empirical results and interprets relevant published results to makes his case that income and gender inequalities pose the greatest internal challenges to these countries today. And there is more: Tausch looks to the future of the Gulf states by situating their societies and economies within the region and the global system as a whole making use of long-term historical data and weighing effects of recent wide-ranging developments from the rise of China to the COVID pandemic to the Abraham Accords. Observers of the Middle East will certainly want this thought-provoking, thoroughly researched study on their shelf."

—Brian M. Pollins, *Associate Professor Emeritus, The Ohio State University*

"An outstanding and topical book by an astute scholar of the MENA region. Arno Tausch's tour de force of the Gulf region confronting economic crisis on the scale of the Great Depression, political upheaval, a seemingly unending pandemic and inequality and restrictive gender norms is a must-read for policy maker and academic. Meticulously researched, expertly contextualized, Tausch explores the future of the Gulf States and the difficult choices their leaders confront."

—Professor Hussein Solomon, *Academic Head of Department, Political Studies and Governance, University of the Free State, South Africa*

"Applying a highly sophisticated mixed-methods framework that includes both quantitative and qualitative approaches, Arno Tausch's brilliant assessment of current and future socio-political trends in the Gulf region is the most comprehensive and insightful study on the subject to date. Of particular interest is the book's careful examination of how the current COVID-19 pandemic will affect internal social dynamics in Gulf states as well as their relationship to a world that is dangerously wavering between globalist expansion and nationalist retrenchment."

—Manfred B. Steger, *Professor of Sociology, University of Hawai'i at Manoa and Global Professorial Fellow, Western Sydney University*

"An impressive review of the Arab Gulf and the Arab World by a leading specialist on quantitative world system theory. The book focuses on beliefs, attitudes, and values. It presents the Gulf as facing new external challenges and basic internal changes, some of them emanating from unexpected social groups, especially women. The book makes use of a highly innovative methodological framework. It effectively shows how, in this region, economic progress and exposure to the outside world have triggered cultural modernization, despite myriad obstacles. This seems to offer the chance of an enduring peace in the Middle East. This region was once a central pillar of Washington's global hegemony—and, to some extent, it still is. However, this book shows how the Gulf is increasingly finding its feet in a more diffuse international system—without losing its distinctive identity."

—Hartmut Elsenhans, *Professor (emeritus) of Political Science, University of Leipzig, FRG*

# Contents

# List of Figures

# List of Tables

# Chapter 1
# Introduction: Factors Affecting the Gulf's Future

**Abstract** How will global economic and political cycles and inequality, still existing restrictive gender norms (UNDP, United Nations Development Programme (UNDP) in Human Development Report. Oxford University Press, Oxford and New York (2019). Available at http://www.hdr.undp.org/), and value change affect the future of the region, which had combined proven reserves of 497 billion barrels of crude oil, which represents approximately 34% of the world's estimated proven crude reserves, and which is the EU's sixth-largest export market (€78.1 bn of exports in 2019, + 5% over 2018)? In this chapter, we present the starting points of our analysis: the global world political and economic environment which the Gulf will most probably face in the near future and the foreseeable emergent real vacuum in world affairs and global leadership; and the multivariate analysis of civil society in the countries of Gulf by international comparison. Here, in the introduction, we present an outline of the social scientific scenery in which this book is located, all characterized by the quest for evidence-driven debates about the realities of the Gulf region by international comparison. The chapter briefly introduces the decisive factors, discussed at length in this book with a wealth of new original, global time series analyses and opinion-survey-co-driven evidence. We highlight, among others, the necessity of the multivariate analyses of opinions on religion, gender, and the desire for democracy, based on World Values Survey data, underlining the general approach of this book in the framework of the recent United Nations Human Development Report, 2019 with its emphasis on inequality and still existing restrictive gender norms.

**Keywords** COVID-19 pandemic · Economic crises · Gulf countries · Global leadership · Opinion surveys · China · USA · Gender norms · Income concentration · Ronald F. Inglehart

## 1.1 What This Book Is All About

The central research question, which we attempt to answer in this book, is how global economic and political cycles, inequality and still existing restrictive gender norms (UNDP, 2019), and value change will affect the future of the region, which had combined proven reserves of 497 billion barrels of crude oil, which represents



A. Tausch, *The Future of the Gulf Region*, Gulf Studies 2,
https://doi.org/10.1007/978-3-030-78299-3_1

approximately 34% of the world's estimated proven crude reserves, and which is the EU's sixth-largest export market (€78.1 bn of exports in 2019, +5% over 2018)?

These days of the global health emergency due to COVID-19 and the profound crisis of democracy in America, which culminated in the 01/06 attack on Capitol Hill, and the global vacuum which follows from the demise of global American leadership also present an urgent necessity for social scientists, especially political scientists, to look anew at the countries of the Gulf region. The establishment of diplomatic relations between Bahrain, Morocco, Sudan, the United Arab Emirates, and the State of Israel is a world historic event which calls for a thorough re-thinking of established social scientific knowledge on the Gulf.

In terms of the chosen methodology of Gulf Studies, the answer is being given rather quickly: combining the approaches of "critical political economy" with its emphasis on dependency, inequality, and globalization (Tausch & Heshmati, 2013) with the analysis of value change in the global periphery and semi-periphery has been already prominently applied over the last decades in Middle East research, among others at Harvard University (see above all, Cammett, 1999, 2014, 2015, 2017, 2018, 2020).

The global international environment which the Gulf countries will face is also quickly described: the current COVID-19 pandemic will further shift the weight of world economic leadership away from the old and in many ways decaying West to the benefit of China and India (see Chaps. 3, 4, and 5 of this book); and the Gulf countries, maybe rather unexpectedly, will be forced to assume a much larger role in global leadership than they did ever before.

The current global economic and social crisis has been aptly described in a recent World Bank document presented in January 2021, where the World Bank predicts the world's future in the following way. This scenario is totally realistic, given the recent mutations of the COVID-19 virus:

> A more protracted pandemic than expected could lead to even larger income losses [...]. In a downside scenario of persistently higher caseloads and delayed vaccination, global output in 2025 would be about 8 percent below earlier expectations and there would be a cumulative loss equivalent to 54 percent of 2019 global output. Delays in vaccine deployment could disappoint financial markets and trigger a repricing of risks [i.e., inflation]. Amid record-high debt, higher borrowing costs could tip many firms into bankruptcy, weakening bank balance sheets, possibly to an extent that could trigger a financial crisis. In such a severe downside scenario, global output could contract by another 0.7 percent in 2021. Cumulative output losses over 2020–25 could amount to 68 percent of 2019 output globally. (World Bank, 2021)

One tool used in Chap. 3 is the comparison of the effects of earlier pandemics on the long cycles of economic development over the last millennia, using the fascinating long-term data provided by Maddison (2007) and his associates, ranging back two thousand years (Bolt et al., 2018; also analyzed by Barro, 2006, Barro & Ursúa, 2008, 2011; Barro et al., 2020). A vast and growing international literature attempts to estimate these short-and long-term effects of pandemic on the social and economic systems of the time, and tries to learn the lessons from these past epidemics (see, among others, Brodeur et al., 2020; Candela, & Geloso, 2020; Doerr & Hofmann,

2020; Furceri et al., 2020; Jordà et al., 2020; Kok, 2020; Morgan et. al., 2021; World Economic Forum, 2020). In its timeline of the most terrible pandemics in human history, the World Economic Forum (2020) highlighted, *inter alia,* the following pandemics and estimated global victim numbers along the Western calendar:

- Antonine Plague 165–180: 5 million,
- Japanese smallpox 735–737: 1 million,
- Plague of Justinian 541–542: 30 to 50 million,
- Black Death 1347–1351: 200 million,
- New World Smallpox: 1520-onwards: 56 million,
- Great Plague of London 1665: 100,000,
- Italian Plague 1629–1631: 1 million,
- Cholera Pandemics 1–6 1817–1923: 1 million+,
- Third Plague 1885: 12 million,
- Yellow Fever Late 1800s: 100,000–150,000,
- Russian Flu 1889–1890: 1 million,
- Spanish Flu 1918–1919: 40–50 million,
- Asian Flu 1957–1958: 1.1 million,
- Hong Kong Flu 1968–1970: 1 million,
- HIV/AIDS 1981–present: 25–35 million.

Inexorably, value change, brought about by decades of modernization and development, will affect the governance structures of the Gulf states, as new generations, which benefitted from a decent tertiary education at the Gulf's world class universities and from social globalization will enter into the arena of public affairs and will have a decisive voice in the future of their countries (see Chaps. 6 to 8 of this book). These changes will not be free of contradictions, and comparative social science can at least estimate the seismographic fault lines of these processes as they emerge from the empirical analysis of the opinion configurations of the populations of the region. Due to the lack of time series opinion data from the *World Values Survey* for the Gulf countries, we have to estimate value change by studying the effects of the age of the respondents on the opinion structures.

With the world now facing the starkest economic depression since the Great Depression of the 1930s in the wake of the current COVID-19 crisis (see Chaps. 3 and 4), we analyze as a starting point the global world political and economic environment that the Gulf will most probably face in the near future. The current depression, like the one which struck the world in the 1930s, can it lead to global war? What are the lessons we have to draw from the past long cycles of global politics and economics?

Recent research, incidentally also having been carried out at Harvard University (Allison, 2017), paid great attention to the mechanisms, how great power conflicts evolve when one dominant power in the international system declines while the other rises inexorably. Owing to the politically destructive forces at work in the current international system, a very substantial space of our book is thus devoted to analyzing this so-called *"Thucydides Trap,"* described in all details in Allison (2017); and in a host of other literature, surveyed in Chaps. 3 and 4. As we will show in Chap. 3 of our book, in ancient Greece, a terrible pandemic coincided with rising tensions between the main military powers of that time and geographical area. Thus, the Peloponnesian War (431–404 BC) has become the father of all Great Power Wars, featured so extensively in the present book, and the staff colleges of the armed forces around the world are well advised to study its manyfold lessons. The ancient historian Thucydides (460 BCE–400 BCE, see Chap. 3 for a detailed debate) already offered a detailed analysis of such processes, leading to a destructive Great Power War, and explained at length how this long, protracted, all-out-war war between the then hegemon (Athens) and its challenger (Sparta) broke out. As we will also show in Chap. 3 of our book, Chinese President Xi Jinping himself seems to be acutely aware of the importance of the ancient author Thucydides for the understanding the events of today and tomorrow.

The amount of attention in our book devoted to the *"Thucydides trap"* also received confirmation by an alarming recent news dispatch, which interestingly enough appeared in the Iranian News Agency *Mehr,* where we could read on March 9, 2021:

> China's top general Mehri Xu Qilian has called for increased military spending to prepare for a war with the United States, Daily Mail reported.[1] Qilian, who is second in command of the country's armed forces after President Xi, said China needs to ready itself for the 'Thucydides Trap'. The term refers to the inevitability of war when a new global power displaces an existing one. Xu said, according to the South China Morning Post,[2] 'In the face of the Thucydides Trap and border problems, the military must speed up increasing its capacity. "We must make breakthroughs in combat methods and ability, and lay a sound foundation for military modernization", he said. He believes China will soon overtake the US as an economic power, with its GDP more than 70 percent that of its rival, he said. Xu made the comments on Friday in a discussion at the annual gathering of China's National People's Congress. The Thucydides Trap is rarely openly discussed by high-ranking officials in China. President Xi said in 2015 while on a visit to Seattle: 'There is no such thing as the so-called Thucydides Trap.[3] 'But should major countries time and again make the mistake of strategic miscalculation, they might create such traps for themselves.' Xu's comments come amid concern in China about its relationship with the US under President Joe Biden. Biden has said Beijing is Washington's most serious competitor', and his administration has indicated it will broadly continue the tough approach taken by Donald Trump. Defense ministry spokesman Wu Qia said some of the 6.8 percent of the increase to the military budget will be spent on trying to catch up with the US to 'fight and win' on the modern battlefield. The rest of the money will be spent on training, weapons, and salaries for China's two million soldiers.[4]

A very central concept in this debate is the idea of long cycles of economics of around half a century duration, forever associated with the name of the Russian economist Nikolai Dmitriyevich Kondratieff (Russian: Никола́й Дм´итриевич

[1] https://www.dailymail.co.uk/news/article-9338759/Chinas-general-calls-increased-military-spending-Thucydides-trap-war-US.html.

[2] https://www.scmp.com/news/china/diplomacy/article/3085321/destined-conflict-xi-jinping-donald-trump-and-thucydides-trap.

[3] https://www.fmprc.gov.cn/mfa_eng/topics_665678/xjpdmgjxgsfwbcxlhgcl70znxlfh/t1305429.shtml.

[4] https://en.mehrnews.com/news/170874/China-top-cmdr-calls-for-military-readiness-for-war-with-US.

Кондра́тьев; 4 March 1892–17 September 1938), who became a victim of Stalin's purges, and whose ideas were taken up by Harvard economist Joseph Alois Schumpeter (February 8, 1883–January 8, 1950; Schumpeter & Elliott, 1983; see also Grinin et al., 2016; Modis, 2017).

Our book presents time series analyses of the long cycles of global politics and economics on a global level in the framework of these debates initiated by Kondratieff and Schumpeter and later taken up by the American political scientist Joshua Goldstein in his Ph.D. work at Yale University in the late 1980s. Prophetically, Goldstein (1985, 1988) highlighted the strong statistical relationships between global warfare (since 1495) and economic depressions (since 1740, see also Chap. 4 of the present book) in the global system. Our data series, especially updated for the purpose of the present book, now suggest that the current COVID-19 depression casts indeed a very long shadow over the world and brings along with it the **rising risk of global confrontations**, fitting in with these global cycle theories (see also the rare joint Russian-NATO study, Devezas, 2006, and the recent volume by Grinin et al., 2016). All this will substantially affect the Gulf's economies and societies in the near future and will be the **first factor to be analyzed here**.

With Chinese power now sweeping across the globe, with which visions of global leadership the political and economic elites of the countries of the Gulf will team up? The author analyses the emergent dangerous **vacuum of global leadership** reflected in current PEW opinion surveys (PEW Spring Survey, 2018) as the **second factor** shaping the Gulf's future. Do the visions offered by world political leaders such as Emmanuel Macron, Angela Merkel, Vladimir Putin, the President of the USA (in the PEW Spring Survey 2018 Donald Trump, now removed from of office by the US presidential election of November 3, 2020), or Xi Xinping command the sufficient support of a significant portion of humanity, and what are the multivariate relationships of these support rates? Or is there an emergent real vacuum in global leadership? The emerging vacuum in global leadership, diagnosed in our analysis on the basis of PEW Spring Survey, 2018 data, is the second factor under scrutiny here, and it will indeed substantially affect the long-term options for the Gulf leadership in the 21st Century. No global leadership–Xi, Trump, Putin, Macron, and Merkel—commanded the sufficient backing of more than 2/3 of the global population, and none seemed to have offered a credible guidance in the current crisis. In the respective chapter of our book (Chap. 5), we pinpoint the underlying trends and contradictions of what the people of our globe really think about these leading global actors and global constellations. For the countries of the Gulf, this global leadership vacuum, exacerbated by an acute value crisis and the loss of trust in the West, also analyzed in Chap. 5 of the present book with the latest data, 2017–2020 from the *World Values Survey* from 79 countries, implies to have second thoughts about "*playing the Chinacard,*" since only the free West guarantees the Gulf states a "*constitution of liberty*" (Hayek) with the sufficient respect for property values and family values. These values are, following Hayek, at the basis of all successful global religions, and of course they play, just as in Judaism and Christianity, the other two Abrahamic religions, also a pivotal role in Islam. This implies that the Gulf countries themselves must assume a positive and strategic role in such institutions as the United Nations, the IMF and the World Bank, the G-20, and hopefully also the OECD.

Our book then focuses on the **multivariate analysis of civil societyin the countries of Gulf.** This is the **third factor under consideration here**, and we devote the remaining portion of our book to the analysis of this factor (Chap. 6 through to Chap. 8). These analyses and value rankings apply global comparisons, as data about the values of the citizens of our planet and from the Arab region emerge from the representative opinion surveys of the *World Values Survey* and also from the *Arab Barometer*.

We look at the differential development patterns of global society and the Arab countries along rising human development and the advancement of world class tertiary education. The great American sociologist Ronald Inglehart and his associates (Inglehart, 2008, 2017; Inglehart & Baker, 2000; Inglehart & Norris, 2003; Inglehart & Welzel, 2010; Norris & Inglehart, 2015, 2016) believes that rising human security, human values change decidedly. Our book, developing approaches in this direction, already tested in Grinin et al. (2018), and Solomon and Tausch (2020a, 2020b, 2021), analyzes not only general value change and value decay, but also such factors as the population's trust in the institutions of the state. And while the civil societies of the Gulf countries increasingly think that—with all its deficiencies—democracies are the best form of government, and other values typical for countries with high development levels and human security increasingly take hold in the Gulf region as well, there are as yet no viable structures of democracy implemented in the countries of the region. The data of this book could help the decision-makers and civil societies of the region toward an orderly and smooth transition to such democratic, constitutional monarchies.

Comparative global value research in the tradition of Inglehart and Norris and associates (see Tausch & Moaddel, 2009; Tausch et al., 2014; Inglehart, 2008, 2017; Inglehart & Baker, 2000; Inglehart & Norris, 2003; Inglehart & Welzel, 2010; Norris & Inglehart, 2011, 2015) has maintained for a long time that manifestations of xenophobia and racism have a strong tendency to be related to the lack of human security, and that with rising human security, which can be measured among others by the UNDP's Human Development Index (UNDP, 2019), these manifestations will diminish. With a growing "thaw" and "détente" under way between the State of Israel and the countries of the Gulf region, the book also analyzes the drivers of the still existing and lamentably high Antisemitism in the region, based on a multivariate analysis of global and Gulf ADL-100 Index data on Antisemitism (Chap. 7). Developing analyses, already presented in Tausch (2019) and Solomon and Tausch (2020a, 2020b), the present analysis now specifically takes the all-important gender policy variables (again UNDP (2019) on still existing restrictive gender norms) and employment variables properly into consideration. These gender policy variables rapidly change in the Gulf region as well, and women in Gulf region play an ever-greater role in the countries' labor markets. *Ceteris paribus,* what is the effect of these gender policy variables on the global rates of Antisemitism and on Antisemitism in Gulf? According to our estimates, their effects on alleviating Antisemitism are considerable and have been hitherto largely neglected in comparative research on the subject. Yes, we optimistically expect that with changing gender policy variables and employment variables Antisemitism in the region will diminish. Also, there is some reason for hope that the Gulf will re-discover its own Jewish heritage.

In Chap. 8, the book offers several multivariate analyses of opinions in Saudi Arabia, the biggest country of the region, on religion, gender, and the desire for democracy, based on *World Values Survey* data, underlining the general and also comparative approach of our book. The debate should be considered as a contribution in the framework of the Saudi Vision 2030, which the present author considers to be a good step in the right direction of focusing on a development perspective of human development, social inclusion, and gender justice, inherent for many years in the analyses of the United Nations Human Development Reports.[5]

What is the exact relationship between opinions on polygamy, gender inequality, inequality, and trust in the political system of this key Gulf country? Our Chap. 8 is the first in-depth and incidentally also the largest multivariate opinion survey analysis of Saudi opinion ever undertaken in the literature, and it attempts comparisons of Saudi society with other countries in the region and internationally as well. Without question, the restrictive gender norms dimension will be decisive for the future of the Kingdom.

In our final Chapter (Chap. 9), we look at the dynamics of value change in the region by international comparison, taking value differences between the generations into account as a proxy for the non-existing time series data. The partially optimistic conclusions of our study testify to the underestimated and incipient maturity of the Gulf's civil society and strongly suggest that the Gulf's future is indeed rather with the free societies of the West and not with any only partially free or unfree society, be it China, Iran, Russia, or Turkey, currently and lamentably the Neo-Ottoman Empire in the making (see also Karoui, 2020, for a debate about the legacy of the Ottoman Empire in the Arab World of today), and proposes such steps as a much closer cooperation of the Gulf countries with the West. In this context, we especially stress the pivotal role of world class tertiary education and global academic exchange with the Gulf's universities and think tanks.

In our Appendix, we present condensed open data compiled from our own research and from international organizations, think tanks, and foreign ministries to provide our readers with concise and readable up-to-date country informations for the countries of the region.

## 1.2 Gulf Studies: General Perspectives on Current Research on the Subject

So, the Gulf Cooperation Council (GCC) countries had combined proven reserves of 497 billion barrels of crude oil. This represents approximately 34% of the world's estimated proven crude reserves. According to OPEC's 2018 data, Saudi Arabia has

[5] https://www.vision2030.gov.sa/en.

the highest share of crude oil reserves in the GCC. Its proven reserves estimate of 267 billion barrels of crude oil represents 54% of the GCC's total proven reserves, followed by Kuwait with 101.5 billion barrels of oil (around 20% of the GCC's total proven reserves).[6]

Many European observers, most notably Bianco (2017, 2018, 2020a, 2020b, 2021), Bianco and Stansfield (2018), seem to start from the normative question what Europe can do to expand its "*soft power*" in this region and in the world. Bianco (2020b) seems to summarize the perceptions of many European political elites when she says:

> Gulf Arab states are at the centre of a new balance of power in the Middle East and north Africa (MENA). With US dominance of the region declining amid the chaos of the past decade, members of the Gulf Cooperation Council (GCC) have expanded their international role like never before. They pursued rivalries and developed alliances across north Africa and the Levant. They used military power to try to determine the outcome of intractable conflicts in Libya and Yemen. And they reached across borders to stifle or support grass-roots Islamist movements, not least the Muslim Brotherhood and Shia resistance groups. Meanwhile, European countries – long used to working under a US umbrella in the region – have struggled to adapt to the emerging multipolar regional order, and have been slow to acknowledge the impact of the political networks and vast financial resources of states such as Qatar, Saudi Arabia, and the United Arab Emirates. For European governments that recognise the many ways in which the MENA region is key to their interests, the situation is unsustainable. (Bianco, 2020b)

and to continue:

> Instability in the region poses a serious threat to Europe's economic, energy, and physical security. This is clear from the dispute between the United States and Iran, which – centring on the Joint Comprehensive Plan of Action (JCPOA), a deal that European governments helped negotiate – could precipitate further military escalation between two major powers. The spread of conflict through countries in which some GCC actors have game-changing capabilities, such as Libya and Yemen, could force Europe to confront a rising terrorist threat and even a new refugee crisis. The erosion of maritime security in the Gulf and off the coast of Yemen threatens to disrupt vital European trade networks. In all these areas, Europe must find new ways to address the destructive competition for influence between GCC states, their allies, and their rivals that stretches across the region. (Bianco, 2020b)

Our approach is much humbler in nature. Much has been achieved in the discipline of Gulf studies in recent years, but a careful analysis of the literature will reveal that as yet, the central research question, which we attempt to answer in this book, i.e., how will inequality and the norms about it, still existing restrictive gender norms (UNDP, 2019), value change, and global economic and political cycles affect the future of the region, as yet has not been adequately dealt with. Most Western observers would agree with the leading Israeli specialist on the Gulf region, Yoel Guzansky, a researcher from the Institute of National Security Studies in Tel Aviv (Guzansky, 2020), in addition on the fact that the Gulf Cooperation Council (GCC) was established as a result of prolonged cooperative processes, mainly in trade and economics. The organization, in Guzansky's opinion (2020), reflected shared interests, the monarchial

[6] https://www.kapsarc.org/file-download.php?i=56797.

nature of the members' regimes, their economies, their religious affiliation as Sunni Muslims, and their common Arab origin, in contrast to revolutionary, Shiite, and non-Arab Iran. It was, following Guzansky (2020), an expression of the attempts to find an agreed formula for security in the Gulf, which began even before the British left the Gulf in 1971. The events in the late 1970s and 1980s accelerated the pace of cooperation and contributed to a change in its character.

> The countries believed that they were capable of putting aside the disputes that had hindered their relations, and of cooperating on the basis of common interests. A number of seminal events contributed to this, especially the revolution in Iran and the outbreak of the Iran-Iraq War, which led the Gulf states to recognize that they had to cooperate in order to confront the potential threats against them more effectively. Yet while steps were taken toward cooperation, progress was especially slow, due to the cautious and calculating nature of the Gulf rulers [...] (Guzansky, 2020)

Despite the obvious advantages of cooperation, due to the countries' military weakness and the threats directed at them (which were inversely proportionate to their size, location, and military capabilities), potential unification with Saudi Arabia, which has a geographic, economic, and military advantage over the smaller other GCC countries, complicated, according to Guzansky (2020), any such initiative.

Since 1981, the year of the foundation of the Gulf Cooperation Council, a truly vast literature has been published on the GCC. We cannot even pretend to adequately review this literature, and can only indicate the basic structures of Gulf integration and some of the major directions and areas of research on it for the readers who are not as yet familiar with the region. Important areas of research on the Gulf countries beyond the realm of Gulf Security Studies, deal with the following issues:

- **Gulfandmigration** (Kibria, 2011; Rahman et al., 2014; Ullah, 2010a, 2010b, 2012; Ullah & Haque, 2020; Ullah & Rahman, 2012).
- **Gulf and ecology** (Jalilvand & Westphal, 2018; Luomi, 2012).
- **Political economy approaches to the modernization process of the Gulf countries and their position in the world economy** (for a very brief debate which is necessary in the context of our focus on economic cycles, see Chap. 4 of the present volume, and Cooke, 2014; Davidson, 2013; Hanieh, 2011, 2013, 2015; Legrenzi & Momani, 2011).
- **State classes in the Arab/Gulf countries and rentier states** (Elsenhans, 1983, 1991, 1996, 2011, 2014; Elsenhans & Babones, 2020; Gray, 2011; Miniaoui, 2020).

In the following, we will first attempt to describe for our readers the general evolution of the Gulf Cooperation Council (GCC). We concentrate merely on a few salient studies, which acquired a certain role in the global social science journals, in the stacks of the global academic libraries and which also became textbooks of the world's leading universities and centers of academic learning.[7]

[7] See SCOPUS with its access of > 40,000 scientific sources from around the world at https://www.scopus.com/freelookup/form/author.uri; OCLC Classify counting the library presence of authors in OCLC Worldcat 15,637 member institutions in 107 countries at http://classify.oclc.org/classify2/,

### *1.2.1 The Gulf Countries—An Attempt at Regional Integration*

Where to start with? The Gulf Cooperation Council (GCC) has been dealt with, among others in Britannica (2021), Gresh (2015), Guzansky (2015), Hossain and Muhammad (2009), Jewish Virtual Library (2021), Jones and Guzansky (2020), Kamrava & Georgetown University (2012), Krieg (2019), Sadig (2019), Sick (2009), Steinberg (2014, 2020), Ulrichsen (2011), Zweiri et al. (2021) ).[8] Also, recent *"Gulf Studies"* publications brought out by the current publisher of this book can and should beconsulted (Cowan, 2018; Dwivedi, 2018; Fromm, 2018; Gremm, 2018; Karolak & Allam, 2020; Khatib, 2021; Krieg, 2019; Miniaoui, 2020; Mohamed Nour, 2013; Ramady, 2012; Sadig, 2019; Ullah & Haque, 2020; Zweiri et al., 2021).

So, the Cooperation Council for the Arab States of the Gulf (GCC) is a regional organization, with six members: The Kingdom of Bahrain, the State of Kuwait, the Sultanate of Oman, the State of Qatar, the Kingdom of Saudi Arabia, and the United Arab Emirates.[9] Set up, as we already said, in 1981, its objectives are to enhance coordination, integration, and inter-connection among its members. All GCC members are also members of the Arab League. Qatar, Saudi Arabia, Kuwait, and the United Arab Emirates are prominent members of OPEC.

Especially our readers in Europe should well remember in addition that the EU established bilateral relations with the GCC countries through the 1988 Cooperation Agreement,[10] intended to:

- strengthen stability in a region of strategic importance,
- facilitate political and economic relations,
- broaden economic and technical cooperation,
- broaden cooperation on energy, industry, trade and services, agriculture, fisheries, investment, science, technology, and environment.

The agreement provides for

- annual joint councils/ministerial meetings between the EU and the GCC foreign ministers, as well as
- meetings of the EU-GCC Regional Directors and Joint Cooperation Committees at senior official level.

---

and Open Syllabus with its access to nearly 8 million classes at more than 7,000 Universities and institutions of other higher learning around the globe at https://opensyllabus.org/.

[8] https://www.gcc-sg.org/en-us/CooperationAndAchievements/Achievements/Pages/Default.aspx.

[9] https://eeas.europa.eu/diplomatic-network/gulf-cooperation-council-gcc/338/gulf-cooperation-council-gcc-and-eu_en.

[10] https://eeas.europa.eu/diplomatic-network/gulf-cooperation-council-gcc/338/gulf-cooperation-council-gcc-and-eu_en.

The agreement[11] allows for the development of closer cooperation on issues such as energy, transport, research and innovation, and the economy, as well as regular political consultations. On June 25, 2020, the 29th meeting of EU-GCC Joint Cooperation Committee took place in a virtual format because of the global COVID-19 restrictions. The EU is diplomatically present in the region with three EU Delegations accredited to the six GCC Countries.[12]

European readers, in addition, should especially notice that the GCC is the EU's sixth-largest export market (€78.1bn of exports in 2019, +5% over 2018), and the EU is the grouping's second-biggest trading partner, with trade flows totalling €121.2 billion, or 13.7% of the GCC's global trade. In return, the GCC represents 3.0% of the EU's total trade.[13]

As we already briefly mentioned above, the purpose of the GCC[14] is to achieve unity among its members based on their common objectives and their similar political and cultural identities. The Presidency of the council rotates annually.[15]

The current literature, published on the subject, would also uniformly agree on the following points (see Britannica, 2021; Gresh, 2015; Guzansky, 2015; Hossain & Muhammad, 2009; Jewish Virtual Library, 2021; Jones & Guzansky, 2020; Kamrava & Georgetown University, 2012; Krieg, 2019; Sadig, 2019; Sick, 2009; Steinberg, 2014, 2020; Ulrichsen, 2011; Zweiri et al., 2021):

- The GCC alliance was formed to strengthen relations among its member countries and to promote cooperation among the countries' citizens.
- The GCC also has a defense planning council that coordinates military cooperation between member countries.
- The highest decision-making entity of the GCC is the Supreme Council, which meets on an annual basis and consists of GCC heads of state.
- Decisions of the Supreme Council are adopted by unanimous approval.
- The Ministerial Council, made up of foreign ministers or other government officials, meets every three months to implement the decisions of the Supreme Council and to propose new policy.
- The administrative arm of the alliance is the office of the Secretariat-General, which monitors policy implementation and arranges meetings.
- In terms of security coordination, policies have included the creation of the Peninsula Shield Force in 1984, a joint military venture based in Saudi Arabia,
- and the signing of an intelligence-sharing pact in 2004.

[11] https://eeas.europa.eu/diplomatic-network/gulf-cooperation-council-gcc/338/gulf-cooperation-council-gcc-and-eu_en.

[12] https://eeas.europa.eu/diplomatic-network/gulf-cooperation-council-gcc/338/gulf-cooperation-council-gcc-and-eu_en.

[13] https://eeas.europa.eu/diplomatic-network/gulf-cooperation-council-gcc/338/gulf-cooperation-council-gcc-and-eu_en.

[14] https://www.gcc-sg.org/en-us/CooperationAndAchievements/Achievements/Pages/Default.aspx.

[15] https://www.britannica.com/topic/Gulf-Cooperation-Council.

- An agreement to launch a single regional currency similar to the euro by 2010 saw little movement apart from the establishment of a monetary council in 2009.
- Coordination in tax policy proved fruitful; however, a customs union was implemented in 2015, and the member states began rolling out a value-added tax of 5 percent in 2018.

### *1.2.2 Gulf Security*

Turning to this sub-field of Gulf Studies, which often catches the imagination of worldwide media, from the *New York Times* to the *BBC* and *Al Jazeera,* and from *TASS* to *Xinhua,* we should emphasize that again, a great number of studies from a Western perspective (Heistein & Guzansky, 2021; Jones & Guzansky, 2020; Guzansky, 2015, 2020; Guzansky & Berti, 2013; Guzansky & Even, 2020; Guzansky & Marshall, 2020; Guzansky & Segal, 2020; Guzansky & Winter, 2020) analyzed international security aspects of the current situation in the Gulf. Readers can and should also refer to the publications on Gulf Security and development, published by the present publishers, which include analyses written by scholars, based in the region (Cowan, 2018; Dwivedi et al., 2018; Fromm, 2018; Gremm, 2018; Karolak & Allam, 2020; Khatib, 2021; Krieg, 2019; Miniaoui, 2020; Mohamed Nour, 2013; Ramady, 2012; Sadig, 2019; Ullah & Haque, 2020; Zweiri et al., 2021). A regional balance in the sources of information is always very advisable.

Since the present study is primarily concerned with inequality, still existing restrictive gender norms (UNDP, 2019), value change, and global economic and political cycles, we make only a few comments on these issues, in order not to unduly detract our readers' attention from our new "cycle analysis" and "world values analysis" focus on Gulf studies, based on the rigorous statistical analysis of open data in the tradition of Karl Wolfgang Deutsch (see Taylor & Russett, 2020, on the general legacy of Karl Deutsch for quantitative political science; Grinin et al., 2016 on the analysis of long cycles of economics and politics; Solomon and Tausch (2020a, 2020b, 2021) on the analysis of global values; and UNDP (2019), on inequality and restrictive gender norms). Readers really interested in Gulf Security Studies, a quickly evolving and often changing subject of its own, very independent from the subject of the present study, are advised to consider the analyses of the world's leading think tanks. For lists of the global and regional ranking of think tanks, readers can consult McGann (2020). In addition, the usually very reliable international country and regional studies by the Konrad Adenauer Foundation and the Friedrich Ebert Foundation, Germany's top analytical institutions, are also recommended to be consulted on the subject of Gulf Security Studies.

With the already mentioned mainstream of the current literature on the Gulf (see above), we can summarize the following main points:

- The immediate reasoning for the Gulf Cooperation Council was security concerns in the wake of the Iran–Iraq war.

- The Jewish Virtual Library (2021) well highlights the following areas of cooperation of the GCC, a point repeated throughout the literature published on the subject:
  - Collaboration in science and technology sectors,
  - The establishment of scientific research facilities for collaboration on research,
  - Military unity and strength,
  - Increased economic cooperation in the private sectors,
  - The establishment of a common currency (this goal was to be met by 2010, and has yet to be accomplished).
- Already in 2016, the GCC officially recognized Hezbollah as a terrorist organization
- Members of the GCC and officials from the USA hold a yearly meeting known as the GCC-US Strategic Cooperation Forum (SCF), in order to coordinate various activities and enhance cooperation between all parties involved.
- The SCF was formed in 2012 and normally occurs in tandem with United Nations General Assembly meetings.
- In December 2000, the GCC states signed a joint defense agreement (Peninsula Shield Force).
- Peninsula Shield Force troops played a vital role during the Bahraini uprising in 2011.
- On December 13, 2013, it was announced that the GCC had formed a joint military command and police forces for the Gulf.
- The establishment of a powerful think tank at the service of the decision-makers in the region, which will be named *"Gulf Academy for Strategic and Security Studies":* In the relevant Riyadh declaration of December 9, 2018, we read: *"In the field of joint defense of the GCC countries, the appointment of the commander of the unified military command of the GCC is an important step to complete the joint defense system. Their Majesties and Highnesses gave directives to accelerate the completion of the procedures for activating the unified military leadership and carrying out its duties, and establishing the Gulf Academy for Strategic and Security Studies in order to establish the unified military command on solid strategic bases and the qualification of the Gulf military leadership to perform these tasks."*[16]
- The Republican administration in the USA, 2016–2020 pushed ahead with naming Qatar as a major non-NATO ally. This status of an especially close strategic relationship with the USA is currently shared by only eighteen countries in the world.[17]

---

[16] https://gulfnews.com/world/gulf/saudi/read-full-text-of-riyadh-declaration-1.1544384331348.

[17] See https://www.aljazeera.com/news/2020/9/17/us-hopes-to-name-qatar-as-major-non-nato-ally-official. On the eighteen countries, allied with the USA, see https://en.wikipedia.org/wiki/Major_non-NATO_ally. On the re-normalization of relations between the other Gulf States and Qatar, see also, just to name a few, https://www.gco.gov.qa/en/focus/gcc-crisis/; ; and https://www.aljazeera.com/news/2020/12/8/uae-official-lauds-efforts-to-strengthen-gulf-unity.

- Yildiz (2021) highlighted the relationship between Turkey and the Gulf region after the recent reconciliation between Qatar and its neighbors following the crisis events of 2017, having been dealt with, among others, in Zweiri et al. (2021). Yildiz (2021) concludes that the resolution of the crisis and the opening of borders and airspace will make Qatar less dependent on Turkey. Qatar Petroleum joined ExxonMobil in signing exploration and production-sharing contracts with Cyprus in 2017, which contradicts Turkey's Eastern Mediterranean policy. Yildiz (2021) predicts that it is plausible that Turkey loses its influence over Qatar as the latter's relations with its neighbors revive, leaving Ankara further isolated in the region.
- In accordance with these perspectives, a very senior Israeli observer of the Middle East, Neriah (2018), also highlighted the following trends:
  - Turkey's projection of its military presence in the Middle East has become a source of worry to the "moderate" Arab states and specifically to Egypt and Saudi Arabia.
  - There are Turkish forces deep inside Syrian and Iraqi territory.
  - Turkey's military intelligence had been blamed by Egyptian authorities for being involved in the Sinai insurgency, siding with the Islamists, while Lebanese sources have mentioned the activities of Turkish agents in destabilizing Lebanon.
  - In 2014, Turkey deployed a 4,000-strong military contingent in Qatar.
  - In 2018, Turkey completed constructing a $50 million training base in the Somalian capital of Mogadishu to be run by the Turks for 5,000–10,000 recruits from Somalia and other African countries.
  - Moreover, the Sudanese government agreed to lease the port of Suakin, along the Red Sea shore, to Turkey for restoration as a major military and civilian port. Suakin was once the naval headquarters of the Ottoman fleet in the Red Sea.
- The "tacit security regimes" between Israel and the Gulf states (Jones & Guzansky, 2020) existed long before the establishment of full diplomatic relations between Israel, Bahrain, and the United Arab Emirates.
  - Trade between Israel and the Gulf states topped $1 billion in 2017.
  - The tacit security regime (TSR) is underway because it allows security cooperation to be pursued between the actors involved (most notably over Iran) but without compromising sensitive political positions that might give rise to internal opposition (Jones & Guzansky, 2020). Three factors strengthen this TSR: Iranian adventurism, jihadi militancy, and American retrenchment.
  - Oman coordinates with Israel via a joint desalination center, but it does not share the perceived threat of Iran. Oman maintains close relations with Tehran and shares control of the Straits of Hormuz with Iran. Relations between Oman and Iran warmed especially after Hassan Rouhani was elected President of Iran in 2013, and Oman has worked to take advantage of this, in contrast with Saudi Arabia and the UAE, who are trying to induce Oman to follow their example

- In the late 1990s, Qatar hosted an Israeli trade delegation, all the while supporting Hamas. Israel and Qatar have stepped up their cooperation since Operation Protective Edge in providing humanitarian aid to the Gaza Strip and in mediation with Hamas. Israel has a clear interest in the continuation of Qatar's aid, because it believes that improvement of the humanitarian situation in the Gaza Strip will help ward off a conflict with Hamas. The director of Israel's intelligence service, Mossad Yossi Cohen, reportedly visited Doha in February 2020 in order to ensure that Qatar's aid to the Gaza Strip, which has totaled over $1 billion since 2012, would continue. Qatar believes that giving aid to the Gaza Strip reinforces its regional status and contributes to strengthening its relations with the US administration, thereby improving its bargaining position in the dispute within the GCC (Guzansky, 2020).
- Internal constraints on all sides (Israel and the Gulf states) determine the type and intensity of external engagement.

### 1.2.3 *The GCC—The Balance Sheet of the Integration Process*

So, what is then the positive balance sheet of four decades of Gulf integration? In Western Gulf Studies research, a very prominent approach (Guzansky, 2020) drew the following rather positive balance:

- Qatar has enjoyed better ties with the USA in the past three years. The two countries have had a security cooperation agreement since 1992, under which 10,000 American soldiers are stationed in Qatar. Qatar is now paying for renovations on the huge US al-Udeid Air Base in its territory and is substantially increasing its procurement of weapons from the USA (Guzansky, 2020)
- On the positive side, the GCC is also the most successful model of inter-Arab cooperation. The history of the organization's six member countries proved that as the threats to them increased, it became easier for them to put their disputes aside.
- The organization's institutions have continued to function even during crises and have been driven by joint economic ventures, such as a shared electrical grid and railway network, while the trade barriers between the countries have been gradually lowered. Coordination on political and economic matters is also underway, as well as cooperation on border controls, the war against terrorism *et cetera* (Guzansky, 2020).
- For nearly 40 years and in light of its special conditions, the GCC, has displayed a considerable degree of coordination and cooperation, together with the almost inherent disagreement between its members (Guzansky, 2020).

### 1.2.4 The Gulf and Comparative Regional Integration

Apart from focusing on the rigorous empirical analysis of inequality, still existing restrictive gender norms, overall value change, and economic and political cycles, a certainly promising other hard-core future empirical research direction on the countries of the Gulf could be to assess the success or failure of regional integration by analyzing the economic and social convergence process of the region and to systematically compare indicators of Gulf convergence with those of other regional integration processes across the globe. In what has been described as *"comparative regionalism," "intersecting interregionalism," "comparative regional integration,"* or similar terms, the social sciences have begun studying the comparative success balances or failures of the ever growing and now already nearly endless list of regional integration processes such as NAFTA, MERCOSUR in Latin America, the European Union, or ASEAN in Asia. The rigorous methodology applied in such studies was described, among others, in Baert et al. (2014), Börzel (2013), Börzel and Risse (2019), Closa and Casini (2016), Ehteshami (2014), Hänggi (2000), Laursen (2018), Sadiki (2020), Simionescu (2014), Siljak (2015) and Te Velde, 2011. The methodology of analyzing convergence processes is standard modern macroeconomic research and was developed, among others, in Barro (2015), Barro and Sala-i-Martin (1992), Barro et al. (1991). In his analysis of the convergence process in the European Union over the last decades, Siljak (2015), neatly summarized these contemporary econometric procedures to measure convergence as follows:

> Beta convergence measures the speed at which poor countries approach rich countries in real GDP per capita terms in a specified time interval. Sigma convergence indicates whether the cross-sectional variation of the real GDP per capita among a group of countries decreases over time. Beta convergence is a necessary but not a sufficient condition for sigma convergence. (Siljak, 2015)

Such a theoretically and also practically rewarding comparison of the convergence process of the Gulf countries with the integration processes in other world regions as yet would have to be written. With Siljak (2015), the following basic three points would have to be remembered:

- Sigma convergence is a simple way of measuring convergence using the standard deviation or the coefficient of variation. Often, the coefficient of variation of purchasing power-adjusted GDP per capita is used as well as the minimums and maximums of GDP per capita relative to the simple average. The coefficient of variation is calculated as the statistical standard deviation divided by the mean. A declining coefficient of variation indicates convergence, while an increase in this measure indicates divergence.
- Beta convergence: There are two types of beta convergence; absolute/unconditional and conditional. When it is assumed that the countries converge to the same terminal point or steady states, convergence is called absolute. It is analyzed by regression with one dependent and one independent variable, where the dependent variable is the growth rate of per capita GDP and the independent variable is the initial level of per capita GDP in purchasing power terms.

- When the economies have different structures, they converge to a different steady-state point, and convergence is called conditional. Conditional convergence is analyzed by the multiple regression with the same dependent variable, but includes various economic, social, and political variables as independent variables in addition to the initial level of per capita GDP. The beta coefficient represents the rate at which a country's real GDP per capita approaches the steady-state rate of growth or the speed of convergence. A negative beta coefficient indicates convergence, while a positive rate indicates divergence.

Setting the scene for our own venturing into the terrains of Gulf studies.

For Guzansky and Segal (2020), it is clear that the Gulf monarchies are ruled by extended families that control most of the centers of power in their countries. Guzansky and Segal highlight also that as long as these families succeeded in cultivating consensus among their different branches, this centralized power has, over the years, contributed to the relative stability of the monarchies:

> Struggles within the ruling families, especially regarding monarchial succession, constitute a weak point with the potential to endanger regime stability, particularly when the family in question is a large one. In some countries, the designated heir is not young or is not healthy, which could shorten the duration of his rule and spark an ongoing struggle over succession. In any event, in addition to the fear of Iranian aggression, the rise of a younger generation of leaders in the Gulf states appears to explain their policies, which are more assertive than traditional policies, and include an increasing openness and willingness to cooperate with Israel. (Guzansky & Segal, 2020)

For a number of years now, public opinion surveys, like the *Arab Opinion Index,* and the *Arab Barometer Survey,* both with a strong relationship to the Qatari scientific scenery, play an increasing role in the region and document basically three trends: a desire for democracy, a certain distance to such players as Iran and Turkey, a more tolerant vision of the relationship between religion and the state, and a more detached and incipient rational view of the role of Israel in the region. With all the talk that we have had in the international media and in the international scientific debate ever since Samuel P. Huntington published his classic on the *"Clash of Civilizations"* (Huntington, 1993, 2000), an astonishing 65% of the entire Arab World now thinks that no religious authority is ever entitled to declare followers of other religions infidels (see Chaps. 6 to 8, below).

For political science, venturing into the terrains of the study of the Gulf monarchies implies many methodological problems. Since there are no electorates or parliamentarians to be interviewed, as usually happens in political science in established and new democracies, should we interview Princes, Emirs, and Consultative Assembly deputies? Should we undertake content analyses of the press and the social media in the region? Should we run experiments with randomly chosen citizens of the countries or with the elites concerned? Or should we rather rely on representative opinion surveys of the inhabitants of the countries of the region?

Standard methodologies well-known from general sociology and political science do simply not apply in these relatively closed systems, so the approach chosen in this volume is based on the data of comparative public opinion survey research and analyses of economic cycles and processes with well-known methodologies and data, used in econometrics and in comparative economics, sociology, and political science. Such methods are surely an alternative to gain at least reasonably objective knowledge as a second-best solution to field studies on the ground. The *World Values Survey* now offers free individual level representative data from around three-fourth of the inhabitants of our globe, and these surveys contain also a reasonable representation of the Gulf countries. In turn, the *Arab Barometer* data contains a wealth of representative individual level opinion survey data from around 80% of the inhabitants of the Arab World.

Our empirical results are totally independent from the world view the present author holds. This statement is worthwhile being pronounced as clearly as possible, and it is also expressed as an invitation to those, who do not share this world view, to look at the empirical results of our study from their own perspective. With an appropriate multivariate statistics computer software, our analyses based on open sources data should be retrievable at any point around the world.

Our world view on the Middle East, we frankly admit this, is the world view, expressed in the yearly report by the Institute for National Security Studies, in Tel Aviv, Israel (INSS, 2020), called the *"Strategic Survey for Israel 2019–2020."* So, with the INSS, we see the following political forces at work in the Gulf region and the greater Middle East, a general political map of the region, which we are further specifying with our own empirical data and background informations:

- **The radical Shiite axis**: This cluster is led by Iran and includes Bashar al-Assad's Syria, Hezbollah, the Houthis in Yemen, the Shiite militias operating in various arenas throughout the Middle East, and Palestinian Islamic Jihad (despite its Sunni identity). A direct estimate of the strength of support for this camp among Arab publics can be concluded from the latest version of the *Arab Opinion Index* (2020). 25% of all interviewed inhabitants in thirteen Arab countries and territories have a (very) positive view of Iranian foreign policy. Opposition to this camp among Arab publics today is truly considerable; 58% hold a negative view of Iranian foreign policy, and 12% think that today Iran poses the greatest security threat to the Arab World.
- **The pragmatic Sunni states**: This bloc includes Egypt, Jordan, Morocco, Saudi Arabia, the United Arab Emirates, and other Arab Gulf states. These actors are advancing a pro-Western, anti-Iranian, anti-Islamist, and nationalist vision. In this book, we analyze data from the University of Qatar, anchored *Arab Barometer* (see Chap. 2 of this work)[18] and suggest that 41% of all Arabs now hold a negative view of the Muslim Brotherhood, which is the strongest and most cohesive force of the Sunni Islamists (see below).
- **The Sunni Islamists**: This group includes supporters of Muslim Brotherhood-style *"Political Islam"*: Turkey, and surely Hamas and remnants of the Brotherhood and its derivative movements throughout the region, such as *Ennahda*, the dominant political party in Tunisia, belong to this camp (see also Solomon

[18] https://www.arabbarometer.org/about/

& Tausch, 2020, 2021). One could emphasize in this context as well the recent political science work by Freer (2018), on the actions of the Muslim Brotherhood in the Gulf region, an Oxford University Press publication, which received *nota bene* a wide positive coverage in the London and Qatar-based news channel "*The New Arab.*"[19] Data from the *Arab Barometer* suggest that 49% of all Arabs still hold positive views of the leadership figure of contemporary Sunni Islamism, Turkish President Recep Tayyip Erdoğan. General support for the Muslim Brotherhood and other Islamist political forces has sharply declined in recent years: *"Hezbollah, the Muslim Brotherhood and Hamas have fallen in public opinion in the years since the Arab uprisings, research commissioned by BBC News Arabic found. More than 25,000 Arabs in 11 states and territories—Morocco, Algeria, Tunisia, Libya, Egypt, Palestine, Yemen, Jordan, Iraq, Sudanand Lebanon—were asked their views on everything from religionto mental healthand homosexuality to women's role in society. A sharp drop in trustfor groups espousing political and radical Islam was among the key findings. Trust in political Islam has declined significantly overall. [...] This pattern continues a general trend of decline in trust for Islamists in Mena, which has been taking place across the region since the Arab uprisings [...] Despite the initial success of Islamist movements in Egyptand Tunisia, there is increasing evidence that Islamism has been in decline over the last eight years." The National (UAE), June 24, 2019, available at* https://www.thenationalnews.com/trust-in-radical-islamist-movements-plummets-major-survey-finds-1.878578.

- **The jihadists**: This camp includes the Islamic State (ISIS) and al-Qaeda and the terrorist organizations associated with them. According to the latest *Arab Opinion Index*, 3% of all Arabs now openly and strongly support ISIL (Daesh), and 2% support it to some extent. Support for ISIL (Daesh) declined since 2014, when 4% of all inhabitants of the Arab World had a very positive view of ISIL (Daesh) and a further 7% had a to some extent positive view of ISIL (Daesh).

In our Fig. 1.1, we further set out to show the strength of support for the foreign policies of different actors in the Middle East among Arab publics.

The recent UNDP *Human Development Report,* 2019 (UNDP, 2019) highlighted what in our opinion are the real issues at stake, and which a good part of academic writing on the Middle East tended to overlook in recent years: the dire issues of global values on inequality and gender norms (Tausch, 2019). The UNDP *Human Development Report,* 2019 (UNDP, 2019) explicitly applied comparative global value research based on the *World Values Survey* in an exemplary way to these two most pressing issues of global development: income and wealth inequality and the way people think about these inequalities, and still existing restrictive gender norms, as

[19] https://english.alaraby.co.uk/english/comment/2018/9/12/rentier-islamism-politics-and-religion-in-arab-gulf-societies. On the importance of this news-channel, see Khatib (2014); on Al Jazeera, see the quantitative content analysis by AlSamrin (2017).

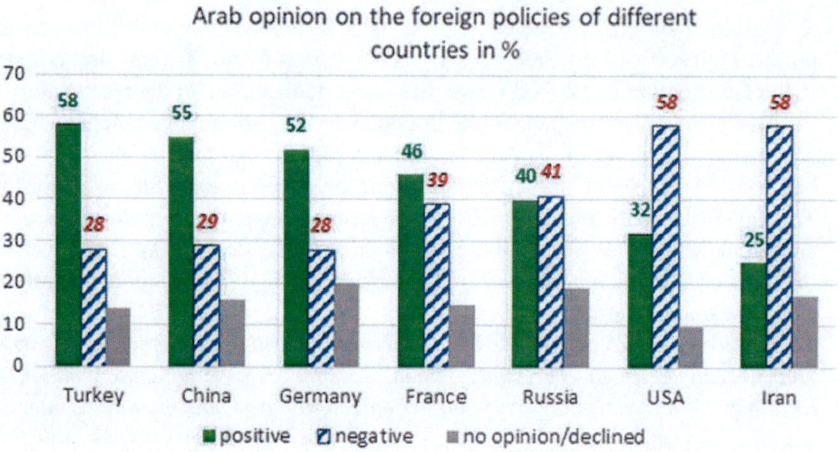


**Fig. 1.1** Opinions of Arab publics on different foreign policy actors in the Middle East according to the latest edition of the *Arab Opinion Index*

they change along development and with the advancement of human security. Our book faces up to these challenges, and features on the multivariate relationships between *World Values Survey* data on opinions on inequality and gender norms, and how these opinions and norms interrelate with more general value perspectives including the role democracies should perform.

## 1.3 Introducing a "UNDP Perspective": Focusing on Inequality and Still Existing Restrictive Gender Norms

The present work thus continues our focus on Middle Eastern countries featuring an approach characterized by comparative global value research and the analysis of long economic cycles already started in Grinin et al. (2016, 2018), Solomon and Tausch (2020a, 2020b, 2021).The availability of an ever-growing number of reliable comparative opinion surveys, including from the Muslim world, revolutionized the way in which research about the region of the *"Middle East"* is possible nowadays (see also Tausch & Moaddel, 2009; Tausch et al., 2014). Also, the art of global developing accounting (see Tausch & Heshmati, 2013; UNDP, 2019) more and more combines data from cross-national opinion surveys with "hard" economic data, allowing insights into social and political processes unthinkable decades ago. The works of two leading global economists of our time, the late Alberto Alesina (1957–2020) and Robert J. Barro, are here a role model to follow for the social sciences in general how to combine macroeconomic data with results from the *World Values Survey*.

The recent data about global inequality and the share of the richest 1% in total national incomes could not be more dramatic and are of pivotal importance for the analysis, performed by the recent flagship of global development analysis, the UNDP Human Development Report, 2019 (UNDP, 2019), see also Chap. 3. The UNDP mentions—in descending order—the twenty-five global record holders of national income concentration in the hands of the richest 1% (UNDP, 2019),[20] and of these twenty-five countries characterized by very high inequality, a breathtaking none the less than fifteen countries are countries with a solid Muslim majority population. Even more important for our research endeavor is that all the Gulf countries (Qatar, United Arab Emirates, Kuwait, Saudi Arabia, Oman, and Bahrain) are among this really lamentable league.[21]

As yet, there are hardly any systematic studies how attitudes on inequality interact with other social and political attitudes in the Arab World and especially in the Gulf region. The mentioned UNDP Human Development Report 2019 (UNDP, 2019) also highlights in an impressive way that gender inequality has long been associated with persistent discriminatory social norms, prescribing social roles and power relations between men and women in society. For the UNDP (2019), it is crystal clear that social norms held by individuals and their reference groups are values, beliefs, attitudes, and practices that assert preferred power dynamics for interactions between individuals and institutions. As broader constructs, norms are operationalized through beliefs, attitudes, and practices.

The Gender Social Norms Index, proposed in the UNDP (2019) Human Development Report, comprises four dimensions—political, educational, economic, and physical integrity—and is constructed based on responses to seven questions from the *World Values Survey.* Among the record holders of countries with a high share of at least two biases regarding the role of women in society,[22] theUNDP names the following Muslim majority countries: Pakistan; Jordan; Nigeria; Qatar; Azerbaijan; Mali; Libya; Iran, Islamic Republic of; Palestine (West Bank and Gaza); Yemen; Kuwait; Iraq; Malaysia; Uzbekistan; Algeria; Turkey, Kyrgyzstan, and Tunisia. Qatar and Kuwait are the Gulf countries among this League (UNDP, 2019).

After the impetus which global research now has received from this UNDP Human Development Report (2019), it is important to recall that the specialists in the field of political sociology are already well acquainted with the massive empirical literature, written by such authors as Ronald F. Inglehart, Pippa Norris, and Mansoor Moaddel, which already established for a long time the validity of analyzing the social and

[20] Qatar; Brazil; Chile; Turkey; Lebanon; United Arab Emirates; Iraq; India; Colombia; Russian Federation; Thailand; USA; Kuwait; Saudi Arabia; Oman; South Africa; Egypt; Bahrain; Côte d'Ivoire; Iran (Islamic Republic of); Jordan; West Bank and Gaza; Yemen; Syrian Arab Republic; and Malaysia

[21] Qatar; Turkey; Lebanon; United Arab Emirates; Iraq; Kuwait; Saudi Arabia; Oman; Egypt; Bahrain; Iran (Islamic Republic of); Jordan; Palestine (West Bank and Gaza); Yemen; and the Syrian Arab Republic).

[22] Pakistan; Jordan; Nigeria; Qatar; Azerbaijan; Mali; Libya; Haiti; Ghana; Iran, Islamic Republic of; West Bank and Gaza; Yemen; Kuwait; Iraq; Rwanda; Malaysia; Uzbekistan; Algeria; Philippines; Burkina Faso; Turkey; Kyrgyzstan; Zimbabwe; Tunisia; and India.

political values of global publics in a growing number of countries in a comparative perspective. Does this type of research, as suggested by the UNDP data on inequality and gender norms, holds in store only pessimistic tendencies for the Gulf region?

Looking at international opinion survey data reveals also some more optimistic results about the Gulf region by international comparison. Some Muslim countries, most notably Morocco, are now among the global top three for the *World Values Survey* items *"democracy may have problems, but it is better"* and *"it is (very) good to have a democratic system."* In several East European new democracies, but also even in key Western countries like Germany and the USA, there seems to be now a real fatigue with democracy, while many Arab countries long for democracy. This fatigue with democracy in many countries of the West is an alarming trend, and combines with the Tsunami force of the gathering storm of the global economic crisis in the wake of COVID-19, and it bodes very badly for the future of democracy in the world.

For the Gulf countries, available data seem to suggest that *"Arab Springs"* are still distant from them. Saudi Arabia is rank 55 of 66 nations with complete data for the variable *"democracy may have problems but it is better,"* and Bahrain is rank 90 of the 96 nations with complete data for the indicator *"it is good to have a democratic system."* But still, Saudi Arabia's civil society is in the vicinity of the most vital and solid Western ally in Asia, Japan, and comparable to some of the new democracies in Eastern Europe and ahead of the Philippines; and Bahrain—fasten your seatbelts—is in the vicinity of Finland, and in the vicinity of success stories of the democratic transformation since the 1980s such as Brazil and the Philippines, and even ahead of South Korea and also Poland, which was the pioneer in the transition from "state socialism" and the crumbling of the political walls that separated Europe in 1989 in the first place.

So, far from being a mere "green box" of potential support for radicalism and extremist Islamist political forces, which was the tenor of academic writing on Saudi Arabia after 9/11 (see Solomon & Tausch, 2021), there are countertendencies at work, which also begin to affect the way in which people begin to interpret gender roles and their relationship to religion. Thus, there is real hope for what is called here *"the pragmatic Sunni states,"* including Egypt, Jordan, Morocco, Saudi Arabia, the United Arab Emirates, and the other Arab Gulf states. The overwhelming majority of the inhabitants of Qatar, interviewed in our surveys, is light years away in their opinions from the radical Shia axis and the Sunni radicals. Our comprehensive data, hopefully, could contribute to the success in creating a cohesive strategy against the radical Shia axis and the Sunni radicals.

As we show in the following first analytical maps to be presented in this book, which set the social scientific scene of our study, Human Development and satisfaction with life in the region is often higher than in many other parts of the world, including in numerous countries of the European Union, where already the crisis 2008/2009 left deep scars in the social fabric, and where we must be all the more preoccupied about the future of democracy. We start here from the robust hypothesis that COVID-19, if anything, will deepen this crisis aspect of democracies in Europe (see Solomon & Tausch, 2021). So, our Figs. 1.2 and 1.3 establish the extent

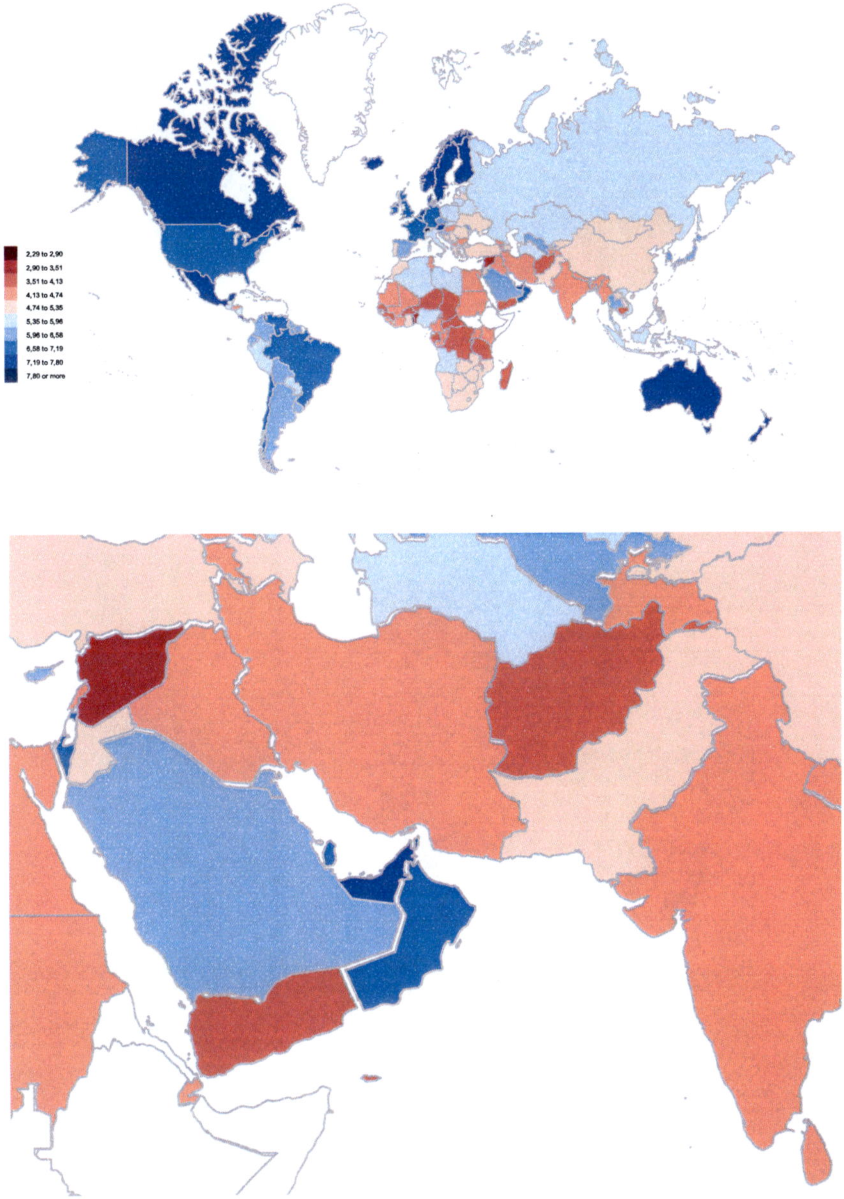


**Fig. 1.2** Overall life satisfaction according to UNDP/Gallup poll (Tausch, 2019)

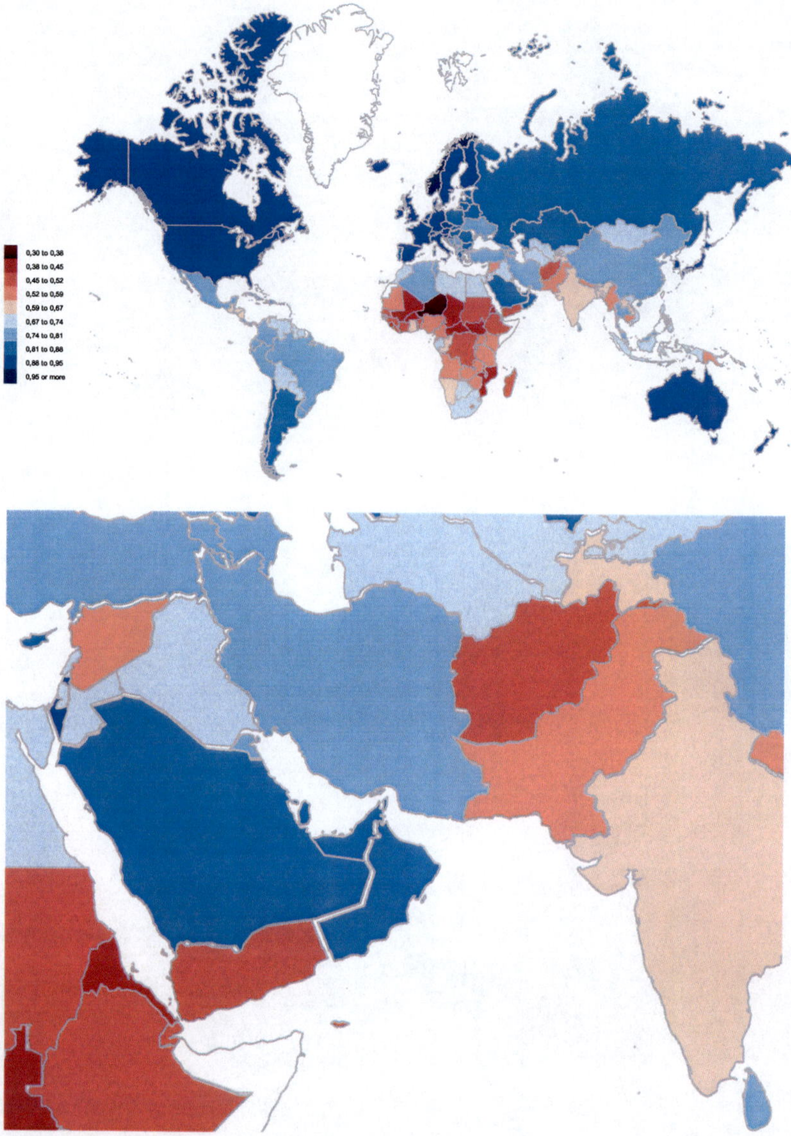


**Fig. 1.3** UNDP Human Development Index, 2018

of Human Development and overall life satisfaction in the region by international comparison.[23]

Figures 1.4 and 1.5 portray the extent of support or lack of support for democracy in a global perspective. "1" is *"it is good to have a democratic political system";* and the scale ends with *"4": "it is bad to have a democratic political system."* The rejection of democracy in core Western countries like the USA, Japan, Poland, and several new democracies in Eastern Europe is really amazing. The skepticism about democracy in Mexico, Brazil, Chile, South Africa, South Korea, and the Philippines is equally disturbing, while Fig. 1.4 also reflects the hopes for democracy in large parts of the Muslim world:

In Fig. 1.5, "1" corresponds to agreement to the statement: *"democracy may have problems, but it is better,"* while "4" implies disagreement to this opinion, i.e., in the opinion of the representative respondents that democracy is not better than other forms of government. Again, the disappointing results for countries like Japan, Australia, New Zealand, Mexico, and several "new democracies," which emerged in the 1970s, 1980s, and 1990s after years of dictatorship, like Chile, Poland, and several other East European countries, South Korea and the Philippines have to be noted, while at the same time, and as in Fig. 1.4, results for several countries from the Muslim world are more optimistic:

Figure 1.6 highlights how the international reverse ranking of the dynamics of Human Development, 2010–2018, have shifted the weights of the global economy away from the hitherto developed regions of our globe. In much of Africa and Asia, there were spectacular improvements in Human Development, while the traditional West fell dramatically behind. The weight of global Human Development clearly shifted away from the West (Table 1.1).

Figures 1.7 and 1.8 highlight the two main problem areas of global development, as identified by the UNDP's Human Development Report, 2019: inequality and restrictive social gender norms. Figure 1.7 features the share of the richest 1% in total incomes, while Fig. 1.8 portrays the UNDP's results about the global Gender Social Norms Index (version GSNI2, expressing the share of people with at least two biases against women in the political, economic, education sphere, and against the physical integrity of women).

Elsenhans (2016) described the dilemma of income concentration in peripheral countries in the following way:

> There are markets, there are people who want to make "profits", there are people who exploit the labor of others, there is competition on markets, and obviously private property which its owners use for increasing their wealth. But the mechanisms which limit the possibilities of capitalists of enriching themselves, which exist in a capitalist economy with fairly high levels of competition, are absent. The basic Keynesian and Kaleckian principle that profit depends on investment spending, that "capitalists get what they spend" and not what they

[23] See Tausch (2019). Migration from the Muslim world to the West: Its most recent trends and effects. Jewish Political Studies Review, 30(1–2), 65–225, available at http://jcpa.org/article/migration-from-the-muslim-world-to-the-west-its-most-recent-trends-and-effects/ (with data definitions and sources). Free data download available from https://www.academia.edu/37568941/Migration_from_the_Muslim_World_to_the_West_Its_Most_Recent_Trends_and_Effects].

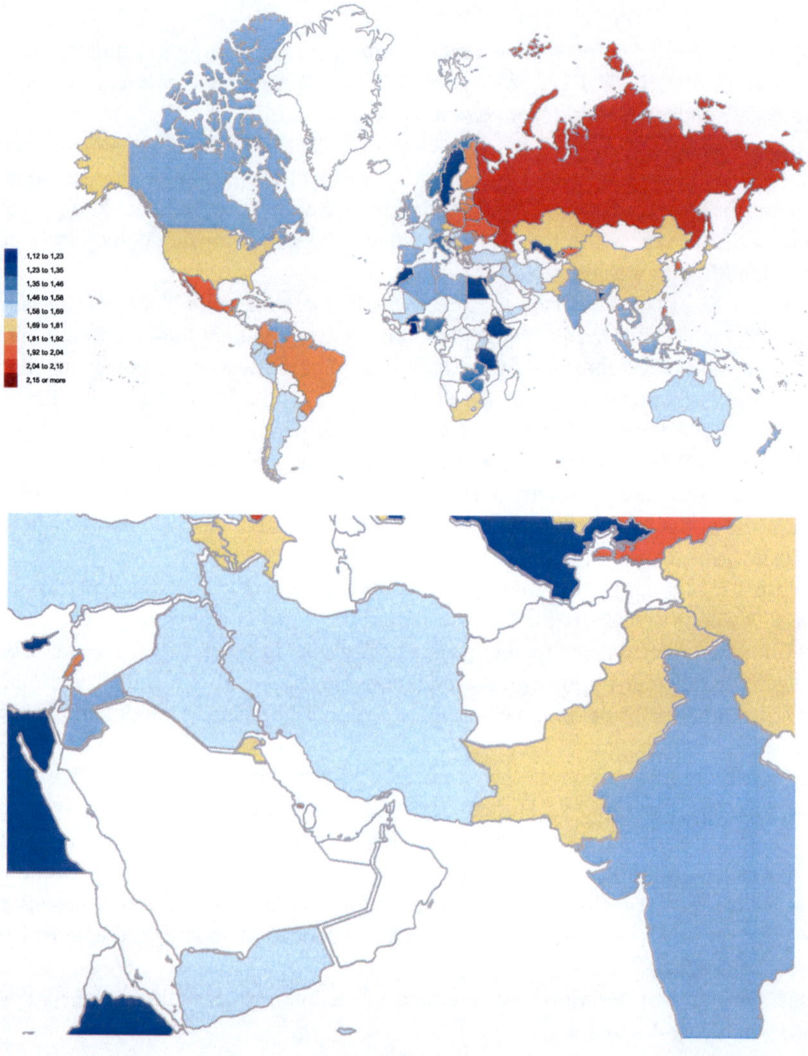


**Fig. 1.4** Support for democracy/lack of support for democracy *("Political system: Having a democratic political system")*

can appropriate by political means does not exist in an economy where there are not high levels of employment, where increasing investment does not lead to higher mass incomes, that in the long run only higher incomes lead to additional investment opportunities via the market etc. The economic structure of non-capitalist and pre-capitalist societies does not empower labor but only increases the possibilities for the rich to get richer, with some of the (rising) rich enriching themselves via imperfect markets. In some historical narratives, the rich are called capitalists [...]. (Elsenhans, 2016)

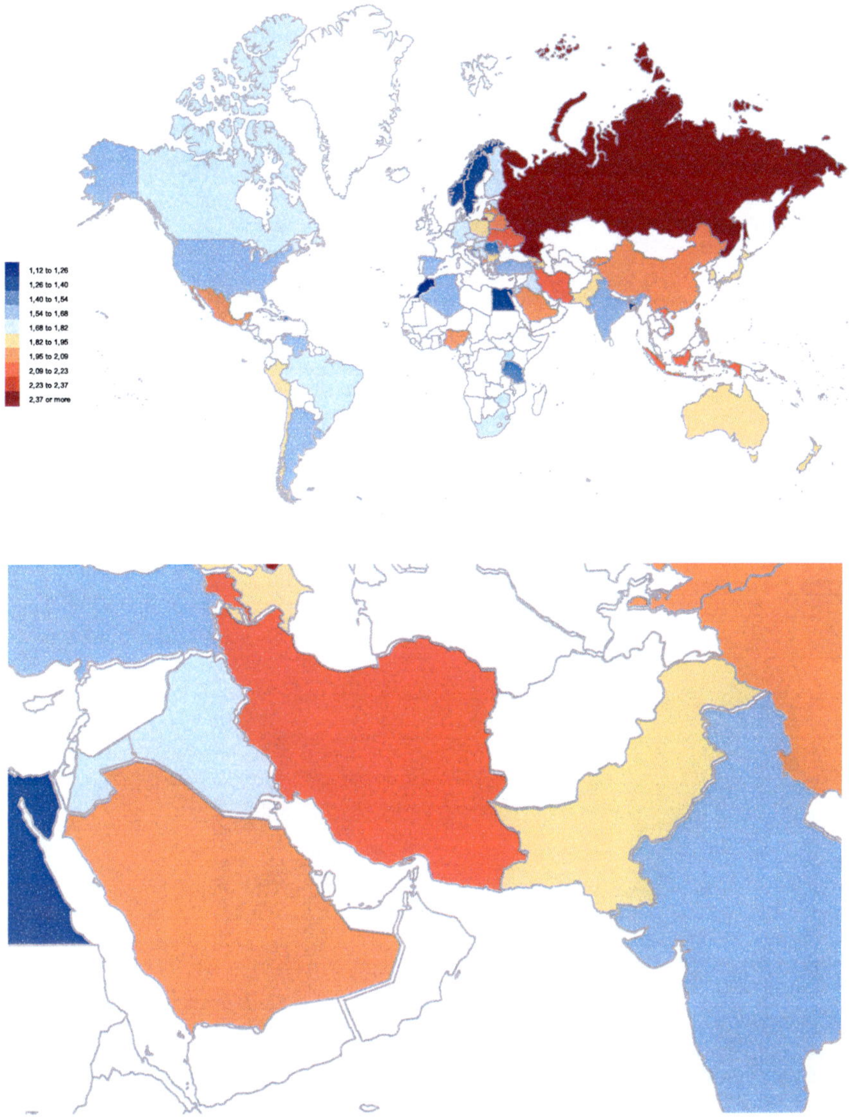


**Fig. 1.5** Support for democracy/lack of democracy ("*Democracy may have problems, but it is better*")

Why do we dare now to hold nevertheless some optimism about changing societies in the Gulf, which now seems to affect even such long-standing issues as the normality of diplomatic relations between the State of Israel and its Arab neighbors more than seven decades after the foundation of the State of Israel in 1948? The works of the American sociologist Ronald F. Inglehart are full of evidence about the

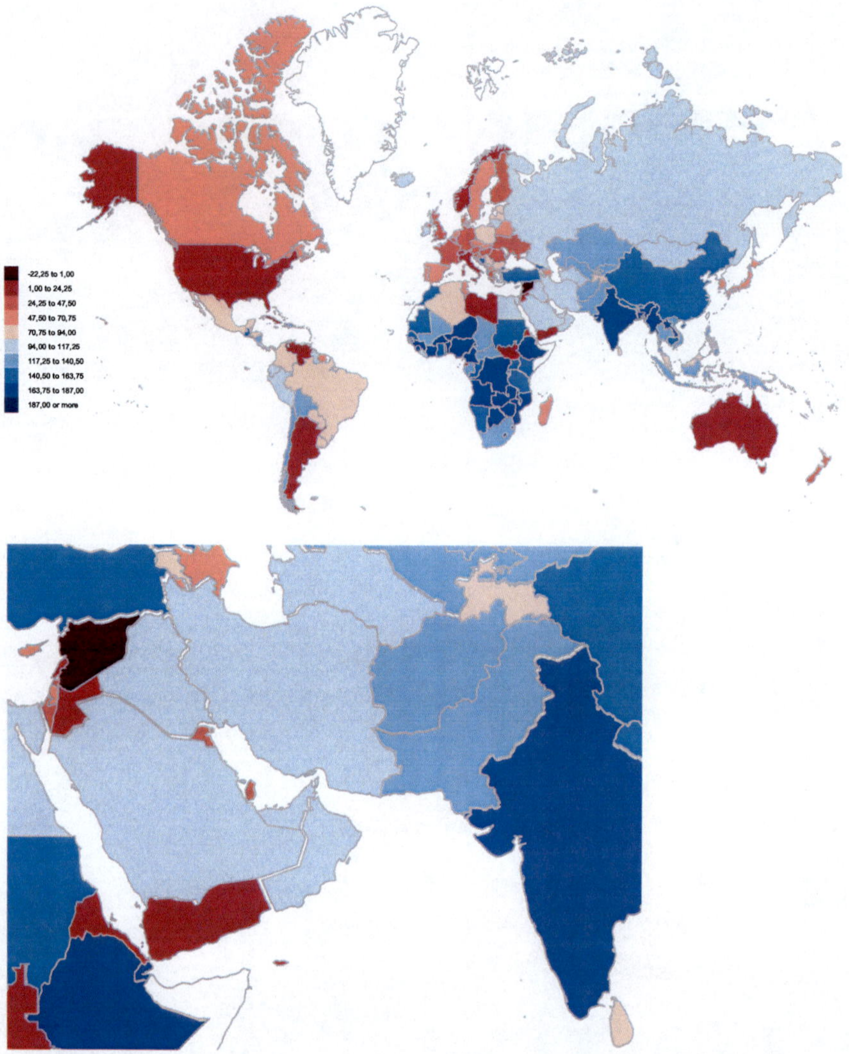


**Fig. 1.6** Dynamics of Human Development, 2010–2018 shifting away from the West (reverse international ranking)

simple and compelling fact that human values change with Human Development (as understood by the UNDP), and Figs. 1.9 and 1.10 indicate again the strength of these relationships. Over 60% of life satisfaction are explained by the UNDP's Human Development Index (HDI), and although restrictive gender social norms first rise with a higher Human Development Index at low levels of the UNDP's HDI, there is a persuasive and strong tendency for the waning of support for restrictive gender

**Table 1.1** Gulf countries on the data radar of the UNDP Human Development Report, 2019

| Country | Restrictive gender norms UNDP GSNI2 | UNDP richest 1% | Global Rank low GSNI2 (75 countries) | Global Rank low share richest 1% (70 countries) |
|---|---|---|---|---|
| Bahrain | xx | 18.00 | xx | **53** |
| Kuwait | 91.56 | 19.93 | **63** | **58** |
| Oman | xx | 19.49 | xx | **56** |
| Qatar | 94.90 | 29.03 | **72** | **70** |
| Saudi Arabia | xx | 19.70 | xx | **57** |
| United Arab Emirates | xx | 22.81 | xx | **65** |

social norms with higher Human Development. More than 62% of restrictive gender social norms are explained in such a way.

Further setting the social scientific scene of this book, Fig. 1.11 features on the country-level relationship between the UNDP's flagship variables, income concentration and restrictive social gender norms, and Fig. 1.12 attempts to answer the simple question whether or not income concentration will diminish with higher Human Development. More than one-third of restrictive social gender norms are explained by income concentration, while more and more countries with a high human development are now characterized by high income concentration.

With these perspectives, we hope to have described sufficiently the general scientific focus of our study. In Chap. 2, below, we thoroughly debate the methodological aspects of our study. In the following three Chapters (Chaps. 3, 4, and 5), we will then define the global environment affecting the future trajectory of the Gulf states. Chapters 6 to 8 analyze the political, social, and religious values, affecting the Gulf region.

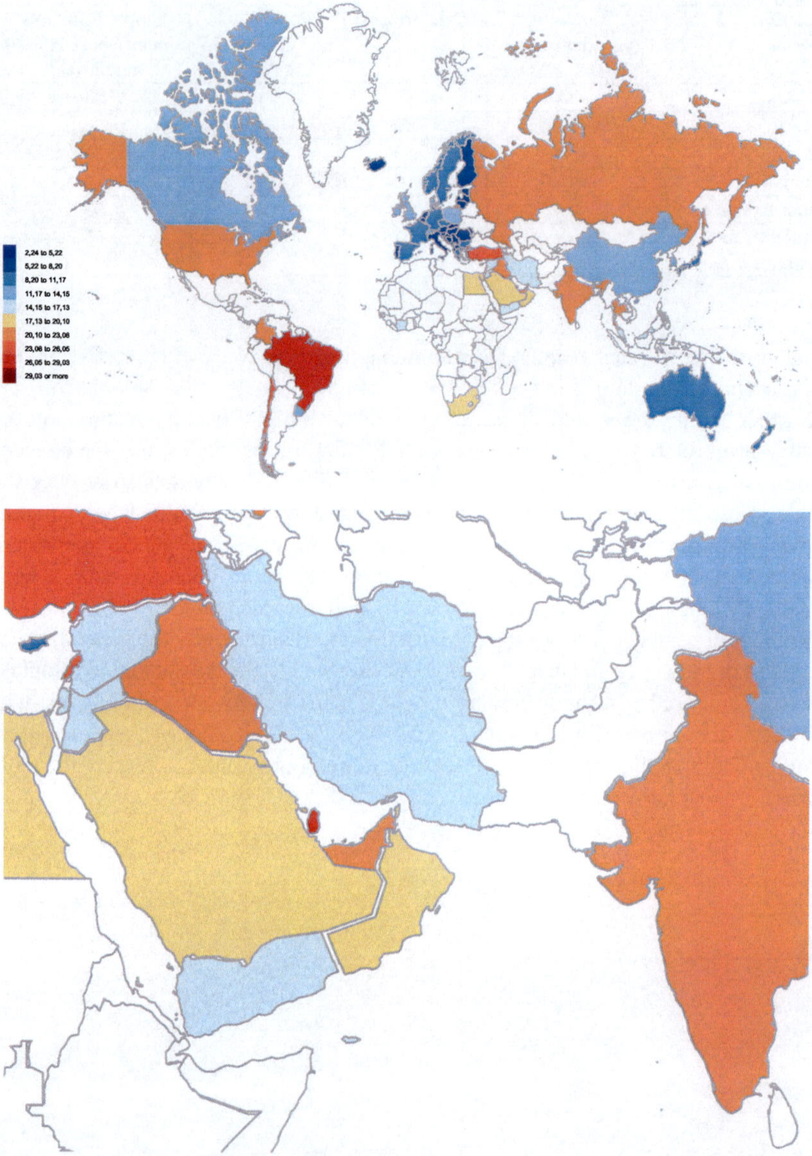


**Fig. 1.7** Share of richest 1% in total incomes

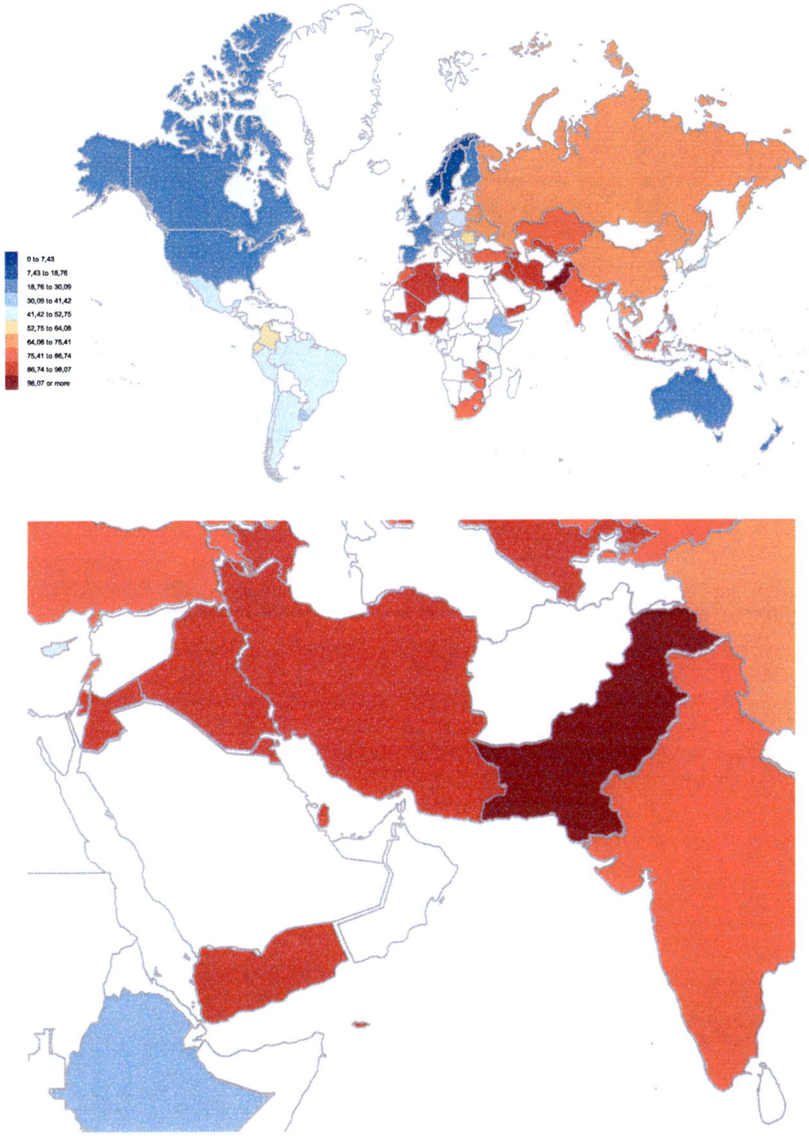


**Fig. 1.8** UNDP Index of restrictive gender norms (percentage of people supporting at least two restrictive gender norms) (UNDP GSNI2)

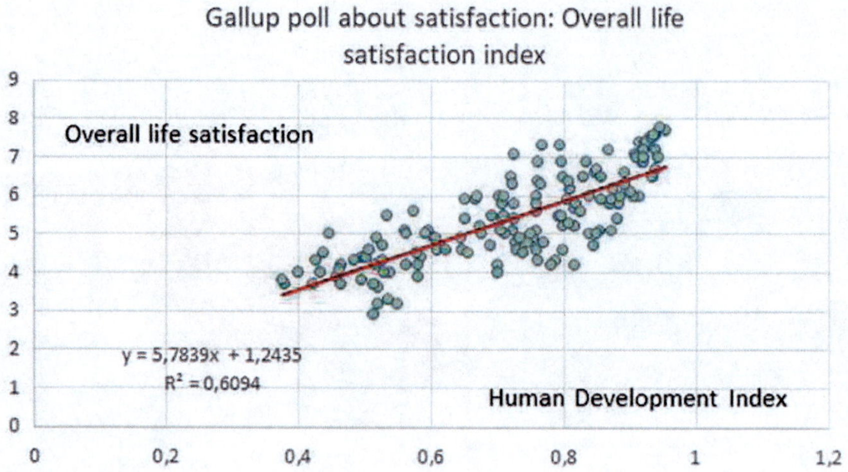


Fig. 1.9 Human Development and life satisfaction

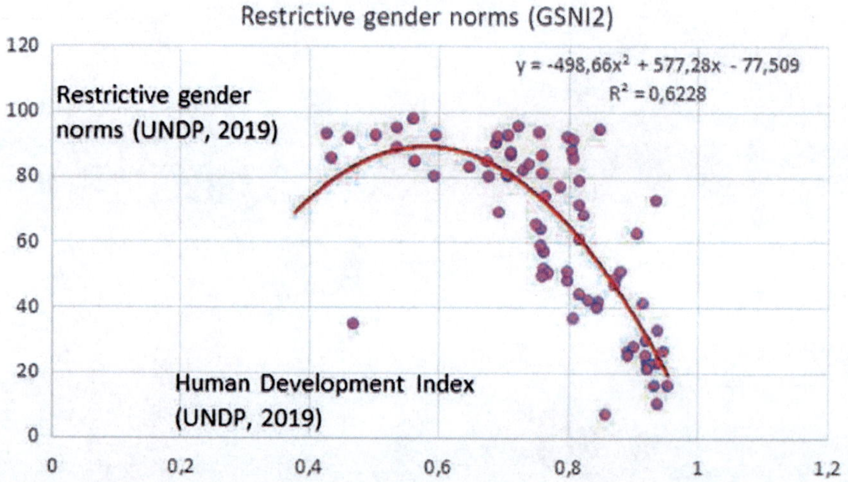


Fig. 1.10 Human Development and restrictive gender norms

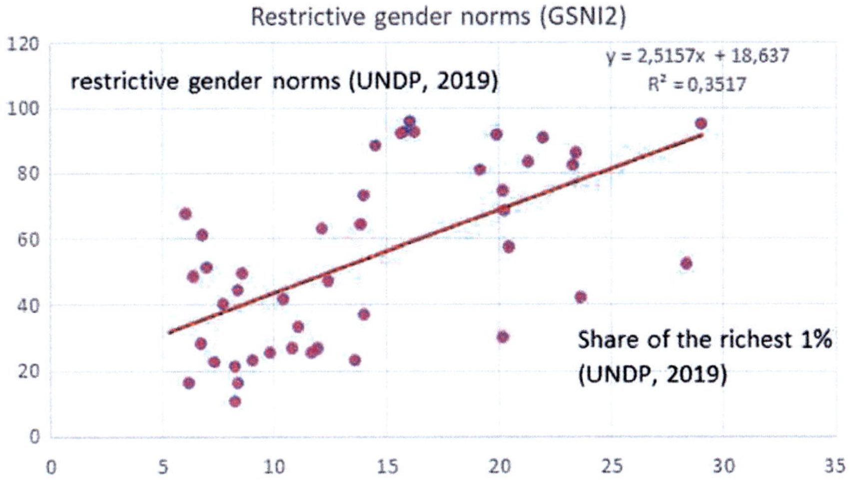


**Fig. 1.11** Income concentration and restrictive gender norms (UNDP, 2019)

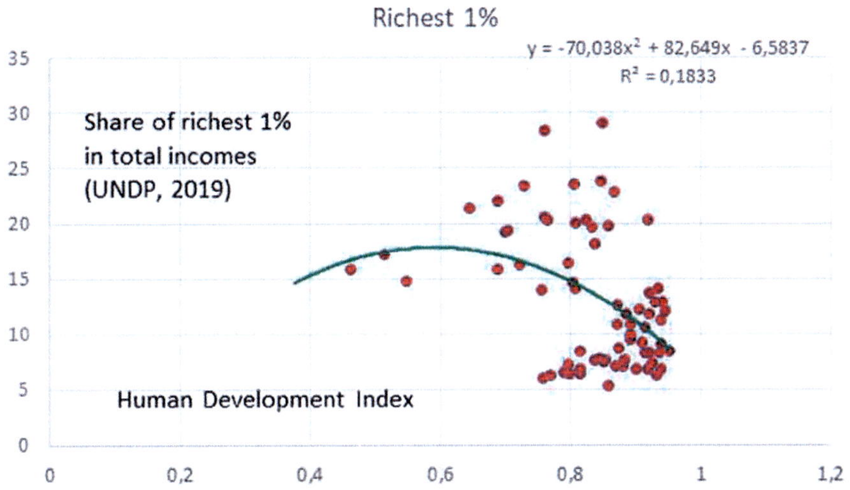


**Fig. 1.12** Human Development and income concentration

## References


Allison, G. T. (2017). *Destined for war? Can America and China escape Thucydides' trap?* Houghton Mifflin Harcourt.

AlSamrin, H. (2017). A comparison of news coverage on the Arabic and English versions of the Al Jazeera websites (Doctoral dissertation, Drexel University). Available at https://core.ac.uk/download/pdf/190333228.pdf.

*Arab Barometer*. (2015). *Arab Barometer*. http://www.arabbarometer.org/.

*Arab Opinion Index*. (2020). Available at https://www.dohainstitute.org/en/Lists/ACRPS-PDFDocumentLibrary/Arab-Opinion-Index-2019-2020-Inbreef-English-Version.pdf.

Baert, F., Scaramagli, T., & Söderbaum, F. (2014). *Intersecting interregionalism.* Springer.

Barro, R. J. (2006). Rare disasters and asset markets in the twentieth century. *The Quarterly Journal of Economics, 121*(3), 823–866.

Barro, R. J. (2015). Convergence and modernisation. *The Economic Journal, 125*(585), 911–942.

Barro, R. J., & Sala-i-Martin, X. (1992). Convergence. *Journal of Political Economy, 100*(2), 223–251.

Barro, R. J., & Ursúa, J. F. (2008). Macroeconomic Crises since 1870. Brookings Papers on Economic Activity, Spring (pp. 255–335). Available at: http://www.people.fas.harvard.edu/~jfursua/papers/bpea.pdf (Download: April 10, 2019.

Barro, R. J., & Ursúa, J. F. (2011). *Barro-Ursúa Macroeconomic Data.* Available at: https://scholar.harvard.edu/barro/publications/barro-ursua-macroeconomic-data.

Barro, R. J., Sala-i-Martin, X., Blanchard, O. J., & Hall, R. E. (1991). *Convergence across states and regions. Brookings papers on economic activity* (pp. 107–182).

Barro, R. J., Ursúa, J. F., & Weng, J. (2020). The Coronavirus and the Great Influenza Epidemic - Lessons from the 'Spanish Flu' for the Coronavirus's Potential Effects on Mortality and Economic Activity (2020). CESifo Working Paper No. 8166). Available at SSRN: https://ssrn.com/abstract=3556305.

Bianco, C. (2017). *The Intra-GCC crisis: domestic, regional and international layers.* Instituto Affari Internazionali. July 2017.

Bianco, C. (2018). Gulf security after 2011: A threat analysis. *Middle East Policy, 25*(2), 27–41.

Bianco, C. (2020a). The GCC monarchies: Perceptions of the Iranian threat amid shifting geopolitics. *The International Spectator, 55*(2), 92–107.

Bianco, C. (2020b). A Gulf apart: How Europe can gain influence with the Gulf Cooperation Council. European Council on Foreign Relations, February 2020. Available at https://ecfr.eu/archive/page/-/a_gulf_apart_how_europe_can_gain_influence_with_gulf_cooperation_council.pdf.

Bianco, C. (2021). Can Europe Choreograph a Saudi-Iranian Détente? European University Institute, Robert Schuman Center for Advanced Studies, Middle East Directions. Available at: https://cadmus.eui.eu/bitstream/handle/1814/70351/PB_2021_10-MED.pdf?sequence=1.

Bianco, C., & Stansfield, G. (2018). The intra-GCC crises: Mapping GCC fragmentation after 2011. *International Affairs, 94*(3), 613–635.

Bolt, J., Inklaar, R., de Jong, H., & Luiten van Zanden, J. (2018). Rebasing 'Maddison': new income comparisons and the shape of long-run economic development. Maddison Project Working Paper, nr. 10. Available at www.ggdc.net/maddison.

Börzel, T. A. (2013). Comparative regionalism: European integration and beyond. *Handbook of International Relations, 2*, 503–530.

Börzel, T. A., & Risse, T. (2019). Grand theories of integration and the challenges of comparative regionalism. *Journal of European Public Policy, 26*(8), 1231–1252.

Britannica, The Editors of Encyclopaedia. (2021). Gulf Cooperation Council. Encyclopedia Britannica, 13 Jan 2021. https://www.britannica.com/topic/Gulf-Cooperation-Council. Accessed 13 April 2021.

Brodeur, A., Gray, D. M., Islam, A., & Bhuiyan, S. (2020). A Literature review of the economics of COVID-19. Available at: https://www.econstor.eu/bitstream/10419/222316/1/GLO-DP-0601.pdf.

Cammett, M. (1999). Defensive integration and late developers: The Gulf Cooperation Council and the Arab Maghreb Union. *Global Governance a Review of Multilateralism and International Organizations, 5*(3), 379–402.

Cammett, M. (2014). Is there an Islamist political advantage? *Annual Review of Political Science., 17*, 187–206.

Cammett, M. (2017). Development and Underdevelopment in the Middle East and North Africa. In C. Lancaster & N. van de Walle (Eds.) *The Oxford handbook of the politics of development,*

*political economy*. Online Publication Date: Jan 2017. https://doi.org/10.1093/oxfordhb/9780199845156.013.25.
Cammett, M. (2018). Popular grievances in the Arab region: Evaluating explanations for discontent in the lead-up to the uprisings. *Middle East Development Journal, 10*(1), 64–96.
Cammett, M. (2020). Insecurity and political values in the Arab World. *Democratization, 27*(5), 699–716.
Cammett, M., et al. (2015). *A political economy of the middle east* (4th ed.). Westview Press.
Candela, R., & Geloso, V. (2020). *Economic freedom, pandemics and robust political economy* (December 11, 2020). Available at SSRN: https://ssrn.com/abstract=3747216 or https://doi.org/10.2139/ssrn.3747216.
Closa, C., & Casini, L. (2016). *Comparative regional integration: governance and legal models* (Vol. 10). Cambridge University Press.
Cooke, M. (2014). *Tribal modern: Branding new nations in the Arab Gulf*. University of California Press.
Cowan, D. (2018). *The coming economic implosion of Saudi Arabia: A behavioral perspective.* Palgrave Macmillan.
Davidson, C. (2013). *After the Sheikhs: The coming collapse of the Gulf monarchies.* Oxford University Press.
Devezas, T. C. (Ed.). (2006). Kondratieff waves, warfare and world security. In: *Proceedings of the NATO Advanced Research Workshop on The Influence of Chance Events and Socioeconomic Long Waves in the New Arena of Asymmetric Warfare*, Covilhã, Portugal, 14–18 February (2005]. IOS Press.
Doerr, S., & Hofmann, B. (2020). *Recessions and mortality: A global perspective*. Bank for International Settlements. Available at: https://www.bis.org/publ/work910.pdf.
Dwivedi, Y. K., et al. (Eds.). (2018). *Emerging markets from a multidisciplinary perspective: Challenges, opportunities and research agenda.* Springer (Advances in Theory and Practice of Emerging Markets Ser).
Ehteshami, A. (2014). The Union Moment for the GCC. GRC; Gulf Research Center. Available at https://www.files.ethz.ch/isn/184050/Unity_Ehteshami_5260.1.pdf
Elsenhans, H. (1983). Rising mass incomes as a condition of capitalist growth: Implications for the world economy. *International Organization*, 1–39.
Elsenhans, H. (1991). *Development and underdevelopment: The history, economics, and politics of north-south relations.* Sage.
Elsenhans, H. (1996) State, class, and development. Radiant.
Elsenhans, H. (2011). *The rise and demise of the capitalist world system.* Leipziger Universitätsverlag.
Elsenhans, H. (2014). *Saving capitalism from the capitalists: world capitalism and global history.* Sage.
Elsenhans, H. (2016). Overcoming Political Isolation: the Relevance of Keynesian Fundamentals for Moral Economy Approaches in an Increasingly Multipolar International System. In: *13th Post-Keynesian Conference*, University of Missouri, 15 to 18 September 2016, Panel: "Perspectives on Keynesian Economics", 10:30 AM on Thursday, Sept. 15th, 2016. Available at: https://hartmutelsenhans.net/paper/
Elsenhans, H., & Babones, S. (2020). *BRICS or Bust?* Stanford University Press.
Freer, C. J. (2018). *Rentier Islamism: The influence of the Muslim Brotherhood in Gulf monarchies.* Oxford University Press.
Fromm, N. (2018). *Constructivist niche diplomacy: Qatar's Middle East diplomacy as an illustration of small state norm crafting*. Springer VS (Politik und Gesellschaft des Nahen Ostens).
Furceri, D., Loungani, P., Ostry, J. D., & Pizzuto, P. (2020). Will Covid-19 affect inequality? Evidence from past pandemics. *COVID-19 Economics, 12*(1), 138–157.
Goldstein, J. S. (1985). Kondratieff waves as war cycles. *International Studies Quarterly, 29*(4), 411–444.
Goldstein, J. S. (1988). *Long cycles: Prosperity and war in the modern age.* Yale University Press.

Gray, M. (2011). A Theory of 'Late Rentierism' in the Arab States of the Gulf (August 18, 2011). CIRS Occasional Papers, 2011, Available at SSRN: https://ssrn.com/abstract=2825905 or https://doi.org/10.2139/ssrn.2825905

Gremm, J., et al. (2018). *Transitioning towards a knowledge society: Qatar as a case study*. Springer. https://doi.org/10.1007/978-3-319-71195-9

Gresh, G. F. (2015). *Gulf security and the U.S. military: regime survival and the politics of basing*. Stanford University Press (Stanford security studies).

Grinin, L., Korotayev, A. V., & Tausch, A. (2018). Islamism, Arab Spring, and the future of democracy: World system and world values perspectives Ser. Perspectives on development in the Middle East and North Africa (Mena region). Springer.

Grinin, L., Korotayev, A., & Tausch, A. (2016). *Economic cycles, crises, and the global periphery*. Springer.

Guzansky, Y. (2015). *The Arab Gulf states and reform in the Middle East: Between Iran and the "Arab Spring"* . Palgrave Macmillan.

Guzansky, Y. (2020). Engulfed in Dispute? The Future of the Gulf Cooperation Council. *Policy Analysis, 23*(4), INSS, Tel Aviv. Available at: https://strategicassessment.inss.org.il/en/articles/engulfed-in-dispute-the-future-of-the-gulf-cooperation-council/

Guzansky, Y., & Berti, B. (2013). Is the New Middle East Stuck in Its Sectarian Past? The Unspoken Dimension of the "Arab Spring". Orbis, 57135–151). doi:https://doi.org/10.1016/j.orbis.2012.10.009.

Guzansky, Y., & Even, S. (2020). *The economic crisis in the Gulf States: A challenge to the "contract" between rulers and ruled*. INSS Insight No. 1327, June 1, 2020. Available at https://www.INSS.org.il/publication/gulf-states-economy/?offset=7&posts=201&outher=Yoel%20Guzansky.

Guzansky, Y., & Marshall, Z. A. (2020). The Abraham accords: Immediate significance and long-term implications. *Israel Journal of Foreign Affairs*, 1–11.

Guzansky, Y., & Segal, E. (2020). *All in the family: Leadership changes in the Gulf*. INSS Insight No. 1378, August 30, 2020. Available at: https://www.INSS.org.il/publication/gulf-royal-families/?offset=1&posts=201&outher=Yoel%20Guzansky

Guzansky, Y., & Winter, O. (2020). Apolitical Normalization: A New Approach to Jews in Arab States. INSS Insight No. 1332, June 8, 2020. Available at: https://www.INSS.org.il/publication/judaism-in-the-arab-world/?offset=5&posts=201&outher=Yoel%20Guzansky.

Hänggi, H. (2000). *Interregionalism: Empirical and theoretical perspectives* (pp. 1–14). University of St. Gallen.

Hanieh, A. (2011). *Capitalism and Class in the Gulf Arab States*. Palgrave Macmillan.

Hanieh, A. (2013). *Lineages of Revolt: Issues of Contemporary Capitalism in the Middle East*. Haymarket Books.

Hanieh, A. (2015). *Capitalism and class in the Gulf Arab states*. Palgrave Macmillan.

Heistein, A., & Guzansky, Y. (2021). Could Gulf Reconciliation Herald a Broader Regional Realignment? INSS Insight No. 1430, January 20, 2021. Available at https://www.inss.org.il/wp-content/uploads/2021/01/no.-1430.pdf

Hossain, I., & Muḥammad, S. M. (2009). American foreign policy & the Muslim world (1st edn.). al-Zaytouna Centre for Studies & Consultations.Available at http://eng.alzaytouna.net/product/american-foreign-policy-the-muslim-world/#.YG1IYegzayI.

Huntington, S. P. (1993). The third wave: Democratization in the late twentieth century (Vol. 4). University of Oklahoma press.

Huntington, S. P. (2000). The clash of civilizations? In*culture and politics* (pp. 99–118). Palgrave Macmillan US.

Inglehart, R. F. (2008). Changing values among western publics from 1970 to 2006. *West European Politics, 31*(1–2), 130–146.

Inglehart, R. F. (2017). Changing values in the Islamic world and the West. *Values, Political Action, and Change in the Middle East and the Arab Spring*, 3–24.

Inglehart, R. F., & Baker, W. E. (2000). Modernization, cultural change, and the persistence of traditional values. *American Sociological Review, 65*(1), 19–51.

Inglehart, R. F., & Norris, P. (2003). *Rising tide: Gender equality and cultural change around the world*. Cambridge University Press.

Inglehart, R. F., & Welzel, C. (2010). Changing mass priorities: The link between modernization and democracy. *Perspectives on Politics, 8*(02), 551–567.

INSS (Institute for National Security Studies, Tel Aviv. (2020). Strategic Survey for Israel 2019–2020. Available at https://www.INSS.org.il/publication/strategic-survey-for-israel-2019-2020/.

Jalilvand, D. R., & Westphal, K. (Eds.). (2018). *The political and economic challenges of energy in the Middle East and North Africa*. Routledge.

Jewish Virtual Library. (2021). *Arab/Muslim World: The Gulf Cooperation Council (GCC)*. Available at: https://www.jewishvirtuallibrary.org/the-gulf-cooperation-council.

Jones, C., & Guzansky, Y. (2020). *Fraternal enemies: Israel and the Gulf monarchies*. Oxford University Press.

Jordà, Ò., Singh, S. R., & Taylor, A. M. (2020). *Longer-run economic consequences of pandemics* (No. w26934). National Bureau of Economic Research. Available at https://www.frbsf.org/economic-research/files/wp2020-09.pdf.

Kamrava, M., & Georgetown University. Center for International and Regional Studies. (2012). *The political economy of the Persian Gulf*. Hurst.

Karolak, M., & Allam, N. (Eds.). (2020). *Gulf cooperation council culture and identities in the new millennium: Resilience, transformation, (re)creation and diffusion*. Palgrave Macmillan.

Karoui, H. (2020). *Overcoming the Darkness of Ottoman Rule. T.E. Lawrence and the Arab Revolution*. Gulf Futures Center, London. Available at https://gulffutures.org/en/archives/3253.

Khatib, D. K. (ed.) (2021). *The Syrian crisis: effects on the regional and international relations*. Springer (Perspectives on Development in the Middle East and North Africa (MENA) Region).

Khatib, L. (2014). *Qatar and the recalibration of power in the Gulf*. Carnegie Middle East Center. Available at https://carnegieendowment.org/files/qatar_recalibration.pdf.

Kibria, N. (2011). *Muslims in motion: Islam and national identity in the Bangladeshi diaspora*. Rutgers University Press.

Kok, J. (2020). Short-term trade-off between stringency and economic growth. *COVID Economics, 60*, 172–189. Available at https://www.researchgate.net/profile/Jason_Kok/publication/346643101_Short-term_trade-off_between_stringency_and_economic_growth/links/5fcad9f645851568d13f07ec/Short-term-trade-off-between-stringency-and-economic-growth.pdf

Krieg, A. (ed.) (2019). *Divided Gulf: the anatomy of a crisis*. Palgrave Macmillan (Contemporary Gulf studies).

Laursen, F. (2018). *Comparative regional integration: Theoretical perspectives*. Routledge.

Legrenzi, M., & Momani, B. (2011) *Shifting geo-economic power of the Gulf: oil, finance and institutions*. Ashgate (The international political economy of new regionalisms series).

Luomi, M. (2012). *The Gulf monarchies and climate change: Abu Dhabi and Qatar in an era of natural unsustainability*. Hurst Company (Power and politics in the Gulf).

Maddison A. (2007). *Contours of the world economy, 1–2030 AD: essays in macro-economic history*. Oxford University Press.

McGann, J. (2020). *2019 Global go to think tank index report*. University of Pennsylvania. Available at https://repository.upenn.edu/think_tanks/17/

Miniaoui H. (Ed.) (2020). *Economic development in the Gulf Cooperation Council countries: from rentier states to diversified economies*. Springer (Gulf Studies, Volume 1). https://doi.org/10.1007/978-981-15-6058-3.

Modis, T. (2017). A hard-science approach to Kondratieff's economic cycle. *Technological Forecasting and Social Change, 122*, 63–70.

Mohamed Nour, S. (2013). *Technological change and skill development in Arab Gulf countries*. Springer (Contributions to Economics).

Morgan, A. K., Awafo, B. A., & Quartey, T. (2021). The effects of COVID-19 on global economic output and sustainability: Evidence from around the world and lessons for redress. *Sustainability: Science, Practice and Policy, 17*(1), 77–81.

Neriah, J. (2018). *Turkey's expansionist military policies in the Middle East*. Jerusalem Center for Public Affairs. Available at https://jcpa.org/article/turkeys-expansionist-military-policies-in-the-middle-east/.

Norris, P., & Inglehart, R. F. (2011). *Sacred and secular: Religion and politics worldwide*. Cambridge University Press.

Norris, P., & Inglehart, R. F. (2015). *Are high levels of existential security conducive to secularization? A response to our critics. In The changing world religion map* (pp. 3389–3408). Springer Netherlands.

Rahman, M., Tan, T. Y., & Ullah, A. K. M. A. (Eds.). (2014). *Migrant remittances in South Asia: Social, economic and political implications*. Palgrave Macmillan.

Ramady, M. A. (2012). *The GCC economies: Stepping up to future challenges*. Springer Science Business Media.

Sadig, H. B. (Ed.) (2019). *Al Jazeera in the Gulf and in the world is it redefining global communication ethics?* Palgrave Macmillan (Contemporary Gulf studies).

Sadiki, L. (2020). Regionalism in crisis: GCC integration without democracy. *The International Spectator, 55*(2), 17–33.

Schumpeter, J. A., & Elliott, J. E. (1983). *The theory of economic development: An inquiry into profits, capital, credit, interest, and the business cycle* (R. Opie, Trans.) (Ser. Social science classics series). Transaction.

Sick, G. (2009). The United States and the Persian Gulf in the twentieth century. In *The Persian Gulf in history* (pp. 295–310). Palgrave Macmillan.

Siljak, D. (2015). Real Economic Convergence in Western Europe from 1995 to 2013. *International Journal of Business and Economic Development (IJBED), 3*(3). Available at https://ijbed.org/cdn/article_file/i-9_c-94.pdf

Simionescu, M. (2014). Testing sigma convergence across EU-28. *Economics & Sociology, 7*(1), 48–60.

Solomon, H., & Tausch, A. (2020). *Islamism, crisis and democratization: Implications of the World Values Survey for the Muslim world (Perspectives on development in the Middle East and North Africa (Mena region)*. Springer.

Solomon, H., & Tausch, A. (2020b). The demise of the Muslim Brotherhood in the Arab World? *Jewish Political Studies Review, 31*(1/2), 171–209.

Solomon, H., & Tausch, A. (2021). *Arab MENA countries: Vulnerabilities and constraints against democracy on the eve of the global Covid-19 crisis* (Perspectives on development in the Middle East and North Africa (Mena region). Springer.

Steinberg, G. (2014). *Leading the counter-revolution: Saudi Arabia and the Arab Spring*. (German Institute for International and Security Affairs). Available at https://www.swp-berlin.org/en/publication/saudi-arabia-and-the-arab-spring/

Steinberg, G. (2020). *Regional Power United Arab Emirates*. SWP (German Institute for International and Security Affairs). Available at https://www.swp-berlin.org/en/publication/regional-power-united-arab-emirates/

Tausch A., & Moaddel, M. (2009). *What 1.3 Billion Muslims really think. An answer to a recent Gallup study*, based on the '***World Values Survey***. Nova Science Publishers.

Tausch, A., Heshmati, A., & Karoui, H. (2014). *The Political algebra of global value change. General models and implications for the Muslim world*. Nova Science Publishers.

Tausch, A. (2019). Migration from the Muslim world to the West: Its most recent trends and effects. *Jewish Political Studies Review, 30*(1–2), 65–225. Available at http://jcpa.org/article/migration-from-the-muslim-world-to-the-west-its-most-recent-trends-and-effects/ (with data definitions and sources. Free data download available from https://www.academia.edu/37568941/Migration_from_the_Muslim_World_to_the_West_Its_Most_Recent_Trends_and_Effects

Tausch, A., & Heshmati, A. (2013). *Globalization, the human condition, and sustainable development in the twenty-first century: Cross-national perspectives and European implications*. Anthem Press (with data definitions and sources). Free data download available from https://www.academia.edu/35044095/Globalization_the_human_condition_and_sustainable_development_in_the_21st_Century._Cross-national_perspectives_and_European_implications_Codebook_and_EXCEL_data_file.

Taylor, C. L., & Russett, B. M. (Eds.). (2020). *Karl W. Deutsch: Pioneer in the theory of international relations*. Springer.

Te Velde, D. W. (2011). Regional integration, growth and convergence. *Journal of Economic Integration*, 1–28.

Ullah, A. K. M. A. (2010). *Population migration and Asia: theories and practice*. Nova Science.

Ullah, A. K. M. A. (2010). *Rationalizing migration decisions: Labour migrants in East and South-East Asia*. Ashgate Pub.

Ullah, A. K. M. A. (2012). *Divergence and convergence in the nation state: the roles in religion and migration*. Nova Science.

Ullah, A. K. M. A., & Haque, S. (2020). *The migration myth in policy and practice: Dreams, development and despair*. Springer.

Ullah, A. K. M. A., & Rahman, M. (Eds.). (2012). *Asian migration policy: South, Southeast and East Asia*. Nova Science Publishers.

Ulrichsen, K. (2011). *Insecure Gulf: The end of certainty and the transition to the post-oil era*. Columbia University Press.

United Nations Development Programme (UNDP). (2019). *Human development report*. Oxford University Press. Available at http://www.hdr.undp.org/.

World Bank. (2021). Global economic prospects. A World Bank Flagship Report, January 2021. World Bank, Washington D.C. Available at https://www.worldbank.org/en/publication/global-economic-prospects.

World Economic Forum. (2020, March). *A visual history of pandemics*. World Economic Forum. https://www.weforum.org/agenda/2020/03/a-visual-history-of-pandemics/

Yildiz, G. (2021). GCC/Qatar Reconciliation: Good or Bad News for Turkey? Point of View, 18.01.2021, SWP Berlin. Available at https://www.swp-berlin.org/en/publication/gccqatar-reconciliation-good-or-bad-news-for-turkey/

Zweiri, M., Rahman, M. M., & Kamal, A. (Eds). (2021). *The 2017 Gulf crisis: an interdisciplinary approach*. Springer (Gulf Studies, Vol. 3). https://doi.org/10.1007/978-981-15-8735-1.

# Chapter 2
# Methodology and Data for Assessing the Effects of Global Cycles and Value Change on the Future of the Gulf Region

**Abstract** In this chapter, we debate the data sources of our analysis and the statistical methods used. We mention the methodologies and data in the order of their occurrence in the present book. Our used data sources are the best available opinion surveys on the Arab countries in the world today, i.e., the Arab Opinion Index of the Arab Center for Research and Policy Studies in Doha, Qatar; the Arab Barometer, and the World Values Survey. For global opinion comparisons, we used the freely available PEW Spring 2018 Survey of the PEW Institute in Washington, D.C. For the analysis of Antisemitism, we rely on the ADL (2014) ADL-100 study by the Anti-Defamation League of Antisemitism in more than 100 countries. Our research also uses myriads of cross-national development data, made freely available online in EXCEL format at https://www.academia.edu/44295690/GULF_2020_Covid_19_PUBLIC_ACCESS_Data_base_for_the_book_Factors_affecting_the_Gulfs_future_Global_cycles_and_value_change_Springer_Publishing_Series_https_www_springer_com_series_16417_Gulf_Studies_edited_by_Prof_Mizanur_Rahman_Qatar_University. Our analysis was written in the well-established tradition of macro-quantitative and time series development research. We debate error margins and other statistical concepts and highlight such methods as factor analysis and econometric time series analysis, available via our used statistical program, the IBM-SPSS, XXIV.

**Keywords** Opinion surveys in the Arab countries · Arab Opinion Index · Arab Barometer · World Values Survey · PEW Spring Survey · ADL-100 Index · Cross-national development data · Macro-quantitative and time series development research · Error margins · Factor analysis · Econometric time series analysis · IBM-SPSS

 
A. Tausch, *The Future of the Gulf Region*, Gulf Studies 2,
https://doi.org/10.1007/978-3-030-78299-3_2

## 2.1 Introduction to the Methodology and Data for the Study of the Factors Affecting the Gulf's Future: Global Cycles and Value Change

In this chapter, we debate the data sources of our analysis and the statistical methods used. We mention the methodologies and data in the order of their occurrence in the present book.

Our used data sources are the best available opinion surveys on the Arab countries, i.e., the ***Arab Opinion Index*** of the Arab Center for Research and Policy Studies in Doha, **Qatar**; the ***Arab Barometer***, and the ***World Values Survey***.

Our research attempt is of course guided by the vast traditions of mathematical-statistical analysis in opinion survey research(Abdi, 2003; Basilevsky, 2009; Braithwaite & Law, 1985; Brenner, 2016; Browne, 2001; Davidov et al., 2008; Fabrigar et al., 1999; Hedges & Olkin, 2014; Inglehart, 1988, 2006; Inglehart & Baker, 2000; Inglehart & Norris, 2012, 2016; Inglehart & Welzel, 2003, 2009; Kim, 2010; Kline, 2014; Knippenberg, 2015; McDonald, 2014; Minkov & Hofstede, 2011, 2013; Minkov, 2014; Mulaik, 2009; Norris & Inglehart, 2002, 2012, 2015; Suhr, 2012; Yeşilada & Noordijk, 2010).

Our methodological approach is within a more general framework to study global values with the methodology of comparative and opinion-survey-based political science (Brenner, 2016; Knippenberg, 2015; Norris & Inglehart, 2015). Our methodology of evaluating the opinions of global publics from global surveys is in addition based on recent advances in mathematical statistical factor analysis (Basilevsky, 2009; Cattell, 2012; Hedges & Olkin, 2014; Kline, 2014; McDonald, 2014; Mulaik, 2009; Tausch et al., 2014). Such studies allow to project the underlying structures of the relationships between the variables.

Current methodology of the social sciences makes it clear that besides factor analysis, there are also other powerful tools of multivariate analysis available to test complex relationships between an independent variable and independent variables (Blalock, 1972; Tabachnick & Fidell, 2001). In our case, we used standard OLS multiple regression analysis and partial correlation analysis. Throughout this work, we use standard partial regression analysis to come to terms with the often-underestimated omitted variable bias in quantitative social science, always cross-checking significant correlations with already existing literature on the subject. Omitted variable bias indeed is a serious problem in the discipline (see Clarke, 2005, 2009; Jargowsky, 2005).

For global opinion comparisons, we used the freely available PEW ***Spring 2018 Survey*** of the PEW Institute in Washington, D.C. For the analysis of Antisemitism, we rely on the ADL (2014) ***ADL-100 Index*** study by the Anti-Defamation League of Antisemitism in more than 100 countries. Our research also uses myriads of cross-national development data and was written in the well-established tradition of macro-quantitative and time series development research (see below). We especially mention what can be termed "development accounting" (see: Grinin et al., 2016;

Tausch, 2010a, 2010b; Tausch & Heshmati, 2012, 2013). We project our research results also at the level of choropleth maps.[1]

In Chaps. 3 and 4 of the present book, we analyze the world political and economic environment which the Gulf countries will face in the coming years, relying on time series analyses of the cyclical movements of world political and economic conditions since 1495 and 1740, respectively, using the advanced long cycle methodologies presented in Grinin et al. (2016), such as spectral analysis and analyses of autocorrelations, well established in econometrics. So, we present the necessary methodology for Chaps. 3 and 4.

In general, our entire work firmly shares the established methodology of global values and comparative opinion research (Davidov et al., 2008; Inglehart, 2006; Norris & Inglehart, 2015; Tausch et al., 2014). Illuminating earlier opinion studies on the Arab countries, making proper use of the systematic study of comparative opinion surveys, were published, among others, in Achilov (2013), Al-Ississ and Diwan (2016), Ciftci and Bernick (2015), Fox et al. (2016), Gorman (2015), Grim (2014), Hoffman and Jamal (2012), Nisbet et al. (2004), Sjoberg and Whooley (2015), and Tessler (2004). Consequently, these articles serve as a methodological guiding post for the present book. The mentioning of these articles could inspire interested readers to take up the challenge of working with the presented open data and the IBM-SPSS statistical software package of their own choice to produce new and even more complete insights on the questions under debate here.

## 2.2 Statistical Methods for the Study of the Factors Affecting the Gulf's Future

The aim of this book is to present reliable and comparable analyses and data on the environment and the internal conditions which the Gulf countries find in the twenty-first century. Based on comparable cross-national and time series macro-societal and opinion survey data, we have to rely on statistical methods to calculate the results.

Our main statistical calculations relied on simple cross-tables, comparisons of means, bi-variate and partial correlation analyses, factor analysis (oblique factor rotations based on promax factor analysis) and standard multiple regressions (OLS) (for a survey, see also Tausch et al., 2014).

Concerning factor analysis and the so-called oblique rotation of the factors, which are underlying the correlation matrix, re also refer our readers to important literature on the subject (Abdi, 2003; Browne, 2001; Kim, 2010). The IBM-SPSS routine chosen in this context was the so-called *promax* rotation of factors (Braithwaite & Law, 1985; Browne, 2001; Fabrigar et al., 1999; Minkov, 2014; Suhr, 2012; Yeşilada & Noordijk, 2010), which in many ways must be considered to be the best suited

[1] The choropleth maps of this book were drawn using the free software developed by Robert Mundigl, available at https://www.clearlyandsimply.com/.

rotation of factors in the context of our research today.[2] Formulated in plain everyday language, the mathematical procedures of the rotation of factors which best represent the dimensions underlying a correlation matrix are necessary to make the structure simpler and more reliable.

The problem which factor analysis is solving can be described as follows: can the variables under consideration here be represented in mathematically reduced dimensions, and what percentages of the total reality are thus reproduced, and how are these dimensions related to each other? And what is the relationship of the underlying variables with these dimensions? Is there indeed such a "factor" or "dimension" as religiosity, and how does it affect phenomena like "trust in the police" or "Antisemitism"? Is there, apart from it, also something like "feminism," and also something like "class" or "status," which influences "trust in the police" or "Antisemitism," independent from the other "factors"? Promax factor analysis is a well-established multivariate and mathematical variety among the general techniques of factor analysis, which extracts the underlying dimensions from the matrix of correlations between the variables and precisely answers the questions just raised above.[3] It was amply described in recent literature (Finch, 2006; Tausch et al., 2014, see, furthermore Gorsuch, 1983; Harman, 1976; Rummel, 1970). As already stated, Promax factor analysis is considered to be the most appropriate technique of factor analysis in public opinion survey studies today (Ciftci, 2010, 2012, 2013; Ciftci & Bernick, 2015; Finch, 2006). Factor analysis—in our case promax factor analysis—also allows the researcher to use the mathematical model for the development of a new measurement scale for the new dimensions, derived in the research process (Tausch et al., 2014). In modern social indicators research, such new scales are called "*parametric indices*."

For the calculation of error margins of the representative opinion survey, readers are referred to the easily readable introduction to opinion survey error margins, prepared by Cornell University Roper Center (2017). Readers more interested in the details are also being referred to Langer Research Associates n.d.[4] On the basis of the methodological literature on opinion surveys, this Web site makes available a direct opinion survey error margin calculator. It is important to recall that, for example at a 5% distrust rate in the police, error margins for our chosen samples of around 1,000 representative interview partners for each country are +− 1.4%. A 10% distrust rate, the error margin is +− 1.9%: and at a distrust rate of 15% the error margin is +− 2.2%; see Langer Research Associates n.d. That error margins differ according to

[2] Older approaches often assumed that there is no correlation between the factors, best representing the underlying dimensions of the variables. But for example, in attempting to understand the pro-Brexit vote in the UK it would be ridiculous to assume that, say, there is no correlation between anti-immigration attitudes and the feeling to be among the losers of globalization.

[3] The mathematical algorithm is described in https://www.ibm.com/support/knowledgecenter/en/SSLVMB_22.0.0/com.ibm.spss.statistics.algorithms/alg_factor_promax.htm. Interested readers are also referred to materials used at the University of Texas in Dallas, available at https://www.utdallas.edu/~herve/Abdi-rotations-pretty.pdf.

[4] https://www.langerresearch.com/moe/.

reported rates of, say, distrust in the police, is an important fact of opinion survey research theory, often forgotten to be mentioned in the public debate.

Keeping in line with standard traditions of empirical opinion survey research (Tausch et al., 2014), for all analyzed groups and sub-groups, a minimum sample size of at least 30 respondents per country had to be available to be able to attempt reasonable predictions (Clauß & Ebner, 1970).

## 2.3 Time Series Analysis in Chaps. 3 and 4

Our book also analyzes long cycles of both global and Gulf countries development. Providing the adequate methods for the fascinating totality of such an approach is not an easy task.

Our data for world industrial production growth since 1740–1975 were provided by Goldstein (1988). From 1975 onwards, these data were updated by Tausch and Ghymers (2007), relying on UNIDO data on worldwide industrial production growth from the mid-1970s to the turn of the millennium. From 1995 onwards, we rely on World Bank Open Data figures for world industrial production growth.

Our figures on annual battle fatalities from major power wars (= permanent current members of the UN Security Council plus Germany) from 1495 onwards were first presented by Goldstein (1988), and were updated by PRIO data (major power wars) from 1975 until 2008 (PRIO Battle Deaths Dataset, 3.1; Lacina & Gleditsch, 2005; Lacina et al., 2006; Peace Research Institute, Oslo (PRIO), 2021; Sarkees, 2000; see also Tausch & Ghymers, 2007).

In considering long economic and political waves, we also considered in this book also data from a variety of other sources. Our standard data on real purchasing power per capita for the 31 countries under scrutiny here are exclusively based on Angus Maddison's widely received research; as freely downloadable in Bolt and van Zanden (2013), for further debates see also Maddison (2003, 2007). These countries currently make up approximately 40.8% of global population and 57.8% of global purchasing power. At the Web site https://www.academia.edu/3742045/Background_materials_to_the_article_The_hallmarks_of_crisis_A_new_center-periphery_perspective_on_long_cycles, readers can download the most important data and also hundreds of spectral density graphs, rolling correlations and regressions, autocorrelation analyses at the level of the world system, and at the level of the 31 analyzed countries.

**Annual battle fatalities from major power wars** (= permanent current members of the UN Security Council plus Germany) from 1495 onwards (Peace Research Institute, Oslo (PRIO), 2021).

**Barro datasets**: http://rbarro.com/data-sets/.

**IMF World Economic Outlook Database**: http://www.imf.org/external/pubs/ft/weo/2014/01/weodata/index.aspx.

**Inequality**: Estimated Household Income Inequality Dataset (EHII)—is a global dataset, derived from the econometric relationship between UTIP-UNIDO, other conditioning variables, and the World Bank's Deininger & Squire dataset http://utip.gov.utexas.edu/data.html.

**Maddison time series data**. Bolt and Van Zanden (2013). The First Update of the Maddison Project; Re-Estimating Growth Before 1820, Maddison Project Working Paper 4. http://www.ggdc.net/maddison/maddison-project/abstract.htm?id=4.

**Top Income data**: http://piketty.pse.ens.fr/fr/capitalisback.

**Unemployment**: unemployment as % of the civilian labor force: http://stats.oecd.org/Index.aspx.

**War cycle since 1495 Data**: Goldstein, J. S. 1988. *Long Cycles: Prosperity and War in the Modern Age*. New Haven: Yale University Press, and—after 1945—PRIO Oslo, see: Tausch A., & Ghymers Ch. 2007. *From the "Washington" towards a "Vienna Consensus"? A quantitative analysis on globalization, development and global governance*. Hauppauge, N.Y.: Nova Science Publishers. The data are available from our electronic appendix.[5]

**World Bank data**: http://data.worldbank.org/indicator.

**World industrial production growthsince 1750**. 1740–1975: based on Goldstein, 1988; 1975–1994, based on UNIDO, as documented in Tausch and Ghymers (2007); 1995–2019, taken from World Bank data, indicator: NV.IND.TOTL.KD.ZG, which refers to industry (including construction), value added (annual % growth), annual growth rate for industrial value added based on constant local currency. Aggregates are based on constant 2010 US dollars. Industry corresponds to ISIC divisions 10–45 and includes manufacturing (ISIC divisions 15–37). It comprises value added in mining, manufacturing (also reported as a separate subgroup), construction, electricity, water, and gas. Value added is the net output of a sector after adding up all outputs and subtracting intermediate inputs. It is calculated without making deductions for depreciation of fabricated assets or depletion and degradation of natural resources. The origin of value added is determined by the International Standard Industrial Classification (ISIC), revision 3. The data are available from our electronic appendix.[6]

**World industrial production growthin 2020**: The International Monetary Fund now already anticipates a global recession of minus 2.9% due to the "corona crisis".[7] United Nations UNIDO research, in a "meta-analysis" of all the published estimates

[5] https://www.academia.edu/45321329/Factors_affecting_the_Gulfs_future_Global_cycles_and_value_change_Electronic_Appendix.

[6] https://www.academia.edu/45321329/Factors_affecting_the_Gulfs_future_Global_cycles_and_value_change_Electronic_Appendix.

[7] https://www.imf.org/en/Topics/imf-and-covid19/Policy-Responses-to-Covid-19 and https://www.nzz.ch/wirtschaft/der-imf-erachtet-die-corona-krise-als-ernsthaftes-risiko-fuer-die-weltwirtschaft-ld.1544427.

by international organizations now estimates[8] that COVID-19 will cause a substantial global output contraction. The costs of uncertainty are comparable to those generated by the pandemic itself. The accelerated spread of COVID-19 has brought the world economy to a standstill. UNIDO underlines that there is near unanimous consensus that the world economy is facing the most serious challenge of the post-war era due to the sudden halt in economic activity in both advanced and developing countries. The WTO's latest forecasts are the most pessimistic, projecting a negative growth rate of nearly 9% for 2020 in the worst-case scenario. The projections for the volume of trade of goods and services are even more dire. The WTO predicted a decline of between 13 and 32%. Capital flows are flooding back to advanced economies. UNCTAD's latest Global Investment Trend Monitor expects that foreign direct investment will shrink between – 30 and – 40% in 2020–2021.

## 2.4 Methods of Time Series Analyses in Chaps. 3 and 4

Our book is based on a standard IBM-SPSS XXIV time series analysis of the data, and all the used methods and their algorithms are fully available to the international public (IBM, 2014). For that reason, we refrain from reproducing the mathematical formula, which interested readers might easily download from the freely available IBM Internet documentation, 2014 (if they are not very familiar with the mathematical formulas anyway):

- As to **correlations**, we used standard Pearson-Bravais correlation coefficients. In calculations not reported here also, multivariate analyses (principal components) were used (Blalock, 1972; Dziuban & Shirkey, 1974; Harman, 1976; Rummel, 1970). ***"Rolling"*** **regressions and correlations** are quite a powerful and straightforward instrument of the analysis of time series and became more popular in recent times in the framework of financial market trend analysis and the necessity to have easily interpretable and reliable instruments of analysis at hand (Smith & Taylor, 2001; Tang, 2010; Zivot & Wang, 2006, furthermore: Cook, 1977; Dempster, 1969; Velleman & Welsch, 1981).
- Throughout this work, we use 25 year (Kondratieff cycles, 2 × 25 years = 50 years) and 75-year periods (war cycles, 2 × 75 years = 150 years) for the moving time window of regression/correlation analysis. We also used shorter windows to reproduce the Barro, the Kuznets, and the Juglar cycles.
- Our analyses of **autocorrelation** and **cross-correlation** are based on the standard **IBM-SPSS ACF** and **CCF algorithm**, which are based on Bartlett (1946), Box and Jenkins (1976), Cryer (1986), Quenouville (1949) (**autocorrelation**) and Box and Jenkins (1976) (**cross-correlation**). Our graphs allow also for the inspection of longer time series. In presenting the graphs, we also took care of the better visibility of the significant results.

[8] https://www.unido.org/stories/coronavirus-economic-impact.

- The **IBM-SPSS spectral density routine**, which is based on the methodological developments, presented by Bloomfield (1976), and Fuller (1976) was performed by using the IBM-SPSS default options; the chosen window was most of the time the so-called Tukey-Hamming window with three periods. We also tested the validity of our main results with longer windows as well. As we demonstrate, however, in our non-mathematical primer on spectral density analysis, available at https://www.academia.edu/3742045/Background_materials_to_the_article_The_hallmarks_of_crisis._A_new_center-periphery_perspective_on_long_cycles, longer windows seem to distort the results even of simulated time series, where we really know the length of the oscillations, and they do not really help us to discover the real periodicity of the oscillations in question, especially when we are confronted with the already mentioned "Devezas paradox" of longer cycles and "nested" shorter cycles (Devezas, 2006, 2010, 2012). We use both the standard IBM-SPSS periodograms and the spectral density graphs and add that standard econometric methodological literature usually maintains that the periodograms are not the best estimator of the spectrum because it is not convergent (Diebolt, 2012).

In the following, we start from a simulated 190-year cycle, in order to demonstrate the potentials of the methodology (Fig. 2.1).

Applying SPSS XXIV Spectral Analysis and Autocorrelation Analysis, we get Figs. 2.2, 2.3, and 2.4 which clearly demonstrate the existing 190 years long cycle of Fig. 2.1:

Following Devezas (2006, 2010, 2012), we now introduce the notion of a "nested cycle" of 63 years duration, which we simulate for the aims of demonstration of the methodology of long cycle analysis all along with the 190-year cycle. As we will see

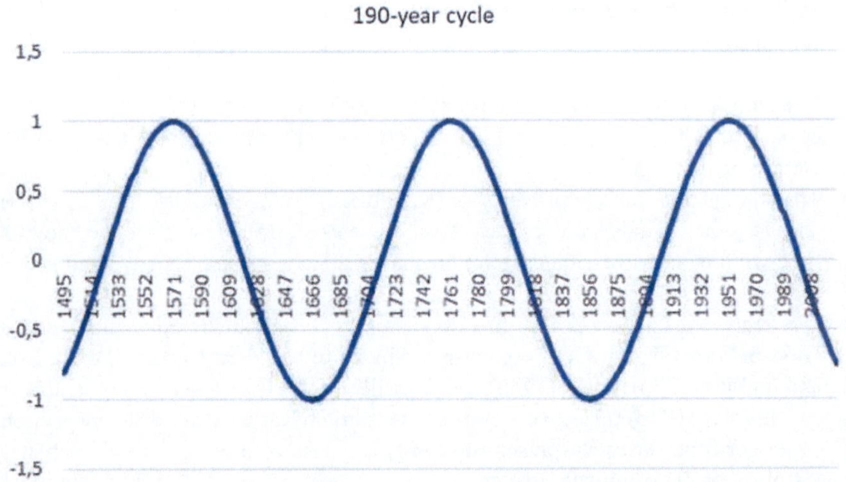


**Fig. 2.1** Simulated 190-year cycle. *Note*: y = sin (time axis * 0.033); troughs: 1666 and 1856; peaks: 1571, 1761, 1952

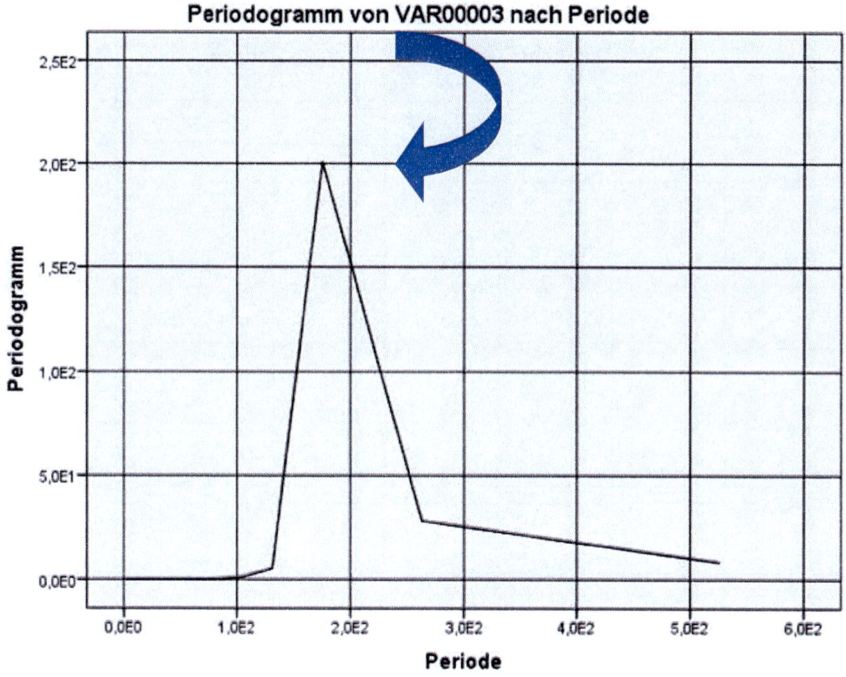


**Fig. 2.2** Periodogram of the data from Fig. 2.1 in the SPSS-XXIV "spectral analysis"

in Fig. 2.5 with real data from the growth of the US economy below, such a simulated structure is not too distant from the empirical realities.

The powerful tools of the periodograms of Spectral Analysis now allow us to identify in a crystal-clear fashion the existence of these two simulated cycles (Fig. 2.6).

The spectral density diagram (Fig. 2.7), making use of the usual window in the SPSS XXIV routine (on the function of these windows, see Diebolt, 2012), looks like the following, suggesting that for the identification of cyclical swings and waves, the periodogram will deliver more clear-cut results:

The autocorrelation procedure in SPSS XXIV delivers the following image (Fig. 2.8) from the data of Fig. 2.5. The two peaks from the 190-year cycle with a "nested" 63-year cycle are clearly visible, but the procedure produces other peaks as well. Thus, the periodogram of the "spectral analysis" SPSS program identifies in the most clear-cut fashion the length of the cycles analyzed from a very large mass of data.

To show the potentials of the time series methodology, available from the SPSS, we show a simulated 190-year cycle and a lagged 190-year cycle, completely depending on the first cycle with a time-lag of 30 years. Such analyses are necessary, for example, in our endeavor to study the question, whether or not there are significant relationships

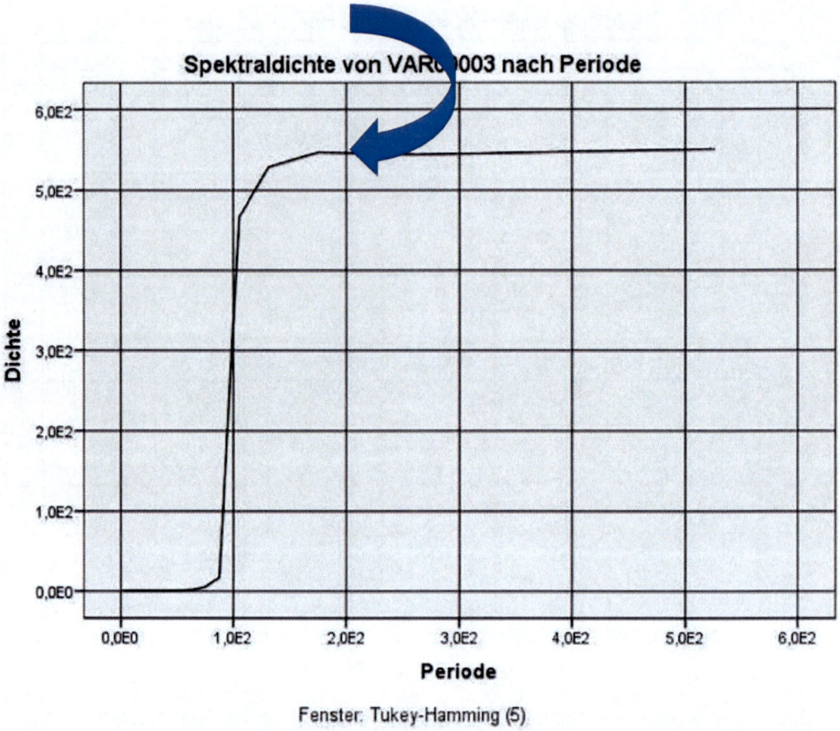


**Fig. 2.3** Spectral density diagram of the data from Fig. 2.1 in the SPSS-XXIV "spectral analysis"

between the economic cycles in the world economy and the cycles of global warfare in world politics (Fig. 2.9).

Figure 2.10 shows the results of the cross-correlation procedure from the SPSS autocorrelation program library. The cross-correlation diagram clearly shows the dependency of cycle 2 from cycle 1 and the time period of 30 years.

In Fig. 2.11, we now let the real world enter our presentation of the time series methodology. Figure 2.11 is based on the Barro-Ursúa dataset, which can be regarded today as the most sophisticated and reliable dataset of per capita income growth over very long periods. The often, quite heavy fluctuations of the US economy emerge from Fig. 2.11. The first half of the 20th Century, the period of the cataclysmic conflicts from 1914 to 1945 and the Influenza Pandemic of 1918–1920, had the effect of especially sharps ups and downs for America.

Figure 2.12 identifies the existence of various cycles, among them a 23-year Kuznets cycle, a 36-year Barro cycle, and a 54-year Kondratieff cycle in the United States economy, although the Barro cycles and the Kondratieff cycles are much weaker than the Kuznets cycles.

Figure 2.13 shows the results of the spectral analysis for the data of Fig. 2.11, suggesting that the existence of the 23-year Kuznets cycle in the US is beyond doubt.

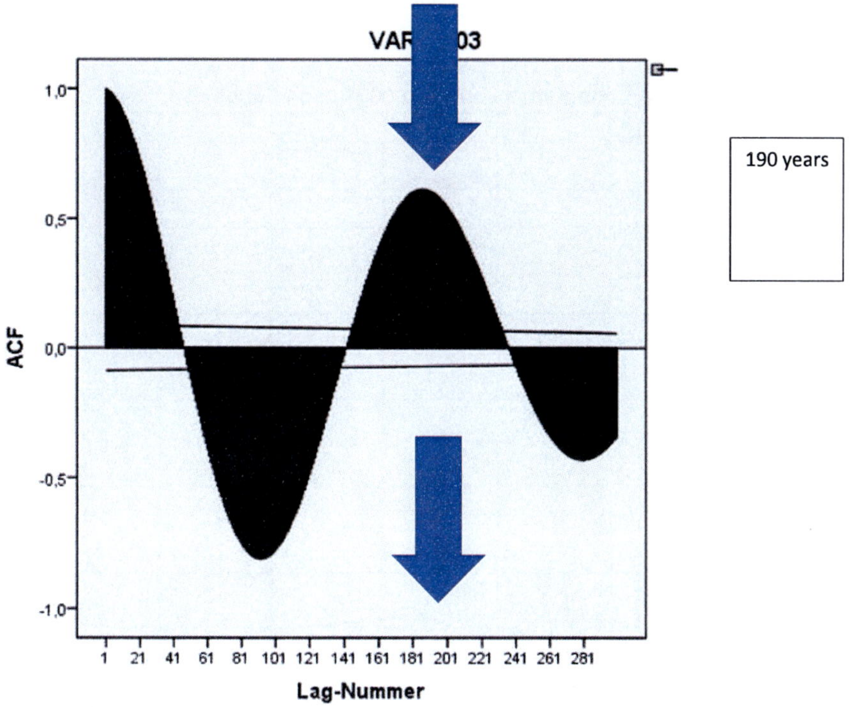


**Fig. 2.4** Autocorrelation diagram of the data from Fig. 2.1 in the SPSS-XXIV "autocorrelation analysis"

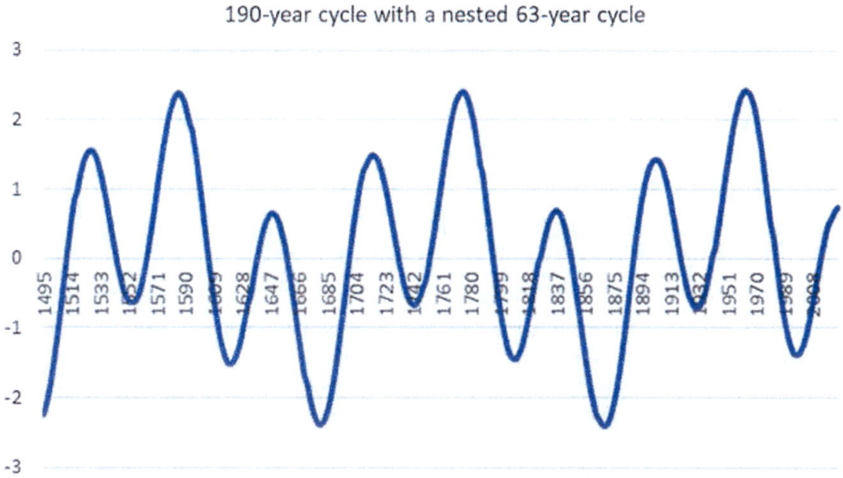


**Fig. 2.5** Simulated 190-year cycle with a "nested" 63-year cycle. *Note* Cycle 1: $y = \sin$ (time axis * 0.033); Cycle 2: $y = \sin$ (time axis * 0.1); Cycle 3 = Cycle 1 + ((Cycle 2) * 1.5))

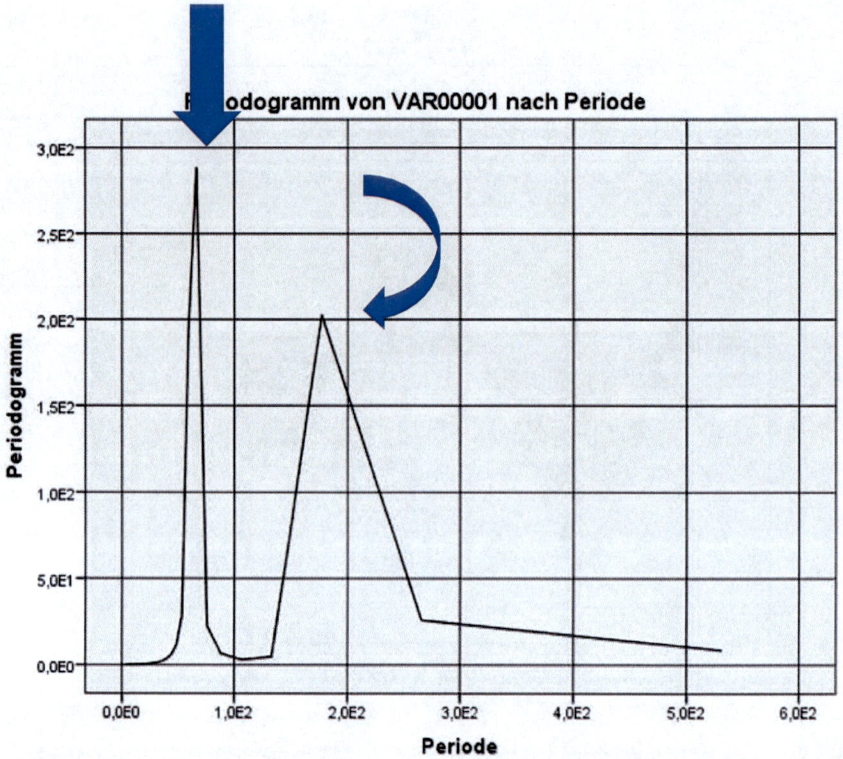


**Fig. 2.6** Periodogram of the data from Fig. 2.5 in the SPSS-XXIV "spectral analysis"

Finally, Fig. 2.14 portrays the results of the autocorrelation analysis for the time series of Fig. 2.11. The image suggests that the cycles, referred to above, are not too strong, and that only the Kondratieff cycles achieve a significance in the statistical sense.

## 2.5 The Survey Data Used from Chaps. 5–8

As already stated, the present book relies on the statistical analysis of open survey data and aggregate country statistical data and country and global time series data and is based on the commonly used statistical software IBM-SPSS XXIV, utilized at many universities and research centers around the world.[9] The program contains the entire array of modern multivariate statistics, and any researcher should be able to arrive at the same results as we do here when she or he uses the same open data and the IBM-SPSS. Clearly, our analysis provides only a first attempt to measure

[9] IBM SPSS 2007.

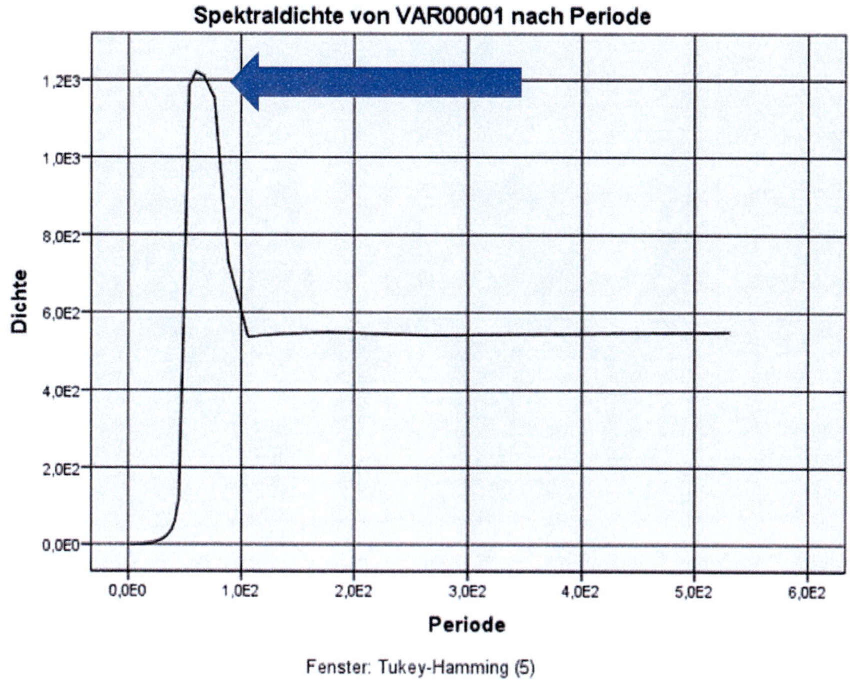


**Fig. 2.7** Spectral density diagram of the data from Fig. 2.5 in the SPSS-XXIV "spectral analysis"

values in the Gulf and is understood as an invitation to further research, especially also along generational, regional, and gender lines.

**Global value studies** are made possible by the availability of systematic and comparative opinion surveys over time under the auspices of leading representatives of the social science research community, featuring the global populations with a fairly constant questionnaire for several decades now. More and more, the countries of the Gulf region form an integral part of these surveys.

So, the sources of our opinion data, mentioned in the order of their occurrence in the present book, are:

## 2.5.1 *Arab Opinion Index*

- First, the  of the Arab Center for Research and Policy Studies in Doha, **Qatar** (***Arab Opinion Index***) 2017/2018. The ***Arab Opinion Index*** is currently the largest public opinion survey in the Arab World. The 2017–2018 ***Arab Opinion Index*** was the sixth in a series of yearly public opinion surveys across the Arab World, always conducted by Arab Center for Research and Policy Studies in

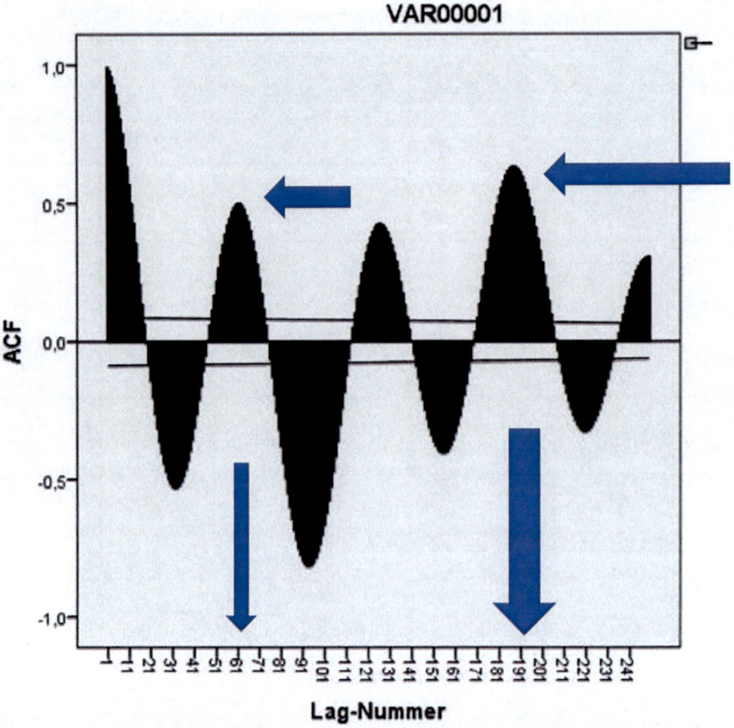


**Fig. 2.8** Autocorrelation diagram of the data from Fig. 2.5 in the SPSS-XXIV "autocorrelation analysis"

Doha, **Qatar**. The first survey was conducted in 2011, with surveys following in 2012–2013, 2014, 2015, and 2016. The 2017–2018 ***Arab Opinion Index*** is based on the findings of face-to-face interviews conducted with 18,830 individual respondents in 11 Arab countries: Egypt, Iraq, Jordan, **Kuwait**, Lebanon, Mauritania, Morocco, Palestine (i.e., Westbank and Gaza), **Saudi Arabia**, Sudan, and Tunisia. Sampling followed a randomized, stratified, multi-stage, self-weighted clustered approach, giving an overall margin of error between ± 2 and 3% for the individual country samples. The overall samples guarantee probability proportional to size (PPS), ensuring fairness in the representation of various population segments. With an aggregate sample size of 18,830 respondents, the ***Arab Opinion Index*** clearly is the largest public opinion survey in the Arab World. The fieldwork was carried out by an overall team of 840 individuals, equally balanced by gender, who conducted 45,000 h of face-to-face interviews. The team covered a total of 700,000 km across the population clusters sampled. The 2019–2020 ***Arab Opinion Index*** is the seventh in a series of yearly public opinion surveys across the Arab World, based on the findings from face-to-face interviews conducted with 28,288 individual respondents in 13 Arab countries: Saudi Arabia, Kuwait, Qatar,

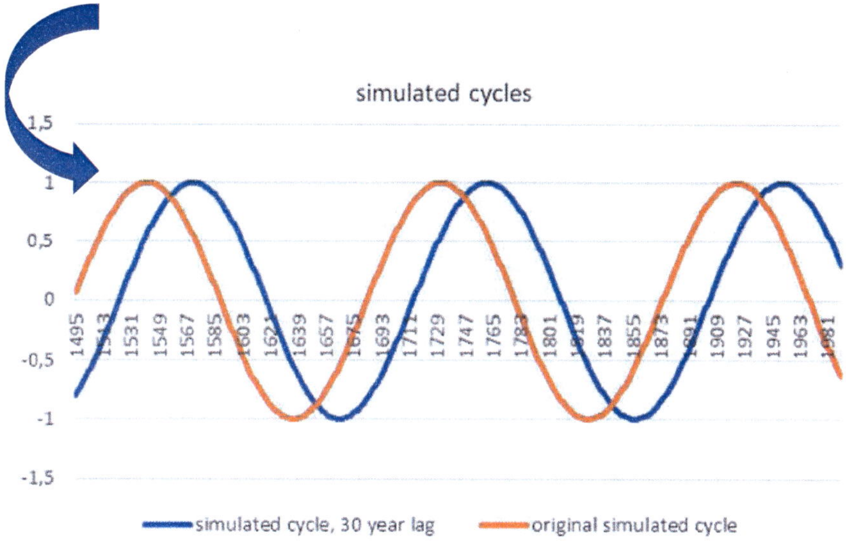


**Fig. 2.9** Simulated 190 cycle and a lagged 190-year cycle (30 years distance to cycle 1)

Iraq, Jordan, West Bank and Gaza, Lebanon, Egypt, Sudan, Tunisia, Morocco, Algeria, and Mauritania. Sampling followed a randomized, stratified, multi-stage, self-weighted clustered approach, giving an overall margin of error between ± 2 and 3% for the individual country samples. The overall samples guarantee probability-proportional-to-size (PPS), ensuring fairness in the representation of various population segments. A total of 69,000 h of actual work was needed to carry out these surveys by an overall team of 900 individuals, equally balanced on gender. The team covered a total of 820,000 km across the population clusters sampled.

Since the original interview data are not available for secondary analysis, we had to work with the research results, as they were published by the ACRPS. It is lamentable that this unique research instrument is so little known in the current scholarly debate on the Middle East. Of the 93 items listed on *Google Scholar,* mentioning the ***Arab Opinion Index*** since 2016, none the less than seven were published by the present author, and five were published in the ACRPS flagship journal "AlMuntaqa." A leading scholar of the ACRPS, Ms. Dana El Kurd, from the Doha Institute for Graduate Studies,[10] authored six of these studies (El Kurd, 2017–2020).

[10] https://scholar.google.com/citations?user=3vs_e5QAAAAJ&hl=en&oi=sra.

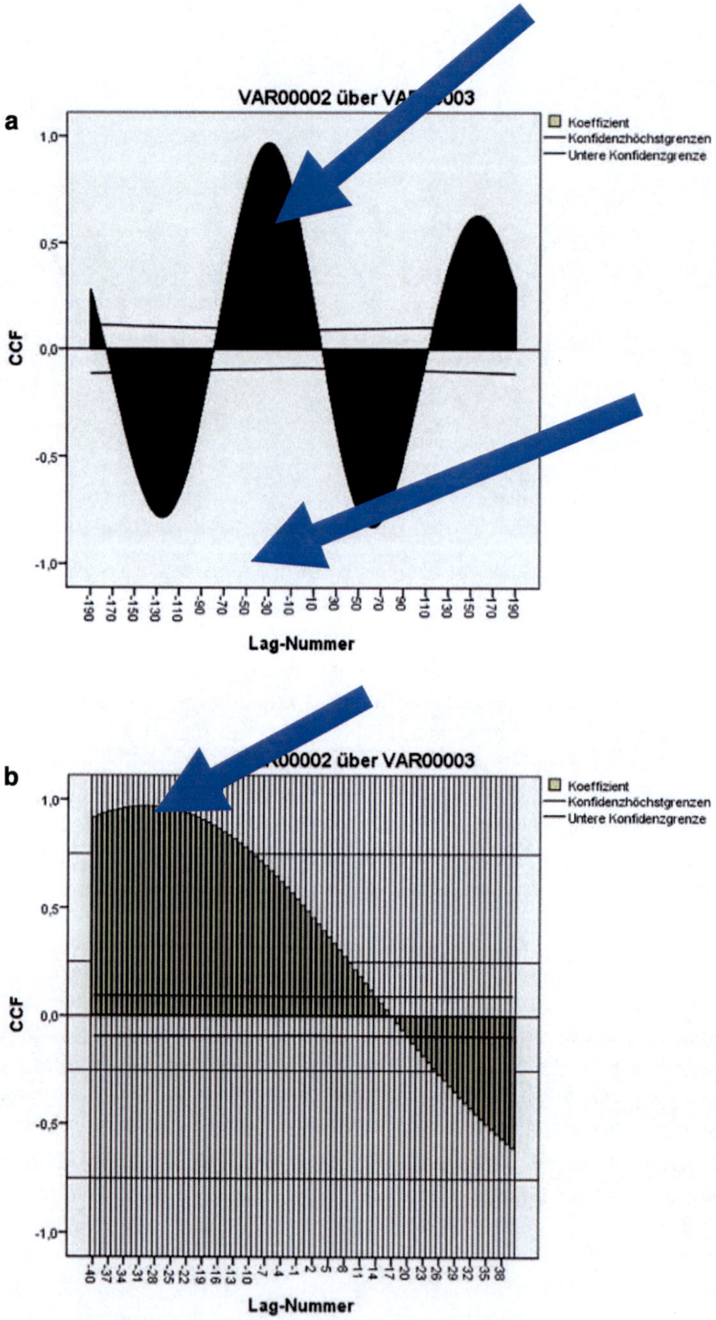


**Fig. 2.10** **a** SPSS CCF—cross-correlation clearly shows that the lagged variable is absolutely dependent on the simulated variable, and that the effect takes thirty-years. **b** SPSS CCF—cross-correlation clearly shows that the lagged variable is absolutely dependent on the simulated variable, and that the effect takes thirty-years, close-up on the first 30 years

**Fig. 2.11** Enter the real world: US economic growth. *Note* Calculated from the Barro dataset about GDP pc Indexes, 2006 = 100; available at: https://scholar.harvard.edu/barro/data_sets; Barro-Ursúa Macroeconomic Data (2010), available in Excel format

## *2.5.2 PEW Spring Surveys*

- Secondly, for our study of the current vacuum in global leadership (Chap. 5) which the countries of the Gulf are now facing, we relied on data of global opinion on world leaders from the PEW Institute in Washington, D.C., USA. The PEW ***Spring 2018 Survey*** (PEW Research Center 2018)[11] was conducted in 2018 and is based on face-to-face and telephone interviews in 27 countries with adults 18 and older (https://www.pewresearch.org/methodology/international-survey-research/international-methodology/global-attitudes-survey/all-country/2018). The survey covers around 47% of the global citizens on earth. The countries are to be seen in our publicly available electronic data appendix.[12]

The data of the survey are freely available for download.[13] We used the following variables for the multivariate analysis[14] in Chap. 5 of the present book:

**Box 2.1: Global leadership preferences and their determinants**

[11] See: https://www.pewresearch.org/global/dataset/spring-2018-survey-data/.

[12] https://www.academia.edu/45321329/Factors_affecting_the_Gulfs_future_Global_cycles_and_value_change_Electronic_Appendix.

[13] https://www.pewresearch.org/global/dataset/spring-2018-survey-data/.

[14] Using PEW data in multivariate analysis, users have to make sure that the don't knows and missing values are properly excluded.

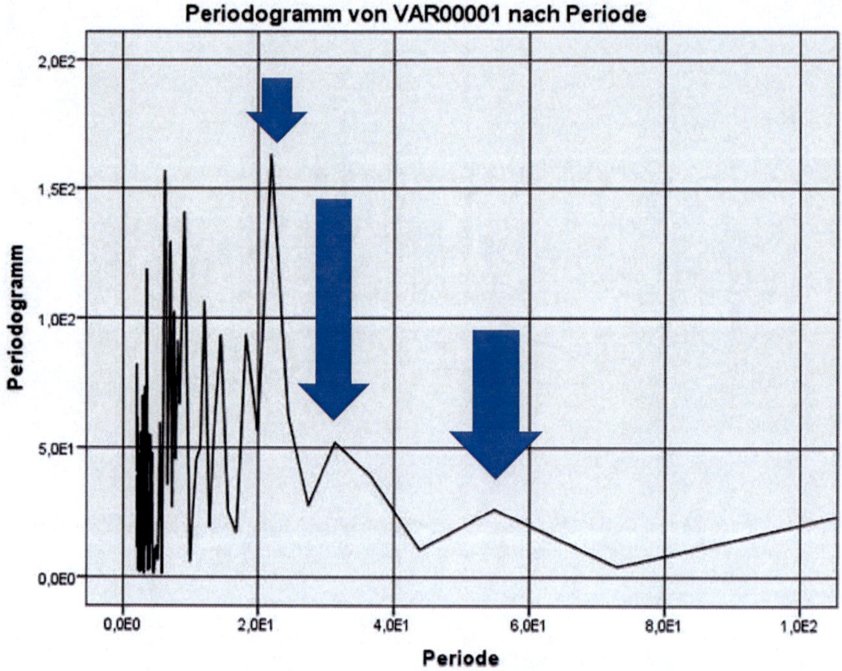


**Fig. 2.12** Periodogram of the data from Fig. 2.11 in the SPSS-XXIV "spectral analysis"—the cycles of the US economy. *Note* Long-wave research currently identifies the following longer economic cycles or waves: the 16-year Juglar cycle; the 23-year Kuznets cycle; the 36-year Barro cycle; the 54-year Kondratieff cycle; and the 140-year Wallerstein cycle; see Grinin et al. (2016)

Q4. How satisfied are you with the way democracy is working in our country—very satisfied, somewhat satisfied, not too satisfied, or not at all satisfied?

Q6. Compared with 20 years ago, do you think the financial situation of average people in (survey country) is better, worse, or do you think there has been no change?

Q17a. Please tell me if you have a very favorable, somewhat favorable, somewhat unfavorable, or very unfavorable opinion of ____. a. The USA

Q17b. Please tell me if you have a very favorable, somewhat favorable, somewhat unfavorable, or very unfavorable opinion of ____. b. China

Q17c. Please tell me if you have a very favorable, somewhat favorable, somewhat unfavorable, or very unfavorable opinion of ____. c. Russia

Q22a. I would like your opinion about some possible international concerns for (survey country). Do you think that ____ is a major threat, a minor threat, or not a threat to (survey country)? a. China's power and influence

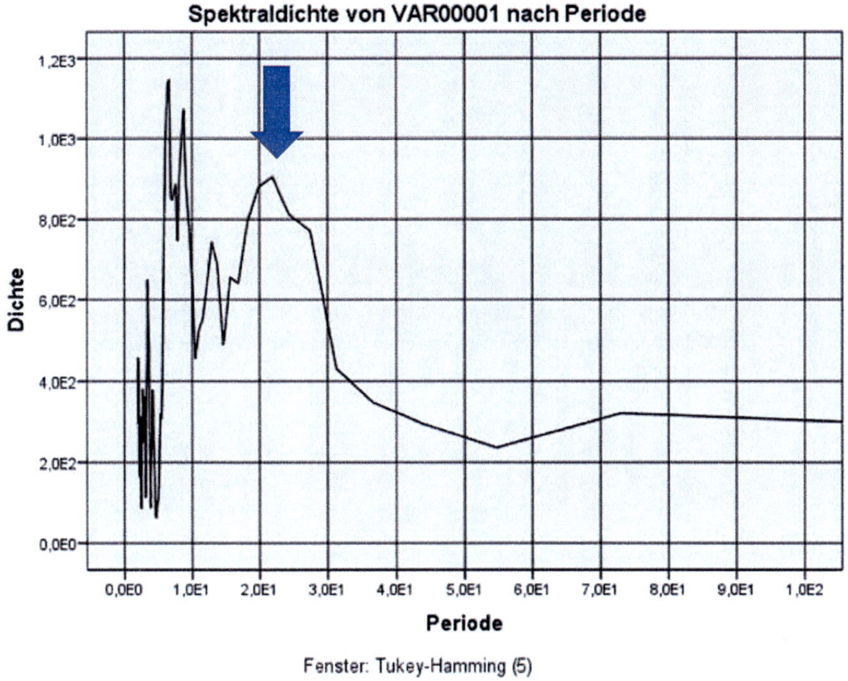


**Fig. 2.13** Spectral density diagram of the data from Fig. 2.11 in the SPSS-XXIV "spectral analysis"—cycles in the US economy. *Note* Long-wave research currently identifies the following longer economic cycles or waves: the 16-year Juglar cycle; the 23-year Kuznets cycle; the 36-year Barro cycle; the 54-year Kondratieff cycle; and the 140-year Wallerstein cycle; see Grinin et al. (2016)

Q22d. I would like your opinion about some possible international concerns for (survey country). Do you think that ____ is a major threat, a minor threat, or not a threat to (survey country)? d. Global climate change

Q22e. I would like your opinion about some possible international concerns for (survey country). Do you think that ____ is a major threat, a minor threat, or not a threat to (survey country)? e. The Islamic militant group known as ISIS

Q22g. I would like your opinion about some possible international concerns for (survey country). Do you think that ____ is a major threat, a minor threat, or not a threat to (survey country)? g. North Korea's nuclear program

Q30a. Do you think the government of ____ respects the personal freedoms of its people or do not you think so? a. China

Q30b. Do you think the government of ____ respects the personal freedoms of its people or do not you think so? b. The United States

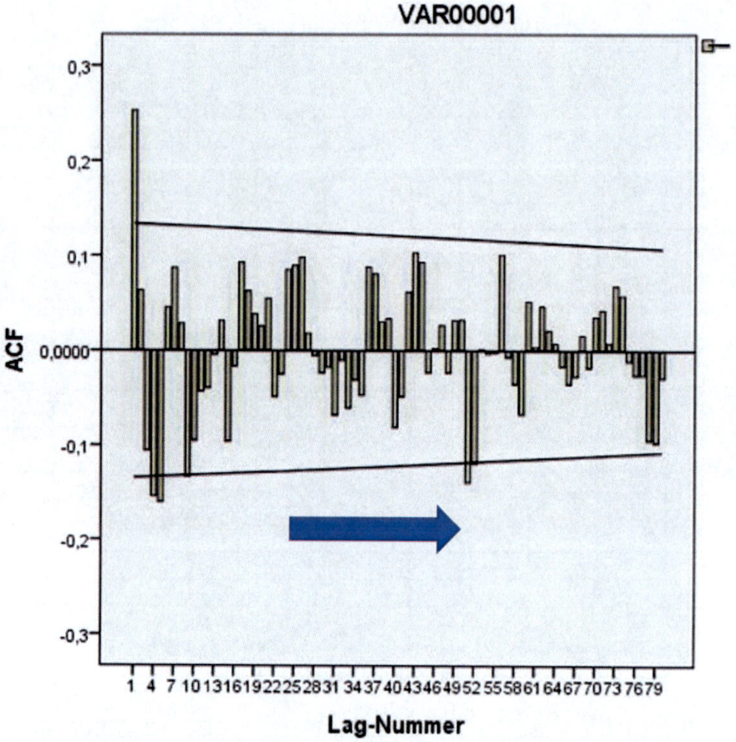


**Fig. 2.14** Autocorrelation diagram of the data from Fig. 2.11 in the SPSS-XXIV "autocorrelation analysis." *Note* Long-wave research currently identifies the following longer economic cycles or waves: the 16-year Juglar cycle; the 23-year Kuznets cycle; the 36-year Barro cycle; the 54-year Kondratieff cycle; and the 140-year Wallerstein cycle; see Grinin et al. (2016)

Q34a. Does this statement describe (survey country) very well, somewhat well, not too well, or not well at all? a. Elected officials care what ordinary people think

Q34c. Does this statement describe (survey country) very well, somewhat well, not too well, or not well at all? c. Most politicians are corrupt

Q34d. Does this statement describe (survey country) very well, somewhat well, not too well, or not well at all? d. Most people live in areas where it is dangerous to walk around at night

Q35b. For each, tell me how much confidence you have in each leader to do the right thing regarding world affairs—a lot of confidence, some confidence, not too much confidence, or no confidence at all. b. Chinese President Xi Jinping

Sex. Gender

Age. How old were you at your last birthday?

Q95. How important is religion in your life: very important, somewhat important, not too important, or not at all important?

### 2.5.3 World Values Survey

- Third, for purposes of comparison and predictions of Gulf population values, we worked with data from the ***"World Values Survey."*** The data for the analysis of the vacuum of global leadership (Chap. 5) were taken from Haerpfer et al. (2020).These data covered the period 2017–2020 and comprise 79 countries and territories. The chosen IBM-SPSS data files from the ***World Values Survey*** database for the analysis of the Gulf countries was "***World Values Survey***_Longitudinal_1981_2014_spss_v2015_04_18.sav" and "***World Values Survey***_Longitudinal_1981_2016_Spss_v20180912.sav." Our samples cover a vast number of countries across the globe. In Chap. 6, for example, our study covers more than 73% of the global population. The Gulf country research designs, based on the *World Values Survey,* are available from the electronic appendix.[15] The countries of comparison are to be seen our publicly available electronic data appendix.[16]

In the case of the ***World Values Survey***, the original data were made freely available to the global scientific publics and render themselves for systematic, multivariate analysis of opinion structures on the basis of the original anonymous interview data.[17] Our data thus are from this set of reliable and regularly repeated global opinion surveys (see Davidov et al., 2008; Inglehart, 2006; Norris & Inglehart, 2015; Tausch et al., 2014).

In the social sciences, there is a rich and evolving debate on the conclusions to be drawn from these comparable and freely available *"omnibus surveys."* For a number of years now, also some leading economists became interested in studying global comparative opinion data, especially from the ***World Values Survey*** (Alesina et al., 2015). The interest of the economics profession in the relationship between religion and economic growth certainly was a factor contributing to the rise of the methodological approach, which we share with many other social scientists in this study (Barro, 2003a, 2003b; McCleary & Barro, 2006). Prejudice is the anti-thesis to societal trust, which is vital for long-run economic success. Racism, Antisemitism,

[15] https://www.academia.edu/45321329/Factors_affecting_the_Gulfs_future_Global_cycles_and_value_change_Electronic_Appendix.

[16] https://www.academia.edu/45321329/Factors_affecting_the_Gulfs_future_Global_cycles_and_value_change_Electronic_Appendix.

[17] http://www.europeansocialsurvey.org/.

religious intolerance, and xenophobia are thus an anti-thesis to societal trust. The majority of the major economic studies using ***World Values Survey*** data concluded that trust is an important factor for long-run economic growth (Alesina & Giuliano, 2015; Alesina et al., 2015; Zak & Knack, 2001). Trust is also an important factor in the political stability of a nation. Some of the countries with very high rates of Antisemitism, religious intolerance, and xenophobia, like Iraq, are also countries with extreme problems of political stability and very low interpersonal trust (Tausch, 2016; Tausch et al., 2014).

In the present book, we feature on societal values in the framework of what is called in political science the *"civic culture"* of the respective societies where the people holding religious beliefs live (Almond & Verba, 1963) and the role played by religiosity in this *"civic culture"* (Inglehart, 1988; Silver & Dowley, 2000). More and more data also from the Muslim countries became available in global social science (Tausch & Moaddel, 2009).

Sociologists, working with the unique comparative and longitudinal opinion survey data from the ***World Values Survey*** have discovered *inter alia* that there are pretty constant and long-term patterns of value change (Inglehart, 2006; Inglehart & Norris, 2003a, 2003b; Norris & Inglehart, 2011). Inglehart and his associates firmly believe that in the West, the ability of the representatives of the Roman Catholic hierarchy, which is still the most numerous of the Christian denominations, to tell people how to live their lives is declining steadily.[18]

As already hinted at above, in the electronic appendix[19] we give an overview of our research designs, based on ***World Values Survey*** data.

In Chap. 5 of our book, we use the latest edition of the *World Values Survey,* 2017–2020. These data are ideally fit to test our hypothesis of the vacuum in global leadership which naturally and inexorably affects the Gulf region, although the dataset itself does not cover a single Gulf country, and is based on 79 countries. It is based on 127,358 interviews. The countries of the survey are to be seen in our publicly available electronic data appendix.[20]

### 2.5.4 Arab Barometer

- Fourth, the ***Arab Barometer.*** This openly available original survey data allows researchers free direct access to the original data for multivariate analysis (***Arab Barometer*** III-V[21]). The ***Arab Barometer*** Wave I was fielded in 2007, Wave 2 in

[18] http://ur.umich.edu/0405/Apr11_05/11.shtml.

[19] https://www.academia.edu/45321329/Factors_affecting_the_Gulfs_future_Global_cycles_and_value_change_Electronic_Appendix.

[20] https://www.academia.edu/45321329/Factors_affecting_the_Gulfs_future_Global_cycles_and_value_change_Electronic_Appendix.

[21] The available codebook describes the survey and sampling techniques for each country covered by the survey. The data download page is https://www.arabbarometer.org/survey-data/data-downloads/.

**Table 2.1** Arab population assumptions

| | Total population | Population share in % |
|---|---|---|
| Algeria | 42,545,964 | 12.81 |
| Egypt | 97,664,971 | 29.40 |
| Iraq | 39,339,753 | 11.84 |
| Jordan | 10,252,700 | 3.09 |
| **Kuwait** | 4,226,920 | 1.27 |
| Lebanon | 6,855,713 | 2.06 |
| Libya | 6,470,956 | 1.95 |
| Morocco | 38,803,500 | 11.68 |
| Sudan | 40,875,455 | 12.31 |
| Tunisia | 11,446,300 | 3.45 |
| West Bank and Gaza | 4,780,978 | 1.44 |
| Yemen | 28,915,284 | 8.70 |
| Subtotal | 332,178,494 | 100.00 |
| Total 22 Arab countries | 423,400,000 | xx |
| Percentage of Arab World covered in our survey | 78.46% | xx |

2011, Wave III in 2013, Wave IV in 2016, and Wave V in 2018. Currently, the participating institutions are the Center for Strategic Studies of the University of Jordan, the Center for Political Studies at the University of Michigan, the Palestinian Center for Policy and Survey Research, Princeton University, One to One for Research and Polling, and SESRI, the Social and Economic Survey Research Institute. Funding for the Project was received from MEPI, the US-Middle East Partnership Initiative, the BBC, the **Qatar** National Research Fund, Princeton University, the University of Michigan, and the United Nations Development Program.[22]

The ***Arab Barometer*** seeks to measure and track civilian attitudes, values, and behavior patterns relating to pluralism, freedom, tolerance, and equal opportunity; social and interpersonal trust; social, religious, and political identities; conceptions of governance and the understanding of democracy; and civic engagement and political participation. Data from the last wave used in this became publicly available in the fall of 2019.[23] Our data on Arab population numbers were based on https://en.wikipedia.org/wiki/List_of_Arab_countries_by_population (Table 2.1).

22 https://www.arabbarometer.org/about/.

23 As stated on the website, "*the operational base for the third wave is the Center for Strategic Studies in Jordan(CSS). This third phase was funded by the Canadian International Research and Development Centre (IDRC) and the United StatesInstitute for Peace (USIP).*"

Data about the sampling are available for each country in the respective technical report.[24] The original file name for our analysis was "***Arab Barometer*** *Wave V EN2.sav*".[25]

The surveys were based on the following number of respondents:

| | |
|---|---|
| Algeria | 1766 |
| Egypt | 1727 |
| Iraq | 2273 |
| Jordan | 2108 |
| Lebanon | 1171 |
| Libya | 1739 |
| Morocco | 1518 |
| West Bank and Gaza | 2030 |
| Sudan | 1508 |
| Tunisia | 1971 |
| Yemen | 2182 |

We present valid country values, the population-weighted results for the entire region, and our aggregate nonparametric indicators. The population data used in our work relied on the following sources:

1. https://en.wikipedia.org/wiki/List_of_Arab_countries_by_population
2. https://www.cia.gov/library/publications/the-world-factbook/
3. https://www.worldometers.info/world-population/state-of-palestine-population/

The available countries and territories with complete data for this analysis were:

Algeria
Egypt
Jordan
**Kuwait**
Lebanon
Morocco
Sudan
Tunisia
West Bank and Gaza
Yemen

[24] https://www.arabbarometer.org/wp-content/uploads/ABIV_Methods_Report.pdf.

[25] Using PEW data in multivariate analysis, users have to make sure that the don't knows and missing values are properly excluded.

### 2.5.5 *The ADL-100 Index of Global Antisemitism*

- **Fifth, we used the ADL-100 Index for the study of Antisemitism in Chap. 7.** The analysis of Antisemitism now can look back on more than a century of fruitful studies with a vast and still growing literature (Jikeli & Allouche-Benayoun, 2012; Rosenfeld, 2013; von Bieberstein, 1977; Wistrich, 2010; see also Bauer, 1993; Wistrich, 2004, 2007, 2010). Here, our study relies on the ***ADL-100 Index***. The data provided by the ADL (2014) is based on survey researched attitudes and opinions toward Jews in more than 100 countries around the world. Parameters of the survey included. Sample size: 53,100 total interviews among citizens aged 18 and over across 101 countries and the Palestinian Territories in the West Bank and Gaza. Sampling error: between ± 4.4 and ± 3.2%. Sampling methodology: all respondents were selected at random, based on random-digit dial sampling; with geographically stratified, randomly-selected sampling points in each country and at the household level. Telephone interviewing was only conducted in countries where the combined telephone density exceeded 90%. Samples were adapted to the landline and mobile phone density in the total population. Representative samples at the national level: In an overwhelming majority of the countries and territories, the samples are fully nationally representative. In some countries (China, India, Ghana, Indonesia, Nigeria, Bangladesh, Laos, Malaysia, Mauritius, Uganda, and Vietnam), national coverage was not complete. Even there, sampling points were selected, and the data was weighted to ensure the interviews from those countries reflected the national perspective. Index Scores and question wording: created by asking whether the negative stereotypes (11 stereotypes) are "probably true" or "probably false." Respondents who said at least 6 out of 11 statements are "probably true" are considered to hold anti-semitic attitudes.

Our analysis also deals with Antisemitism in the Gulf countries by comparative perspective (Chap. 7). It further and substantially develops the approach already presented in Solomon and Tausch (2020) and Tausch and Obirek (2020). In analyzing Antisemitism, we introduce in each statistical calculation the dimensions of gender and employment, hitherto lacking in the before-mentioned approaches. The negative stereotypes used in the ADL Survey ADL-100 (see above) are:

1. Jews are more loyal to Israel than to [this country/the country they live in].
2. Jews have too much power in the business world.
3. Jews have too much power in international financial markets.
4. Jews do not care about what happens to anyone but their own kind.
5. Jews have too much control over global affairs.
6. People hate Jews because of the way Jews behave.
7. Jews think they are better than other people.
8. Jews have too much control over the United States government.
9. Jews have too much control over the global media.
10. Jews still talk too much about what happened to them in the Holocaust.
11. Jews are responsible for most of the world's wars.

The countries with complete data covered in the analysis are to be seen in our publicly available electronic data appendix.[26]

Further answering the methodological issues, potentially raised by our study, one would be tempted to say that the ***World Values Survey*** theoretically is still the ideal and single best choice to be the raw data base for the scientific analysis of our theme under consideration in Chap. 7, that of Antisemitism.[27] In practice, however, it is evident that the managers of the ***World Values Survey*** data base hitherto were much more interested in issues of general sociology, the sociology of religion, and the sociology of the norms of family life and sexuality than in the hard-core issues of identifying Antisemitism. As already mentioned in Solomon and Tausch (2020) and Tausch and Obirek (2020), they presented only one item which we can use in our study, i.e., the rejection of Jewish neighbors. Compared to the richness of the theoretical literature, especially Wistrich (2004, 2007, 2010), this operationalization might look even primitive, but what would be the feasible global empirical alternative? We apologize to our readers for the simplicity of the ***World Values Survey*** indicator used; however, we would like to remark that for the comparisons of the opinions of different denominations and different rates of religious activities in as many countries as possible, our chosen indicators are the only information available.

In order to be able to rely in any reasonable measure on our results, at least we had to look into the empirical relationship between the ADL 100 variable and our own Antisemitism rates on a country-to-country level, in order to calibrate our estimates. This relationship was developed in Solomon and Tausch (2020) and Tausch and Obirek (2020), and for the readers of Chap. 7, we should—for reasons of simplicity—explain this important argument. For 23 countries, we can calibrate our results accordingly and calculate at least a Pearson-Bravais correlation coefficient between the available two comparative data series on global Antisemitism, i.e., the ***World Values Survey*** data on the rejection of Jewish neighbors, and the ADL (2014) data (see also Solomon & Tausch, 2020; Tausch & Obirek, 2020). The countries with complete data for both variables are: Argentina, Bangladesh, Belarus, Bosnia, Canada, Chile, Czech Republic, Egypt, India, Iran, Iraq, Japan, Korea (South), Mexico, Moldova, Nigeria, Russia, South Africa, Spain, Uganda, USA, Uruguay, and Venezuela, representing a fair mix of global religions and cultures. Indeed, the two measurement scales have 56.16% of the variance in common, i.e., our second-best solution using ***World Values Survey*** data indeed captures more than half of the variance of the far superior measurement scale, the ADL (2014) data series. We can assume from these results that there is a constant of around 14% Antisemitism in any society around the globe, and that for every 10% of people saying that they reject Jewish neighbors according to the ***World Values Survey*** data, there is an increase of 6.6% in the Antisemitism score according to the ADL, 2014 criteria (Fig. 2.15).

---

[26] https://www.academia.edu/45321329/Factors_affecting_the_Gulfs_future_Global_cycles_and_value_change_Electronic_Appendix.

[27] http://www.worldvaluessurvey.org/WorldValues/Survey.jsp.

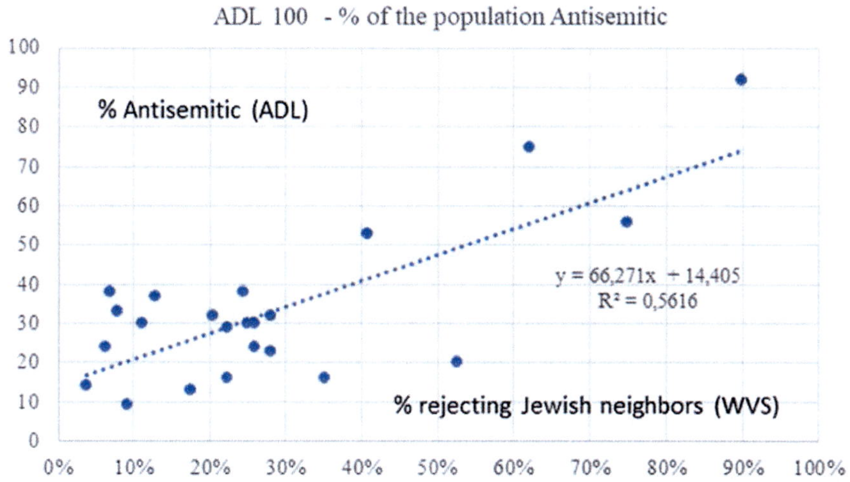


**Fig. 2.15** Correlation between the *World Values Survey* data on rejecting Jewish neighbors and the ADL (2014) rates of Antisemitism. *Note World Values Survey* scale ranging from 0.0 (= 0%) to 1.0 (= 100%); see also Solomon and Tausch (2020) and Tausch and Obirek (2020)

## 2.6 UNDP-type Index Construction

Another methodological note should be allowed about the Index construction in Chaps. 6–8. What we are doing there can be called in everyday plain language the combination of opinion data and the calculation of a single opinion yardstick, comparable to the United Nations Human Development Index. For the calculation of these so-called "nonparametric indicators," we relied on the well-established methodology of the United Nations Human Development Programme and its *UNDP Human Development Index* (UNDP, 2017). Since the 1990s, the widely and internationally highly recognized "*Human Development Index*" equally weights life expectancy, education, and real income. Life expectancy, education, and real incomes are projected onto a scale from 0 (worst value) to 1 (best value). Taking the example of the life expectancy "component" in the *Human Development Index,* let us assume that Italy, according to UNDP (2017) data, has an average life expectancy of 83.3 years and Senegal 66.9 years. Let us also assume that the world's best life expectancy (in 2017 in Hong Kong, China) was 84.2 years, and the global worst value (Central African Republic) was 51.5 years. The life expectancy component for Italy then will be Italy's value minus the global worst observed value, i.e. (83.3 – 51.5), divided by the value of the best placed country minus the value of the worst placed country, i.e. (84.2 – 51.5), which results in a life expectancy component of 0.972 for Italy. The corresponding life expectancy component for Senegal would then be (66.9 – 51.5)/(84.2 – 51.5), i.e., 0.471. The simplified form of a *Human Development Index* for Italy will then be the average of its component indices for life expectancy, education, and real income, and likewise, the simplified form of a *Human Development Index* for Senegal will be the average of its component indices for life expectancy, education. Although UNDP calculation methods have become more complicated in recent years, the simple rationale remains. Much to the detriment of inter-temporal comparability, the variables used in the computation of the Human Development Index changed over the years, and the same can be said about the now startling variety of other UNDP Human Development Report Indices: the Human Development Index (HDI), the Inequality-Adjusted Human Development Index (IHDI), the Gender Development Index (GDI), the Gender Inequality Index (GII), and the Multidimensional Poverty Index (MPI).[28] These indices are available not only at the country-to-country world level, but also for an immense variety of regions of our globe, contained in the macro-regional reports like Africa or Latin America and the Caribbean and the country reports.[29]

In the vast literature, surveyed in Tausch et al. (2014), there are two ways to add together the results from the different components, making up an UNDP-type of performance index indicator being mentioned: simply adding the results together, or first grouping them together to various subcomponents, and only from there to arrive at the final results. In Chap. 6, we present results from the first of the mentioned two possible methodologies.

## 2.7 Indicators Based on Advanced Multivariate Methodology

In Chaps. 6–8, we present much more advanced indicators, called "parametric indicators," which—in everyday plain language—combine the data with the aim of multivariate statistical analysis (see Tausch et al., 2014). Such parametric indicators usually rely on advanced statistical methods, such as principal components analysis (see, again, Tausch et al., 2014). Such an analysis extracts an overriding indicator, mathematically best representing the component variables and their correlation matrix. We eliminated the missing or refused responses and base our results on the vast majority of the valid responses. Our parametric index thus relies on the original survey respondents of the survey and calculates the country results, based on principal components factor scores. In Table 2.2, we present the example of the Gulf country Bahrain, taken from Chap. 6.

The treatment of highest numerical values, missing values and do not knows in the open access data based on survey research.

It is important to mention that in our empirical work, we took great care to properly take into account for the fact that in the PEW and ***Arab Barometer*** projects, very often the **highest numerical values of a variable express something else than the original PEW and *Arab Barometer* variable labels do.**

[28] http://hdr.undp.org/en/countries.

[29] http://hdr.undp.org/en/global-reports.

**Table 2.2** Example of a parametric index construction—Bahrain Overall Civil Society Index (Chap. 6)

| Promax factor | Bahrain—factor scores | Re-interpreted factor | Eigenvalues | Multiplier for the calculation of the final Overall Civil Society Index | Component of the Overall Civil Society Index |
|---|---|---|---|---|---|
| The violent and lawless society | – 0.214 | The law-abiding society | 4.263 | – 4.263 | 0.912 |
| Democracy movement | – 1.178 | Democracy movement | 2.574 | 2.574 | – 3.032 |
| Climate of personal violence | 0.098 | Climate of personal non-violence | 2.260 | – 2.260 | – 0.221 |
| Lack of trust in institutions | – 0.513 | Trust in institutions | 1.929 | – 1.929 | 0.989 |
| Unhappiness, poor health | 0.077 | Happiness, good health | 1.864 | – 1.864 | – 0.143 |
| Redistributive religious fundamentalism | – 0.261 | No redistributive religious fundamentalism | 1.554 | – 1.554 | 0.405 |
| Rejecting the market economy | 1.102 | Accepting the market economy | 1.434 | – 1.434 | – 1.580 |
| Feminism | – 0.697 | Feminism | 1.245 | 1.245 | –0.868 |
| Distance to politics | – 0.547 | Involvement in politics | 1.197 | – 1.197 | 0.655 |
| Nihilism | 1.125 | Optimism and engagement | 1.141 | – 1.141 | – 1.284 |
| Welfare mentality, rejection of the Calvinist work ethics | 0.240 | No welfare mentality, acceptancy of the Calvinist work ethics | 1.075 | – 1.075 | – 0.258 |
| Overall Civil Society Index | xx | xx | xx | xx | – 4.426 |

## 2.8 Value and Development Accounting for the Gulf Region

Our macro-comparative methodology of value and "development accounting" is oriented by recent advances in this field (see: Babones, 2014; Grinin et al., 2016; Holland & Campbell, 2005; Tausch et al., 2014).

General datasets, used in the present book, are mentioned in alphabetical order here below. The complete list of variables and their sources are available from our electronic appendix,[30] and we also made an EXCEL file with the data available at academia.edu.[31]

- **Barro, R. J. and Ursúa, R. F. 2011. Barro-Ursúa Macroeconomic Data**. Available at: https://scholar.harvard.edu/barro/publications/barro-ursua-macroeconomic-data.
- **Global Development Network Growth Database**, https://wp.nyu.edu/dri/resources/global-development-network-growth-database/ based on IMF.
- **IMF World Economic Outlook** data base, freely available from https://www.imf.org/external/pubs/ft/weo/2019/02/weodata/index.aspx
- **KOF Globalization Index**. Data definitions and free download available from: https://kof.ethz.ch/prognosen-indikatoren/indikatoren/kof-globalisierungsindex.html.[32]
- **Maddison Project Database**, version 2018. Bolt et al. (2018) (Bolt, Jutta, Robert Inklaar, Herman de Jong, and Jan Luiten van Zanden (2018), "Rebasing "Maddison": new income comparisons and the shape of long-run economic development" Maddison Project Working Paper, nr. 10, available for download at www.ggdc.net/maddison). The original Maddison data are available from http://www.ggdc.net/maddison/oriindex.htm.
- **UNDP Human Development Data**. Data definitions and free download available from: http://hdr.undp.org/en/data.
- **UTIP Inequality Index**. Data definitions and free download available from: https://utip.lbj.utexas.edu/data.html.
- **World Bank Open Data**. Data definitions and free download available from: https://data.worldbank.org/.
- **World Inequality Data Base**. Available at https://wid.world/.

Due emphasis was also given to dependency and world system approaches to development, which received a large-scale empirical confirmation in Tausch and Heshmati (2013) (see also Bornschier, 1980a, 1980b, 1983; Bornschier & Ballmer-Cao, 1979; Bornschier & Chase-Dunn, 1985; Tausch, 2010a, 2010b, 2012a, 2012b; Tausch & Heshmati, 2012). Earlier, well-known datasets for these investigations were

- Bornschier & Ballmer-Cao, 1978

[30] https://www.academia.edu/45321329/Factors_affecting_the_Gulfs_future_Global_cycles_and_value_change_Electronic_Appendix.

[31] https://www.academia.edu/44295690/GULF_2020_Covid_19_PUBLIC_ACCESS_Data_base_for_the_book_Factors_affecting_the_Gulfs_future_Global_cycles_and_value_change_Springer_Publishing_Series_springer_com_series_16417_Gulf_Studies_edited_by_Prof_Mizanur_Rahman_Qatar_University.

[32] Data definitions for the current version of the KOF Index are available from: https://ethz.ch/content/dam/ethz/special-interest/dual/kof-dam/documents/Globalization/2018/Definitions_2018_2.pdf.

- Müller & Bornschier, 1988
- Tausch, 2012a, 2012b [**International Macro-quantitative Data**. Faculty of Economics, Corvinus University of Budapest (with data definitions and sources). Free download available from http://www.uni-corvinus.hu/index.php?id=47854&tx_ttnews%5Btt_news%5D=0&tx_ttnews%5BbackPid%5D=31638&tx_ttnews%5BcalendarYear%5D=2012&tx_ttnews%5BcalendarMonth%5D=6&cHash=af8ef6888f7c9922b83b113f71c1ca32].

More recent databases, integrating dependency and world system approaches with conventional economic theories and the sociology of world values research are:

- Tausch, 2019a, 2019b [**Migration from the Muslimworld to the West: Its most recent trends and effects.** Jewish Political Studies Review, 30(1–2), 65–225, available at http://jcpa.org/article/migration-from-the-muslim-world-to-the-west-its-most-recent-trends-and-effects/ (with data definitions and sources). Free data download available from https://www.academia.edu/37568941/Migration_from_the_Muslim_World_to_the_West_Its_Most_Recent_Trends_and_Effects]
- Tausch & Heshmati, 2013 [**Globalization, the human condition, and sustainable development in the twenty-first century: cross-national perspectives and European implications.** London, New York and Delhi: Anthem Press (with data definitions and sources). Free data download available from https://www.academia.edu/35044095/Globalization_the_human_condition_and_sustainable_development_in_the_21st_Century._Cross-national_perspectives_and_European_implications_Codebook_and_EXCEL_data_file].

Here again, our statistical calculations were performed by the routine and standard IBM-SPSS statistical program (IBM-SPSS XXIV),[33] and relied on standard correlation analyses, partial correlation analyses, and standard ordinary least square (OLS) multiple regression analyses (Tausch et al., 2014).[34] Since both our data and the statistical methods used are available around the globe, any researcher can repeat our research exercise with the available open data and should be able to reproduce the same results as we did.

## 2.9 Predictors of Antisemitism

In Chap. 7, we use the data from Tausch (2019a, 2019b), and the data from the ***World Values Survey*** to arrive at comparative analyses of Antisemitism in the Gulf region. The regression analyses based on the ***ADL-100*** survey and the Tausch (2019a, 2019b), dataset yielded country predictions, and we could analyze the residuals from these regressions. The variables from Tausch (2019a, 2019b), were tested for their

[33] https://www-01.ibm.com/software/at/analytics/spss/.

[34] https://www.ibm.com/analytics/spss-statistics-software. The authors express thanks to the Department of Political Science, Innsbruck University, https://www.uibk.ac.at/politikwissenschaft/index.html.en for the opportunity to be able to use the software.

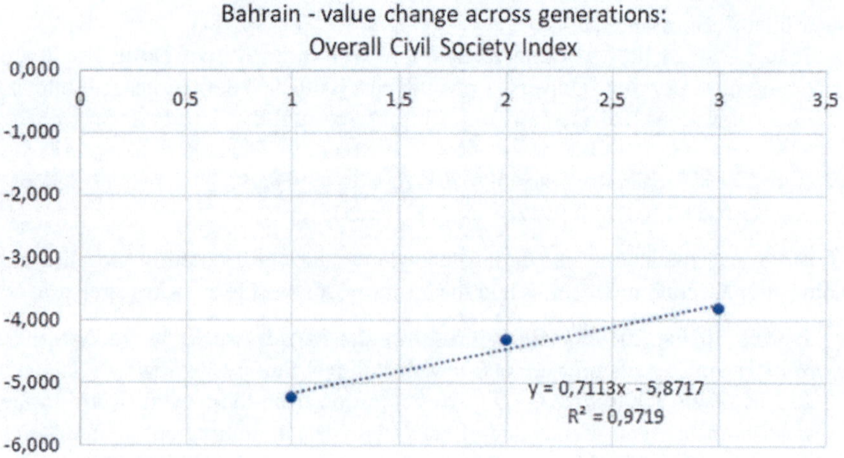


**Fig. 2.16** Value change across generations: Overall Civil Society Index in Bahrain

influence on Antisemitism in partial correlation analysis[35] (see also Tausch et al., 2014) in either direction.

## 2.10 Accounting for Value Change Across Generations

In our data base (see Appendix D of this book[36] and the freely available EXCEL data at academia.edu),[37] we calculated for each country with available data in the *World Values Survey* the value indices of Chap. 6 for each generation, i.e., the populations aged > 50, the populations aged 30–49, and the populations aged < 30. Then, to calculate single value change indices based on correlations for each country with complete data, we assigned for each country for the populations aged > 50 the value 1, the value 2 for the populations aged 30–49, and the value 3 for the populations aged < 30. We then calculated simple Pearson-Bravais correlations between the three value indices of each generation and the numbers 1, 2, and 3, yielding a simple, admittedly somewhat crude measure of a unified internationally comparable yardstick of value change over time in every country of the world with available *World Values Survey* data. Figure 2.16 explains, how for example the dynamics of the intergenerational value change in Bahrain for the Overall Civil Society Index can be assumed to be very

---

[35] Income per capita 2010 (n log EU = 100) and income per capita 2013 (n log EU = 100)$^2$ constant.

[36] https://www.academia.edu/45321329/Factors_affecting_the_Gulfs_future_Global_cycles_and_value_change_Electronic_Appendix.

[37] https://www.academia.edu/44295690/GULF_2020_Covid_19_PUBLIC_ACCESS_Data_base_for_the_book_Factors_affecting_the_Gulfs_future_Global_cycles_and_value_change_Springer_Publishing_Series_https_www_springer_com_series_16417_Gulf_Studies_edited_by_Prof_Mizanur_Rahman_Qatar_University.

**Table 2.3** Value change between the generations—the example of Bahrain

| | Bahrain—generation | Bahrain—Overall Civil Society Index |
|---|---|---|
| Population aged > 50 | 1 | – 5.230 |
| Population aged 30–49 | 2 | – 4.309 |
| Population aged < 30 | 3 | – 3.807 |

strong and positive, i.e., younger people in Bahrain will far more strongly identify with the overall goals of an open civil society than the older generations (Table 2.3).

This procedure was repeated for each of the promax factors resulting from the calculations of Chap. 6:

DYN GEN The law-abiding society
DYN GEN Democracy movement
DYN GEN Climate of personal non-violence
DYN GEN Trust in institutions
DYN GEN Happiness, good health
DYN GEN No redistributive religious fundamentalism
DYN GEN Accepting the market economy
DYN GEN Feminism
DYN GEN Involvement in politics
DYN GEN Optimism and engagement
DYN GEN No welfare mentality, acceptancy of the Calvinist work ethics

Thus, we have single yardsticks of intergenerational value change at our disposal.

## References

Abdi, H. (2003). Factor rotations in factor analyses. *Encyclopedia for research methods for the social sciences* (pp. 792–795). Sage.

Achilov, D. (2013). Social capital, Islam, and the Arab Spring in the Middle East. *Journal of Civil Society, 9*(3), 268–286. https://doi.org/10.1080/17448689.2013.816541.

Alesina, A., Algan, Y., Cahuc, P., & Giuliano, P. (2015). Family values and the regulation of labor. *Journal of the European Economic Association, 13*(4), 599–630.

Alesina, A., & Giuliano, P. (2015). Culture and institutions. *Journal of Economic Literature, 53*(4), 898–944.

Alexander, A. C., Inglehart, R. F., & Welzel, C. (2012). Measuring effective democracy: A defense. *International Political Science Review, 33*(1), 41–62.

Al-Ississ, M., & Diwan, I. (2016). Preference for democracy in the Arab World. *Politics and Governance, 4*(4), 16–26. https://doi.org/10.17645/pag.v4i4.753.

Almond, G. A., & Verba, S. (1963). *The civic culture: Political attitudes and democracy in five nations.* Princeton University Press.

*Arab Opinion Index.* (2015). Arab Public Opinion Programme. Arab Center for Research and Policy Studies. http://english.dohainstitute.org/file/Get/6ad332dc-b805-4941-8a30-4d28806377c4. Accessed May 10, 2017

*Arab Barometer* III. (2017a). Center for Strategic Studies in Jordan (CSS). http://www.arabbarometer.org/content/arab-barometer-iii-0. Accessed May 10, 2017

*Arab Opinion Index*. (2017b). Arab Center for Research and Policy Studies. http://english.dohainstitute.org/portal. Accessed May 10, 2017

Babones, S. (2014). *Methods for quantitative macro-comparative research*. Sage.

Barro, R. J. (2003). Economic growth in a cross section of countries. *International Library of Critical Writings in Economics, 159*(1), 350–386.

Barro, R. J. (2003). *Religion adherence data*. Harvard University, Department of Economics. https://scholar.harvard.edu/Barro/publications/religion-adherence-data

Barro, R. J. (2006). Rare disasters and asset markets in the twentieth century. *The Quarterly Journal of Economics, 121*(3), 823–866.

Barro, R. J., Nakamura, E., & Ursúa, J. F. (2011). *Crises and recoveries in an empirical model of consumption disasters* (unpublished). Harvard University, Department of Economics.

Bartlett, M. S. (1946). On the theoretical specification of sampling properties of autocorrelated time series. *Journal of Royal Statistical Society, Series B, 8*, 27–27.

Basilevsky, A. T. (2009). *Statistical factor analysis and related methods: Theory and applications* (Vol. 418). Wiley.

Bauer, Y. (1993). Antisemitism as a European and world problem. *Patterns of Prejudice, 27*(1), 15–24.

Blalock, H. M. (1972). *Social statistics*. McGraw-Hill.

Bloomfield, P. (1976). *Fourier analysis of time series*. Wiley.

Bolt, J., Inklaar, R., de Jong, H., & Luiten van Zanden, J. (2018). *Rebasing 'Maddison': New income comparisons and the shape of long-run economic development* (Maddison Project Working Paper, No. 10). Available at www.ggdc.net/maddison

Bolt, J., & van Zanden, J. L. (2013). *The first update of the Maddison project; re-estimating growth before 1820* (Maddison Project Working Paper 4). http://www.ggdc.net/maddison/maddison-project/home.htm

Bornschier, V. (1980). Multinational corporations and economic growth: A cross-national test of the decapitalization thesis. *Journal of Development Economics, 7*(2), 191–210.

Bornschier, V. (1980). Multinational corporations, economic policy and national development in the world system. *International Social Science Journal, 32*(1), 158–172.

Bornschier, V. (1983). World economy, level development and income distribution: An integration of different approaches to the explanation of income inequality. *World Development, 11*(1), 11–20.

Bornschier, V., & Ballmer-Cao, T. H. (1978). *Multinational corporations in the world economy and national development: An empirical study of income per capita growth (1960–1975)*. Department of Sociology, University of Zurich.

Bornschier, V., & Ballmer-Cao, T. H. (1979). Income inequality: A cross-national study of the relationships between MNC-penetration, dimensions of the power structure and income distribution. *American Sociological Review*, 487–506.

Bornschier, V., & Chase-Dunn, C. (1985). *Transnational corporations and underdevelopment*. Frederic Praeger.

Bornschier, V., Chase-Dunn, Ch. K., & Rubinson, R. (1978). Cross-national evidence of the effects of foreign investment and aid on economic growth and inequality: A survey of findings and a reanalysis. *American Journal of Sociology, 84*(3), 651–683.

Box, G. E. P., & Jenkins, G. M. (1976). *Time series analysis: Forecasting and control* (rev. ed.). Holden-Day.

Braithwaite, V. A., & Law, H. G. (1985). Structure of human values: Testing the adequacy of the Rokeach Value Survey. *Journal of Personality and Social Psychology, 49*(1), 250.

Brenner, P. S. (2016). Cross-national trends in religious service attendance. *Public Opinion Quarterly*, nfw016.

Browne, M. W. (2001). An overview of analytic rotation in exploratory factor analysis. *Multivariate Behavioral Research, 36*(1), 111–150.

Cattell, R. (Ed.). (2012). *The scientific use of factor analysis in behavioral and life sciences*. Springer.

Ciftci, S. (2010). Modernization, Islam, or social capital: What explains attitudes toward democracy in the Muslim world? *Comparative Political Studies, 43*(11), 1442–1470.

Ciftci, S. (2012). Islamophobia and threat perceptions: Explaining anti-Muslim sentiment in the west. *Journal of Muslim Minority Affairs, 32*(3), 293–309. https://doi.org/10.1080/13602004.2012.727291.

Ciftci, S. (2013). Secular-Islamist cleavage, values, and support for democracy and Shari'a in the Arab World. *Political Research Quarterly, 66*(4), 781–793. Available at SSRN https://ssrn.com/abstract=2920382

Ciftci, S., & Bernick, E. M. (2015). Utilitarian and modern: Clientelism, citizen empowerment, and civic engagement in the Arab World. *Democratization, 22*(7), 1161–1182. https://doi.org/10.1080/13510347.2014.928696.

Clarke, K. A. (2005). The phantom menace: Omitted variable bias in econometric research. *Conflict Management and Peace Science, 22*(4), 341–352.

Clarke, K. A. (2009). Return of the phantom menace: Omitted variable bias in political research. *Conflict Management and Peace Science, 26*(1), 46–66.

Clauß, G., & Ebner, H. (1970). *Grundlagen der Statistik für Psychologen.* Volk und Wissen Volkseigener Verlag.

Cook, R. D. (1977). Detection of influential observations in linear regression. *Technometrics, 19*, 15–18.

Cornell University Roper Center. (2017). *Polling fundamentals—Total survey error*. https://ropercenter.cornell.edu/support/polling-fundamentals-total-survey-error/. Accessed May 10, 2017

Cryer, J. D. (1986). *Time series analysis.* Duxbury Press.

Davidov, E., Schmidt, P., & Schwartz, S. H. (2008). Bringing values back in: The adequacy of the European social survey to measure values in 20 countries. *Public Opinion Quarterly, 72*(3), 420–445.

Dempster, A. P. (1969). *Elements of continuous multivariate analysis.* Addison-Wesley.

Devezas, T. C. (2010). Crises, depressions, and expansions: Global analysis and secular trends. *Technological Forecasting and Social Change, 77*(5), 739–761.

Devezas, T. C. (2012). The recent crisis under the light of the long wave theory. In L. E. Grinin, T. C. Devezas, & A. V. Korotayev (Eds.), *Kondratieff waves. Dimensions and prospects at the dawn of the 21st century* (pp. 138–175). Uchitel.

Devezas, T. C. (Ed.). (2006). Kondratieff waves, warfare and world security. In *Proceedings of the NATO Advanced Research Workshop on The Influence of Chance Events and Socioeconomic Long Waves in the New Arena of Asymmetric Warfare*, Covilhã, Portugal, February 14–18, 2005. IOS Press.

Diebolt, C. (2012). Cliometrics of economic cycles in France. In L. E. Grinin, T. C. Devezas, & A. V. Korotayev (Eds.), *Kondratieff waves. Dimensions and prospects at the dawn of the 21st century* (pp. 120–137). Uchitel.

Dziuban, C. D., & Shirkey, E. C. (1974). When is a correlation matrix appropriate for factor analysis? *Psychological Bulletin, 81*, 358–361.

El Kurd, D. (2017). Policing expression: A survey experiment on the efficacy of repression in Saudi Arabia. *Journal of Arabian Studies, 7*(2), 265–282.

Fabrigar, L. R., Wegener, D. T., MacCallum, R. C., & Strahan, E. J. (1999). Evaluating the use of exploratory factor analysis in psychological research. *Psychological Methods, 4*(3), 272).

Finch, H. (2006). Comparison of the performance of Varimax and Promax rotations: Factor structure recovery for dichotomous items. *Journal of Educational Measurement, 43*(1), 39–52.

Fox, A. M., Alzwawi, S. A., & Refki, D. (2016). Islamism, secularism and the woman question in the aftermath of the Arab Spring: Evidence from the *Arab Barometer*. *Politics & Governance, 4*(4), 40–57. https://doi.org/10.17645/pag.v4i4.767.

Fuller, W. A. (1976). *Introduction to statistical time series.* Wiley.

Goldstein, J. S. (1988). *Long cycles: Prosperity and war in the modern age.* Yale University Press.

Gorman, B. (2015). The case for cognitive interviewing techniques in the post-Arab spring environment. *Middle East Law and Governance, 7*(1), 50–60. https://doi.org/10.1163/18763375-00701011.

Gorsuch, R. L. (1983). *Factor analysis.* Erlbaum.

Grim, B. J. (2014). Growing religion and growing religious restrictions: A global overview. *International Area Studies Review, 17*(2), 134–145. https://doi.org/10.1177/2233865914537054.

Grinin, L., Tausch, A., & Korotayev, A. (2016). *Economic cycles, crises, and the global periphery.* Springer.

Haerpfer, C., Inglehart, R., Moreno, A., Welzel., Kizilova., Diez-Medrano J., M. Lagos, P. Norris, E. Ponarin, & B. Puranen et al. (Eds.). (2020). *World Values Survey*: Round Seven–Country-Pooled Datafile. Madrid, Spain & Vienna, Austria: JD Systems Institute & WVSA Secretariat. doi.org/https://doi.org/10.14281/18241.1

Harman, H. H. (1976). *Modern Factor Analysis* (3rd ed.). University of Chicago Press.

Hedges, L. V., & Olkin, I. (2014). *Statistical methods for meta-analysis.* Academic press.

Hoffman, M., & Jamal, A. (2012). The youth and the Arab Spring: Cohort differences and similarities. Middle East Law and Governance, 4(1), 168–188). doi:https://doi.org/10.1163/187633712X632399).

Holland, J., & Campbell, J. (2005). Methods in development research: Combining qualitative and quantitative approaches. ITDG.

Hosseini, S. H., & Saha, L. J. (2018). How 'Critically Open-Minded' Are We? An Australian Perspective Through the ***World Values Survey***. *Social Indicators Research, 136*(3), 1211–1236.

IBM (2014). IBM SPSS Statistics 21 Documentation. Available at http://www-01.ibm.com/support/docview.wss?uid=swg27024972#en .

Inglehart, R. F. (1988). The renaissance of political culture. *American Political Science Review, 82*(04), 1203–1230.

Inglehart, R. F. (1990). *Culture shift in advanced industrial countries.* Princeton University Press.

Inglehart, R. F. (2006). Mapping global values. *Comparative Sociology, 5*(2), 115–136.

Inglehart, R. F., & Baker, W. E. (2000). Modernization, cultural change, and the persistence of traditional values. American Sociological Review, 65(1), (19–51). http://scholaR.F.google.at/citations?view_op=view_citation&hl=de&user=r3vC6IAAAAAJ&citation_for_view=r3vC6IAAAAAJ:9yKSN-GCB0IC. Accessed May 10, 2017

Inglehart, R. F., & Norris, P. (2012). The four horsemen of the apocalypse: Understanding human security. *Scandinavian Political Studies, 35*(1), 71–95.

Inglehart, R. F., & Norris, P. (2003a). Rising tide: Gender equality and cultural change around the world. Cambridge University Press.

Inglehart, R. F., & Norris, P. (2003b). The true clash of civilizations. *Foreign Policy*, 63–70.

Inglehart, R. F., & Norris, P. (2016). Trump, Brexit, and the rise of populism: Economic have-nots and cultural backlash (HKS Working Paper No. RWP16–026). http://ssrn.com/abstract=2818659). Accessed May 10, 2017

Inglehart, R. F., & Welzel, C. (2003). Political Culture and Democracy: Analyzing Cross-Level Linkages. *Comparative Politics, 36*(1), 61–79.

Inglehart, R. F., & Welzel C. (2009). How development leads to democracy. What we know about modernization. Foreign Affairs, March, April. http://www.foreignaffairs.com/articles/64821/ronald-Inglehart-and-christian-welzel/how-development-leads-to-democracy. Accessed May 10, 2017

Inglehart, R. F., Haerpfer, C., Moreno, A., Welzel, C., Kizilova, K., Diez-Medrano, J., Lagos, M., Norris, P., Ponarin, E., & Puranen, B., et al. (Eds.). (2014). *World Values Survey: Round six—Country-pooled datafile version.* JD Systems Institute. https://www.worldvaluessurvey.org/WVSDocumentationWV6.jsp.

Institute for Economics and Peace. (2014). *Global Terrorism Index 2014.* Institute for Economics and Peace. Available at http://www.visionofhumanity.org/sites/default/files/Global%20Terrorism%20Index%20Report%202014_0.pdf. Download: April 10, 2019

Jargowsky, P. A. (2005). Omitted variable bias. *Encyclopedia of Social Measurement, 2*, 919–924.

Jikeli, G., & Allouche-Benayoun, J. (Eds.). (2012). *Perceptions of the Holocaust in Europe and Muslim communities: Sources, comparisons and educational challenges* (Vol. 5). Springer Science, & Business Media.

Kim, S. Y. (2010). Do Asian values exist? Empirical tests of the four dimensions of Asian values. *Journal of East Asian Studies*, 315–344.

Klein, C., Lühr, M., & Streib, H. (2018). Extant empirical research on religiosity and prejudice. In *Xenosophia and religion. Biographical and statistical paths for a culture of welcome* (pp. 23–84). Springer.

Kline, P. (2014). *An easy guide to factor analysis*. Routledge.

Knippenberg, H. (2015). Secularization and transformation of religion in post-war Europe. In *The changing world religion appendix map* (pp. 2101–2127). Springer

Lacina, B., & Gleditsch, N. P. (2005). Monitoring trends in global combat: A new dataset of battle deaths. *European Journal of Population, 21*(2–3), 145–166.

Lacina, B., Gleditsch, N. P., & Russett, B. (2006). The declining risk of death in battle. *International Studies Quarterly, 50*(3), 673–680.

Langer Research Associates. (n.d.). Moe. http://www.langerresearch.com/moe/. Accessed May 10, 2017

Maddison A. (2003). The world economy: Historical statistics. Development Centre of the Organisation for Economic Co-operation and Development.

Maddison A. (2007). Contours of the world economy, 1–2030 AD: essays in macro-economic history. Oxford University Press.

McDonald, R. P. (2014). *Factor analysis and related methods*. Psychology Press.

Minkov, M. (2014). The K factor, societal hypometropia, and national values: A study of 71 nations. *Personality and Individual Differences, 66*, 153–159.

Minkov, M., & Hofstede, G. (2011). *Cultural differences in a globalizing world*. Emerald.

Minkov, M., & Hofstede, G. (2013). *Cross-cultural analysis: the science and art of comparing the world's modern societies and their cultures*. Sage.

Mulaik, S. A. (2009). *Foundations of factor analysis*. CRC Press.

Müller, G. P., & Bornschier, V. (1988). *Comparative world data: A statistical handbook for social science*. Baltimore: Johns Hopkins University Press.

Nisbet, E. C., Nisbet, M. C., Scheufele, D. A., & Shanahan, J. E. (2004). Public diplomacy, television news, and Muslim opinion. *Harvard International Journal of Press/Politics, 9*(2), 11–37.

Norris, P., & Inglehart, R. F. (2002). Islamic culture and democracy: Testing the 'clash of civilizations' thesis. *Comparative Sociology, 1*(3), 235–263.

Norris, P., & Inglehart, R. F. (2011). *Sacred and secular: Religion and politics worldwide*. Cambridge University Press.

Norris, P., & Inglehart, R. F. (2012). Muslim integration into Western cultures: Between origins and destinations. *Political Studies, 60*(2), 228–251.

Norris, P., & Inglehart, R. F. (2015). Are high levels of existential security conducive to secularization? A response to our critics. In *The changing world religion map* (pp. 3389–3408). Springer.

Peace Research Institute, Oslo (PRIO). (2021). The Battle Deaths Dataset Version 3.0 and 3.1. Available at https://www.prio.org/Data/Armed-Conflict/Battle-Deaths/

Quenouville, M. H. (1949). Approximate tests of correlation in time series. *Journal of the Royal Statistical Society, Series B, 11*, 68–68.

Rosenfeld, A. H. (2013). *Resurgent antisemitism: Global perspectives*. Indiana University Press.

Rummel, R. J. (1970). *Applied factor analysis*. Northwestern University Press.

Sarkees, M. R. (2000). The correlates of war data on war: An update to 1997. *Conflict Management and Peace Science, 18*(1), 123–144.

Silver, B. D., & Dowley, K. M. (2000). measuring political culture in multiethnic societies reaggregating the *World Values Survey*. *Comparative Political Studies, 33*(4), 517–550.

Sjoberg, L., & Whooley, J. (2015). The Arab Spring for women? Representations of women in Middle East politics in 2011. *Journal of Women, Politics and Policy, 36*(3), 261–284. https://doi.org/10.1080/1554477X.2015.1050902.

Smith, R. J., & Taylor, A. M. R. (2001). Recursive and rolling regression-based tests of the seasonal unit root hypothesis. *Journal of Econometrics, 105*(2), 309–336.

Solomon, H., & Tausch, A. (2020). Islamism, crisis and democratization: Implications of the ***World Values Survey*** for the Muslim world. *Perspectives on development in the Middle East and North Africa (MENA) region.* Springer.

Suhr, D. (2012). Exploratory factor analysis with the *World Values Survey*. In *Proceedings of the SAS Global Forum 2012 Conference*, Orlando, FL.

Tabachnick, B. G., & Fidell, L. S. (2001). *Using multivariate statistics.* Allyn & Bacon.

Tang, C. F. (2010). The money-prices nexus for Malaysia: New empirical evidence from the time-varying cointegration and causality tests. *Global Economic Review, 39*(4), 383–403.

Tausch, A. (2010). Globalization and development: the relevance of classical "dependency" theory for the world today. *International Social Science Journal, 61*(202), 467–488.

Tausch, A. (2010). Paul Boccara's analysis of global capitalism. The return of the Bourbons, and the breakdown of the Brussels/Paris neo-liberal consensus. *Entelequia. Revista Interdisciplinar, 12*, 105–147.

Tausch, A. (2012). A globalization-oriented perspective on health, inequality and socio-economic development. *The International Journal of Health Planning and Management, 27*(1), 2–33.

Tausch, A. (2012b). *International macroquantitative data.* Faculty of Economics, Corvinus University of Budapest. Available at http://www.uni-corvinus.hu/index.php?id=47854&tx_ttnews%5Btt_news%5D=0&tx_ttnews%5BbackPid%5D=31638&tx_ttnews%5BcalendarYear%5D=2012&tx_ttnews%5BcalendarMonth%5D=6&cHash=af8ef6888f7c9922b83b113f71c1ca32

Tausch, A. (2016). Muslim immigration continues to divide Europe: A quantitative analysis of European Social Survey data. *Middle East Review of International Affairs, 20*(2).

Tausch, A. (2019). Migration from the Muslim world to the West: Its most recent trends and effects. *Jewish Political Studies Review, 30*(1–2), 65–225. Available at http://jcpa.org/article/migration-from-the-muslim-world-to-the-west-its-most-recent-trends-and-effects/ (with data definitions and sources). Free data download available from https://www.academia.edu/37568941/Migration_from_the_Muslim_World_to_the_West_Its_Most_Recent_Trends_and_Effects

Tausch, A. (2019). Multivariate analyses of the global acceptability rates of male intimate partner violence (IPV) against women based on *World Values Survey* data. *International Journal of Health Planning and Management*. https://doi.org/10.1002/hpm.2781

Tausch, A., & Ghymers, Ch. (2007). *From the 'Washington' towards a 'Vienna Consensus'? A quantitative analysis on globalization, development and global governance.* Nova Science Publishers.

Tausch, A., & Heshmati, A. (2016). *Islamism and gender relations in the Muslim world as reflected in recent World Values Survey data* (IZA Discussion Paper No. 9672). Available at SSRN http://ssrn.com/abstract=2725033. Download April 10, 2019

Tausch, A., & Heshmati, A. (2012). The effects of Multinational Corporation (MNC) penetration on the global political economy. A re-analysis of a recurrent sociological proposition with contemporary data. *Sociologia, 44*(3), 314–347.

Tausch, A., & Heshmati, A. (2013). *Globalization, the human condition, and sustainable development in the twenty-first century: Cross-national perspectives and European implications.* Anthem Press (with data definitions and sources). Free data download available from https://www.academia.edu/35044095/Globalization_the_human_condition_and_sustainable_development_in_the_21st_Century._Cross-national_perspectives_and_European_implications_Codebook_and_EXCEL_data_file

Tausch, A., & Heshmati, A., Karoui, H. (2014). *The political algebra of global value change: General models and implications for the Muslim world.* Nova Science Publishers. https://www.scopus.com/inward/record.uri?eid=2-s2.0-84952781328&partnerID=40&md5=7559153e0e895584b810889f47d26721

Tausch, A., & Moaddel M. (2009). What 1.3 billion Muslims really think. An answer to a recent Gallup study, based on the '*World Values Survey*'. Nova Science Publishers.

Tausch, A., & Obirek, S. (2019). *Global Catholicism, tolerance and the open society. An empirical study of the value systems of Roman Catholics.* Springer. https://doi.org/10.1007/978-3-030-23239-9

Tessler, M. (2004). The view from the street: The attitudes and values of ordinary Algerians. *Journal of North African Studies, 9*(2), 184–201.

UNDP. (2014). *Human development report*. Oxford University Press.

UNDP. (2017a). *Human development data (1990–2015)*. http://hdr.undp.org/en/data

UNDP. (2017b). *Human development data (1990–2015)*. http://hdr.undp.org/en/data. Download April 10, 2019

UNDP. (2017). *Human development report*. Oxford University Press.

UNDP (current issues). UNDP Human Development Report. Available at http://www.hdr.undp.org/

Ursúa, J. F. (2010). *Long-run volatility* (unpublished). Harvard University. Available at http://www.people.fas.harvard.edu/~jfursua/papers.html

Ursúa, J. F. (2010). *Macroeconomic archaeology: Unearthing crises and long-run trends*. Harvard University. Available at http://www.people.fas.harvard.edu/~jfursua/papers.html

Velleman, P. F., & Welsch, R. E. (1981). Efficient computing of regression diagnostics. *American Statistician, 35*, 234–242.

Vernengo, M., Perez-Caldentey, E., & Rosser, B. J. (2020). The New Palgrave dictionary of economics. Online-Version, available via Scopus.

von Bieberstein, J. R. (1977). The story of the Jewish-Masonic conspiracy, 1776–1945. *Patterns of Prejudice, 11*(6), 1–21.

Wistrich, R. S. (2004). Anti-Zionism and anti-Semitism. *Jewish Political Studies Review, 2004*, 27–31.

Wistrich, R. S. (2007). *Anti-Semitism and multiculturalism: The uneasy connection*. Vidal Sassoon International Center for the Study of Antisemitism, The Hebrew University of Jerusalem.

Wistrich, R. S. (2010). *A lethal obsession: Anti-Semitism from antiquity to the global jihad*. Random House.

World Values Survey. (2021). http://www.worldvaluessurvey.org/wvs.jsp

Yeşilada, B. A., & Noordijk, P. (2010). Changing values in Turkey: Religiosity and tolerance in comparative perspective. *Turkish Studies, 11*(1), 9–27.

Zak, P. J., & Knack, S. (2001). Trust and growth. *The Economic Journal, 111*(470), 295–321.

Zivot, E., & Wang, J. (2006). *Modeling financial time series with S-PLUS, second edition, with Jiahui Wang*. Springer.

# Chapter 3
# COVID-19 and the Contours of the Impending Global Crisis: Western Meltdown, Gulf Fallout?

**Abstract** This chapter attempts to analyze the current COVID-19 pandemic from a historic-comparative and cross-national analytical perspective. Considering the gravest political, social, economic, and public health consequences of the present crisis, we must assume that the West, for all practical purposes, could be relegated from the landscape of the twenty-first century as a major strategic player in the Middle East and beyond. Our findings are based on the comparison of the effects of earlier pandemics on the long cycles of economic development over the last two thousand years and the multivariate analysis of COVID-19 death rates per 1 million population and COVID-19 fatality rates in a partial correlation analysis on a country-to-country basis. We debate at length the economic consequences of the past more than two millennia of pandemics and also analyze what has become known as the "Thucydides trap" of the hegemonial challenge of a rising power, inexorably leading to Great Power War, all this combined with a pandemic. So, is the ancient writer Thucydides (460 BCE–400 BCE) the prophet of events to come on a global scale? The present chapter also risks a prognosis on the overall societal effects of the pandemic, based on the Piketty time series data of the share of the richest 1% in total incomes, and estimates on the effects of the severity of the 1918–1920 pandemic on international income convergence/divergence over time, made possible by the Maddison/Barro data series. The severity of an epidemic negatively influences international income convergence over time. Equally stark lessons must be drawn from the long-term effects of the crisis of 2008. We show that already before the tsunami of the present COVID-19 crisis the crisis of 2008 hit so many countries with full force, and none the less than 35 countries did not even fully recover from the 2008 crisis, when the current COVID-19 emergency set in. With Germany quenching the Latin and Mediterranean European partner countries to the wall, European austerity measures and the economic crisis of 2008 also led to a significant weakening of national security on the European continent. Three factors will be decisive for the trajectory of the Gulf. These factors are inequality, a pattern of high economic globalization occurring with a relatively low or intermediate de jure social globalization, and a relatively high past fluctuation of economic growth, tied to export prices, which leads to quite strong economic fluctuations over time.



A. Tausch, *The Future of the Gulf Region*, Gulf Studies 2,
https://doi.org/10.1007/978-3-030-78299-3_3

**Keywords** COVID-19 · Coronavirus · Coronanomics · Pandemic · Economic impacts

## 3.1 Introduction to COVID-19 and the Contours of the Impending Global Crisis

How can we study the effects of the current COVID-19 pandemic as social scientists? And which conclusions can we draw from such an analysis for the Gulf countries? In Chap. 1 of the present book, above, we highlighted the overwhelming importance of inequality and gender norms as the current frame of reference of the UNDP's development research. In a recent high-profile article, Elgar et al. (2020) entitled *"The trouble with trust"* performed a time series analysis of social capital, income inequality, and COVID-19 deaths in 84 countries. In this article, the authors empirically tested the hypothesis that social cohesion, public confidence in government sources of health information, and general concern for the welfare of others support health advisories during a pandemic and save lives. Elgar et al. (2020) tested this assumption through a time series analysis of cross-national differences in COVID-19 mortality during the early phase of the pandemic. Country data on income inequality and four dimensions of social capital (trust, group affiliations, civic responsibility, and confidence in public institutions) were linked to data on COVID-19 deaths in 84 countries. Associations with deaths were examined using Poisson regression with population-averaged estimators. Mortality was positively related to income inequality, trust, and group affiliations and negatively related to social capital from civic engagement and confidence in state institutions. These associations held with controls for population size, age, and wealth. According to Elgar et al. (2020), the results indicate that societies that are more economically unequal and lack capacity in some dimensions of social capital experienced more COVID-19 deaths. Elgar et al. (2020) conclude from this that social trust and belonging to groups were associated with more deaths, possibly due to behavioral contagion and incongruence with physical distancing policy. Some countries require a more robust public health response to contain the spread and impact of COVID-19 due to economic and social divisions within them.

The present chapter further develops this approach on the effects of COVID-19 in the Arab World, already presented in Solomon and Tausch (2021). Venturing into this new terrain of the social sciences, which we could call "comparative epidemiology," three names of researchers should be named here at the beginning, whose contributions served as roadmaps in our analysis in this chapter. First, Harvard economist Robert Barro and the Italian economist Giovanni Andrea Cornia, presently at Firenze University, and a longtime associate of UNICEF, provided important insights on how to deal with rare economic disasters and with excess mortality crises, respectively, which we have to expect now on a massive global level (see also Barro, 2006, Barro & Ursúa, 2008, 2011; Barro et al., 2020; and Cornia, 2019; Cornia & Paniccià, 2000). And then at the time of writing this analysis, the well-known M.I.T. economist Daron Acemoglu already ventured into the field of offering a proper framework of analysis of the pandemic.

In fact, his thoughts (Acemoglu, 2020) as those of Barro and Cornia offer to us a theoretical and methodological guidance in the study of the current pandemic. Their ideas greatly influenced the way, we look at these issues. Acemoglu assumes that the COVID-19 pandemic will have

- Major public health implications,
- Major economic implications (possibility of the most severe global recession in at least the last 90 years),
- Major political implications (Acemoglu, 2020).

We also agree with Acemoglu that in the context of COVID-19, guesses and imperfect knowledge abound. With Zhang et al. (2020), we also assume that global infection rates in the northern hemisphere will increase in winter 2020/21, thus aggravating not only the catastrophic health consequences of the pandemic, but also the economic fallout from all these lockdowns around the world. As this book goes to the press, there are nearly 100 million global infections and 2 million deaths.[1]

In current public health research, some high-profile articles (Ahmed et. al., 2020; Douglas et al., 2020; Kawohl & Nordt, 2020) began to look into the probable social consequences of the pandemic. Of equal importance for the question writing of our chapter is the attempt to look more deeply into the effects which earlier pandemics had on the economies and social systems of their time and to attempt to draw prudent conclusions from this for the next years (Ceylan et al., 2020). Kilic (2020) attempted to link the current COVID-19 crisis with the issue of the larger cycles of the world economy.

The World Bank, in its January 2021 flagship report, forecasts:

> Global activity is estimated to have contracted 4.3 percent in 2020 as a result of COVID-19, making it the fourth most severe global recession of the past 150 years, exceeded only by the first World War, the Great Depression, and the Second World War. Given the unprecedented nature of the pandemic, prospects for the global economy are uncertain, and several growth outcomes are possible. In the baseline forecast, global growth is expected at 4 percent in 2021, and is projected to moderate further in 2022, to 3.8 percent. Even by 2022, global GDP is forecast to be 4.4 percent below pre-pandemic projections. (World Bank, 2021)

The World Bank, in its outlook for the Middle East and North Africa (MENA) region, offers an equally dire perspective:

> Output in the region is estimated to have contracted 5 percent in 2020, as countries struggled with significant disruptions from COVID-19 and a sharp fall in oil demand. This contraction adds to already slowing growth in the region and compounds pre-pandemic per capita income losses. Growth is expected to improve to a modest 2.1 percent in 2021, as the pandemic is brought under control and lockdown restrictions are eased, global oil demand rises, and policy support continues. The pandemic is expected to leave lasting economic scars on the region, which will likely dampen potential growth going forward. Disruptions related to geopolitical tensions and political instability, renewed downward pressure on oil prices, and additional balance of payments stress are key downside risks to the outlook. (World Bank, 2021)

[1] https://www.worldometers.info/coronavirus/.

In this chapter, we attempt to deal with the strategic implications of the current crisis and leave the hard work of politometric time series analysis of global conflict data ranging back to 1495 and global world industrial production growth data ranging back to 1740 in Chap. 4, hopefully being able to risk some predictions about the cycles of global conflict and global depressions. Even from the current simple statistics of the COVID-19 disease, it emerges crystal clear that more than three quarters of a century after the end of the Second World War and victory against Nazi Germany, and the NATO countries with the highest relevance for the European and Mediterranean theater are also among the most seriously affected COVID-19 countries: the USA, Spain, Italy, the UK, France, and Germany. In our introductory statistical finding of the COVID-19 death rates per 1 million inhabitants here below we highlight them in turquois color. Considering the gravest political, social, economic, and public health consequences of the present crisis, we must assume that the West, for all practical purposes, could be relegated from the landscape of the twenty-first century as a major strategic player in the Middle East and beyond. According to the mentioned Johns Hopkins data, the following countries had COVID-19 death rates of more than 1.0 per 1000 population[2]:

| | |
|---|---|
| San Marino | 1.914 |
| Belgium | 1.738 |
| Slovenia | 1.477 |
| Bosnia and Herzegovina | 1.337 |
| Italy | 1.330 |
| Liechtenstein | 1.283 |
| Czechia | 1.274 |
| North Macedonia | 1.271 |
| United Kingdom | 1.245 |
| Bulgaria | 1.196 |
| USA | 1.176 |
| Montenegro | 1.173 |
| Peru | 1.156 |
| Hungary | 1.135 |
| Spain | 1.127 |
| Andorra | 1.125 |
| Croatia | 1.093 |
| France | 1.053 |
| Panama | 1.048 |
| Mexico | 1.046 |

Acemoglu highlights the necessity to study more in depths the dynamics of infections when epidemics interact with economic and social behavior, and what kinds of institutions and states–society relations led to the failures and delays in government responses in places such as Italy, the US, the UK, and Turkey. Most important still,

[2] https://www.worldometers.info/coronavirus/ and https://www.worldometers.info/coronavirus/#countries.

Acemoglu thinks that we need to know what the pandemic will imply for the future of global economic and political institutions.

The present chapter will thus evaluate the evidence available on the questions, raised by Acemoglu (2020), on the basis of the *"Worldometers"* COVID-19 country data, contained in https://www.worldometers.info/coronavirus/ (Johns Hopkins University), as of March 31, 2021.

Our data might still be unsatisfactory, given the wealth of other theoretically available data, but for the aims of a first and thorough cross-national assessment of the issues involved, the chosen approach is certainly sufficient and hopefully will be especially of use for the decision-makers in the Gulf region itself.

In global public health research, running epidemiological multivariate analyses about socio-economic conditions which led to outcomes which are under the domain of the medical profession, such as life expectancies, are rather common nowadays, showing the ever more interdisciplinary character of public health scholarship (on global inequality and life expectancy; on globalization and life expectancy and overall public health conditions, see, Tausch, 2012).

So, our aim is to study precisely what Acemoglu, 2020, calls the *"dynamics of infections when epidemics interact with economic and social behavior, and what kinds of institutions and states-society relations led to the failures and delays in governmentresponse in places such as Italy, the US, the UK and Turkey."* Our specially created cross-national *"Jerusalem dataset"* already mentioned in Chap. 2 of this book was augmented for the aims of the present chapter, wherever necessary, by new data from sources such as the IMF World Economic Outlook Database, the KOF Globalization Index by the ETH Zurich, the UNDP Human Development Data, and World Bank Open Data, and was made freely available online in Excel format at academia.edu.[3]

Acemoglu (2020) underlines that most important still, we need to know what the pandemic will imply for the future of economic and political institutions. We partially dare to venture into these grounds, based on the knowledge we have about the interaction of pandemics and economic depressions in the past (see also Barro, 2006, Barro & Ursúa, 2008, 2011; Barro et al., 2020; and Cornia, 2019; Cornia & Paniccià, 2000). For such comparisons, the World Bank Open global data time series for each country of the world from 1960 to 2018, and the GDP per capita in constant 2011 $ in real purchasing power parities, offered by the "Maddison database" at the University of Groningen in the Netherlands (Bolt et al., 2018) for the countries of the world and the time period from the year 1 of our era to the great financial crisis of 2008 and the year 2016, as well as the well-established historical and medical history accounts of earlier pandemics, such as Raoult et al. (2013), are important sources.

[3] https://www.academia.edu/44295690/GULF_2020_Covid_19_PUBLIC_ACCESS_Data_base_for_the_book_Factors_affecting_the_Gulfs_future_Global_cycles_and_value_change_Springer_Publishing_Series_https_www_springer_com_series_16417_Gulf_Studies_edited_by_Prof_Mizanur_Rahman_Qatar_University.

In a rather prophetic article, which appeared already before the outbreak of COVID-19, the University of California-based world systems scholar Christopher K. Chase-Dunn remarked:

> The post-World War II wave of trade globalization and financialization faltered in 2008 but has recovered since then. A future trough of trade deglobalization like what happened in the 1930s could happen again if a perfect storm of calamities and more powerful resistance to further economic globalization should emerge (as seems now to be the case). The declining economic and political hegemony of the United States poses huge challenges for global governance. Newly emergent national economies such as India and China are competing for insertion into the global structure of power.

Barro recently remarked:

> The best source of a worst-case scenario for the ongoing coronavirus pandemic likely comes from the Great Influenza Pandemic of 1918-1920. Our research found from data on 48 countries that the flu death rate averaged 2.1 percent, corresponding to 40 million worldwide deaths. When applied to current population, this death rate implies a staggering 150 million deaths worldwide. The 1918-1920 pandemic was associated with recessions in many countries. The average size of the contraction in GDP was about six percent, lasting around two years; that is, the lost output was about 12 percent of a year's GDP. To put this outcome in a broader context, my research on rare macroeconomic disasters showed that the three worst global macroeconomic crises were WWII, the Great Depression of the early 1930s, and WWI. The Great Influenza may have been the next worst global adverse event going back to 1870. Around 12 or so countries out of 42 with data were estimated to have suffered declines in real per capita GDP by 10% or more with troughs in 1919-21.[4]

Facing these grim realities, we attempt to offer our analysis, based on relatively well-established cross-national and long time series data, hoping that it finds its ways to scientific publics and decision-makers and the analysts in the security establishments of the free democratic West and of the countries of the Gulf.

## 3.2 COVID-19 and the Current Global Crisis

As we already specified in Chap. 1, there are increasing attempts to estimate the short- and long-term effects of the pandemic on the social and economic systems of our time and try to learn the lessons from the past epidemics (see, among others, Benmelech & Tzur-Ilan, 2020; Brodeur et al., 2020; Candela & Geloso, 2020; Doerr & Hofmann, 2020; Furceri et al., 2020; Jordà et al., 2020; Kok, 2020; Morgan et al., 2021.

In a breathtaking parallel to the world of today, in ancient Greece a terrible pandemic coincided with rising tensions between the main military powers of that time and geographical area (see Holladay & Poole, 1979; Maas, 2005; and Langmuir et al., 1985). The pandemic raged in Athens around 430 BCE–427 BCE. The Great Power tensions in ancient Greece led to the Peloponnesian War, 431 BCE–404 BCE, between Athens and its rival Sparta (Brzezinski, 2014; Gries, 2019; Moore, 2017).

[4] https://english.phbs.pku.edu.cn/2020/news_0717/2452.html.

The ancient historian Thucydides (460 BCE–400 BCE) offers a detailed analysis of the interaction between this pandemic of the time, the *PlagueofAthens,* the overall decay of the geostrategic capabilities of the first, though limited democracy on earth, Athens, and the onset of the Peloponnesian War (Ober, 1989, 2008; furthermore Arvanitidis, 2013; McCannon, 2012; Orwin, 1988; Page, 1953).

His accounts are an important lecture for any of today's analysts in the state chancelleries of the Middle East and North Africa, and the global security establishments and think tanks even more than 2400 years afterward. Thucydides himself says about his conviction that events in Athens will be repeated in history:

> To hear this history rehearsed, for that there be inserted in it no fables, shall be perhaps not delightful. But he that desires to look into the truth of things done and which (according to the condition of humanity) may be done again, or at least their like, he shall find enough herein to make him think it profitable. And it is compiled rather for an everlasting possession than to be rehearsed for a prize. Thucydides (460 BCE–400 BCE)[5]

In the evolving literature on this subject, Brzezinski (2014), Gries (2019), and Moore (2017) all have analyzed what has become known as the *"Thucydides trap"* of the hegemonial challenge of a rising power, inexorably leading to Great Power War, all this combined with a pandemic. So, is the ancient writer Thucydides the prophet of events to come on a global scale?

Surprisingly and significantly, Chinese President Xi Jinping,[6] himself seems to be acutely aware of the importance of Thucydides for the understanding the events of today and tomorrow, and already said in a speech in November 2013 at a conference, *"Understanding China,"* hosted by the Berggruen Institute on Governance,[7] that

> we all need to work together to avoid the Thucydides Trap - destructive tensions between an emerging power and an established power, or between established powers themselves.

From Thucydides and also from another historian of antiquity, Procopius (Procopius of Caesarea 500 CE–565 CE) and his accounts on the *PlagueofJustinian* (541–542 AD, with recurrences until 750), we can learn another simple and frightening additional lesson for today: Pandemics not only kill, but have a deeply corrosive effect on the social and geostrategic fabric of the affected countries.

Procopius of Caesarea, 500 CE–565 CE, features, among others, on the effects of *Justinian's Plague* on the capabilities of the then existing Byzantine state and the army. Born in Caesarea on the soil of the land of Israel in 500 CE, Procopius and his analyses could not be more modern for the analysts of international politics around 1500 years after his death: How badly the corrupt elites of the Byzantine Empire reacted to the crisis, how the Plague depleted Byzant's military capabilities, and what were the relations of the then leading power in the region, the Byzantine Empire, to Iran *et cetera.*

---

[5] http://www.perseus.tufts.edu/hopper/text?doc=Thuc.%201.22&lang=original.

[6] President Xi Jinping's speeches often make reference to European antiquity; Thucydides is mentioned none the less than nine times in his speeches; see https://www.fmprc.gov.cn/wjb/eng_search.jsp.

[7] https://www.berggruen.org/about/; furthermore: https://www.berggruen.org/?s=thucydides. Nicolas Berggruen is a German billionaire, businessman, and investor. The official Web site is: https://www.berggruen.org/

In the spirit of these lessons that we can learn from these two great spirits of antiquity, the aim of this chapter is to provide a longer-term strategic assessment of the implications of the current COVID-19 pandemic with a special view to the strategic environment that the Gulf countries will find in the world in a few years to come. The changes, which the pandemic presumably will bring about, affect the very foundations of the world system, and the economic and social well-being of the main Western powers, which traditionally were the anchor of security for the region. The analytical perspective we offer combines macro-quantitative analyses of the drivers of the current international infection and death rates of the pandemic with more long-term economic cyclical analyses of the socio-economic effects of earlier pandemics and the conclusions we can draw from these insights for reasonably founded prognoses for developments to come.

Current events with its presumed catastrophic effects especially for the five main Western players in the Mediterranean, France,[8] Italy,[9] and Spain,[10] and of course the USA with its VI. Fleet,[11] and the UK with its still existing naval presence in Gibraltar and Cyprus,[12] unfold at a time, when the economic crisis of 2008 already left deep wounds in the economic and social fabric of the European South and greatly reduced not only the intensive care capabilities of their hospitals (the effects we saw on international TV channels for the entire year 2020 in their reports from Madrid, Milano, and Rome), but also the financial capabilities of these states to maintain their naval presence in the Mediterranean, guarding the safety of the seaways, which are so vital for the moderate states in the region. Quite symbolically, more than 1000 of the personnel of France's and Europe's pride of the Sea, the Aircraft Carrier *Charles de Gaulle,* were infected by the virus.[13] Similar dramatic events aboard the US Navy Aircraft Carrier *Theodore Roosevelt* in 2020 must equally be mentioned in this context.[14]

As Christopher S. Chivvis, associate director of the RAND International Security and Defense Policy Center, warned, NATO risks coming up dangerously short on the threats that matter to most of Europe and thus to NATO. If NATO fails to define a strategy for its southern challenges, it could slip, Chivvis argues, into strategic irrelevance.[15]

The current crisis, which will also further aggravate severe already existing economic and political tensions between the European South and Germany and

---

[8] https://www.defense.gouv.fr/marine.

[9] http://www.marina.difesa.it/Pagine/default.aspx.

[10] https://armada.defensa.gob.es/ArmadaPortal/page/Portal/ArmadaEspannola/iniciohome/prefLang-es/.

[11] https://www.c6f.navy.mil/.

[12] https://www.royalnavy.mod.uk/.

[13] https://www.france24.com/en/20200417-france-reports-40-of-aircraft-carrier-group-crew-test-positive-for-covid-19.

[14] See https://www.rand.org/blog/2020/04/pandemics-and-the-us-military-lessons-from-1918.html; and https://en.wikipedia.org/wiki/Covid-19_pandemic_on_USS_Theodore_Roosevelt.

[15] https://www.rand.org/blog/2016/04/natos-southern-exposure-the-real-threats-to-europe.html.

other richer countries in the European Union, not only threatens the functioning of NATO, but also threatens to unravel the European Union itself, producing cataclysmic economic and social effects of their own (see also Solomon & Tausch, 2021). Needless to say, the current crisis also led to the diffusion of another global virus—Antisemitism.[16]

One tool used here in this analysis is the comparison of the effects of earlier pandemics on the long cycles of economic development over the last two thousand years, using the data provided by Maddison (2007), and his associates (Bolt et al., 2018; also analyzed by Barro, 2006; Barro & Ursúa, 2008, 2011; Barro et al., 2020).

Are we then witnessing the end of US hegemony in the world system and the rise of China in the wake of COVID-19? Among the social scientists of our time, the Italian–American political economist Giovanni Arrighi (1937–2009), in his prophetic book "*The Long Twentieth Century*" (Arrighi, 1995), exactly predicted today's situation: a rising China, a declining America, which—like all other hegemonial powers in the world economy—now leaves the stage. Arrighi's work shows that since the birth of capitalism in the Italian city-states of the fifteenth century, our present global social order developed over a 700-year period and had up to now four hegemonic powers which provided leadership for the global economy (Genoa, the Dutch Republic, the UK, and the USA).[17] For Arrighi, the ups and downs of hegemonies are deeply connected with the succession of longer economic fluctuations and cycles (Kondratieff cycles, see Grinin et al., 2016; see Chap. 4 of the present book) of "material expansion" and, alternatively, of "financial expansion." The current waning period, like all periods of financial expansion before it, was characterized by rising income inequality. Real wages for American workers have stagnated, and populism now responded to this dissatisfaction. The end of every global hegemony is marked by a "*terminal crisis*"—like the ones in the middle of the eighteenth century and the Great Depression of the 1930s (see Chap. 4). Terminal crises always escalated into the 30-year conflicts: The Thirty Years' War, the Napoleonic Wars, and the World Wars of the twentieth century (see, again, Chap. 4).

The economic and social analysis of the effects of major economic depressions which repeatedly occurred in history quickly becomes "*economics as a dismal science.*" We are also reminded of perspectives like the one presented by Karl Polanyi (1886–1964) in his "*Great Transformation*" (Polanyi & MacIver, 1944) accounting for the devastating effects of the *Great Depression,* 1929, and the end of the *Century of Peace*, 1815–1914, and the global liberal market of the nineteenth century. Thanks to the work of the British economist Angus Maddison (1926–2010) and subsequent work by Robert Barro (see above), social science is now at least being able to work with reasonable and recognized estimates of the evolution of the world economy and GDP (per capita) in many parts of the world from the beginnings of our era to the present time. Of course, the further we move back in history, the more these data

[16] See especially https://jcpa.org/article/anti-semitism-and-the-coronavirus/ and https://www.inss.org.il/publication/coronavirus-and-antisemitism/. On Antisemitism and the Black Death Pandemic in the fourteenth century.

[17] For an interesting evaluation of Arrighi's theory, see also http://freedomlab.org/the-end-of-the-fourth-hegemonic-cycle/.

become mere estimates and even worse, only well-informed guesses, but Maddison's and Barro's data can at least help us to understand the devastating economic effects of earlier pandemics (Maddison, 2007) in order to make reasonable predictions in the current crisis.

A great number of international relations scholars have already pondered on whether there is an "inherent" tendency of the world system to produce major power confrontations of around 30 years' duration between a land-based hegemonic challenger and a dominant sea power, as happened in the Thirty Years War, 1618–1648; in the French/Napoleonic Wars, 1789/1803–1815; and in the German quest for global hegemony, 1914–1945 (see also Chap. 4, for the relevant econometric and politometric time series analyses). Ever since such a statistical analysis, put forward by Goldstein (1985, 1988) aroused so many debates in the discipline of international relations, such analyses have attempted to measure these contentions by statistical and time series analysis of the figures about the fatalities from Great Power Wars since the late fifteenth century, all documented in Goldstein (1988).[18] At a time of rising tensions between China and the USA in the wake of the COVID-19 pandemic, we start our description of the background of this chapter by highlighting that although Goldstein later (Goldstein, 2011; shared also by Gat, 2017) came to an absolutely different conclusion—the disappearance of major power wars in the future—other scholars have continued investigations about long cycles of war in the global system (Tausch, 2007, 2010). Gat, in an analysis published in the widely circulated journal "*Foreign Affairs*," linked his analysis with an unbroken belief in the leadership provided by the USA, now so severely affected by the pandemic and facing the gravest economic crisis since the Great Depression, 1929, not to talk even about the internal divisions of America and the mess that President Donald Trump left with his presidency (see also Norris & Inglehart, 2019):

> For all the criticism leveled against it, the United States – and its alliance with Europe – stands as the single most important hope for the future of liberal democracy. Despite its problems and weaknesses, the United States still commands a global position of strength and is likely to retain it even as the authoritarian capitalist powers grow. Not only are its GDP and productivity growth rate the highest in the developed world, but as an immigrant country with about one-fourth the population density of both the European Union and China and one-tenth of that of Japan and India, the United States still has considerable potential to grow – both economically and in terms of population – whereas those others are all experiencing aging and, ultimately, shrinking populations. China's economic growth rate is among the highest in the world, and given the country's huge population and still low levels of development, such growth harbors the most radical potential for change in global power relations. But even if China's superior growth rate persists and its GDP surpasses that of the United States by the 2020s, as is often forecast, China will still have just over one-third of the United States' wealth per capita and, hence, considerably less economic and military power. Closing that far more Challenging gap with the developed world would take several more decades. Furthermore, GDP alone is known to be a poor measure of a country's power, and evoking it to celebrate China's ascendency is highly misleading. As it was during the twentieth century, the U.S. factor remains the greatest guarantee that liberal democracy will not be thrown on the defensive and relegated to a vulnerable position on the periphery of the international system. (Gat, 2007)

[18] As of January 14, 2021, Google Scholar lists 1242 quotations of Goldstein (1988).

Chinese scholars themselves, it seems, are acutely aware of historical parallels between the Chinese ascendancy to power and that of other hegemonical challengers in the past of human history. Yang (2018), and Lee (2019), like the Chinese President Xi Xinping, speak about the "*Thucydides Trap*" (see above) and warn the Western world about falling into the "*Churchill Trap,*" with Yang (2018) putting the blame for the start of the Cold War on the West and hoping, so Yang (2018), that the West will not "reignite" a Cold War, this time against China. Questionable, as Yang's reading of Cold War History is, his article reveals the pre-COVID-19 Chinese Party ideology of "striving for [global] achievement" (in Chinese "*fen fa you wei*") and stressing the perceived need to calm widespread Western fears about the Chinese ascent.

The rise of Asia, and the decline of the West, has also been reflected on by two late very influential scholars of the "*world system approach*" to international relations, Andre Gunder Frank (1929–2005) (Frank, 1998) in conjunction with the already mentioned Giovanni Arrighi (Arrighi, 2007; Dobbin, 2009).

Frank (1998) examined in a book which is as prophetic as Arrighi's work (1995), the global structure and flow of trade, starting in the Americas and going eastward literally around the globe. Frank (1998), examined the pattern of trade imbalances and their settlement through payment in money, which also flowed predominantly eastward. A dozen regions and their relations with each other are examined, going from the Americas, via Africa and Europe, to and through West, South, and Southeast Asia, to Japan and China and from there both across the Pacific and also back across Central Asia and Russia. Although Frank admits that hard data are hard to come by, but one section of his book devotes efforts to assembling and comparing at least some worldwide and regional dimensions of population, production, trade, and consumption, as well as their respective rates of growth, especially in Asia and Europe. Frank (1998) contends that historically not only were various parts of Asia economically far more important in and to the world economy than all of Europe. The evidence, Frank (1998) maintains, also demonstrates that Asia grew faster and more than Europe and maintained its economic lead over Europe in all these respects until at least 1750. Frank (1998) believes that in the nineteenth century, the "victory" of the West is only temporary. There is a continued dominance of Asia in the world economy through the seventeenth and into the eighteenth century. The effects of Asia's strength led to its decline after 1750 and that Europe's actions reflected the weakness of its previously marginal position in the world economy and led to its ascendance after 1800. The twenty-first-century reversion of the balance of economic, political, and cultural power to Asia has already begun.

As Dobbin correctly emphasizes, the geostrategic implication of Arrighi's analysis, 2007, is that while America recovered from the crisis of the 1960s and 1970s, the failure of the "*Project for a New American Century*" and President George W. Bush's failure in Iraq hastened the decline of the USA as a world political–economic hegemon. Arrighi's thesis is that China is the real winner of Bush's war on terror. For Arrighi, China is a new development model—a Beijing consensus to challenge the "*Washington consensus*" (see also Tausch, 2007). For Arrighi, the Western financial system built on massive borrowing from China is absolutely volatile.

In their comparison of the effects of the Spanish flu in 1918 and COVID-19 today, Barro et al. (2020) conclude:

> Both the implications of our findings from the Great Influenza Epidemic for the ongoing coronavirus epidemic are unsettling. (…) The flu death Rate of 2.0 percent out of the (…) total population in 1918–1920 translates into 150 million deaths worldwide when applied to the world's population of around 7.5 billion in 2020. Further, this death rate corresponds in our regression analysis to declines in the typical country by 6 percent for GDP and 8 percent for consumption. These economic declines are comparable to those last seen during the global Great Recession of 2008–2009. The results also suggest substantial short-term declines in real returns on stocks and short-term government bills. Thus, the possibility exists not only for unprecedented numbers of deaths but also for major global economic dislocation. Although these outcomes for the coronavirus are only possibilities, corresponding to plausible worst-case scenarios, the large potential losses in lives and economic activity justify substantial outlays to attempt to limit the damage. However, extreme mitigation efforts - such as widespread cancellations of travel, meetings, and major events - will themselves contribute to the depressed economic. activity." (Barro et al., 2020)

In Europe, the nationwide victim numbers from the flu epidemic in 1918–20 have a relatively large correlation with today's corona numbers.[19] As if the susceptibility to such disasters had been dislocated forever to Southern Europe! With 1.07% per total population, Egypt's death rate during the flu epidemic of 1918–20 was among the world's higher rates, and about equal to those of Austria, China, Hungary, Italy, Korea, Malaysia, Spain, and Turkey, but still lower than that of India, Indonesia, and Mexico, which were at the lamentable global top. We can roughly say that a fatality rate of around 1% corresponded to the countries of what was then the semi-periphery, while higher rates were typical in the world system's periphery (as to long-run time series data of global development ranging well into the nineteenth century, justifying such assumptions, see Grinin et al., 2016).

Taubenberger and Morens (2006), highlighted that the "Spanish flu" hit the world in three waves, with the first one in summer 1918 and a death rate of around 5 per 1000 persons in the UK, while the deadliest, second wave occurred more than a year later in November 1918, with a death rate of around 25 per 1000 persons, and a third peak having been reached in March, 1919, with a death rate of 10 per 1000 persons. That logic of death today seems to apply only partially, because by April 30, 2020, we had more than 3.2 million global Coronavirus cases and more than global 228,000 deaths,[20] while today (January 2021) we already have around 100 million infections and 2 million deaths.

What the leading global medical literature in a crystal-clear fashion and against all current conspiracy theories calls "*a pneumonia outbreak associated with a new coronavirus of probable bat origin*" (Zhou, 2020; furthermore Chan, 2020; Chen, 2020; Guan, 2020; Huang, 2020; Lu, 2020) is going to change the fate of the world for the coming years to come. Confronted by the emergency of the pandemic, economic literature has begun to establish "*corononomics*" to analyze the most probable economic impacts of the pandemic, benefitting from earlier attempts of *"depression economics"* and *"pandemic economics"* (Aiginger, 2009; Barro et al., 2020; Barua, 2020; Bell & Lewis, 2005; Dayaratna & Michel, 2020; Fernandes, 2020; Ozili & Thankom, 2020; Rubin, 2020; Ruiz Estrada, 2020).

---

[19] Simple Microsoft Excel correlations between the mentioned historical data on the Spanish flu and current COVID-19 infection data, as reported in https://www.worldometers.info/coronavirus/..

[20] https://www.worldometers.info/coronavirus/..

The already mentioned Italian economist Giovanni Andrea Cornia (Cornia, 2019; Cornia & Paniccià, 2000) has proven that not only epidemics kill, but also economic crises hit the world economy with the force of a tsunami (Solomon & Tausch, 2021). Cornia's theses on excess mortality led him to conclude that from 1989 to 2014, 12.28 million people in Russia fell victim to the gradual economic downturns and shock therapies. If the world faces a depression like the one which struck Russia, we can safely assume that up to 546 million people will die a premature death on a global scale from the effects of the economic depression to follow the global lockdown (Solomon & Tausch, 2021).

So, what is the nature of the global recession, which we are now facing since the COVID-19 pandemic took its global trail of destruction from the Huanan Seafood Market in Wuhan, China? As the *NY Times* showed on March 22, 2020 (see Wu et al., 2020), thousands of people from Wuhan did not stop traveling abroad even after the Chinese authorities were aware of the serious pandemic.

But already before the COVID-19 pandemic and recession, the West was weak and internally divided, above all European Union, affected by a plurality of internal and external crises, such as Brexit, the lack of real democracy at a European Union level, the malfunctioning of European monetary union, and the lack of a coherent European foreign policy (Tausch & Heshmati, 2013).

For Acemoglu (2020), the huge variation in the response to the pandemic across countries might have been caused by atrophying in the capacity of state institutions, because of the erosion in public trust in state institutions (various reasons but related to inequality, perceived or real capture of politics by banks and business elites, and responses to the global financial crisis); the weakening of technocratic, autonomous civil service as a political strategy (e.g., Italy, the US, the UK, Turkey); and the general underinvestment in social safety net and public health systems and already very weak state institutions in the developing world.

In the following images (maps), we will highlight some of the variables which were mentioned in the global political debate in the context, regardless of whether or not these variables—as our empirical tests reveal—have a direct statistical connection with the severity of the outbreak of the pandemic in the countries of the world or not. Our images (maps) are thus not to be regarded as a "theory" of what causes—*ceteris paribus*—higher or lower COVID-19 infection or death rates around the world, but only maps reproducing the global geographical distribution of phenomena mentioned in the debate about the virus:

- **The intensity of Chinese immigration**[21]

Former US President Trump said the outbreak had hit the US harder than the Japanese bombing of Pearl Harbor in the Second World Waror the 9/11 attacks two decades

[21] See, among others, https://www.nytimes.com/2020/03/31/world/asia/coronavirus-china-hong-kong-singapore-south-korea.html and https://www.nature.com/articles/d41586-020-00154-w as well as https://www.aljazeera.com/news/2020/01/timeline-china-coronavirus-spread-200126061554884.html. On President Trump's use of the term *"Chinavirus,"* see https://time.com/5807376/virus-name-foreign-history/ and https://www.bbc.com/news/world-us-canada-52568405 as well as https://time.com/5807376/virus-name-foreign-history/.

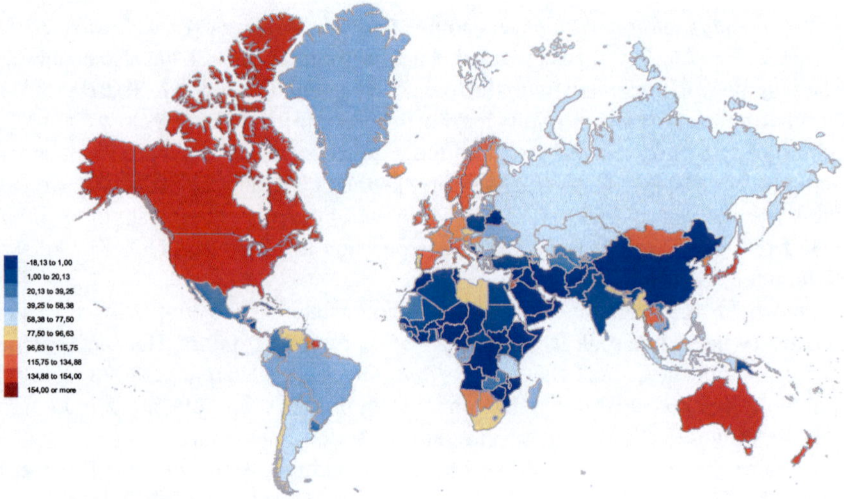


**Fig. 3.1** Intensity of Chinese immigration

ago. Under the Trump administration, punitive actions against China over its early handling of the global emergency were on the table. On the last day of 2019, December 31, reports emerged that China is investigating an "*outbreak of respiratory illness in the central city of Wuhan.*" Already per January 31, 2020, President Trump blocked travel from China (Fig. 3.1).[22]

- **Overall migrationintensity**[23]

President Trump, on April 21, 2020, 02:06 AM, sent a Tweet, saying that in "*light of the attack from the Invisible Enemy, as well as the need to protect the jobs of our GREAT American Citizens, I will be signing an Executive Order to temporarily suspend immigrationinto the United States!*".

In view of the now existing travel bans throughout the world, one reads the following quotation from "*Politico*" which reveals the difficulty of European political culture to come to terms with the harsh realities of COVID-19:

> The free circulation of goods and people, immigration policies and weak controls at the borders obviously allow the exponential spread of this type of virus," exclaimed Aurélia Beigneux, an MEP from France's right-wing National Rally, during a European Parliament debate in early February. Beigneux then asked Andrea Ammon, the director of the European

[22] https://www.aljazeera.com/news/2020/04/timeline-trump-administration-coronavirus-actions-200414131306831.html.

[23] See https://www.bbc.com/news/world-us-canada-52363852. China in turn itself also introduced a rigorous travel ban; see https://www.washingtonpost.com/gdpr-consent/?next_url=https%3a%2f%2fwww.washingtonpost.com%2fhealth%2fchina-restricts-travel-for-35-million-people-to-halt-spread-of-deadly-virus%2f2020%2f01%2f24%2f80086bde-3ee4-11ea-8872-5df698785a4e_story.html.

> Centre for Disease Prevention and Control (ECDC), whether Europe would be better off reintroducing land border checks and stopping immigration, at least temporarily. Delivering a diplomatic "no," Ammon said that closing the Schengen zone lacks scientific justification. The World Health Organization has also advised against restricting travel and trade, while EU health ministers meeting in Brussels last week said that any travel measures meant to contain the spread of the virus should safeguard free movement within the EU.[24]

The International Organization for Migration (IMO), in one analysis,[25] in a way also highlighted why it is important to look into the statistical relationships between immigration and COVID-19. IMO said that in countries all over the world, migrants may live in overcrowded environments without adequate access to water and hygiene products, where respecting social distancing and other basic prevention practices, such as self-isolating in case of illness, is difficult. Formal and informal displacement sites, transit sites, and reception centers may present the most acute challenges, but workers' quarters in industrial and rural areas, and the low-income, underserved locations migrants often share with other marginalized groups of citizens show similar issues. The IOM highlighted that in the absence of systematic screening and tracing, the risk of a rapid spread of the disease is especially heightened in and around these sites. Migrants are also often overrepresented in some of the industries hardest hit by the crisis, such as food services and nonessential retail or domestic work—or in the most precarious positions in any sector. The unprotected, undeclared positions will represent a substantial share of the millions of people COVID-19 has pushed and will push toward underemployment and unemployment—a group in which migrant workers are overrepresented (Figs. 3.2 and 3.3).

- **Current health expenditure(% of GDP)**[26]

Well before COVID-19, the years of the "*savings packages*" in the wake of the 2008 economic crisis depleted the medical resources of the European South. To quote from just one excellent recent analysis[27] explaining well the catastrophe:

[24] https://www.politico.eu/article/populists-cite-coronavirus-outbreak-to-advance-anti-immigration-agenda/ All that, of course, is history now, as a look at the global map of existing travel bans, published by Al Jazeera, https://www.aljazeera.com/news/2020/03/coronavirus-travel-restrictions-border-shutdowns-country-200318091505922.html, or Wikipedia, reveals: https://en.wikipedia.org/wiki/Travel_restrictions_related_to_the_Covid-19_pandemic.

[25] https://publications.iom.int/system/files/pdf/mrs-60.pdf.

[26] See especially https://www.imf.org/en/Topics/imf-and-covid19/Policy-Responses-to-Covid-19; https://www.ceps.eu/eu-solidarity-in-exceptional-times/; https://www.ncsl.org/research/fiscal-policy/state-fiscal-responses-to-covid-19.aspx; https://www.wifo.ac.at/en/news/effects_of_the_corona_crisis_on_public_budgets .

[27] On the health crisis in Southern Europe well before COVID-19, brought about by years of "savings packages" in the wake of the 2008 economic crisis, see https://www.intereconomics.eu/contents/year/2020/number/3/article/decades-of-tight-fiscal-policy-have-left-the-health-care-system-in-italy-ill-prepared-to-fight-the-covid-19-outbreak.html.

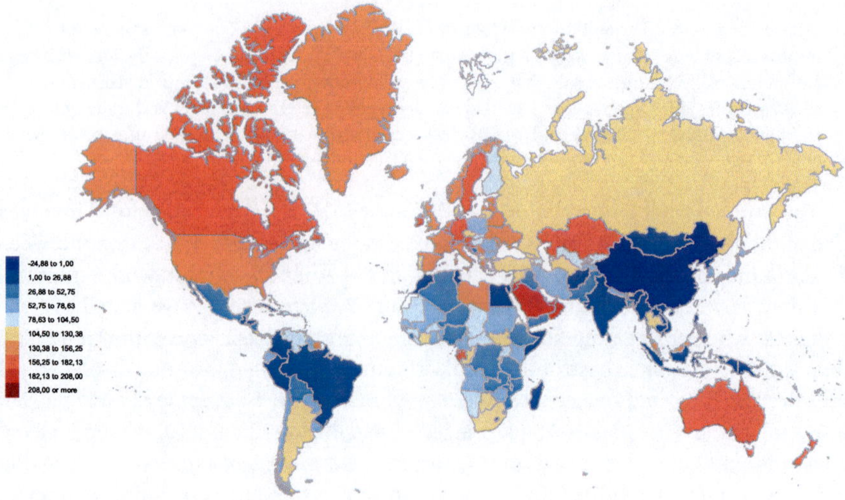


**Fig. 3.2** Overall migration intensity

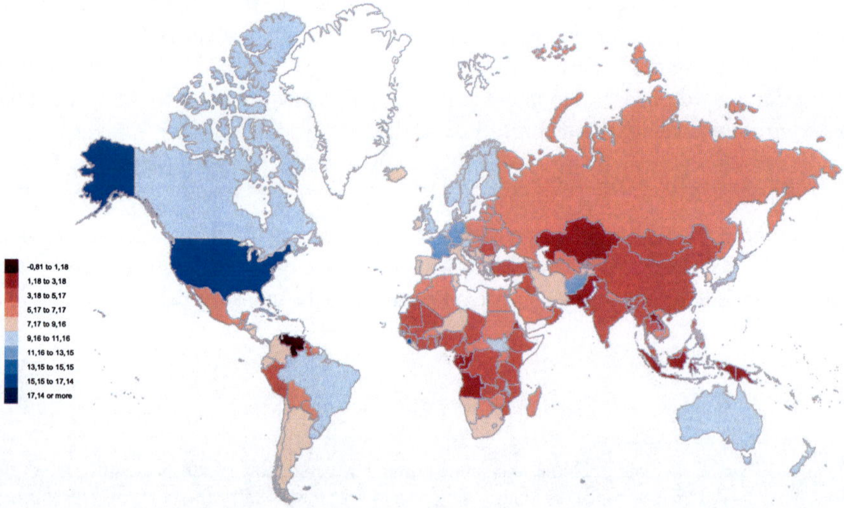


**Fig. 3.3** Current health expenditure (% of GDP), 2017

> The availability of acute care beds was reduced even more drastically than hospital capacity [...] Although a pronounced trend towards reducing acute care beds can be observed in many European countries, few European countries have reduced the number of available beds as much and to such a low level as Italy. In 1990, Italy had seven beds per 1,000 inhabitants, a value close to Germany and above the EU average. In 2017, the number of acute care beds had dropped to 2.6 per 1,000 inhabitants, significantly lower than in Germany with six beds available per 1,000 persons and much closer to the historically low value of Spain. Thus, within a rather short time period, Italy found itself at the lower end of the spectrum in Europe.

- **Economic Globalization(ETH Zurich, KOF Index)**[28]

As *Foreign Policy* highlighted, many see COVID-19 not as a cause around which the world's governments should rally, but rather as the most dramatic example of an already broken globalized system. "*In a global public healthemergency, the U.S. is alone,*" said Peter Navarro, one of former U.S. President Donald Trump's former economic advisor, to the *Financial Times.* This is "*precisely why it is important for the Trump administration to bring home its manufacturing capabilities and supply chains for essential medicines and thereby simultaneously reduce America's foreign dependencies,*" he said.[29]

Ross Douthat noted in *The New York Times* on March 31, 2020, that COVID-19 is a "*world crisistailor-made for an anti-globalization, anti-deep-state worldview—a crisisin which Chinalit the fuse, the World HealthOrganization ran interference for Beijing, the American public healthbureaucracy botched its one essential job, pious anti-racism inhibited an early public-healthresponse, and outsourcing and offshoring left our economy exposed.*"[30]

Pope Francis, reminiscent of the spirit of the social scientist Karl Polanyi (1886-1964), criticized not only globalization,[31] but also, in a Polanyian turn, against the market principle itself (Polanyi & MacIver, 1944; Sarkis et al., 2020) and stated:

> The battle against hunger and malnutrition will not end as long as the logic of the market prevails.[32]

A neo-Marxist commentator, Alfredo Saad Filho, pointed out in a similar vein:

> The pandemic hit after four decades of neoliberalism had depleted state capacities in the name of the 'superior efficiency' of the market, fostered deindustrialization through the

---

[28] On the relationship between globalization and COVID-19, see, among others, https://www.euronews.com/2020/03/24/Globalization-is-no-longer-business-as-usual-coronavirus-has-changed-the-world-works-view; https://www.rand.org/blog/2020/04/why-covid-19-will-not-stop-globalization.html; https://www.voanews.com/covid-19-pandemic/will-covid-19-kill-globalization; https://www.bbc.com/news/business-52104978; https://www.project-syndicate.org/commentary/will-covid19-remake-the-world-by-dani-rodrik-2020-04.

[29] https://foreignpolicy.com/2020/04/17/globalization-trade-war-after-coronavirus-pandemic/.

[30] https://www.nytimes.com/2020/03/31/opinion/covid-conservatism.html.

[31] https://www.ncronline.org/news/opinion/distinctly-catholic/francis-continues-critique-capitalism-globalization-indifference.

[32] http://www.vatican.va/content/francesco/en/messages/food/documents/papa-francesco_20191016_messaggio-giornata-alimentazione.html.

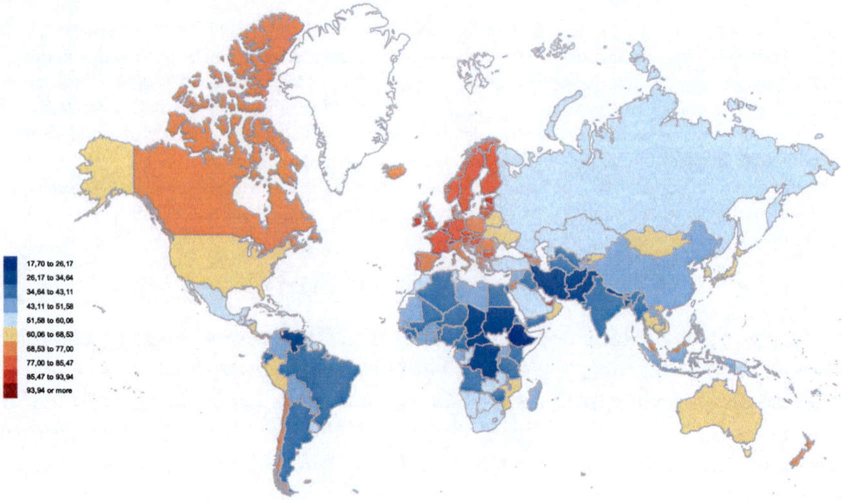


**Fig. 3.4** Economic globalization (ETH Zurich, KOF Index)

> 'globalization' of production, and built fragile financial structures secured only by the state, all in the name of short-term profitability. The disintegration of the global economy left the most uncompromisingly neoliberal economies, especially the UK and the USA, exposed as being unable to produce enough face masks and personal protective equipment for their health personnel, not to speak of ventilators to keep their hospitalized population alive. At the same time, service provision has been transformed beyond recognition, with online work becoming the norm in countless areas in a matter of days, rather than the years that this transition would have normally taken, while the neoliberal worship of consumption dissolved into undignified scrambles for hand sanitizer, pasta and sardines, and fistfights for toilet paper.[33]

Here, we present a global map based on the so-called KOF Index of Globalization, described in all detail in Solomon and Tausch (2021) (Fig. 3.4).

## 3.3 "The Pandemic Is Expected to Leave Lasting Economic Scars" (World Bank, 2021): What the World Can Learn from Past Pandemics

The World Bank, in its January 2021 global forecast, says (see Chap. 1) that we must expect a cumulative output loss that is equivalent to 36% of the world's 2019 output by 2025 (World Bank, 2021).

---

[33] http://ppesydney.net/coronavirus-crisis-and-the-end-of-neoliberalism/.

Today, research can focus on some salient classical studies, which analyzed the most terrible pandemics which struckhumankindinhistory(Barquet, 1997; Benedictow, 2009; Bruun, 2007; Büntgen, 2016; Dols, 2019; Duncan-Jones, 1996; Fenner, 1993; Findlay & Lundahl, 2017; Gottfried, 2010; Little, 2007; McCormick, 2012; Pamuk, 2007; Pamuk & Shatzmiller, 2014; Patz, 2005; Raoult et al., 2013; Rosen, 2010; Sabbatani & Fiorino, 2009; Scheidel, 2009; Schmid, 2015; Wagner et al., 2014; Ziegler, 2013).

First of all, there seems to be an emerging consensus in the literature that pandemics cannot be separated from climate change. Some pandemics, like the Plagues, occurred when there was a significant cooling of the global climate, like the Justinian Plague and the Black Death in the fourteenth century. Other pandemics happened in times of global warming.

In the face of COVID-19, the words of Thucydides should be remembered:

> In the very beginning of summer, the Peloponnesians and their confederates, with two-thirds of their forces as before, invaded Attica under the conduct of Archidamus the son of Zeuxidamas, king of Lacedaemon, and after they had encamped themselves, wasted the country about them. They had not been many days in Attica when the plague first began amongst the Athenians, [...] but so great a plague and mortality of men was never remembered to have happened in any place before. For at first neither were the physicians able to cure it through ignorance of what it was but died fastest themselves, as being the men that most approached the sick, nor any other art of man availed whatsoever. All supplications to the gods and enquiries of oracles and whatsoever other means they used of that kind proved all unprofitable; insomuch as subdued with the greatness of the evil, they gave them all over.[34]

The "*Black Death,*" 1346–53, killed up to 200 million people. Going back in human history, a pandemic, which we should list here as well, was the *"Antonine Plague,"* 165–180, CE, with up to 10 million victims in the Roman Empire, which decisively contributed to its weakening. The "*Plagueof Justinian,*" 541–542, CE, killed up to 100 million people in Europe and West Asia (40–50% of the population of Europe at that time). It sealed the end of the socio-economic order of antiquity.

The Maddison (2007), data, updated by Bolt et al.(2018), offer at least a fairly complete dataset for Sweden and France as well as for Poland for much of the fourteenth century, so we can at least guess the income depletion brought about by the Black Death (Fig. 3.5):

Some of the deadliest other pandemics in human history also left their trace on the economic fluctuations of the world.[35] The Maddison, 2007, data, updated by Bolt et al. (2018), offer some observations about real incomes from France and Sweden, as well as Poland, Portugal, and Mexico, so that we might at least in theory venture to say something about the income depressing effects of the smallpox pandemic in the sixteenth century (with around 50 million deaths) with its clear effects in Sweden, as well as the seventeenth-century plague (with around three million deaths) (Fig. 3.6).

---

[34] http://www.perseus.tufts.edu/hopper/text?doc=Perseus%3Atext%3A1999.01.0247%3Abook%3D2%3Achapter%3D47.

[35] https://www.visualcapitalist.com/history-of-pandemics-deadliest/.

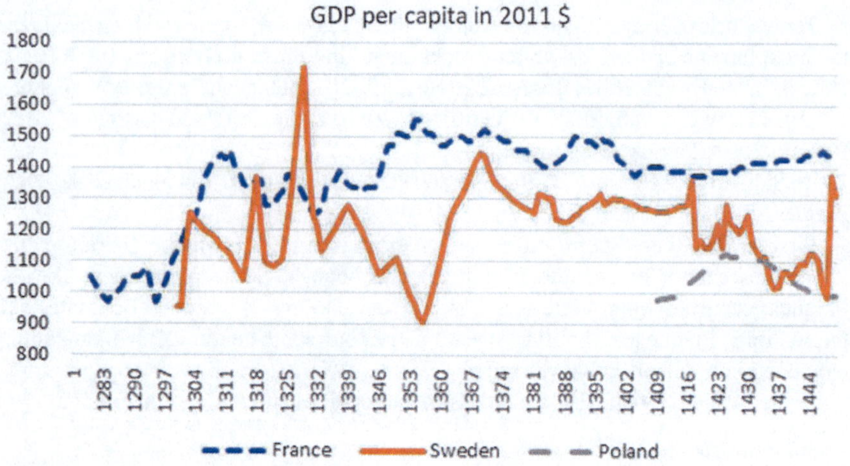


**Fig. 3.5** Effects of the Black Death on the GDP per capita in the fourteenth century, calculated from Bolt et al. (2018)

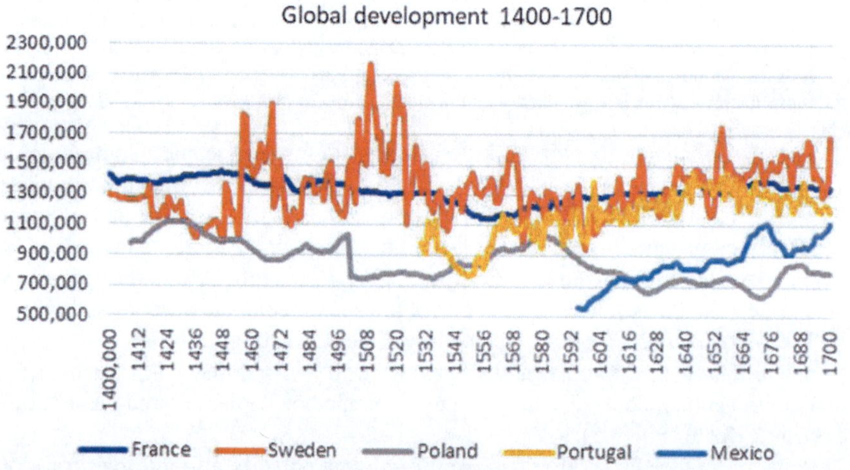


**Fig. 3.6** Global development 1400–1700 (GDP per capita in 2011$) as reflected in the data compiled by Bolt et al. (2018)

Based on the works of of Alfani, 2013; Alfani, & Percoco 2019; Armelagos et al., 2005, we also can attempt to apply the Maddison data to Italy's development 1300–1800. The Black Death and the seventeenth-century and eighteenth-century plagues devastated Italy, the home of the Renaissance and modern European culture, in an incredible way (Fig. 3.7).

Also, the cholera pandemics of the nineteenth century again devastated Italy and other countries, and greatly contributed to their subsequent economic depressions

**Fig. 3.7** Drama of Italian development 1300–1800 as reflected in the data compiled by Bolt et al. (2018)

(see Anderson, 2019; Barnett, 2019; Brody, 2000; Colwell, 1996; Ghose, 2011; Lafferty, 2009; Lim, 2005; Morens, 2008; Patz, 2005; Pinke, 2019; Steere-Williams, 2018; Tatem, 2006). The literature generally distinguishes between the following cholera waves in the nineteenth century:

First, 1817–1824,
Second, 1829–1837,
Third, 1846–1860,
Fourth, 1863–1875,
Fifth, 1881–1896.

Future research could link the occurrences of the cholera pandemics with the time series data contained in Bolt et al. (2018) (Fig. 3.8).

## 3.4 What the World Can Learn from the Great Depression

We already highlighted that the International Monetary Fund (IMF), in a 2020 prediction,[36] expected the global economy to contract by 3% in 2020, much worse than during the 2008–09 financial crisis. This scenario still assumed totally unrealistically that the pandemic fades in the second half of 2020 and containment efforts are effective. The advanced economies would, according to the IMF, shrink by −6.1%, while China and India still would experience some sort of economic growth by < 2.0%, and the Middle East and North Africa would shrink by around −3.3%. The

[36] https://www.imf.org/en/Publications/WEO/Issues/2020/04/14/weo-april-2020.

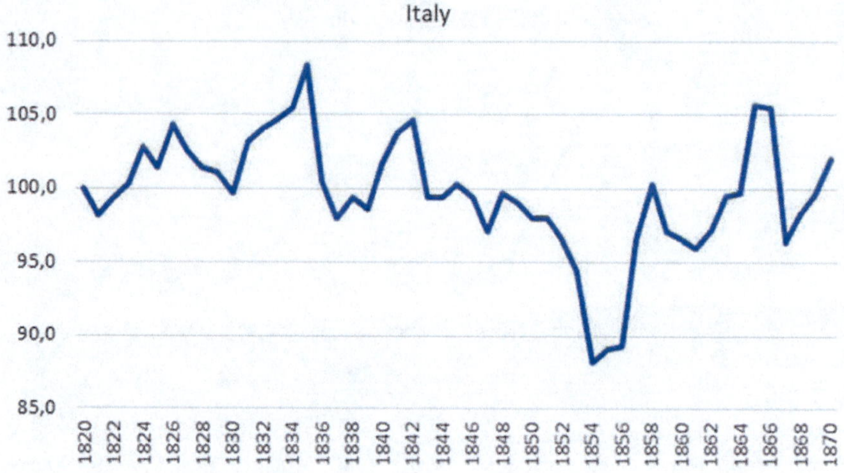


**Fig. 3.8** Deep economic depressions of the nineteenth century as reflected in the data compiled by Bolt et al. (2018): real GDP per capita in 2011US$ (1820 = 100) —the case of Italy

world would thus be confronted with the largest economic depression since the Great Depression of 1929. Everything what we have seen in 2020 and early 2021 dispels this still relatively mild view of events to come.

Thanks to the work of Barro and Ursúa (2011), and Barro et al. (2020), economic science knows much more about the effects of both the Spanish influenza pandemic from 1918 onward as well as the Great Depression from 1929 onward. Since the large literature on long economic cycles, depressions, and their devastating effects has been reviewed extensively elsewhere (Grinin et al., 2016), we rather concentrate here on analyzing what can be expected from the double crisis of a pandemic and a global depression for the countries of the world system in the coming years.

The robust lesson of the "Great Depression" accounting, made possible by the Barro and Ursúa, 2011, data is that there are very considerable differences in the way, and the tsunami of the Great Depression hit the countries of the world system in 1929. To recover their pre-depression income level by around + 5%, it took Denmark, Iceland, Korea, and Russia only a short lapse of time, and for all practical purposes they recovered already in 1930/31. By 1933/34, Colombia, Greece, Brazil, Finland, Japan, Norway, South Africa, Sweden, and Turkey also recovered again.

But only by 1935/36, Italy, Singapore, Taiwan, UK, Australia, Germany, and New Zealand recovered and had an income level of around + 5% above the pre-depression level. In 1937, Portugal and Venezuela followed, while Indonesia only made it in 1938. Even more dramatic, Austria recovered later and had an income level of around + 5% above the pre-depression level only in 1939, after the Anschluss, and thus shared the economic misfortunes of Mexico and the Philippines, which also recovered only in 1939. Even worse, Canada and the United States only could recover their income levels in 1940, after the beginning of the Second World War in 1939.

A still more belated plus of around 5% or more in the real income levels compared to 1929 was recorded in the following unfortunate nations in the following years (Table 3.1).

1944: Argentina
1945: Chile
1946: Peru, Switzerland, and Uruguay
1947: Belgium
1949: France
1950: Netherlands
1956: China and Spain
1959: Egypt
1960 or after: India, Malaysia, and Sri Lanka.

The world ahead of us is also to be seen, when we correlate the data about relative GDP per capita in PPP (UK, 1913 = 100) of the countries of the world system, contained in Barro and Ursúa (2011), and Grinin (2016), with the country death rates of the Great Influenza Epidemic, 1918, documented in Barro et al. (2020). The suggestion is that poorer countries were hit harder by the influenza epidemic of 1918 than countries of the center, although the effect is nonlinear due to the higher death rates in some richer countries. The IBM-SPSS XXIV calculation of the relationship between the natural logarithm of GDP per capita in the last peace year before the First World War in 1913 (UK = 100) and its square and the influenza death rate per country in 1918–20 in the 31 countries with available data is significant at the 1% level, and the F ratio is 8.46; the adjusted $R^2$ is 33.2%, and the predicted influenza death rate was 13,8860 − 6,1790*ln (GDP per capita PPP, UK, 1913 = 100) + 0.7160**ln (GDP per capita PPP, UK, 1913 = 100)$^2$.

Finally, we risk a prognosis on the overall societal effects of the pandemic. A lesson of the 1918–1920 pandemic is that the severity of the pandemic, as it impacts the different countries of the world, contributes to rising income inequality over time. We measured this by calculating a Pearson correlation of income inequality with the time axis, 1921–1929, based on the Piketty time series data of the share of the richest 1% in total incomes, discussed in Grinin et al. (2016), and the death rates from the Great Influenza Epidemic, 1918–1920 (Barro et al., 2020). We also attempted to estimate the effects of the severity of the pandemic on international income convergence/divergence over time. To estimate this, we calculated the Pearson correlation of income convergence with the time axis, 1921–1929, based on the Maddison data of a country's income per capita in PPP, world average = 100, documented in Grinin et al. (2016), and the severity of the influenza epidemic 1918–1920, measured by the flu death rates (Barro et al., 2020). The result is that the severity of an epidemic negatively influences international income convergence over time.

We should finally look at another lesson of history, which might be deduced from the Barro and Ursúa (2011), data. In the UK, it took practically a decade to recover the

**Table 3.1** Recovery from the Great Depression, 1929

| | GDP pc in 2011$ 1918 | GDP pc in 2011$ 1928 | GDP pc in 2011$ 1929 | GDP pc in 2011$ 1938 | Recovery from the Influenza 1918: 1918–1928 1918 = 100 | Recovery from the Great Depression 1929–1938: 1929 = 100 |
|---|---|---|---|---|---|---|
| Spain | 4290 | 5625 | 6009 | 3539 | 131.12 | 58.89 |
| Nicaragua | No data | 1710 | 1883 | 1157 | No data | 61.44 |
| Honduras | No data | 2165 | 2097 | 1545 | No data | 73.68 |
| Malaysia | 1658 | 2376 | 2877 | 2328 | 143.31 | 80.92 |
| Cuba | 3351 | 3314 | 3653 | 3012 | 98.90 | 82.45 |
| Canada | 7250 | 8987 | 8856 | 7754 | 123.96 | 87.56 |
| Chile | 3443 | 4195 | 4248 | 3793 | 121.84 | 89.29 |
| Sri Lanka | 1289 | 1463 | 1554 | 1426 | 113.50 | 91.76 |
| United States | 8648 | 10,040 | 10,543 | 9797 | 116.10 | 92.92 |
| Argentina | 5564 | 7351 | 7481 | 6975 | 132.12 | 93.24 |
| El Salvador | No data | 971 | 958 | 900 | No data | 93.95 |
| India | 1234 | 1492 | 1544 | 1466 | 120.91 | 94.95 |
| Indonesia | 1263 | 1560 | 1576 | 1526 | 123.52 | 96.83 |
| France | 3348 | 5433 | 5693 | 5620 | 162.28 | 98.72 |
| Ecuador | 1327 | 1555 | 1601 | 1600 | 117.18 | 99.94 |
| Bolivia (Plurinational State of) | 1062 | 1273 | 1311 | 1313 | 119.87 | 100.15 |
| Panama | 2115 | 2058 | 2095 | 2111 | 97.30 | 100.76 |
| Netherlands | 4150 | 5751 | 5588 | 5634 | 138.58 | 100.82 |
| Peru | 1070 | 1557 | 1693 | 1709 | 145.51 | 100.95 |
| Singapore | 1350 | 1612 | 2269 | 2296 | 119.41 | 101.19 |
| Former Yugoslavia | No data | 1016 | 1054 | 1068 | No data | 101.33 |
| China | No data | No data | 950 | 965 | No data | 101.58 |
| Austria | 2767 | 3962 | 4006 | 4080 | 143.19 | 101.85 |
| Philippines | 1574 | 1704 | 1729 | 1763 | 108.26 | 101.97 |
| Mexico | 1741 | 1935 | 1829 | 1870 | 111.14 | 102.24 |
| Uruguay | 3502 | 4626 | 4583 | 4713 | 132.10 | 102.84 |
| Belgium | 4152 | 5746 | 5482 | 5640 | 138.39 | 102.88 |
| Romania | No data | 746 | 787 | 810 | No data | 102.92 |

(continued)

**Table 3.1** (continued)

| | | | | | | |
|---|---|---|---|---|---|---|
| Poland | No data | No data | 2718 | 2801 | No data | 103.05 |
| Thailand | No data | No data | 963 | 1003 | No data | 104.15 |
| Australia | 7993 | 10,690 | 10,437 | 11,280 | 133.74 | 108.08 |
| Ireland | No data | 3679 | 3795 | 4113 | No data | 108.38 |
| Portugal | 1527 | 1953 | 2138 | 2320 | 127.90 | 108.51 |
| Italy | 2458 | 2664 | 2741 | 3010 | 108.38 | 109.81 |
| Switzerland | 4692 | 5672 | 5799 | 6467 | 120.89 | 111.52 |
| Brazil | 721 | 1034 | 1015 | 1139 | 143.41 | 112.22 |
| Greece | 1963 | 3066 | 3214 | 3674 | 156.19 | 114.31 |
| United Kingdom | 6888 | 7150 | 7380 | 8456 | 103.80 | 114.58 |
| Taiwan, Province of China | 1399 | 1813 | 1868 | 2161 | 129.59 | 115.69 |
| Denmark | 4756 | 6111 | 6431 | 7507 | 128.49 | 116.73 |
| Costa Rica | No data | 2451 | 2300 | 2729 | No data | 118.65 |
| Japan | 2228 | 2661 | 2705 | 3259 | 119.43 | 120.48 |
| Bulgaria | No data | 2146 | 2172 | 2653 | No data | 122.15 |
| Colombia | 1335 | 2065 | 2086 | 2555 | 154.68 | 122.48 |
| New Zealand | 6789 | 7107 | 7275 | 8926 | 104.68 | 122.69 |
| Sweden | 4294 | 4880 | 5061 | 6236 | 113.65 | 123.22 |
| Norway | 4382 | 4660 | 4850 | 6048 | 106.34 | 124.70 |
| Hungary | No data | 1748 | 1792 | 2316 | No data | 129.24 |
| Guatemala | No data | 1591 | 1758 | 2273 | No data | 129.29 |
| South Africa | 2421 | 3192 | 3117 | 4073 | 131.85 | 130.67 |
| Jamaica | No data | No data | 1118 | 1554 | No data | 139.00 |
| Finland | 1488 | 2942 | 2952 | 4112 | 197.72 | 139.30 |
| Republic of Korea | 950 | 945 | 888 | 1266 | 99.47 | 142.57 |
| Venezuela (Bolivarian Republic of) | 977 | 1373 | 1410 | 2055 | 140.53 | 145.74 |
| Turkey | 924 | 1145 | 1265 | 1898 | 123.92 | 150.04 |
| Germany | 4184 | 4682 | 4533 | 6968 | 111.90 | 153.72 |
| Former USSR | 1327 | 2737 | 2769 | 4294 | 206.25 | 155.07 |

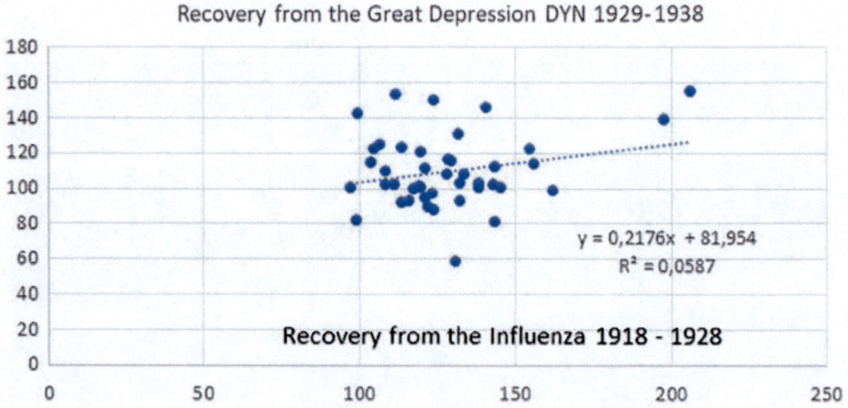


**Fig. 3.9** Recovery from the Influenza Pandemic, 1918–1920, and recovery from the Great Depression, 1929

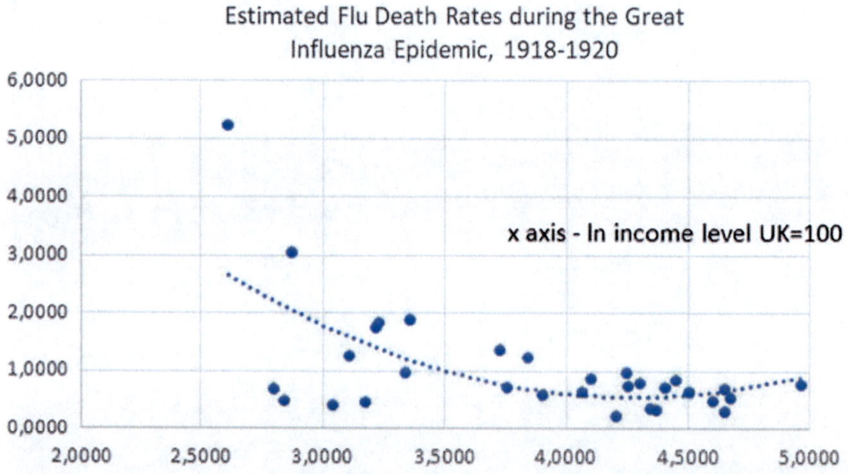


**Fig. 3.10** Estimated flu death rates during the Great Influenza Epidemic, 1918, predicted by real natural logarithm of GDP per capita in PPP (UK, 1913, = 100)

pre-1918 pandemic income levels; in Italy, the crisis was even more severe, Italian democracy collapsed, and Benito Mussolini rose to power (Fig. 3.9).

The world ahead of us is also reflected in Fig. 3.10, which correlates the data about relative GDP per capita in PPP (UK, 1913 = 100) of the countries of the world system, contained in Barro and Ursúa (2011), and Grinin (2016), with the country death rates of the Great Influenza Epidemic, 1918. The suggestion is that poorer countries were hit harder by the influenza epidemic of 1918 than countries of the center, although the effect is nonlinear due to the higher death rates in some richer countries (see also Table 3.2).

Table 3.2 now mentions the original data.

**Table 3.2** GDP per capita in PPP (UK = 100) (Grinin et al., 2016) and estimated flu death rates during the Great Influenza Epidemic, 1918-1920 (Barro et al., 2020)

| | GDP per capita in PPP (UK, 1913 = 100) | Estimated flu death rates during the Great Influenza Epidemic, 1918–1920 |
|---|---|---|
| Argentina | 77.16 | 0.33 |
| Australia | 104.80 | 0.28 |
| Austria | 70.41 | 0.97 |
| Belgium | 85.75 | 0.83 |
| Brazil | 16.48 | 0.69 |
| Canada | 90.37 | 0.62 |
| Chile | 60.72 | 0.86 |
| Colombia | 17.17 | 0.46 |
| Denmark | 79.50 | 0.31 |
| Finland | 42.90 | 0.71 |
| France | 70.82 | 0.74 |
| Germany | 74.13 | 0.78 |
| Greece | 23.92 | 0.45 |
| India | 13.68 | 5.22 |
| Indonesia | 17.66 | 3.04 |
| Italy | 46.84 | 1.23 |
| Japan | 28.19 | 0.96 |
| Netherlands | 82.28 | 0.71 |
| New Zealand | 104.69 | 0.69 |
| Norway | 49.73 | 0.57 |
| Peru | 20.97 | 0.39 |
| Portugal | 25.40 | 1.81 |
| Russia | 28.73 | 1.87 |
| Spain | 41.78 | 1.36 |
| Sri Lanka | 25.08 | 1.74 |
| Sweden | 58.40 | 0.63 |
| Switzerland | 144.14 | 0.76 |
| UK | 100.00 | 0.46 |
| Uruguay | 67.26 | 0.22 |
| USA | 107.72 | 0.52 |
| Venezuela | 22.43 | 1.25 |

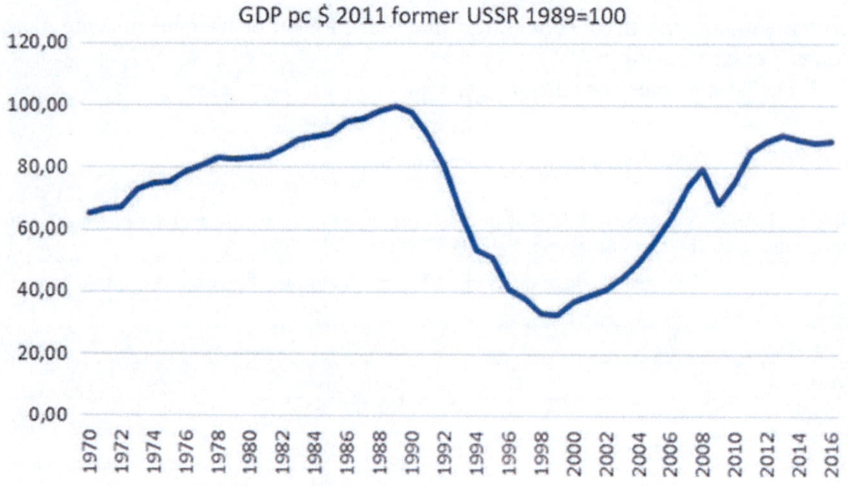


**Fig. 3.11** Great Depression at the end of the Soviet Union

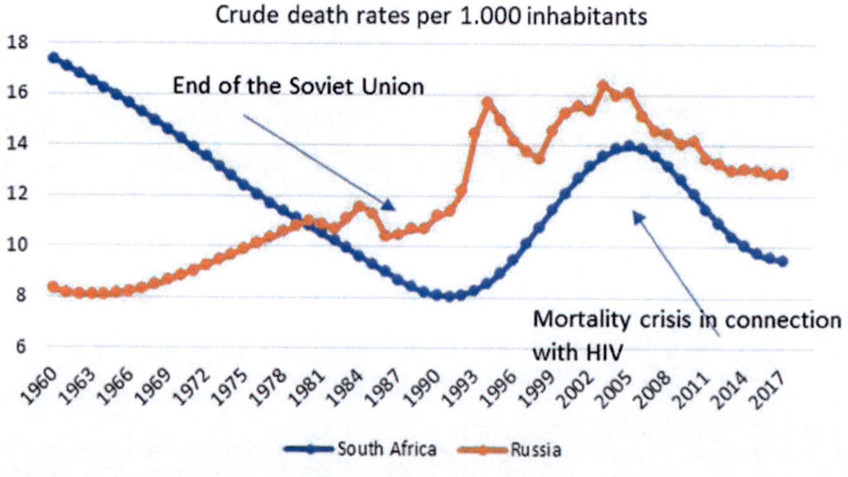


**Fig. 3.12** Mortality crises in Russia and South Africa

## 3.5 What the World Can Learn from the End of the Soviet Union and "Real Socialism" in Eastern Europe

The economic and social *tsunami* that swept across Eastern Europe and the former USSR could also become a realistic image of what we can expect in the Western countries after COVID-19.

Cornia's theses on excess mortality led him to conclude that from 1989 to 2014, 12,28 million people in the region of Eastern Central Europe and the former USSR fell victim to the gradual economic downturns and shock therapies. The hurricane that is now brewing over the world after the expected health and economic crisis 2020/2021 is already recognizable. The neoliberal set of rules, which determined the West's path after the 2008 financial crisis, has caused death rates to climb since 2008 and led to exactly the situation which Cornia describes in his works. The shock therapies are a mirror of things to come.

Downgraded hospitals in Italy, Spain, the reduction of public services and health services to the limits of state failure, such as in Greece in the wake of the European Stability Mechanism (ESM) austerity packages, aggervated the severity of the current COVID-19 crisis. Applied to the population as a whole, we must start from the thesis that the 2008 crisis and the subsequent crises in the Eurozone led to the premature death of 1.76 million people (see Solomon & Tausch, 2021, reporting calculations, based on https://data.worldbank.org/).[37]

If Russia could have stabilized the death rate at 10.0 per 1000 as in the Brezhnev era, there are 14 million deaths to date that could have been avoided. If the gross death rate had stabilized at 12.0 as at the beginning of perestroika, still 8 million lives could have been saved (our own calculations, based on https://data.worldbank.org/) (Figs. 3.11 and 3.12).[38]

---

[37] Crude death rates per 1.000 inhabitants in the European Union, according to the World Bank Open Data, Available from: https://data.worldbank.org/data, have a clear and straight linear downward trend, which Microsoft Excel regression routine (OLS-regression) identifies as being dependent on the time axis with -0,04 as the unstandardized regression coefficient and 86,85 as the constant. Without using more sophisticated models, and predicting crude death rates from this trend after 2008 yields us a first realistic estimate of the crude death rates per 1000 inhabitants in the European Union, if this good health policy trend would have continued. We compared then these predicted crude death rates per 1000 inhabitants with the crude death rates after the crisis of 2008, which really exploded, considering the long-term European advances in public health policy before the crisis of 2008. We then projected for each year the numbers of deaths which would have been avoidable, subtracting the crude death rates per 1.000 inhabitants after 2008 from the expected numbers, based on the before-mentioned simple linear regression (regression residuals), which we then weighted with the yearly population figures of the European Union. From 2008 onward, these figures add up together to 1.76 million human beings who died a death which would have been avoidable had the European Union not experienced the 2008 crisis.

[38] Crude death rates per 1.000 inhabitants in Russia, according to the World Bank Open Data, available from: https://data.worldbank.org/data. We compared Russian crude death rates per 1000 inhabitants, as they occurred in Russian history, with the crude death rates which were recorded under Brezhnev (variant 1, crude death rate per 1000 inhabitants = 10.0) and during the onset of "perestroika" (crude death rate per 1000 inhabitants = 12.0). A simple linear prolongation of public health policy standards during the waning of the communist system would have at least stabilized crude death rates per 1.000 inhabitants enjoyed then in Russia under Brezhnev and enjoyed today by the inhabitants of countries such as Armenia, the Slovak Republic, Slovenia, North Macedonia, and Bosnia and Herzegovina (variant 1) or at last at the beginning of Perestroika then in Russia or in the region of Central Europe and the Baltics (World Bank terminology) today. What happened, however, ever since was not a stabilization of public health, but one of the biggest mortality crises in modern history after the Second World War. Using the mentioned World Bank data, we arrive at the distressing conclusion that up to 12 million people died prematurely (variant 1) or at least 8 million people (lower variant, 2). Simply comparing crude death rates per 1.000 inhabitants in 2017 with those in 2000 reveals another striking aspect: In descending order, Japan, Lithuania, Bosnia and Herzegovina, Romania, Greece, Syrian Arab Republic, Croatia, Albania, Venezuela, RB, North

## 3.6 What the World Can Learn from the 2008/2009 Crisis

Equally stark lessons must be drawn from the long-term effects of the crisis of 2008. In Table 3.3, we show that already before the tsunami of the present COVID-19 crisis hit so many countries with full force, and none the less than 35 countries did not even fully recover from the last major crisis which began in 2008. That means that for all practical purposes they must brace them now for at least two decades of income loss. Once, hopefully, when the COVID-19 crisis and its expected terrible economic aftermath will be over, their populations will remember that the ordeal of falling real incomes already began with the economic crisis of 2008. The Arab countries among them are Yemen, Rep., Libya, Kuwait, United Arab Emirates, Oman, Jordan, and Qatar. Also, several countries of the European Union of stagnation and downward economic trends are among them: If we take the real income in purchasing power in last year before the 2008 crisis, i.e., 2007, as the benchmark (2007 = 100), World Bank figures for GDP per capita in purchasing power in constant 2011 international dollars indicate that the GDP per capita in Greece was 76,70; in Italy, it was 91,71; in Cyprus, it was 94,67; in Finland, it was 97,46, in Luxembourg, it was 97,76; and in Spain, it was 99,94. So in all these 35 countries, people were already confronted by a "lost decade" of development, and now they are again confronted with the prospect of yet another prolonged recession. Only Bahrain and Saudi Arabia were the Gulf countries to have had GDP per capita PPP (constant 2011 international $) in 2017 which was higher than the GDP per capita PPP (constant 2011 international $) in 2007, while all the other Gulf countries were racing backward in their real income development endeavor already a decade before the present COVID-19 crisis:

At the same time, it is clear that the Gulf countries increased their economic influence over the ailing European continent after the 2008/2009 crisis. Bianco, 2020, with justification, highlights this important point in the context of our study:

> When the crisis hit Europe, GCC countries stepped forward to become significant sources of investment and other financial support in various EU member states. On top of flagship investments – such as those in Paris Saint-Germain, London's Shard, and luxury estates in Italy – GCC countries funded banks and other areas of the private sector in Europe. For example, the Qatar Investment Authority (QIA), a sovereign wealth fund, invested €300m in Spain's troubled savings banks and €1 billion in a joint fund focused on Greek small and medium-sized enterprises (SMEs). Abu Dhabi's Mubadala Investment Company and the QIA both established partnerships with the Caisse des Dépôts et Consignations to invest €300 m and €800 m in French SMEs respectively. Similarly, in 2014, Qatar and Kuwait agreed to invest €1 bn and €500 m to Italy's Fondo Strategico Italiano respectively. Mubadala also acquired 5 percent of Italy's largest bank, UniCredit, in 2017. In 2018 Saudi Arabia offered to revitalise an agreement it signed with Spain in 2007 for an infrastructure and energy fund worth more than €1 billion. Gulf Arab countries also stepped up their economic engagement with more robust European economies, such as Germany's. The QIA has invested in major German entities such as Volkswagen, Hochtief, Siemens, and Deutsche Bank, while around 700 German companies are active in Saudi Arabia – which is now Germany's third-largest trade partner in the Middle East. The balance of trade between Europe and the GCC

---

Macedonia, Cuba, Puerto Rico, Armenia, Georgia, Latvia, Mexico, Germany, Serbia, and Poland suffered from an increase in their crude death rates per 1000 inhabitants over the last 20 years.

**Table 3.3** GDP per capita PPP (constant 2011 international $) in the world since 2007

| Country | GDP pc in 2007 | GDP pc in 2017 | GDP pc in 2017, 2007 = 100 |
|---|---|---|---|
| Yemen, Rep. | 4285 | 2404 | 56.12 |
| Libya | 28,600 | 17,321 | 60.56 |
| Equatorial Guinea | 35,901 | 22,299 | 62.11 |
| San Marino | 90,468 | 57,297 | 63.33 |
| Kuwait | 96,850 | 65,167 | 67.29 |
| Timor–Leste | 9309 | 6741 | 72.41 |
| Greece | 32,074 | 24,602 | 76.70 |
| Central African Republic | 885 | 746 | 84.31 |
| Antigua and Barbuda | 26,517 | 22,400 | 84.48 |
| Brunei Darussalam | 84,481 | 72,524 | 85.85 |
| United Arab Emirates | 77,236 | 66,478 | 86.07 |
| Bahamas, The | 32,884 | 28,318 | 86.11 |
| Cayman Islands | 76,468 | 65,996 | 86.31 |
| Oman | 43,125 | 37,779 | 87.60 |
| Curacao | 28,549 | 25,111 | 87.96 |
| Jordan | 9531 | 8400 | 88.13 |
| Aruba | 40,172 | 35,862 | 89.27 |
| Italy | 38,701 | 35,491 | 91.71 |
| Trinidad and Tobago | 31,067 | 28,596 | 92.05 |
| Ukraine | 8522 | 7902 | 92.72 |
| Barbados | 17,786 | 16,570 | 93.16 |
| Jamaica | 8707 | 8157 | 93.69 |
| Cyprus | 35,523 | 33,628 | 94.67 |
| Burundi | 707 | 671 | 94.81 |
| Madagascar | 1706 | 1648 | 96.62 |
| Gambia, The | 2311 | 2240 | 96.95 |
| Finland | 42,523 | 41,443 | 97.46 |
| Qatar | 11,6023 | 11,3262 | 97.62 |
| Micronesia, Fed. States | 3256 | 3179 | 97.64 |
| Luxembourg | 97,864 | 95,666 | 97.75 |
| Gabon | 16,496 | 16,145 | 97.87 |
| Belize | 7742 | 7592 | 98.07 |
| Spain | 34,146 | 34,126 | 99.94 |
| Vanuatu | 2840 | 2839 | 99.97 |
| Norway | 65,026 | 65,014 | 99.98 |
| Angola | 6044 | 6045 | 100.01 |
| Dominica | 9655 | 9680 | 100.25 |
| Lebanon | 11,522 | 11,647 | 101.08 |
| Chad | 1735 | 1754 | 101.10 |
| Bahrain | 42,754 | 43,321 | 101.32 |
| Puerto Rico | 34,821 | 35,454 | 101.82 |
| Samoa | 5786 | 5904 | 102.03 |

(continued)

**Table 3.3** (continued)

| | | | |
|---|---|---|---|
| South Africa | 11,925 | 12,215 | 102.43 |
| Portugal | 27,579 | 28,257 | 102.46 |
| Mauritania | 3567 | 3657 | 102.52 |
| Denmark | 46,374 | 47,555 | 102.55 |
| Croatia | 22,254 | 22,861 | 102.73 |
| St. Vincent and the Grenadines | 10,392 | 10,710 | 103.06 |
| Austria | 43,939 | 45,421 | 103.37 |
| Switzerland | 56,269 | 58,171 | 103.38 |
| France | 37,671 | 38,956 | 103.41 |
| United Kingdom | 38,783 | 40,229 | 103.73 |
| Netherlands | 46,879 | 48,809 | 104.12 |
| St. Kitts and Nevis | 25,202 | 26,241 | 104.12 |
| Belgium | 41,313 | 43,133 | 104.41 |
| Slovenia | 30,086 | 31,449 | 104.53 |
| Marshall Islands | 3282 | 3437 | 104.72 |
| Argentina | 18,030 | 18,945 | 105.08 |
| St. Lucia | 11,676 | 12,276 | 105.13 |
| Suriname | 12,964 | 13,636 | 105.19 |
| Haiti | 1571 | 1653 | 105.24 |
| Canada | 41,629 | 43,871 | 105.39 |
| Kiribati | 1910 | 2025 | 106.00 |
| Japan | 36,697 | 38,907 | 106.02 |
| Brazil | 13,334 | 14,236 | 106.77 |
| United States | 50,898 | 54,471 | 107.02 |
| Mexico | 16,772 | 17,956 | 107.06 |
| Iceland | 44,624 | 47,840 | 107.21 |
| Congo, Rep. | 4757 | 5103 | 107.28 |
| European Union | 34,493 | 37,010 | 107.30 |
| Sweden | 43,972 | 47,261 | 107.48 |
| Saudi Arabia | 45,252 | 48,804 | 107.85 |
| Estonia | 27,379 | 29,704 | 108.49 |
| Russian Federation | 23,020 | 25,062 | 108.87 |
| New Zealand | 33,102 | 36,046 | 108.90 |
| Iran, Islamic Rep. | 17,507 | 19,098 | 109.09 |
| Comoros | 2341 | 2556 | 109.19 |
| Algeria | 12,622 | 13,822 | 109.51 |
| Australia | 40,733 | 44,781 | 109.94 |
| Maldives | 11,984 | 13,201 | 110.16 |
| Cabo Verde | 5769 | 6368 | 110.39 |
| Guatemala | 6713 | 7423 | 110.58 |
| Grenada | 11,860 | 13,320 | 112.31 |
| Palau | 14,456 | 16,239 | 112.33 |
| Arab World | 13,836 | 15,543 | 112.33 |
| Germany | 40,271 | 45,393 | 112.72 |
| Czech Republic | 28,844 | 32,571 | 112.92 |
| Latvia | 22,011 | 24,859 | 112.94 |
| Mali | 1788 | 2019 | 112.94 |

(continued)

**Table 3.3** (continued)

| | | | |
|---|---|---|---|
| Tunisia | 9692 | 10,950 | 112.98 |
| Honduras | 3951 | 4469 | 113.12 |
| Liberia | 1034 | 1176 | 113.67 |
| Guinea–Bissau | 1383 | 1577 | 114.00 |
| Benin | 1806 | 2069 | 114.60 |
| Hungary | 23,492 | 27,032 | 115.07 |
| Israel | 28,717 | 33,123 | 115.34 |
| Solomon Islands | 1842 | 2125 | 115.38 |
| Tuvalu | 3065 | 3548 | 115.75 |
| Cameroon | 2856 | 3313 | 116.00 |
| Pakistan | 4097 | 4765 | 116.29 |
| El Salvador | 6203 | 7247 | 116.82 |
| Eswatini | 7920 | 9316 | 117.63 |
| Azerbaijan | 13,499 | 15,929 | 118.00 |
| Senegal | 2737 | 3232 | 118.10 |
| Ecuador | 8823 | 10,454 | 118.48 |
| Egypt, Arab Rep. | 8987 | 10,673 | 118.75 |
| Niger | 772 | 921 | 119.19 |
| Serbia | 12,754 | 15,289 | 119.87 |
| Namibia | 8394 | 10,093 | 120.25 |
| Chile | 18,529 | 22,297 | 120.33 |
| Nigeria | 4421 | 5351 | 121.04 |
| Montenegro | 13,603 | 16,477 | 121.13 |
| Hong Kong SAR, China | 45,937 | 56,094 | 122.11 |
| World | 12,723 | 15,558 | 122.28 |
| Malawi | 942 | 1154 | 122.47 |
| Botswana | 13,165 | 16,165 | 122.79 |
| Tonga | 4632 | 5746 | 124.06 |
| Sierra Leone | 1129 | 1404 | 124.30 |
| Uganda | 1422 | 1768 | 124.38 |
| Slovak Republic | 24,104 | 30,059 | 124.70 |
| Armenia | 7012 | 8745 | 124.72 |
| Costa Rica | 12,334 | 15,430 | 125.10 |
| North Macedonia | 10,481 | 13,140 | 125.37 |
| Kenya | 2356 | 2961 | 125.68 |
| Seychelles | 20,968 | 26,416 | 125.98 |
| Burkina Faso | 1339 | 1697 | 126.72 |
| Belarus | 13,560 | 17,193 | 126.79 |
| Singapore | 69,203 | 87,760 | 126.82 |
| Colombia | 10,389 | 13,186 | 126.92 |
| Fiji | 7371 | 9389 | 127.37 |
| Kazakhstan | 18,885 | 24,079 | 127.50 |
| Sao Tome and Principe | 2349 | 3011 | 128.21 |
| Korea, Rep. | 28,014 | 35,938 | 128.29 |
| Morocco | 5845 | 7515 | 128.55 |
| Bulgaria | 14,463 | 18,606 | 128.65 |
| Thailand | 12,610 | 16,286 | 129.15 |

(continued)

**Table 3.3** (continued)

| | | | |
|---|---|---|---|
| Congo, Dem. Rep. | 623 | 808 | 129.71 |
| Guinea | 1654 | 2154 | 130.25 |
| Bosnia and Herzegovina | 9335 | 12,166 | 130.32 |
| Lithuania | 22,738 | 29,668 | 130.48 |
| Nicaragua | 3959 | 5169 | 130.56 |
| Kyrgyz Republic | 2599 | 3394 | 130.56 |
| Ireland | 50,420 | 66,132 | 131.16 |
| Kosovo | 7409 | 9777 | 131.96 |
| Macao SAR, China | 79,634 | 10,5774 | 132.82 |
| Zambia | 2798 | 3718 | 132.87 |
| Togo | 1156 | 1538 | 132.99 |
| West Bank and Gaza | 3495 | 4652 | 133.09 |
| Malta | 27,669 | 37,146 | 134.25 |
| Cote d'Ivoire | 2650 | 3565 | 134.49 |
| Iraq | 11,774 | 15,924 | 135.24 |
| Lesotho | 2102 | 2847 | 135.48 |
| Malaysia | 20,022 | 27,291 | 136.30 |
| Tanzania | 2053 | 2809 | 136.86 |
| Paraguay | 8605 | 11,790 | 137.02 |
| Romania | 17,065 | 23,468 | 137.52 |
| Bolivia | 4912 | 6799 | 138.43 |
| Sudan | 3194 | 4436 | 138.86 |
| Guyana | 5318 | 7401 | 139.16 |
| Papua New Guinea | 2841 | 3957 | 139.26 |
| Albania | 8451 | 11,797 | 139.59 |
| Mozambique | 922 | 1289 | 139.78 |
| Turkey | 17,905 | 25,031 | 139.80 |
| Poland | 19,563 | 27,379 | 139.95 |
| Afghanistan | 1248 | 1758 | 140.93 |
| Zimbabwe | 1810 | 2568 | 141.92 |
| Dominican Republic | 10,453 | 14,880 | 142.35 |
| Moldova | 4344 | 6234 | 143.51 |
| Mauritius | 141,46 | 20,319 | 143.64 |
| Uruguay | 14,365 | 20,658 | 143.81 |
| Peru | 8637 | 12,517 | 144.93 |
| Philippines | 5175 | 7581 | 146.50 |
| Georgia | 6825 | 10,172 | 149.04 |
| Nepal | 1750 | 2613 | 149.35 |
| Indonesia | 7470 | 11,161 | 149.40 |
| Ghana | 2645 | 4051 | 153.19 |
| Tajikistan | 1908 | 2924 | 153.20 |
| Rwanda | 1206 | 1889 | 156.66 |
| Cambodia | 2332 | 3654 | 156.66 |
| Panama | 14,007 | 22,244 | 158.81 |
| Sri Lanka | 7328 | 11,706 | 159.75 |
| Vietnam | 3852 | 6233 | 161.81 |
| Bangladesh | 2216 | 3634 | 164.01 |

(continued)

**Table 3.3** (continued)

| | | | |
|---|---|---|---|
| Mongolia | 7049 | 11,627 | 164.94 |
| Bhutan | 5472 | 9043 | 165.24 |
| Uzbekistan | 4398 | 7348 | 167.10 |
| India | 3849 | 6516 | 169.30 |
| Lao PDR | 3524 | 6310 | 179.08 |
| Myanmar | 2815 | 5610 | 199.32 |
| Ethiopia | 860 | 1724 | 200.41 |
| Turkmenistan | 7815 | 16,390 | 209.73 |
| China | 7264 | 15,254 | 209.99 |
| Nauru | 4049 | 13,670 | 337.63 |

increasingly favours the former, while there are potentially profitable opportunities for European firms in real estate and infrastructure projects under development plans such as Saudi Arabia's Vision 2030. Additionally, defence industries across Europe – especially those in France, Italy, and the UK, but also Spain – have benefited from an increase in lucrative contracts and joint ventures with GCC countries. (Bianco, 2020)

## 3.7 The Geostrategic Implications of the 2008/2009: Now, There Is Worse to Come

On January 6, 2021, a violent mob, instigated by the country's president, stormed Capitol Hill. Across the Atlantic, the crisis is more subtle. The most divisive issue is the cleavage between the European North and South, which already became so acutely evident in the 2008 crisis, which almost led to the disintegration of European Monetary Union. To observers from outside Europe, it is often hard to understand how deeply Monetary Union already affects almost any decision which has to be taken in Europe. What, then, will be the answer of the European Union to this economic and social hurricane, which now gathers over the European Mediterranean? We are afraid that the European Union did not learn the bitter lessons of the crisis in 2008 (for an ample debate on Euro-monetarism, Solomon & Tausch, 2021). The consensus emerging from the literature, critical of German monetarism and neoliberalism , is that at the core of the failure of European Monetary Union lie the German mercantilist economic model and the inability of the other European countries to question this model openly and to convince Germany that it is not even in that country's own interest to opt for competition rather than cooperation between nations, in particular between the members of a currency union. In the following graphs, we would like to analyze the effects of neoliberal and neomonetarist policies for the unity and cohesion and national security of the European continent, based on World Bank Open Data.

Figures 3.13 and 3.14 offer a snapshot of the severe imbalances that characterize the functioning of the European Monetary Union already before the onset of the COVID-19 crisis.

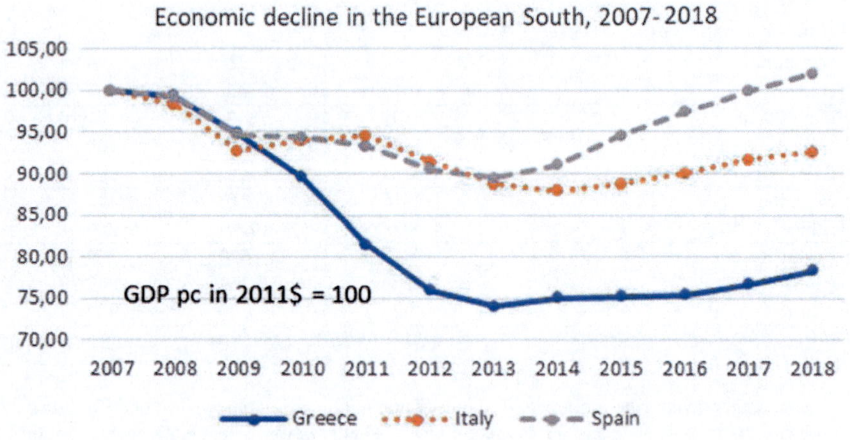


**Fig. 3.13** GDP per capita PPP (constant 2011 international $) in the European South, 2007–2018

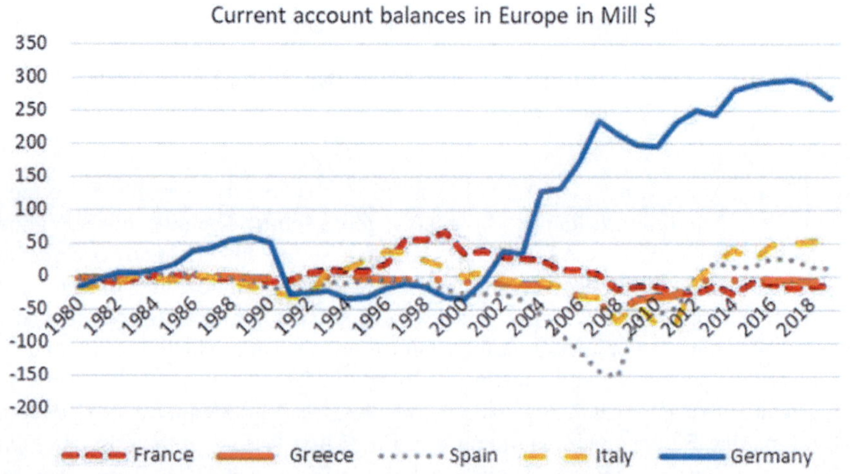


**Fig. 3.14** Current account imbalances in the European Monetary Union

For all practical purposes, the long decade between 2008 and 2020 was for the European South a lost decade, with stagnating and falling real incomes and an exploding current account imbalance with the continents leading economy, Germany:

With Germany quenching the Latin and Mediterranean European partner countries to the wall, European austerity measures and the economic crisis of 2008 also led to a significant weakening of national security on the European continent, as evidenced in Figs. 3.15 and 3.16 of this chapter. Now, of course, the UK was never a member of

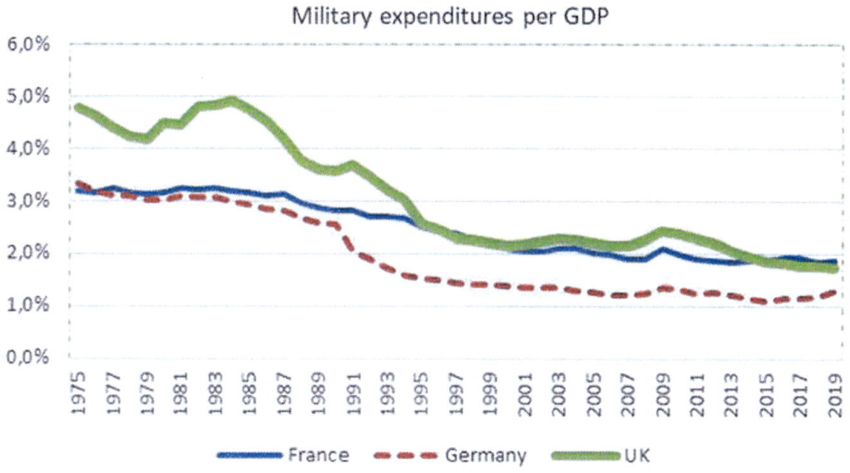


**Fig. 3.15** Military expenditures in France, Germany, and the UK since 1975

**Fig. 3.16** Military expenditures in Greece, Italy, and Spain since 1975

European Monetary Union, and with Brexit, it left the European Union altogether, but none the less its economic and political fortunes were tied to the Union for decades.

## 3.8 How Well Prepared Are the Gulf Countries?

As the world confronts COVID-19, and the Gulf countries as well, according to our analysis three factors will be decisive for the trajectory of the Gulf. These factors are inequality, a pattern of high economic globalization occurring with a relatively low or intermediate de jure social globalization (see also Gygli et al. 2019), and a relatively high past fluctuation of economic growth, tied to export prices, which leads to quite strong economic fluctuations over time.

So, first of all, our analysis reveals the importance of inequality in the Gulf countries as an. The most comprehensive science documentation system in the world, *Scopus,* listed, by January 8, 2021, none the less than 121 research articles featuring "inequality" and the "Arab World," and a simple *Google* search, by the same date, lists important sources such as Brookings,[39] Carnegie,[40] and the World Bank,[41] all speaking about high rates of inequality in the Arab World. One particular analysis, Harry (2007), which received an impact of—currently—94 quotations in the literature, maintained:

> The countries of the GCC use human resources inefficiently because the governments believe that 'rents' (from natural resources) can be used to catch up with industrialized countries without changing their social structures. This, however, has not been a successful strategy as can be seen in the stresses that are occurring in the social structures due to the lack of job opportunities for young people. If the governments will now invest the 'rents' to diversify from their reliance on oil and gas production by creating new industries and boost the small and medium enterprise sector there are opportunities to create wealth instead of spending on consumption. There will still be a need to make changes in the social structures but the GCC is wealthy enough to afford to invest in change. Harry (2007)

Joseph (1996), speaks about the persistence of patriarchy in the Arab World, and in other regions. Patriarchy is an obstacle for women, children, families, and states. It affects health, education, labor, human rights, and democracy. She argues that patriarchy is powerful in the Arab World because age-based kinship values and relationships are crucial socially, economically, politically, ideologically, and psychologically (see also Boutayeb & Serghini, 2006). Also, the World Bank emphasizes wealth inequality as the overriding and most central issue of the region.[42]

This last qualification seems to be an important one. All comparative indicators seem to suggest that in the Arab World wealth concentration and the share of the top 1% in total incomes are very high, but that the share of the poorest 40% in total incomes is also relatively high, thus presenting an interesting puzzle (based on UNDP, 2019). Not to overburden our readers with too much statistics, the UNDP

---

[39] https://www.brookings.edu/blog/future-development/2015/02/04/how-unequal-are-arab-countries-2/.

[40] https://carnegie-mec.org/2018/05/31/arab-1-inequality-and-development-event-6901.

[41] http://documents.worldbank.org/curated/en/303441467992017147/pdf/Inequality-uprisings-and-conflict-in-the-Arab-World.pdf.

[42] http://documents.worldbank.org/curated/en/303441467992017147/pdf/Inequality-uprisings-and-conflict-in-the-Arab-World.pdf.

(2019) data suggest the following linear OLS regression equation between the share of the poorest 40%, $y$, and the share of the richest 1% in total incomes, $x$:

$$y = -0.3037x + 23.03; \; R^2 = 0.2974 \tag{3.1}$$

Looking at the residuals from this simple OLS regression, easily performed on the basis of the UNDP data (UNDP, 2019), one sees immediately that—as a rule—Arab nations belong to the fortunate countries in the world where the poorest 40% are with their share above the regression line of Eq. (3.1). In descending order, the following countries have a share of their poorest 40% in total incomes above the share which can be predicted by income and wealth concentration in the hands of the richest 1%: These nations are:

- **Iraq**, **Egypt**, *Lebanon*, Czech Republic, Denmark, India, Slovenia, Norway, Finland, Malta, Iceland, **Jordan**, Poland, Ireland, Netherlands, Slovakia, Belgium, Sweden, Thailand, Serbia, Russia, Germany, Albania, **West Bank and Gaza**, Korea, South, France, Switzerland, Austria, **Yemen**, Japan, Hungary, UK, and Canada.

In light of Eq. (3.1), the share of the poorest 40% is in the following countries below the regression line. Again, we order our countries by the size of the residual, this time the size of the negative residuals:

- South Africa, Colombia, Romania, Macedonia, Brazil, Lithuania, Italy, *Malaysia*, Bulgaria, Spain, Uruguay, Portugal, Greece, Cote d'Ivoire, China, USA, Mauritius, *Iran*, Australia, Chile, *Bosnia and Herzegovina*, Latvia, Luxembourg, Estonia, Cyprus, Croatia, *Turkey*, and Montenegro.

A plausible explanation would be that either de facto social Keynesian welfare policies or a functioning Muslim wealth tax (the "Zakat," see Dean & Khan, 1997; Javaid & Al-Malkawi, 2018) explains the fact why diverse countries such as Iraq, Egypt, Lebanon, Czech Republic, Denmark, India, Slovenia, Norway, Finland, and Malta have a share of their poorest 40% higher than would have been to be expected by the data on the share of their richest 1% in total incomes.

So, in Fig. 3.17, we contrast the data of Table 3.4 (see also Fig. 1.7 in Chap. 1) with the UNDP data on the share of the poorest 40% in total incomes onto a world map.

Figure 3.18 shows the development of inequality over time in the Gulf region since the late 1960s. Our analysis is based on the University of Texas Inequality Project data (see Chap. 2 of the present book),[43] which are based on the very encompassing and yearly available UN UNIDO data of the inequality of wages between economic sectors and the robust econometric relationship of these data with the less often available World Bank household survey inequality data, thus creating the unique opportunity of time series of inequality over fairly long historical periods. In Kuwait and Qatar, income inequality considerably increased over time.

---

[43] https://utip.gov.utexas.edu/.

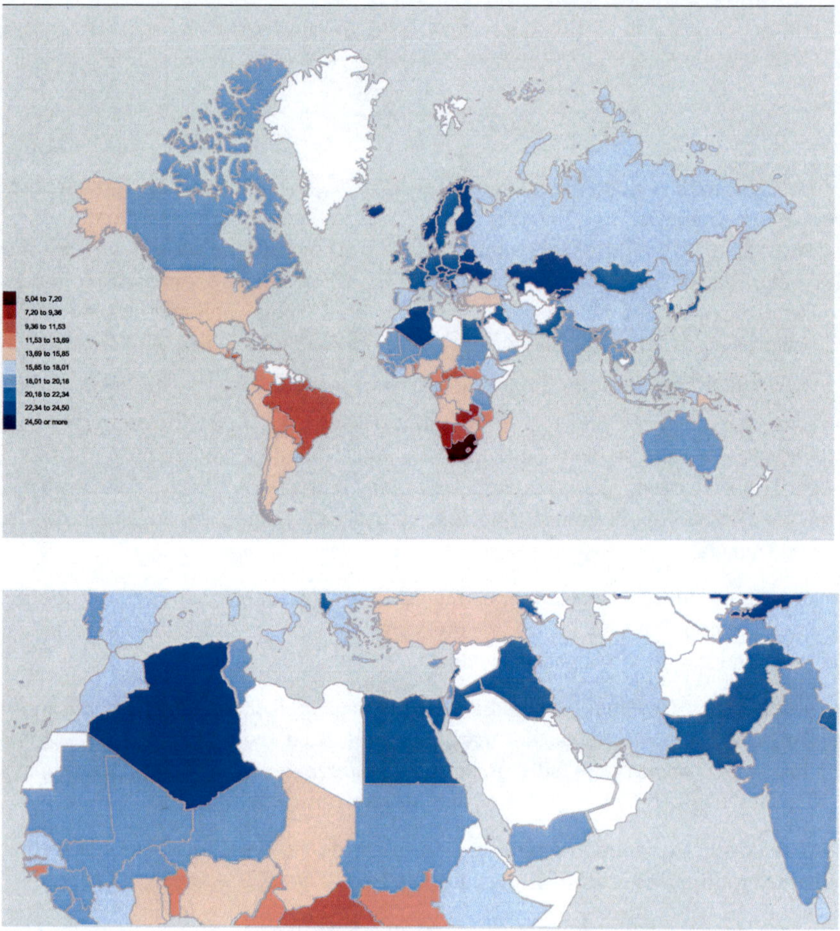


**Fig. 3.17** Inequality on a global scale—UNDP share of the poorest 40% in total incomes

We now move on to analyze the second of our benchmarks proposed here, after showing Gulf inequality rates: the fluctuations of economic development. Table 3.5 reflects the cycles of economic growth in the Gulf region.

Combining these data in a single image (Fig. 3.19), the message of Table 3.5 emerges very clearly: The Gulf region is characterized by very strong economic fluctuations, making economic development very unpredictable.

Cammett (2018) summarizes the growth balance of the recent years as follows:

> Economic liberalization has had a mixed record at best in the MENA region. Throughout the region, economic reforms have not produced sustained high growth rates, while inequality has increased somewhat according to official data […] Growth rates have been erratic, particularly in Jordan, which is especially vulnerable to regional conflict, and Morocco, which is highly sensitive to drought, among other factors. Economic restructuring has generally

**Table 3.4** Inequality—share of the richest 1% in total incomes (UNDP data)

| Country | Richest 1% | Global rank in share of the richest 1% | League of global inequality—country is among the top … % |
|---|---|---|---|
| Qatar | 29.03 | 1 | 1.41 |
| Brazil | 28.35 | 2 | 2.82 |
| Chile | 23.66 | 3 | 4.23 |
| Turkey | 23.44 | 4 | 5.63 |
| Lebanon | 23.36 | 5 | 7.04 |
| United Arab Emirates | 22.81 | 6 | 8.45 |
| Iraq | 22.02 | 7 | 9.86 |
| India | 21.34 | 8 | 11.27 |
| Colombia | 20.45 | 9 | 12.68 |
| Russian Federation | 20.24 | 10 | 14.08 |
| Thailand | 20.23 | 11 | 15.49 |
| United States | 20.20 | 12 | 16.90 |
| Kuwait | 19.93 | 13 | 18.31 |
| Saudi Arabia | 19.70 | 14 | 19.72 |
| Oman | 19.49 | 15 | 21.13 |
| South Africa | 19.21 | 16 | 22.54 |
| Egypt | 19.13 | 17 | 23.94 |
| Bahrain | 18.00 | 18 | 25.35 |
| Côte d'Ivoire | 17.08 | 19 | 26.76 |
| Iran (Islamic Republic of) | 16.26 | 20 | 28.17 |
| Jordan | 16.10 | 21 | 29.58 |
| West Bank and Gaza | 15.81 | 22 | 30.99 |
| Yemen | 15.70 | 23 | 32.39 |
| Syrian Arab Republic | 14.68 | 24 | 33.80 |
| Malaysia | 14.55 | 25 | 35.21 |
| Singapore | 14.02 | 26 | 36.62 |
| Uruguay | 14.00 | 27 | 38.03 |
| China | 13.92 | 28 | 39.44 |
| Canada | 13.62 | 29 | 40.85 |
| Denmark | 12.83 | 30 | 42.25 |
| Ireland | 12.80 | 31 | 43.66 |
| Poland | 12.46 | 32 | 45.07 |
| Korea (Republic of) | 12.16 | 33 | 46.48 |
| Switzerland | 11.93 | 34 | 47.89 |
| United Kingdom | 11.71 | 35 | 49.30 |
| Malta | 11.65 | 36 | 50.70 |
| Germany | 11.13 | 37 | 52.11 |
| Greece | 10.83 | 38 | 53.52 |
| France | 10.80 | 39 | 54.93 |
| Japan | 10.44 | 40 | 56.34 |

(continued)

**Table 3.4** (continued)

| | | | |
|---|---|---|---|
| Spain | 9.82 | 41 | 57.75 |
| Czechia | 9.46 | 42 | 59.15 |
| Luxembourg | 9.15 | 43 | 60.56 |
| Australia | 9.07 | 44 | 61.97 |
| Cyprus | 8.56 | 45 | 63.38 |
| Norway | 8.37 | 46 | 64.79 |
| Bulgaria | 8.36 | 47 | 66.20 |
| Sweden | 8.25 | 48 | 67.61 |
| New Zealand | 8.24 | 49 | 69.01 |
| Austria | 8.17 | 50 | 70.42 |
| Hungary | 7.72 | 51 | 71.83 |
| Croatia | 7.58 | 52 | 73.24 |
| Latvia | 7.56 | 53 | 74.65 |
| Italy | 7.50 | 54 | 76.06 |
| Portugal | 7.40 | 55 | 77.46 |
| Finland | 7.29 | 56 | 78.87 |
| Mauritius | 7.08 | 57 | 80.28 |
| Lithuania | 7.02 | 58 | 81.69 |
| Estonia | 6.99 | 59 | 83.10 |
| Iceland | 6.78 | 60 | 84.51 |
| Romania | 6.76 | 61 | 85.92 |
| Slovenia | 6.70 | 62 | 87.32 |
| Belgium | 6.69 | 63 | 88.73 |
| Albania | 6.40 | 64 | 90.14 |
| Montenegro | 6.38 | 65 | 91.55 |
| Serbia | 6.38 | 66 | 92.96 |
| Netherlands | 6.21 | 67 | 94.37 |
| Bosnia and Herzegovina | 6.15 | 68 | 95.77 |
| Moldova (Republic of) | 6.05 | 69 | 97.18 |
| North Macedonia | 5.85 | 70 | 98.59 |
| Slovakia | 5.22 | 71 | 100.00 |

> come with enormous social costs […]. Political and economic elites, who enjoy close ties to rulers - whether presidents or monarchs - have benefited disproportionately from the new opportunities generated by greater global economic integration and increased emphasis on private sector-led development.

We identify relative restrictions on social globalization, especially its de jure aspects, as the third major conditioning factor for the future of the Gulf's long-run development. Based on the intensive debate about the KOF Globalization Index, developed at the ETH Zurich (Gygli et al., 2019; Solomon & Tausch, 2021), we present our readers the global ranks of the Gulf countries according to the different aspects of the all-important social globalization, which is the real tie between the social system and economic globalization, and without which the imbalances of globalization might increase (Solomon & Tausch, 2021).

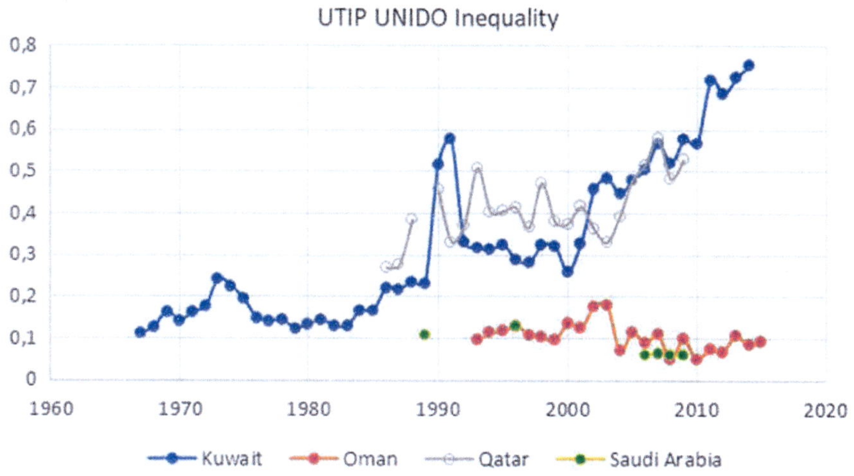


**Fig. 3.18** UTIP UNIDO Inequality Index in the Gulf region since the late 1960s

- Social globalization, de facto, is composed of interpersonal globalization, de facto; informational globalization, de facto; and cultural globalization, de facto.
- Social globalization, de jure, by contrast, is composed of interpersonal globalization, de jure; informational globalization, de jure; and cultural globalization, de jure.

The sub-components of these two indicators are defined as follows:

- Interpersonal globalization, de facto, is measured by the ETH Zurich team by data about international voice traffic, transfers, international tourism, international students, and migration.
- Informational globalization de facto is combined of used Internet bandwidth, international patents, and high technology exports.
- Cultural globalization de facto is measured by trade in cultural goods, trade in personal services, international trademarks, McDonald's restaurants, and IKEA stores.
- Interpersonal globalization, de jure, is measured by telephone subscriptions, freedom to visit, and international airports.
- Informational globalization, de jure, is based on the combination of television access, Internet access, and press freedom.
- Cultural globalization, de jure, is based on the measurement of gender parity, human capital, and civil liberties.

Table 3.6 thus highlights the performance of the Gulf countries.

The conclusion from this table and our data (https://www.academia.edu/44295690/GULF_2020_Covid_19_PUBLIC_ACCESS_Data_base_for_the_book_Factors_affecting_the_Gulfs_future_Global_cycles_and_value_change_Springer_Publishing_Series_https_www_springer_com_series_16417_Gulf_Studies_edited_by_Prof_Mizanur_Rahman_Qatar_University) is that the agenda for the Gulf would consist in increasing the performance in

**Table 3.5** Cycles of economic growth per capita in the Gulf (gross domestic product per capita, constant prices, purchasing power parity; percent change, IMF World Economic Outlook Database)

| | Bahrain | Kuwait | Oman | Qatar | Saudi Arabia | United Arab Emirates |
|---|---|---|---|---|---|---|
| 1980 | 2.91 | −25.02 | −2.76 | −9.67 | −0.40 | −10.53 |
| 1981 | −1.17 | −22.43 | 8.09 | −11.29 | −2.92 | −0.81 |
| 1982 | 2.68 | −13.43 | 11.50 | −14.77 | −24.51 | −12.74 |
| 1983 | 3.47 | 0.62 | 7.67 | −11.64 | −20.05 | −8.38 |
| 1984 | 0.79 | 3.00 | 6.31 | 8.74 | −9.21 | −3.52 |
| 1985 | −4.23 | −9.89 | 14.52 | −17.66 | −14.09 | −7.47 |
| 1986 | −2.94 | 2.53 | −4.24 | −1.26 | 11.44 | −22.63 |
| 1987 | −4.63 | 2.45 | −9.64 | −3.19 | −11.08 | 1.09 |
| 1988 | 2.27 | −14.55 | 5.25 | 0.62 | 7.72 | −18.40 |
| 1989 | −2.50 | 19.90 | −2.74 | 1.35 | −5.25 | 11.38 |
| 1990 | 2.61 | −27.27 | 19.68 | −26.96 | 9.70 | 24.63 |
| 1991 | −1.81 | −7.61 | 2.32 | −3.41 | 9.53 | −2.28 |
| 1992 | 5.38 | 75.09 | 4.72 | 10.31 | −2.15 | −1.19 |
| 1993 | 4.81 | 29.65 | 2.60 | −1.80 | −3.24 | −3.48 |
| 1994 | 0.53 | −12.16 | 0.65 | 0.78 | −1.85 | 0.29 |
| 1995 | −0.72 | −5.07 | 1.96 | 1.16 | −2.19 | −1.42 |
| 1996 | 0.57 | −5.66 | 0.41 | 2.07 | 0.18 | 3.96 |
| 1997 | −0.32 | −3.02 | 3.92 | 25.83 | −1.32 | 2.79 |
| 1998 | 2.12 | 0.82 | 0.74 | 7.00 | 0.43 | −8.24 |
| 1999 | 3.22 | −1.10 | −1.47 | 0.31 | −6.07 | −3.06 |
| 2000 | 4.23 | 6.48 | 4.70 | 4.11 | 3.10 | 13.75 |
| 2001 | −1.19 | −3.78 | 2.72 | 0.82 | −3.58 | −4.22 |
| 2002 | −3.81 | −1.70 | −2.72 | 4.13 | −5.15 | −3.09 |
| 2003 | −1.21 | 11.49 | −4.27 | −1.07 | 8.57 | 2.62 |
| 2004 | −0.71 | 2.44 | −0.45 | 9.27 | 5.36 | 3.46 |
| 2005 | −1.05 | 1.33 | 0.62 | −5.70 | 2.11 | −3.94 |
| 2006 | −1.47 | 1.04 | 3.31 | 7.08 | −0.59 | −10.03 |
| 2007 | 0.08 | −0.76 | 2.30 | −0.94 | −1.50 | −16.80 |
| 2008 | −0.21 | 1.22 | 5.91 | −12.73 | 2.77 | −20.50 |
| 2009 | −3.72 | −8.23 | 2.52 | 6.16 | −5.27 | −6.69 |
| 2010 | −0.41 | −5.02 | 1.88 | 12.86 | 1.60 | 0.80 |
| 2011 | 5.36 | 6.21 | −1.13 | 12.22 | 6.85 | 3.82 |
| 2012 | 2.53 | 3.56 | −0.77 | −1.04 | 2.45 | 1.44 |
| 2013 | 1.70 | −1.06 | −3.77 | −4.49 | −0.03 | 1.99 |
| 2014 | −0.52 | −3.17 | −1.92 | −5.99 | 2.47 | 1.25 |

(continued)

**Table 3.5** (continued)

| | Bahrain | Kuwait | Oman | Qatar | Saudi Arabia | United Arab Emirates |
|---|---|---|---|---|---|---|
| 2015 | −1.32 | −2.86 | 2.97 | −5.77 | 1.69 | 2.05 |
| 2016 | −0.41 | −0.55 | −1.16 | −4.89 | −0.65 | 0.18 |
| 2017 | 1.78 | −7.05 | −2.87 | −2.41 | −3.25 | −2.32 |
| 2018 | −0.04 | −2.78 | 0.86 | 0.19 | −0.02 | −1.11 |
| 2019 | 1.63 | −2.00 | −2.49 | −1.17 | −1.64 | −1.71 |
| 2020 | −5.48 | −3.79 | −5.74 | −4.58 | −4.20 | −6.35 |
| 2021 | 0.97 | 0.63 | −0.12 | 4.78 | 0.91 | 0.21 |

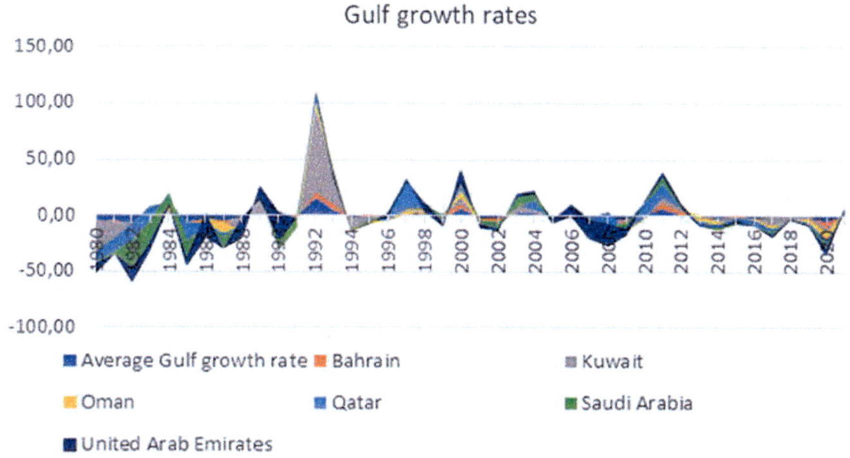


**Fig. 3.19** Gulf economic cycle since 1980 (IMF data)

**Table 3.6** Ranks of the Gulf countries in the global process of social globalization

| | Social globalization, overall, rank | Social globalization, de facto, rank | Social globalization, de jure, rank |
|---|---|---|---|
| Qatar | 65 | 16 | 127 |
| Kuwait | 71 | 27 | 119 |
| Bahrain | 72 | 29 | 117 |
| United Arab Emirates | 73 | 6 | 133 |
| Oman | 80 | 50 | 113 |
| Saudi Arabia | 84 | 30 | 131 |

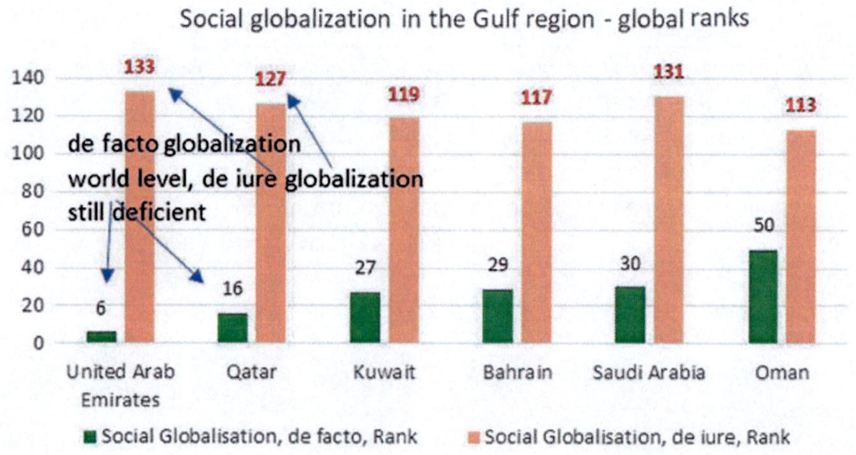


**Fig. 3.20** Even in the most globalized Gulf countries, de jure social globalization is still deficient

- Civil liberties,
- Gender parity,
- Human capital,
- Press freedom.

Figure 3.20 summarizes these tendencies:

## 3.9 "The Pandemic Is Expected to Leave Lasting Economic Scars" (World Bank, 2021): Public Health and Mortality effects

We already mentioned that Taubenberger and Morens (2006), highlighted that the "Spanish flu" hit the world in three waves, with the second wave about 5 times deadlier than the first one and the third wave two times as deadly as the first wave. On a yearly basis, the two million COVID-19 deaths recorded in January 2021 have to be compared with the 56.9 million deaths worldwide,[44] which are recorded nowadays, and more than half (54%) were due to the top 10 causes. Ischemic heart disease and stroke are the world's biggest killers, accounting for a combined 15.2 million deaths in 2016, chronic obstructive pulmonary disease claimed 3.0 million lives in 2016, while lung cancer (along with trachea and bronchus cancers) caused 1.7 million deaths. Diabetes killed 1.6 million people in 2016. Lower respiratory infections remained the deadliest communicable disease, causing 3.0 million deaths

44 https://www.who.int/news-room/fact-sheets/detail/the-top-10-causes-of-death.

**Table 3.7** Main victims of COVID-19

| | % of global Covid–19 cases | % of global Covid–19 deaths | % of global Covid–19 tests |
|---|---|---|---|
| EU–27 | 8.4 | 14.6 | 12.6 |
| Other traditional Western democracies | 24.9 | 26.5 | 23.9 |
| GCC | 2.4 | 0.7 | 3.0 |
| Other Arab countries | 2.4 | 2.2 | 1.2 |
| 32 nations with a low effective democracy index (<10)[a] | 7.4 | 6.4 | 34.7 |
| Rest of the world | 54.5 | 49.6 | 24.6 |
| Total | 100.0 | 100.0 | 100.0 |

[a]Angola, Azerbaijan, Belarus, Burma/Myanmar, Burundi, Cambodia, Cameroon, Central African Republic, Chad, China, Congo, Republic of the, Cote d'Ivoire, Cuba, Equatorial Guinea, Eritrea, Guinea, Haiti, Iran, Kazakhstan, Kyrgyzstan, Laos, Libya, Pakistan, Russia, Rwanda, Swaziland, Tajikistan, Togo, Turkmenistan, Uzbekistan, Vietnam, Zimbabwe

worldwide in 2016. The death rate from diarrheal diseases caused 1.4 million deaths in 2016, and tuberculosis had a death toll of 1.3 million people. HIV/AIDS killed 1.0 million people in 2016, and road injuries killed 1.4 million people in 2016 (Tables 3.7 and 3.8).

Our publicly available electronic appendix to this book[45] summarizes the available information on a country-to-country basis and some of the variables, often mentioned in the context of the spread of the disease. Of the 211 countries and territories with complete data, the rates of COVID-19 deaths per 1 million population indicate that the Gulf countries were not at the forefront of the most seriously affected countries up to now. In that sense, these tables are the "master tables" of this chapter, from which we draw our multivariate conclusions in Tables 3.9 and 3.10. The Gulf countries are ranked the following way on the indicator *"COVID-19deaths per 1 million population*":

43 Oman
52 Bahrain
53 Kuwait
55 Saudi Arabia
75 Qatar
100 United Arab Emirates.

Looking at the global distribution of the pandemic, the following facts emerge: Table 3.7 summarizes the results of our analysis according to the following global breakdown of countries:

- EU-27,
- Other traditional Western democracies,
- GCC,

[45] https://www.academia.edu/45321329/Factors_affecting_the_Gulfs_future_Global_cycles_and_value_change_Electronic_Appendix.

**Table 3.8** Current accounts, unemployment, and economic growth in the Gulf region, 2020/21, as predicted by the IMF. Absolute values and global performance ranks

| Country | Bahrain | Kuwait | Oman | Qatar | Saudi Arabia | United Arab Emirates |
|---|---|---|---|---|---|---|
| Current account 2020 per GDP | −9.64 | −10.24 | −14.19 | −1.92 | −3.08 | 1.52 |
| Current account 2021 per GDP | −7.33 | −7.81 | −11.07 | −1.78 | −3.39 | 4.09 |
| Unemployment 2020 | 4.369 | xx | xx | xx | xx | xx |
| Unemployment 2021 | 3,81 | xx | xx | xx | xx | xx |
| DYN GDP pc PPP 2020 | −5.48 | −3.79 | −5.74 | −4.58 | −4.20 | −6.35 |
| DYN GDP pc PPP 2021 | 0.97 | 0,63 | -0.12 | 4.78 | 0.91 | 0.21 |
| Rank current account 2020 per GDP 191 countries | 142 | 146 | 164 | 62 | 81 | 33 |
| Rank current account 2021 per GDP 190 countries | 141 | 143 | 162 | 68 | 96 | 20 |
| Rank avoiding unemployment 2020 100 countries | 16 | xx | xx | xx | xx | xx |
| Rank avoiding unemployment 2021 100 countries | 12 | xx | xx | xx | xx | xx |
| Rank DYN GDP pc PPP 2020 191 countries | 113 | 72 | 117 | 91 | 79 | 129 |
| Rank DYN GDP pc PPP 2021 190 countries | 163 | 169 | 179 | 58 | 164 | 175 |

- Other Arab countries,
- 32 nations with a low Effective Democracy Index (<10),
- Rest of the world.

It clearly emerges that the West is at the receiving end of the current global crisis. So, what COVID-19 does is nothing more and nothing less than undermining the fabric of the Free World and benefitting totalitarian regimes.

**Table 3.9** Partial correlations of deaths per 1 m population

| | COVID-19 deaths per 1 million population, end September 2020 | Error p | df |
|---|---|---|---|
| Share of the richest 1% in total incomes | 0.396 | 0.001 | 61 |
| Share of Roman Catholics per total population | 0.365 | 0.000 | 158 |
| Years of membership in the EU, 2010 | 0.296 | 0.000 | 150 |
| Quintile share income difference between richest and poorest 20% | 0.260 | 0.005 | 114 |
| MNC outward investments (stock) per GDP | 0.250 | 0.007 | 114 |
| Unemployment 2020 | 0.242 | 0.023 | 87 |
| GINI income | 0.221 | 0.013 | 123 |
| Democracy measure | 0.199 | 0.019 | 137 |
| Gallup poll about satisfaction: standard of living | 0.195 | 0.019 | 143 |
| Tertiary enrollment | 0.182 | 0.034 | 133 |
| Social protection (ILO) | 0.179 | 0.030 | 145 |
| Happy Planet Index, HPI | 0.178 | 0.042 | 129 |
| Share of richest 10% | 0.171 | 0.048 | 132 |
| Chinese immigrants as a percentage of total immigrants | −0.162 | 0.041 | 158 |
| Share of Buddhists per total population | −0.164 | 0.038 | 158 |
| Civil and Political Liberties violations | −0.169 | 0.038 | 149 |
| Nurses and midwives (per 1,000 people), 2010 | −0.175 | 0.046 | 128 |
| Economic Globalization, de facto index | −0.193 | 0.015 | 158 |
| Share of adherents of Eastern religions per total population | – | 0.009 | 158 |
| Gender empowerment index | −0.238 | 0.049 | 67 |
| Share of poorest 40% | −0.244 | 0.004 | 132 |
| Hospital beds per 1000 people 2011 | −0.296 | 0.001 | 111 |
| Trust in institutions | −0.358 | 0.008 | 51 |

**Table 3.10** Partial correlations of COVID-19 fatality rates per total cases

| | COVID-19 fatality rate per total cases | Error p | df |
|---|---|---|---|
| Current health expenditure (% of GDP), 2017 | 0.322 | 0.000 | 167 |
| Years of membership in the EU, 2010 | 0.268 | 0.001 | 160 |
| Earthquakes—average number of earthquakes per year | 0.261 | 0.001 | 160 |
| Global migration senders 2013 | 0.249 | 0.001 | 167 |
| Global migration senders 2017 | 0.224 | 0.003 | 167 |
| Happy life years | 0.221 | 0.010 | 132 |
| Life expectancy (years) (by 2010) | 0.213 | 0.013 | 132 |
| Happy Planet Index, HPI | 0.207 | 0.017 | 132 |
| UNDP education index | 0.202 | 0.010 | 160 |
| Life satisfaction (0–10) | 0.196 | 0.023 | 132 |
| Female survival probability of surviving to age 65 female | 0.195 | 0.015 | 155 |
| Social protection (ILO) | 0.191 | 0.017 | 154 |
| Democracy measure | 0.179 | 0.031 | 143 |
| Years of membership in EMU, 2010 | 0.169 | 0.031 | 160 |
| Mean economic growth rate crisis years 2008/2011 | −0.157 | 0.047 | 159 |
| Net international migration rate, 2005–2010 | −0.174 | 0.028 | 157 |
| Annual population growth rate, 1975–2005 (%) | −0.175 | 0.028 | 155 |
| GDP per capita PPP$ | −0.185 | 0.019 | 158 |
| Net migration rate | −0.192 | 0.013 | 164 |
| Gross national income (GNI) per capita | −0.200 | 0.009 | 167 |
| Gulf Cooperation Council | −0.203 | 0.008 | 167 |
| DYN HDI 2010-2018 | −0.208 | 0.007 | 167 |
| LFPR 55–59 Labor Force Participation Rate of elderly workers | −0.209 | 0.008 | 156 |
| Migrants 2017 percentage of total population | −0.218 | 0.005 | 167 |
| Economic globalization, de facto index | −0.225 | 0.003 | 167 |
| Trust in institutions | −0.313 | 0.022 | 51 |

The data listed in our electronically available appendix[46] peak for themselves.[47] Our choropleth maps summarize the available information. For reasons of data visibility, we work here with inverted ranks. In each case, red colors indicate that a country is severely affected by the disease, while blue colors indicate that a country is not/not yet severely affected by the disease. Figure 3.21 shows just how dramatically the pandemic affected the Americas, some countries of Western Europe, South Africa, the Gulf region, and Russia.

Figure 3.22 shows how deaths per 1 million population are distributed around the world. The Americas, Latin Europe, Sweden, and the UK are at the forefront of those countries, suffering most severely from the pandemic.

In Fig. 3.23, we ask ourselves which countries—given the global infection rates—still handled to protect their populations in the best possible manner from dying from the disease. Many different factors seem to contribute to this. Countries as diverse as Canada, China, Sweden, the UK, Latin Europe, Iran, and some Arab countries and some other nations belong to this unfortunate and very heterogeneous group. The Gulf states did remarkably well according to this scale.

Finally, Fig. 3.24 shows just how dependent COVID-19 testing capacity, which is so vital in the framework of the WHO's strategy to combat the disease,[48] is on overall socio-economic development.

## 3.10 The Economic Fallout from COVID-19

Our research presents here calculations about the already observable economic fallout of the COVID-19 pandemic on the basis of the IMF World Economic Outlook Database.[49] Our main results for the Gulf region are presented in Table 3.9, and other conclusions can be drawn from our maps (Figs. 3.25, 3.26, and Fig. 3.27). Inexorably, the center of gravitation of the world economy is further propelled by COVID-19 toward the two giants which will dominate the twenty-first century, China and India (Table 3.8).

Unemployment is highest in the larger Mediterranean and MENA area (Fig. 3.25), in Brazil and South Africa:

---

46 https://www.academia.edu/45321329/Factors_affecting_the_Gulfs_future_Global_cycles_and_value_change_Electronic_Appendix.

47 Freely available from https://www.academia.edu/44295690/GULF_2020_Covid_19_PUBLIC_ACCESS_Data_base_for_the_book_Factors_affecting_the_Gulfs_future_Global_cycles_and_value_change_Springer_Publishing_Series_https_www_springer_com_series_16417_Gulf_Studies_edited_by_Prof_Mizanur_Rahman_Qatar_University.

48 https://www.who.int/director-general/speeches/detail/who-director-general-s-opening-remarks-at-the-media-briefing-on-covid-19. 16 March 2020.

49 https://www.academia.edu/44295690/GULF_2020_Covid_19_PUBLIC_ACCESS_Data_base_for_the_book_Factors_affecting_the_Gulfs_future_Global_cycles_and_value_change_Springer_Publishing_Series_https_www_springer_com_series_16417_Gulf_Studies_edited_by_Prof_Mizanur_Rahman_Qatar_University.

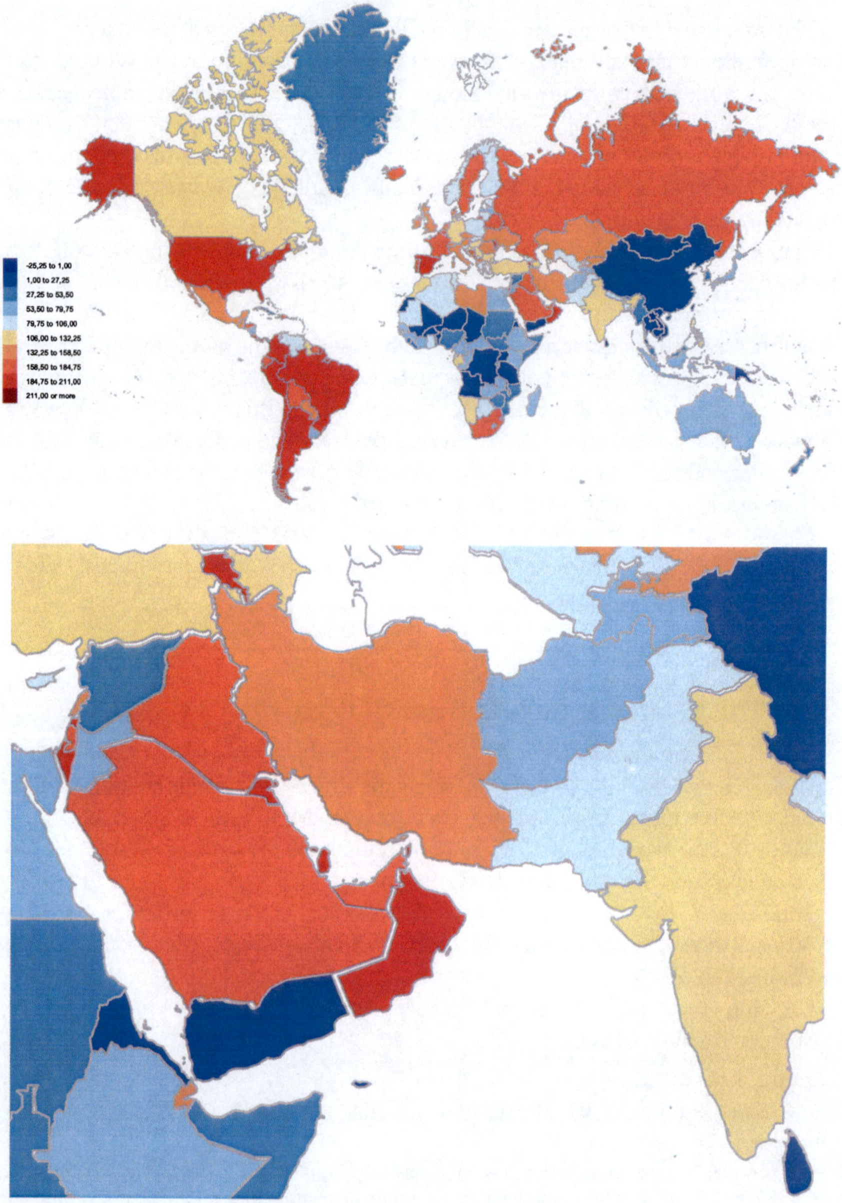


**Fig. 3.21** Inverted global ranks—COVID-19 cases per 1 million population

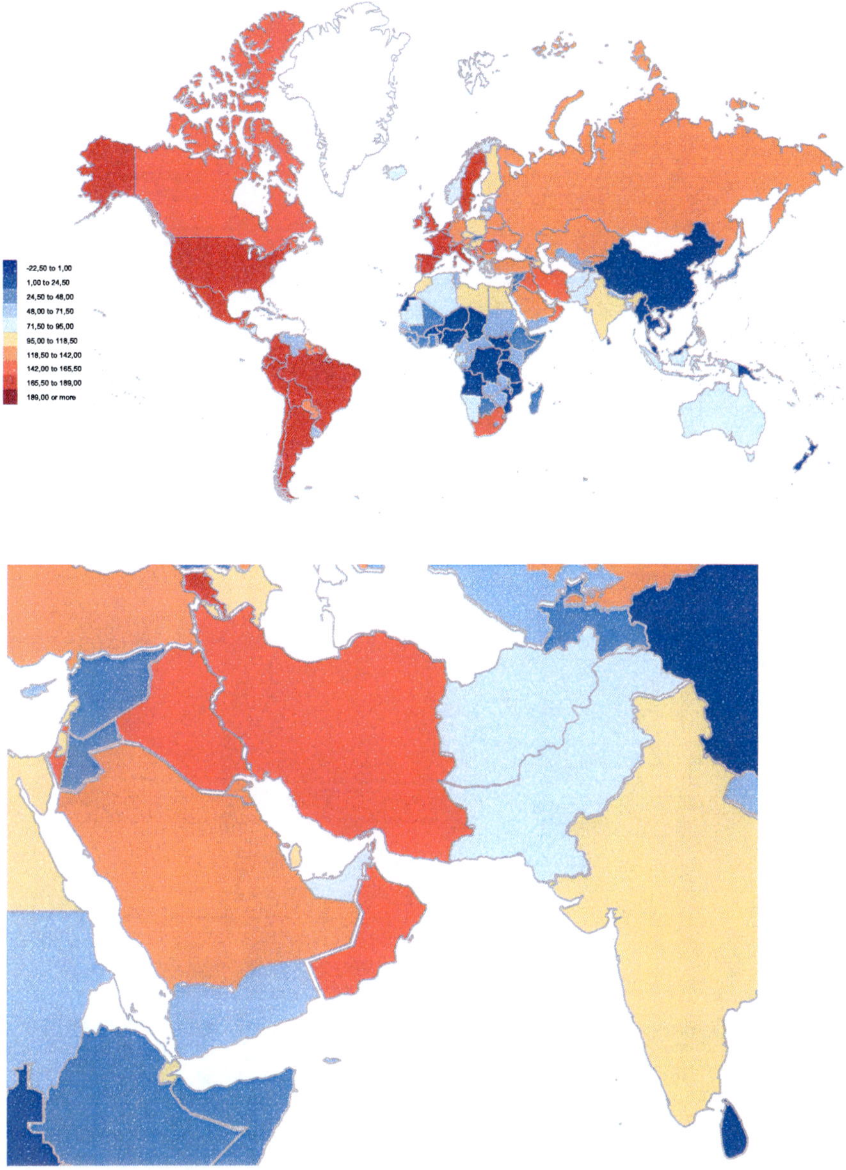


**Fig. 3.22** Inverted global ranks—COVID-19 deaths per 1 m total population

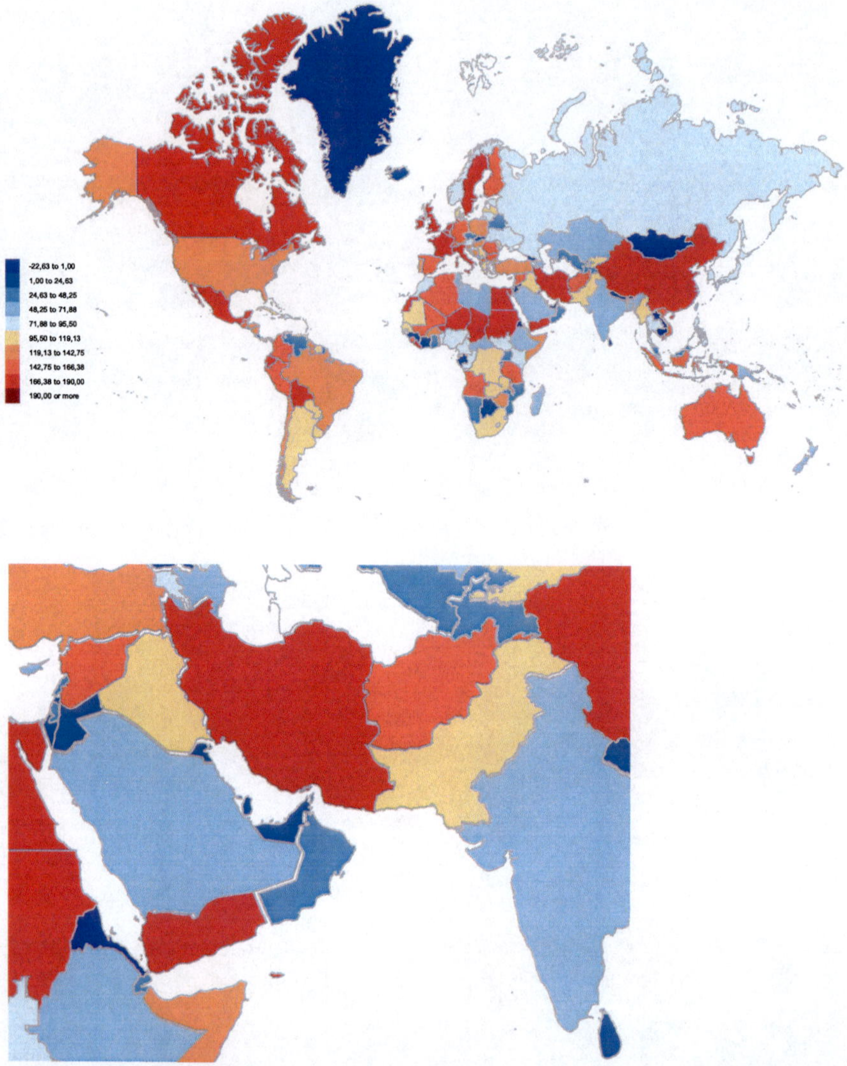


**Fig. 3.23** Inverted global ranks—COVID-19 Deaths per COVID-19 cases (in %)

The implosion of economic growth, due to COVID-19, is scattered around the following group of countries:

- Some, but not all Western democracies,
- Most of the Mediterranean and MENA countries,
- Southern Africa,
- Some former Communist countries.

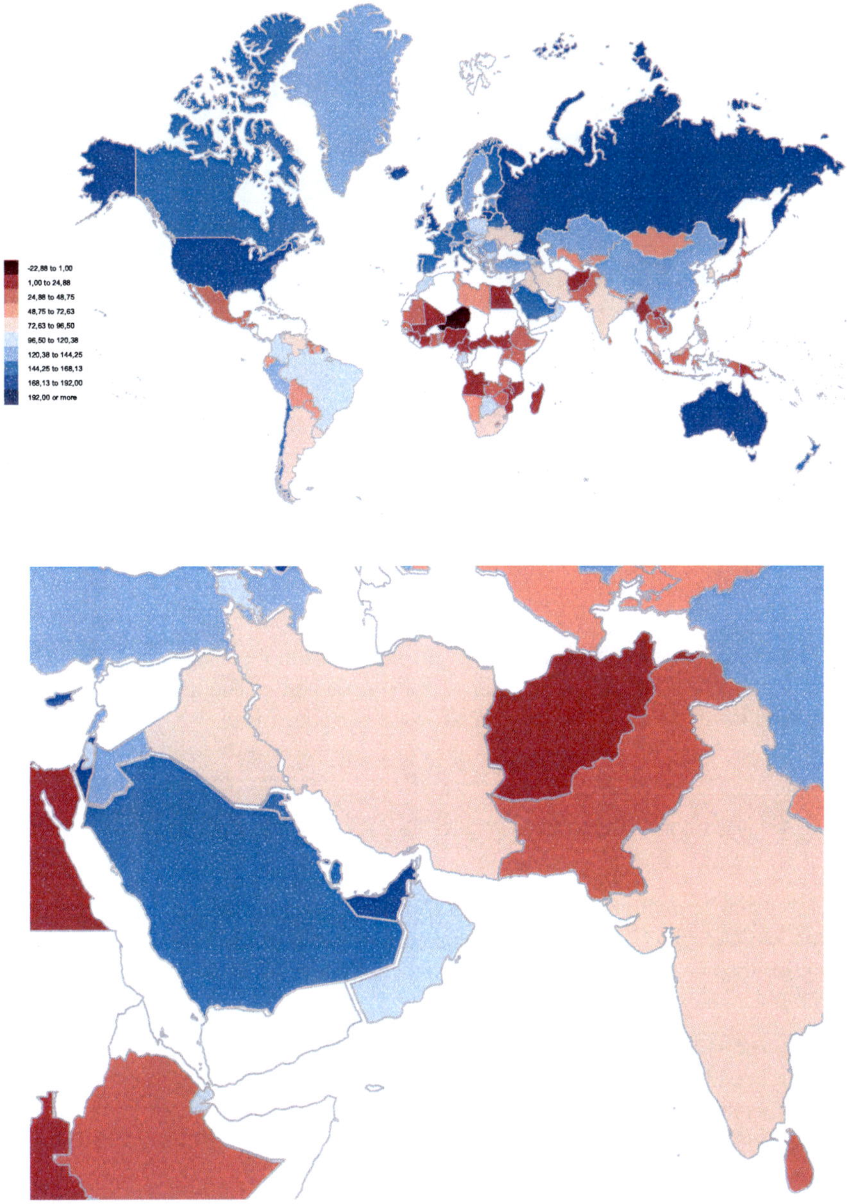


**Fig. 3.24** Tests per 1 m total population

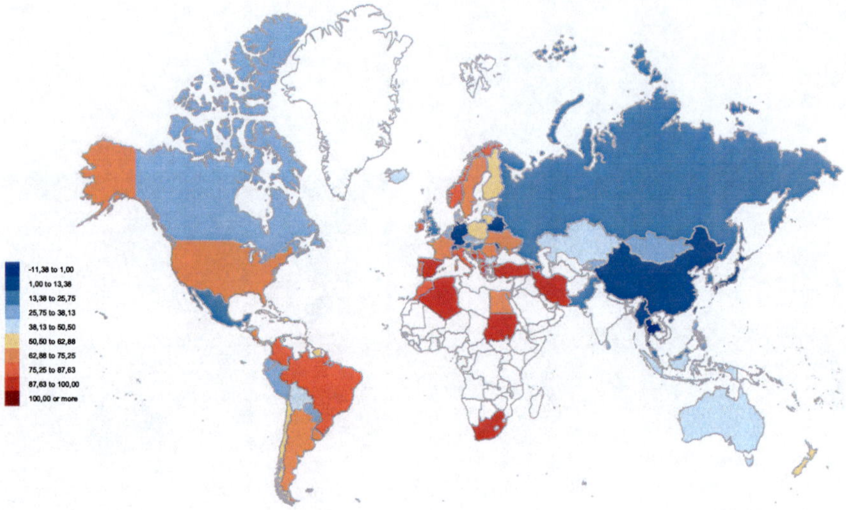


**Fig. 3.25** Unemployment (IMF World Economic Outlook Database, April 2020, inverted ranks of avoiding unemployment)

Figure 3.27 shows the current account balance surpluses or deficits per GDP, clearly reflecting the global center-periphery division. The post-Brexit United Kingdom is—it seems—also on a path toward peripherization.

### 3.10.1 *Drivers of the Severity of the Global Health Impact*

To ascertain the multivariate effects on COVID-19 death rates per 1 million population and COVID-19 fatality rates (see our freely available electronic appendix)[50] in a partial correlation analysis on a country-to-country basis, we kept the following variables constant:

- Share of Muslims per total population,
- Membership in the Arab League,
- Human Development Index (HDI, 2018),
- Human Development Index (HDI, 2018)^2.

[50] https://www.academia.edu/45321329/Factors_affecting_the_Gulfs_future_Global_cycles_and_value_change_Electronic_Appendix.

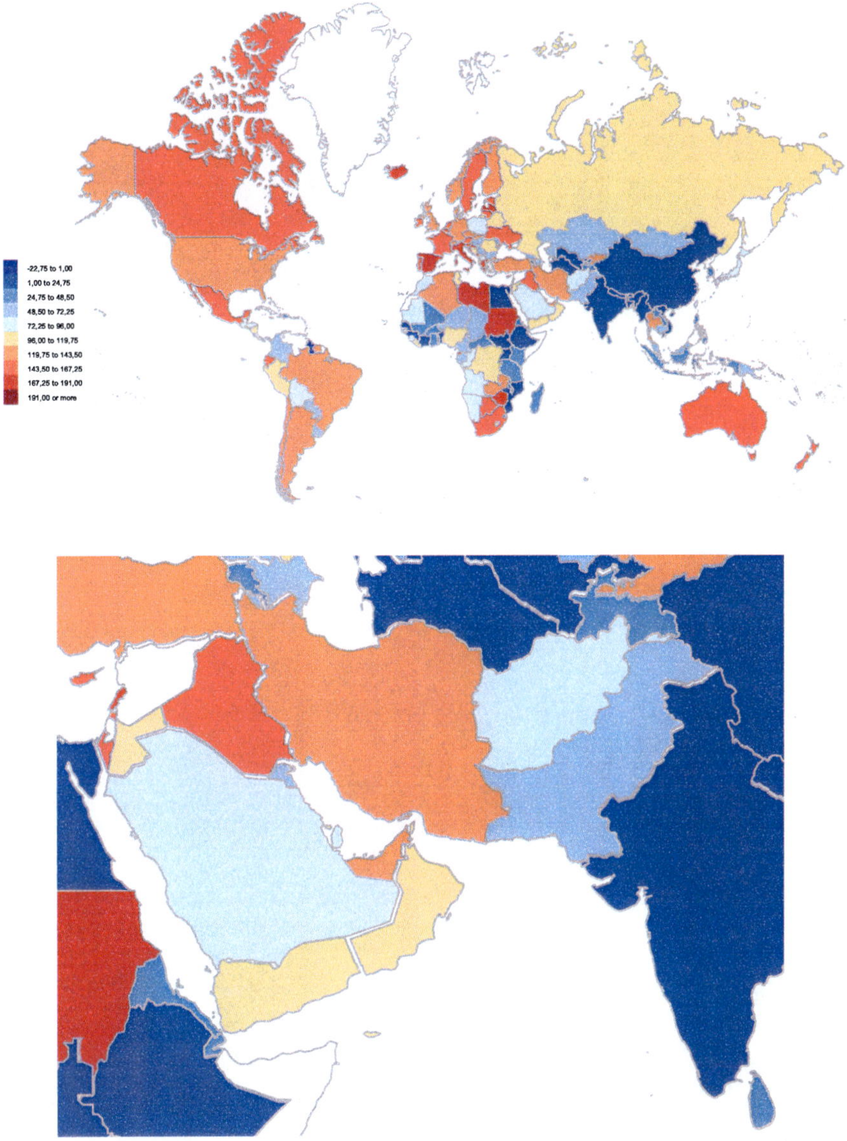


**Fig. 3.26** Recession (IMF World Economic Outlook Database, April 2020, inverted ranks of global economic growth rates)

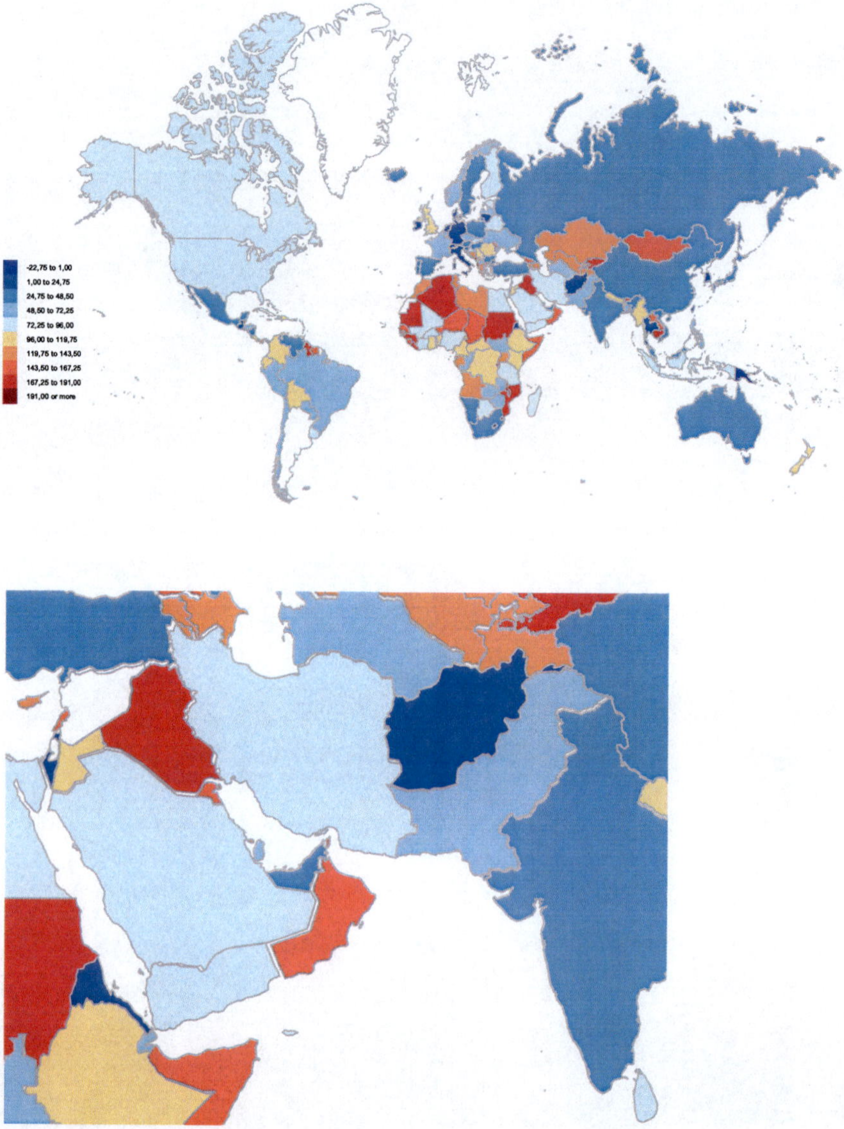


**Fig. 3.27** Current account surplus or deficit per GDP (IMF World Economic Outlook Database, April 2020, inverted ranks)

Our variables are well documented and are freely available to global publics.[51] On the one hand, it is clear that countries with a high share of Roman Catholics per total population, years of membership in the EU, 2010, a high degree of democracy, a former high standard of living, a well-developed system of tertiary enrollment, and an equally well-developed system of social protection as well as a high Happy Planet Index, HPI, were the infection centers in the current pandemic. However, we must also consider that countries with a high social inequality, as well as a high unemployment and a high rate of outward investment of their transnational enterprises, were among the centers of the spread of the disease. Now, which conditions contributed to a relatively lower spread of the disease? In accordance with the hypothesis, voiced by Acemoglu, 2020, and also in accordance with a vast literature in the connection of the late Harvard professor Alberto Alesina (Alesina & Guiliano, 2013, 2015; Alesina & Ferrara, 2000; Alesina et al., 2015), trust in institutions seems to be a decisive factor against a rapid spread of the disease. The availability of hospital beds and a great number of nurses and midwives per total population, a high share of the poorest in total incomes, and a high gender empowerment index all contributed toward the ability of countries to control the spread of the COVID-19 virus. Contrary to the widely debated opinion that economic globalization is to blame for the spread of the virus, the opposite is true, because globalized system could react more rapidly to the emergency. Contrary to the opinion voiced by President Trump, debated above ("China virus"), the presence of Chinese immigrants per total immigrants did not create conditions for a rapid spread of the virus, and also the share of Buddhists and also the share of the adherents of (other) Eastern religions per total population was not a "driver" of high infection rates.

Analyzing COVID-19 fatality rates, the results are at times even more surprising. It emerges again as confirmed by the facts that countries with a relatively high standard of living and a developed democracy were the countries not only where there were many infections but also where those infected had a higher death rate. Especially the countries of the European Union were affected by this, and the significant drivers of high fatality rates were:

- Years of membership in the EU, 2010,
- Happy life years,
- Life expectancy (years) (by 2010),
- Happy Planet Index, HPI,
- UNDP education index,
- Life satisfaction (0–10),
- Female survival probability of surviving to age 65 female,
- Social protection (ILO),
- Democracy measure,
- Years of membership in EMU, 2010.

---

[51] https://www.academia.edu/44295690/GULF_2020_Covid_19_PUBLIC_ACCESS_Data_base_for_the_book_Factors_affecting_the_Gulfs_future_Global_cycles_and_value_change_Springer_Publishing_Series_https_www_springer_com_series_16417_Gulf_Studies_edited_by_Prof_Mizanur_Rahman_Qatar_University.

But, perhaps paradoxically, the following conditions were significantly associated with high death rates:

- Current health expenditure (% of GDP), 2017,
- Global migration senders, 2017.

The following variables show a significant negative partial correlation with COVID-19 fatality rates:

- Trust in institutions,
- Economic globalization, de facto index,
- Migrants 2017 percentage of total population,
- LFPR 55–59 Labor Force Participation Rate of elderly workers,
- DYN HDI 2010-2018,
- Gulf Cooperation Council membership,
- Gross national income (GNI) per capita,
- Net migration rate,
- GDP per capita PPP$,
- Annual population growth rate, 1975–2005 (%),
- Net international migration rate, 2005–2010,
- Mean economic growth rate crisis years 2008/2011 (Table 3.10).

## 3.11 What We Know About Pandemic Severity Rates

With the majority of the populations in the Arab World predictably becoming poorer still by international standards after the pandemic, and with income inequality predictably rising, Solomon and Tausch, 2021, also looked at another lesson of history, which might be deduced from the Barro and Ursúa, 2011, data. Solomon and Tausch analyzed the developments of average per capita incomes in purchasing power parities in Italy and the UK, two European countries which were particularly hard hit by the Influenza Pandemic of 1918–1920. In the UK, it took practically a decade to recover the pre-pandemic income levels; in Italy, the crisis was even more severe, Italian democracy collapsed, and Benito Mussolini rose to power (Figs. 3.28 and 3.29 and Table 3.11).

## 3.12 The Rising Chinese Global Dominance

Rather in accordance with the "cycle of global politics and economics" literature (Goldstein, 1985, 1988; Goldstein and Rapkin, 1991; and Tausch, 2007, 2010), we think that the repetition of the Willhelminian path of a global Chinese challenge to

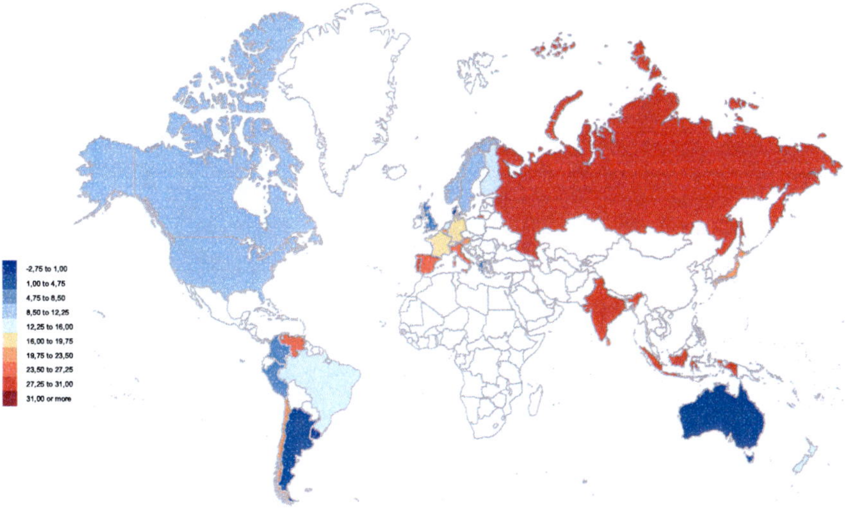


**Fig. 3.28** Severity of the Influenza Pandemic 1918–1920 (death rates as % of the total population, inverse ranks)

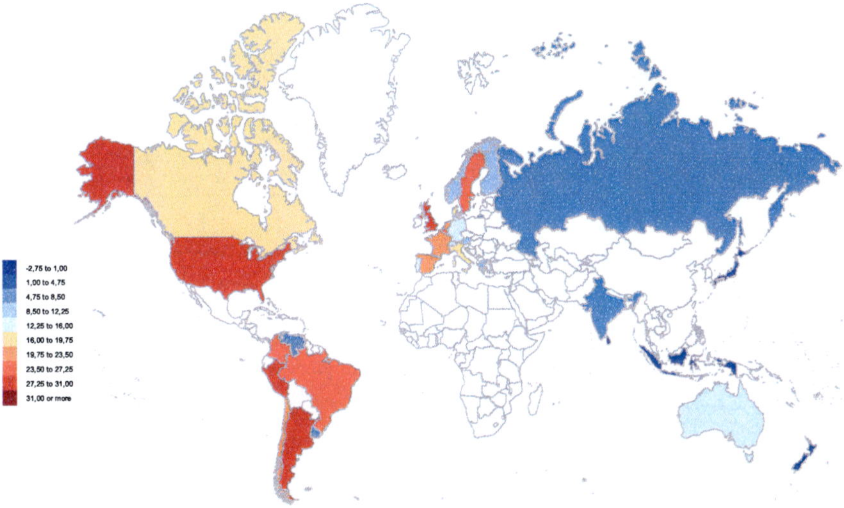


**Fig. 3.29** Severity of the current pandemic as compared to the "Spanish flu," 1918–1920

US global leadership is now inevitable.[52] Not only President Xi Jinping seems to be acutely aware of this, also Chinese Central Television produced a TV series with hundreds of millions of watchers, dealing precisely with the parallels between China

[52] https://foreignpolicy.com/2018/11/09/world-war-is-depressing-lessons-for-asia/.

**Table 3.11** Death rates of the current pandemic in % of the historic influenza death rates, 1918–1920

| | Death rates of the current pandemic in % of the historic influenza death rates, 1918–1920 |
|---|---|
| Sri Lanka | 0.0034 |
| New Zealand | 0.0725 |
| Japan | 0.1250 |
| Indonesia | 0.1283 |
| India | 0.1360 |
| Venezuela | 0.1760 |
| Uruguay | 0.6364 |
| Russia | 0.7540 |
| Greece | 0.8222 |
| Finland | 0.8732 |
| Norway | 0.8772 |
| Austria | 0.9072 |
| Portugal | 1.0663 |
| Australia | 1.2500 |
| Germany | 1.4615 |
| Switzerland | 3.1447 |
| Denmark | 3.6129 |
| Canada | 3.9677 |
| Italy | 4.8293 |
| Spain | 4.9706 |
| Netherlands | 5.2535 |
| France | 6.5946 |
| Chile | 7.7209 |
| Sweden | 9.2381 |
| Brazil | 9.7391 |
| Belgium | 10.3855 |
| Colombia | 11.0000 |
| Argentina | 11.0606 |
| United States | 12.2308 |
| United Kingdom | 13.4565 |
| Peru | 25.1026 |

and Imperial Germany before the First World War.[53] The video is globally available even today. Already, Polybius (200 BE–120 BE), another historian of the ancient world of antiquity, was keenly aware of Great Power rivalries leading to war. Seen

[53] http://www.cctv.com/program/cultureexpress/20070601/104110.shtml.

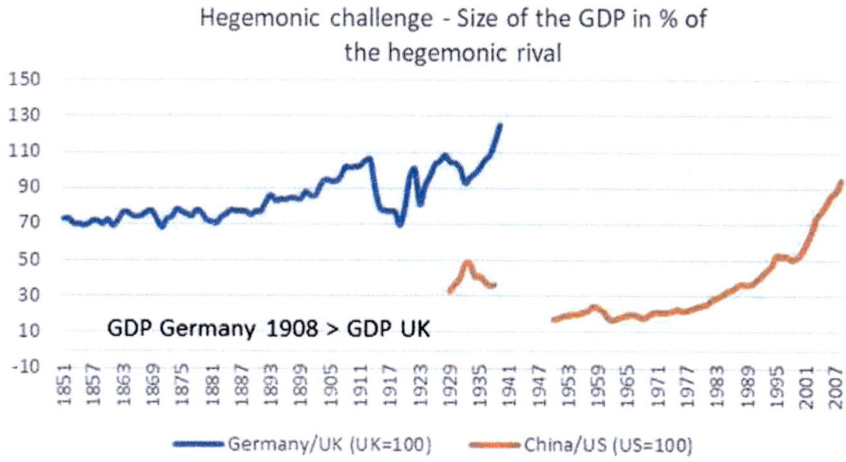


**Fig. 3.30** Germany's and China's GDP over the last 200 years compared to their hegemonic rivals (UK and the USA) according to Maddison, 2007 (Angus Maddison: Historical Statistics of the World Economy: 1–2008 AD, available at http://www.ggdc.net/maddison/Maddison.htm)

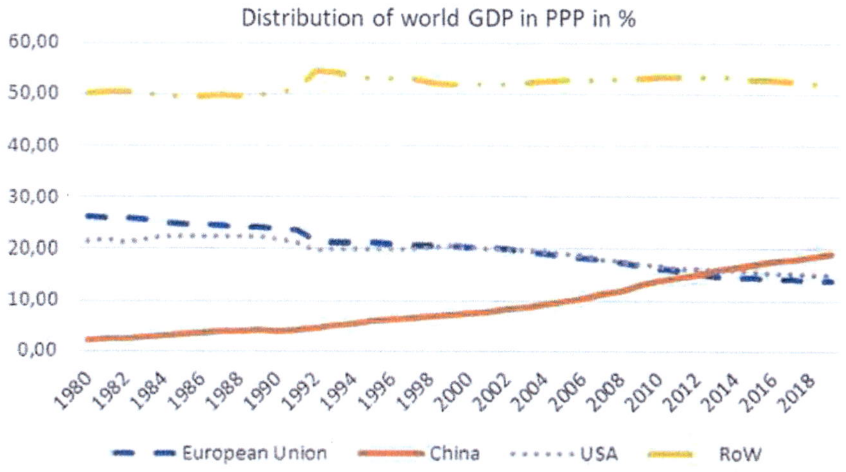


**Fig. 3.31** World power distribution, 1980–2019, according to World Bank Open Data

in the perspective of Fig. 3.30, China already reached in 2007 a level comparable to the year 1900 of the power contest Germany/UK.

According to the World Bank data, the all-decisive point of no return of the Chinese global challenge against the USA and also against the European Union already was reached in 2012, when Chinese GDP measured in purchasing power already surpassed that of its rivals (Fig. 3.31).

**Fig. 3.32** Military expenditures in the USA, China, and Russia since 1975

As shown in Fig. 3.32, Chinese military expenditures are still below 2%, but it is only a question of time when the superior economic strength of China will translate into a superior global military power:

## 3.13 Prospects and Conclusions for the World After the Pandemic

As we highlighted in the preceding analysis, global social science has a lot of analytical insights and data at disposal to draw parallels between the current COVID-19 pandemic and the Influenza Pandemic, 1918–1920. Thanks to the work of Barro and Ursúa, (2011), and Barro et al. (2020), economic science also knows much more about the effects of both the Spanish Influenza Pandemic from 1918 onward and the subsequent Great Depression from 1929 onward. Since the large literature on long economic cycles, depressions, and their devastating effects has been reviewed extensively elsewhere (Grinin et al., 2016) and will be taken up again with the latest data in Chap. 4 of the present book, we rather concentrated here in this chapter on analyzing what can be expected from the double crisis of a pandemic and a global depression for the countries of the Gulf region.

The robust lesson of the "Great Depression" accounting, made possible by the Barro and Ursúa (2011), data is that there are very considerable differences in the way, the tsunami of the Great Depression hit the countries of the world system in 1929. To recover their pre-depression income level by around + 5%, we highlighted the fact that it took Denmark, Iceland, Korea, and Russia only a short lapse of time, and for all practical purposes they recovered already in 1930/31 (see analysis above, and also Solomon & Tausch, 2021). By 1933/34, Colombia, Greece, Brazil, Finland, Japan, Norway, South Africa, Sweden, and Turkey recovered again. Only in 1935/36, Italy, Singapore, Taiwan, UK, Australia, Germany, and New Zealand recovered and had an income level of around + 5% above the pre-depression level. In 1937, Portugal and Venezuela followed, while Indonesia only made it in 1938. Even more dramatic, Austria recovered and had an income level of around + 5% above the pre-depression level in 1939, the year after the Anschluss, and thus shared the economic misfortunes of Mexico and the Philippines, who also recovered only in 1939. Even worse, Canada and the USA only could recover their income levels in 1940, after the beginning of the Second World War in 1939 (Solomon & Tausch, 2021).

As it emerged from the analysis above, a plus of around 5% or more in the real income levels, compared to 1929, was recorded in the following unfortunate nations in the following years:

1944: Argentina
1945: Chile
1946: Peru, Switzerland, and Uruguay
1947: Belgium
1949: France
1950: Netherlands
1956: China and Spain
1959: Egypt
After 1960: India, Malaysia and Sri Lanka.

The World Bank (2021) forcefully reminded international publics and decision-makers what kind of a stage of world development we are entering by saying that past global recessions were associated with highly persistent output losses. By 2025, global output is still expected to be 5% below the pre-pandemic trend—a cumulative output loss that is equivalent to 36% of the world's 2019 output.

The world ahead of us can be guessed by the data and perspectives contained in Barro and Ursúa (2011), Grinin (2016), and Solomon and Tausch (2021). The suggestion is that poorer countries were hit harder by the influenza epidemic of 1918 than countries of the center, although the effect, analyzed in Solomon and Tausch, 2021, is nonlinear due to the higher death rates in some richer countries. The IBM-SPSS 24 calculation of the relationship between the natural logarithm of GDP per capita in the last peace year before the First World War in 1913, UK = 100 and its square and the influenza death rate per country in 1918–20 in the 31 countries reported in Solomon and Tausch, 2021, is explaining 33.2% of the influenza death rate per country in 1918–20.

As Barro et al. 2020, convincingly argued, every global pandemic has terrible effects on the mortality rates especially in poorer countries. Finally (Solomon & Tausch, 2021), and—in a more detailed fashion, the above analysis—risked a prognosis on the overall societal effects of the pandemic. A lesson of the 1918–1920 pandemic is that the severity of the pandemic, as it impacts the different countries of the world, contributes to rising income inequality over time. We measured this by calculating a Pearson correlation of income inequality with the time axis, 1921–1929, based on the Piketty time series data of the share of the richest 1% in total incomes, discussed in Grinin et al. (2016), and the flu death rates during the Great Influenza Epidemic, 1918–1920 (Barro et al., 2020). We also attempted to estimate the effects of the severity of the pandemic on international income convergence/divergence over time. To estimate this, they calculated the Pearson correlation of income convergence with the time axis, 1921–1929, based on Maddison data of a country's income per capita in PPP, world average = 100, documented in Grinin et al. (2016), and the severity of the flu epidemic 1918–1920, measured by the flu death rates (Barro et al., 2020). The result was that the severity of an epidemic negatively influences international income convergence over time.

Barro predicts the following to happen:

> I do not find that the Great Influenza Pandemic caused permanent structural changes in the macroeconomy or in government policies, although the later Great Depression and World War II did have these kinds of effects. I think that the ongoing coronavirus crisis will be similar to the Great Influenza and will have no great impact on the structures of economies. Yet, there will be some permanent changes. For example, countries such as the United States will be less willing to rely on foreign suppliers for goods and services regarded as crucial during a crisis. Part of this change will involve further movement away from China as a source of many products. And some industries, such as airlines, will likely change permanently, notably in terms of maintaining a greater degree of social distancing among passengers. The practice of shaking hands will likely also become a thing of the past.[54]

The rise of China to a global position of dominance is now irreversible and seems to revert the world to the geographical and economic order which existed before the age of industrialization. Arrighi (1995, 2007), as well as Frank (1998) were very clear on this, and in fact, also the Maddison database confirms this perspective (see Fig. 3.27).

## References

Acemoglu, D. (2020). Unknowns, Challenges and Opportunities in the Time of Covid-19). Massachusetts Institute of Technology, MIT, April 9, (2020). Available at https://economics.mit.edu/files/19370.

Ahmed, F., Ahmed, N., Pissarides, C., & Stiglitz, J. (2020). Why inequality could spread COVID-19. *The Lancet Public Health, 5*(5), e240. https://doi.org/10.1016/S2468-2667(20)30085-2

Aiginger, K. (2009). *The current economic crisis: Causes, cures and consequences.* WIFO Working Papers, No. 341, August 2009, Austrian Institute for Economic Research, available at https://karl.aiginger.wifo.ac.at/fileadmin/files_aiginger/publications/2009/WP_2009_341__01.PDF

Alesina A., & Guiliano P. (2013). *Culture and institutions.* NBER Working Paper Series, Working Paper 19750. Available at http://scholar.harvard.edu/alesina/publications.

Alesina, A., & Ferrara, E. L. (2000). *The determinants of trust* (No. w7621). National Bureau of Economic Research.

Alesina, A., & Giuliano, P. (2015). Culture and institutions. *Journal of Economic Literature, 53*(4), 898–944.

[54] https://english.phbs.pku.edu.cn/2020/news_0717/2452.html..

Alesina, A., Algan, Y., Cahuc, P., & Giuliano, P. (2015). Family values and the regulation of labor. *Journal of the European Economic Association, 13*(4), 599–630.

Alfani, G. (2013). *Calamities and the economy in Renaissance Italy: the grand tour of the Horsemen of the Apocalypse*. Springer.

Alfani, G., & Percoco, M. (2019). Plague and long-term development: the lasting effects of the 1629–30 epidemic on the Italian cities. *The Economic History Review, 72*(4), 1175–1201.

Anderson, W. (2019). The history in epidemiology. *International Journal of Epidemiology Official Journal of the International Epidemiological Association, IEA, 48*(3), 672–674.

Armelagos, G. J., Brown, P. J., & Turner, B. (2005). Evolutionary, historical and political economic perspectives on health and disease. *Social Science & Medicine, 61*(4), 755–765.

Arrighi, G. (1995). *The long 20th century. Money, power, and the origins of our times*. Verso.

Arrighi, G. (2007). *Adam Smith in Beijing: Lineages of the twenty-first century*. Verso.

Arvanitidis, P. A. (2013). Democracy and public choice in classical Athens. *Peace Economics, Peace Science and Public Policy /, 19*(2), 213–248.

Barnett, R. (2019). Cholera. *The Lancet, 393*(10168), 218.

Barquet, N. (1997). Smallpox: The triumph over the most terrible of the ministers of death. *Annals of Internal Medicine, 127*(8 I), pp.635–642.

Barro, R. J. (2006). Rare disasters and asset markets in the twentieth century. *The Quarterly Journal of Economics, 121*(3), 823–866.

Barro, R. J., & Ursúa, J. F. (2008). Macroeconomic Crises since 1870. Brookings Papers on Economic Activity, Spring (pp. 255–335). Available at: http://www.people.fas.harvard.edu/~jfursua/papers/bpea.pdf (Download: April 10, (2019).

Barro, R. J., & Ursúa, J. F. (2011). Barro-Ursúa Macroeconomic Data. Available at: https://scholar.harvard.edu/barro/publications/barro-ursua-macroeconomic-data.

Barro, R. J., Ursúa, J. F., & Weng, J. (2020). The coronavirus and the Great Influenza Epidemic—Lessons from the 'Spanish Flu' for the Coronavirus's Potential Effects on Mortality and Economic Activity (2020). CESifo Working Paper No. 8166). Available at SSRN: https://ssrn.com/abstract=3556305

Barua, S. (2020). Understanding Coronanomics: The Economic Implications of the Coronavirus (Covid-19) Pandemic (April 1, 2020). Available at SSRN: https://ssrn.com/abstract=3566477 or /https://doi.org/10.2139/ssrn.3566477

Bell, C. & Lewis, M. (2005). *The economic implications of epidemics old and new*. Working Paper Number 54, February 2005, Center for Global Development. Available at http://ssrn.com/abstract=997387.

Benedictow, O. (2009). The Justinianic Plague pandemic: Progress and problems. *Early Science and Medicine, 14*(4), 543–548.

Benmelech, E., & Tzur-Ilan, N. (2020). *The determinants of fiscal and monetary policies during the COVID-19 crisis* (No. w27461). National Bureau of Economic Research. https://www.nber.org/system/files/working_papers/w27461/w27461.pdf

Bianco, C. (2020). *A Gulf apart: How Europe can gain influence with the Gulf Cooperation Council*. European Council on Foreign Relations, February 2020. Available at https://ecfr.eu/archive/page/-/a_gulf_apart_how_europe_can_gain_influence_with_gulf_cooperation_council.pdf

Bolt, J., Inklaar, R., de Jong, H., & Luiten van Zanden, J. (2018). *Rebasing 'Maddison': new income comparisons and the shape of long-run economic development*. Maddison Project Working Paper, nr. 10. Available at www.ggdc.net/maddison.

Boutayeb, A., & Serghini, S. (2006). Health indicators and human development in the Arab region. *International Journal of Health Geographics, 5*, 1–13.

Brodeur, A., Gray, D. M., Islam, A., & Bhuiyan, S. (2020). *A literature review of the economics of COVID-19*. Available at: https://www.econstor.eu/bitstream/10419/222316/1/GLO-DP-0601.pdf

Brody, H. (2000). Map-making and myth-making in Broad Street: The London cholera epidemic, 1854. *The Lancet, 356*(9223), 64–68.

Bruun, C. (2007). The Antonine plague and the 'third-century crisis'. In *Crises and the Roman Empire* (pp. 201–217). Brill.

Brzezinski, Z. (2014). Can China avoid the Thucydides trap? *New Perspectives Quarterly, 31*(2), 31–33.

Büntgen, U. (2016). Cooling and societal change during the Late Antique Little Ice Age from 536 to around 660 AD. *Nature Geoscience, 9*(3), 231–236.

Candela, R., & Geloso, V. (2020). *Economic freedom, pandemics and robust political economy* (December 11, 2020). Available at SSRN: https://ssrn.com/abstract=3747216 or https://doi.org/10.2139/ssrn.3747216.

Ceylan, R., Ozkan, B., & Mulazimogullari, E. (2020). Historical evidence for economic effects of COVID-19. *The European Journal of Health Economics, 21*(6), 817–823.

Chan, J.F.-W., et al. (2020). A familial cluster of pneumonia associated with the 2019 novel coronavirus indicating person-to-person transmission: A study of a family cluster. *The Lancet, 395*(10223), 514–523. https://doi.org/10.1016/S0140-6736(20)30154-9

Chen, N., et al. (2020). Epidemiological and clinical characteristics of 99 cases of 2019 novel coronavirus pneumonia in Wuhan, China: A descriptive study. *The Lancet, 395*(10223), 507–513. https://doi.org/10.1016/S0140-6736(20)30211-7

Colwell, R. R. (1996). Global climate and infectious disease: The cholera paradigm. *Science, 274*(5295), 2025–2031.

Cornia, G. A. (2019). A theory of why potentially favorable political and economic changes may lead to mortality crises (No. wp2019_23.rdf). Universita' degli Studi di Firenze, Dipartimento di Scienze per l'Economia e l'Impresa.

Cornia, G. A., & Paniccià, R. (2000). *The mortality crisis in transitional economies*. OUP Oxford.

Dayaratna, K., & Michel, N. (2020). *The Challenges of Forecasting the Spread and Mortality of Covid-19*. Heritage Foundation. Available at https://www.heritage.org/public-health/report/the-challenges-forecasting-the-spread-and-mortality-covid-19

Dean, H., & Khan, Z. (1997). Muslim perspectives on welfare. *Journal of Social Policy, 26*(2), 193–209.

Dobbin F. (2009). Review of Giovanni Arrighi, Adam Smith in Beijing: Lineages of the Twenty-first Century' American Journal of Sociology, 115: 293–295. Available at: https://www.researchgate.net/publication/260869819_Review_of_Giovanni_Arrighi_Adam_Smith_in_Beijing_Lineages_of_the_Twenty-first_Century. Accessed 9 May 2020.

Doerr, S., & Hofmann, B. (2020). *Recessions and mortality: A global perspective*. Bank for International Settlements. Available at: https://www.bis.org/publ/work910.pdf.

Dols, M. W. (2019). *The black death in the Middle East*. Princeton University Press.

Douglas, M., Katikireddi, S.V., Taulbut, M., McKee, M., & McCartney, G. (2020). Mitigating the wider health effects of covid-19 pandemic response. *The BMJ, 369*, art. no. m1557. https://doi.org/10.1136/bmj.m1557

Duncan-Jones, R. P. (1996). The impact of the Antonine plague. *Journal of Roman Archaeology, 9*, 108–136.

Elgar, F. J., Stefaniak, A., & Wohl, M. J. (2020). The trouble with trust: Time-series analysis of social capital, income inequality, and COVID-19 deaths in 84 countries. *Social Science & Medicine, 263*, 113365.

Fenner, F. (1993). Smallpox: Emergence, global spread, and eradication. *History and Philosophy of the Life Sciences, 15*(3), 397–420.

Fernandes, N. (2020). Economic effects of coronavirus outbreak (Covid-19) on the World Economy (March 22, 2020). Available at SSRN: https://ssrn.com/abstract=3557504 or https://doi.org/10.2139/ssrn.3557504

Findlay, R., & Lundahl, M. (2017). Demographic shocks and the factor proportions model: from the plague of Justinian to the Black Death. In *The economics of the frontier* (pp. 125–172). Palgrave Macmillan.

Frank, A. G. (1998). *ReOrient: Global economy in the Asian Age*. University of California Press.

Furceri, D., Loungani, P., Ostry, J. D., & Pizzuto, P. (2020). Will Covid-19 affect inequality? Evidence from past pandemics. *COVID-19 Economics, 12*(1), 138–157.

Gat, A. (2007). The return of authoritarian great powers. *Foreign Affairs, 86*(4), 59–69.

Gat, A. (2017). *The causes of war and the spread of peace but will war rebound?* Oxford University Press.

Ghose, A. C. (2011). Lessons from cholera & Vibrio cholerae. *The Indian Journal of Medical Research., 133*(2), 164–170.

Goldstein, J. S. (1988). *Long cycles: Prosperity and war in the modern age.* Yale University Press.

Goldstein, J. S. (1985). Kondratieff waves as war cycles. *International Studies Quarterly, 29*(4), 411–444.

Goldstein, J. S. (2011). *Winning the war on war: The decline of armed conflict worldwide.* Dutton.

Goldstein, J. S., & Rapkin, D. P. (1991). After insularity: Hegemony and the future world order. *Futures, 23*(9), 935–959.

Gottfried, R. S. (2010). *Black death.* Simon and Schuster.

Gries, P. (2019). Are the US and China fated to fight? How narratives of 'power transition' shape great power war or peace. *Cambridge Review of International Affairs, 32*(4), 456–482.

Grinin, L., Tausch, A., & Korotayev, A. (2016). *Economic cycles, crises, and the global periphery.* Springer.

Guan, W.-J., et al. (2020). Clinical characteristics of coronavirus disease 2019 in China. *The New England Journal of Medicine.* https://doi.org/10.1056/NEJMoa2002032

Gygli, S., Haelg, F., Potrafke, N., & Sturm, J. E. (2019). The KOF globalization index—Revisited. *Review of International Organizations, 14*(3), 543–574. https://doi.org/10.1007/s11558-019-09344-2

Harry, W. (2007). Employment creation and localization: The crucial human resource issues for the GCC. *International Journal of Human Resource Management, 18*(1), 132–146.

Holladay, A. J., & Poole, J. C. F. (1979). Thucydides and the Plague of Athens. *The Classical Quarterly, 29*(2), 282–300.

Huang, C., et al. (2020). Clinical features of patients infected with 2019 novel coronavirus in Wuhan, China. *The Lancet, 395*(10223), 497–506. https://doi.org/10.1016/S0140-6736(20)30183-5

Javaid, S., & Al-Malkawi, H.-A.N. (2018). Corporate social responsibility and financial performance in Saudi Arabia: Evidence from Zakat contribution. *Managerial Finance, 44*(6), 648–664.

Jordà, Ò., Singh, S. R., & Taylor, A. M. (2020). *Longer-run economic consequences of pandemics* (No. w26934). National Bureau of Economic Research. Available at https://www.frbsf.org/economic-research/files/wp2020-09.pdf.

Joseph, S. (1996). Patriarchy and development in the Arab World. *Gender and Development., 4*(2), 14–19.

Kawohl, W., & Nordt, C. (2020). COVID-19, unemployment, and suicide. *The Lancet Psychiatry, 7*(5), 389–390. https://doi.org/10.1016/S2215-0366(20)30141-3

Kılıç, S. (2020). Does COVID-19 as a long wave turning point mean the end of neoliberalism? *Critical Sociology.* https://doi.org/10.1177/0896920520980083.ArticleinPress

Kok, J. (2020). Short-term trade-off between stringency and economic growth. *COVID Economics, 60*, 172–189. Available at https://www.researchgate.net/profile/Jason_Kok/publication/346643101_Short-term_trade-off_between_stringency_and_economic_growth/links/5fcad9f645851568d13f07ec/Short-term-trade-off-between-stringency-and-economic-growth.pdf

Lafferty, K. (2009). The ecology of climate change and infectious diseases. *Ecology, 90*(4), 888–900.

Langmuir, A. D., Worthen, T. D., Solomon, J., Ray, C. G., & Petersen, E. (1985). The Thucydides Syndrome: A new hypothesis for the cause of the plague of Athens. *New England Journal of Medicine, 313*(16), 1027–1030. https://doi.org/10.1056/NEJM198510173131618

Lee, J. (2019). Did Thucydides believe in Thucydides' trap? The history of the Peloponnesian War and its Relevance to US-China Relations. *Journal of Chinese Political Science, 24*(1), 67–86.

Lim, M. L. (2005). History of U.S. military contributions to the study of diarrheal diseases. *Military Medicine, 170*(4 SUPPL.), 30–38.

Little, L. K. (Ed.). (2007). *Plague and the end of antiquity: the pandemic of 541–750.* Cambridge University Press.

Lu, R., et al. (2020). Genomic characterisation and epidemiology of 2019 novel coronavirus: Implications for virus origins and receptor binding. *The Lancet, 395*(10224), 565–574. https://doi.org/10.1016/S0140-6736(20)30251-8

Maas, M. (2005). *The Cambridge companion to the Age of Justinian.* Cambridge University Press.

Maddison A. (2007). *Contours of the world economy, 1–2030 AD: essays in macro-economic history.* Oxford University Press.

McCannon, B. C. (2012). The origin of democracy in Athens. *Review of Law & Economics., 8*(2), 531–562.

McCormick, M. (2012). Climate change during and after the Roman Empire: Reconstructing the past from scientific and historical evidence. *The Journal of Interdisciplinary History, 43*(2), 169–220.

Moore, G. J. (2017). Avoiding a Thucydides Trap in Sino-American Relations (…and 7 reasons why that might be difficult). *Asian Security., 13*(2), 98–115.

Morens, D. M. (2008). Emerging infections: A perpetual challenge. *The Lancet Infectious Diseases., 8*(11), 710–719.

Morgan, A. K., Awafo, B. A., & Quartey, T. (2021). The effects of COVID-19 on global economic output and sustainability: Evidence from around the world and lessons for redress. *Sustainability: Science, Practice and Policy, 17*(1), 77–81.

Norris, P., & Inglehart, R. (2019). *Cultural backlash: Trump, Brexit, and the rise of authoritarian populism.* Cambridge University Press.

Ober, J. (1989). *Mass and Elite in democratic Athens: Rhetoric, ideology, and the power of the people.* Princeton University Press.

Ober, J. (2008). *Democracy and knowledge: Innovation and learning in classical Athens.* Princeton University Press.

Orwin, C. (1988). Stasis and plague: Thucydides on the dissolution of society. *The Journal of Politics JOP, 50*(4), 831–847.

Ozili, P., & Thankom, A. (2020). Spillover of Covid-19: impact on the Global Economy. Social Science Research Network, New York. Available at SSRN: https://ssrn.com/abstract=3562570 or https://doi.org/10.2139/ssrn.3562570

Page, D. (1953). Thucydides' Description of the Great Plague at Athens. *The Classical Quarterly, 3*(3/4), 97–119.

Pamuk, S. (2007). The black death and the origins of the 'great divergence' across Europe, 1300–1600. *European Review of Economic History EREH, 11*(3), 289–317.

Pamuk, S., & Shatzmiller, M. (2014). Plagues, wages, and economic change in the Islamic Middle East, 700–1500. *The Journal of Economic History, 74*(1), 196–229.

Patz, J. (2005). Impact of regional climate change on human health. *Nature, 438*(7066), 310–317.

Pinke, Z. (2019). Volcanic mega-eruptions may trigger major cholera outbreaks. Climate research interactions of climate with organisms, ecosystems, and human societies. *International and Multidisciplinary Journal, 79*(2), 151–162.

Polanyi, K., & MacIver, R. M. (1944). *The great transformation.* Beacon press.

Raoult, D., Mouffok, N., Bitam, I., Piarroux, R., & Drancourt, M. (2013). Plague: History and contemporary analysis. *Journal of Infection, 66*(1), 18–26.

Rosen, W. (2010). *Justinian's flea: Plague, empire and the birth of Europe.* Random House.

Rubin, M. (2020). *Capitalism is the solution, not the problem, for the pandemic world.* American Enterprise Institute. Available at https://www.aei.org/op-eds/capitalism-is-the-solution-not-the-problem-for-the-pandemic-world/.

Ruiz Estrada, M. A. (2020). Economic waves: The effect of the Wuhan Covid-19 on the World Economy (2019–2020) (February 28, 2020). Available at SSRN: https://ssrn.com/abstract=3545758 or https://doi.org/10.2139/ssrn.3545758

Sabbatani, S., & Fiorino, S. (2009). The Antonine plague and the decline of the Roman Empire. *Le Infezioni in Medicina: Rivista Periodica Di Eziologia, Epidemiologia, Diagnostica, Clinica e Terapia Delle Patologie Infettive, 17*(4), 261–275.

Sarkis, J., Cohen, M. J., Dewick, P., & Schröder, P. (2020). A brave new world: Lessons from the Covid-19 pandemic for transitioning to sustainable supply and production. *Resources, Conservation, and Recycling*. Available at https://www.ncbi.nlm.nih.gov/pmc/articles/PMC7164912/

Scheidel, W. (2009). Roman wellbeing and the economic consequences of the 'Antonine Plague' (November 10, 2009). Available at SSRN: https://ssrn.com/abstract=1442584 or https://doi.org/10.2139/ssrn.1442584

Schmid, B. V. (2015). Climate-driven introduction of the Black Death and successive plague reintroductions into Europe. *Proceedings of the National Academy of Sciences of the United States of America PNAS, 112*(10), 3020–3025.

Solomon, H., & Tausch, A. (2021). *Arab MENA countries: Vulnerabilities and constraints against democracy on the eve of the global Covid-19 crisis.* (Perspectives on development in the Middle East and North Africa (Mena region). Springer.

Steere-Williams, J. (2018). Edward Ballard and the practice of epidemiology in the 19th-century United Kingdom. *American Journal of Public Health., 108*(12), 1632–1638.

Tatem, A. J. (2006). Global transport networks and infectious disease spread. *Advances in Parasitology, 62*, 293–343.

Taubenberger, J. K., & Morens, D. M. (2006). 1918 Influenza: The mother of all pandemics. *Emerging Infectious Diseases, 12*(1), 15.

Tausch, A. (2007). *War Cycles' Social Evolution and History (moscow), 6*(2), 39–74.

Tausch, A. (2010). Paul Boccara's analysis of global capitalism. the return of the Bourbons, and the breakdown of the Brussels/Paris neo-liberal consensus, Entelequia. *Revista Interdisciplinar, 12*, 105–147.

Tausch, A. (2012). International Macroquantitative Data. Faculty of Economics, Corvinus University of Budapest. Available at http://www.uni-corvinus.hu/index.php?id=47854&tx_ttnews%5Btt_news%5D=0&tx_ttnews%5BbackPid%5D=31638&tx_ttnews%5BcalendarYear%5D=2012&tx_ttnews%5BcalendarMonth%5D=6&cHash=af8ef6888f7c9922b83b113f71c1ca32).

Tausch, A., & Heshmati, A. (2013). *Globalization, the human condition, and sustainable development in the twenty-first century: cross-national perspectives and European implications.* Anthem Press (with data definitions and sources). Free data download available from https://www.academia.edu/35044095/Globalization_the_human_condition_and_sustainable_development_in_the_21st_Century_Cross-national_perspectives_and_European_implications_Codebook_and_EXCEL_data_file.

UNDP (current issues). *United Nations Human Development Report*. Available at http://www.hdr.undp.org/.

Wagner, D. M., Klunk, J., Harbeck, M., Devault, A., Waglechner, N., Sahl, J. W., Enk, J., Birdsell, D. N., Kuch, M., Lumibao, C., & Poinar, D. (2014). Yersinia pestis and the Plague of Justinian 541–543 AD: A genomic analysis. *The Lancet Infectious Diseases, 14*(4), 319–326.

World Bank (2021). Global Economic Prospects. A World Bank Flagship Report, January 2021. World Bank, Washington D.C., available at https://www.worldbank.org/en/publication/global-economic-prospects

Wu, J., Cai, W., Watkins, D., & Glanz, J. (2020, March 22). How the Virus Got Out. *New York Times*. Available at https://www.nytimes.com/interactive/2020/03/22/world/coronavirus-spread.html

Yang, Y. (2018). Escape both the 'Thucydides Trap' and the 'Churchill Trap': Finding a third type of great power relations under the bipolar system. *The Chinese Journal of International Politics, DO*. https://doi.org/10.1093/cjip/poy002,11,2:193-235

Zhang, C. and Liao, H., Strobl, E., Li, H., Li, R., Jensen, S. S., & Zhang, Y. (2020). *Beware that COVID-19 would be worse in winter: A study of a global panel of 1236 Regions* (July 12, 2020). Available at SSRN: https://ssrn.com/abstract=3652202 or https://doi.org/10.2139/ssrn.3652202

Zhou, P., et al. (2020). A pneumonia outbreak associated with a new coronavirus of probable bat origin. *Nature, 579*(7798), 270–273. https://doi.org/10.1038/s41586-020-2012-7

Ziegler, P. (2013). *The black death.* Faber & Faber.

# Chapter 4
# Caught in the "Thucydides Trap?" Revisiting Kondratieff Waves, Warfare, and World Security and Their Implications for the Gulf Region

**Abstract** In Chap. 3, we already analyzed the "Thucydides Trap" of dominant world powers, rising global challengers, and the tendencies for global warfare. In view of rising global tensions in the wake of the COVID-19 depression, this chapter revisits the dire predictions about the future of the international system, which were hotly debated by world systems research for decades now. We test these propositions with new and encompassing data right up to the year 2020, applying advanced time series techniques like spectral analysis, analysis of autocorrelation, and cross-correlation analysis. Is there a danger that the global economic depression, which we now face, will again lead to a global war, just like the Great Depression of the 1930s led up to Second World War? By re-analyzing the latest global conflict data from 1495 onwards and data about the growth of world industrial production from 1740 onwards, we re-investigate propositions from classical world systems analysis and find that instability, and not stability, characterizes the world system, and that there is an indented "W"-shaped pattern of global conflict since 1495, leading from one global war to the other and that this pattern did not end with the end of the Cold War. World hegemonies that characterize the workings of world capitalism arise and they also end. The belle époque of globalization from the 1960s to the COVID-19 crisis did not bring about a more stable, egalitarian, and peaceful world. So, this chapter is a dire synthesis of the predictions and lessons to be drawn from the time series analysis of major power wars since 1495 and global economic cycles since 1740. Our chapter must be regarded as a reconfirmation of the kind of the most dismal world systems literature with advanced method of spectral analysis. Our analysis in particular confirms the existence of Kuznets cycles (22 years); Barro cycles (36 years); Kondratieff cycles (54 years); and Wallerstein cycles (140 years). Our readers should be also reminded of the fact that these results were achieved without any prior transformations of the data and without any omission of time periods to be considered between 1740 and 2020. In this chapter, also the quantitative, statistical global war cycle literature, initiated by Goldstein (Int Stud Q 29:411–444, 1985, Long cycles. Prosperity and war in the modern age. Yale University Press, London, 1988), is again confirmed, with a shorter 30-year war cycle and a longer 130-year war cycle as a result of our time series analysis. Cross-correlation analysis also confirms another world system research hypothesis, mainly a time-lag of about 4 years between a major global



A. Tausch, *The Future of the Gulf Region*, Gulf Studies 2,
https://doi.org/10.1007/978-3-030-78299-3_4

depression and global warfare. For the Gulf countries, this chapter thus implies a clear warning of an impending world political tsunami.

**Keywords** COVID-19 · Coronavirus · Coronanomics · Pandemic · Economic impacts · Economic cycles · Political cycles

## 4.1 Introduction to Kondratieff Waves, Warfare, and World Security and Their Implications for the Gulf Region

In a rather prophetic article, which appeared before the outbreak of COVID-19, the University of California-based world systems scholar Christopher K. Chase-Dunn remarked:

> The post-World War II wave of trade globalization and financialization faltered in 2008 but has recovered since then. A future trough of trade deglobalization like what happened in the 1930s could happen again if a perfect storm of calamities and more powerful resistance to further economic globalization should emerge (as seems now to be the case). The declining economic and political hegemony of the United States poses huge challenges for global governance. Newly emergent national economies such as India and China are competing for insertion into the global structure of power. (Chase-Dunn et al., 2019)

Facing these grim realities, we evaluate a new global data about world industrial production, global warfare, and global alliances, establish their cyclical character, and ask ourselves what these "long waves" of economic crises and world political confrontations imply for the Gulf region.

Confronted by the global COVID-19 crisis (see Chap. 3), the present chapter thus revisits an old debate in global social science which suddenly achieves relevance again with the current rising global tensions between the major powers of the world system. The current constellation of rising global power conflict was exactly already predicted by the German political scientist Dieter Senghaas in an article written four decades ago (Senghaas, 1983). The propensity for international conflicts is likely to increase, Senghaas said at that time, when the predominant position of a once hegemonic power tends to decline.

World political tensions between the global powers have risen to levels unthinkable since the end of the Cold War, and China especially asserts with full might and power its aspirations for world leadership in the face of an America in crisis and decline, accelerated by the Trump Presidency, 2016–2020.

## 4.2 Background on Kondratieff Waves, Warfare, and World Security

In the following, we will shortly present some of the relevant theories and also time series evidence about the Gulf countries in the global economy.

### *4.2.1 Khaleeji Capitalism*

More and more, there seems to emerge a growing "*Gulf power*" in global capitalism. *Khaleeji,* an Arabic word for "Eastern Arab," is a word, more and more used in the analysis, and it seems to be a word, which Westerns should learn more and more. Just take as an example the presence of Qatar Holding[1] in the world economy, which holds the stakes in the following companies, including in the flagship of German industry, Volkswagen. Strategic thinkers, as the managers of Qatar Holding seem to be, they didn't go for "any" company in Germany; no, they targeted the centerpiece of industry of the most powerful country in the European Union. And these strategic moves go on and on:

- Barclays: 5.50% (as of July 2019),
- Credit Suisse: 5.21% (as of August 2019),
- Deutsche Bank: 6.1% (as of August 2019),
- Glencore: 8.73% (as of July 2019),
- Hapag-Lloyd: 14.5% (as of August 2019),
- Harrods Group: Takeover (May 2010),
- Heathrow Airport Holdings: 20% (as of March 2019),
- Lagardère: 13.00% (as of July 2019),
- London Stock Exchange: 10.3% (as of July 2016),
- Pulkovo Airport: 24.99% (as of September 2016),
- Sainsbury's: 21.9% (as of July 2019),
- Siemens: 3.27% (as of August 2019),
- Tiffany & Co.: 9.70% (as of July 2019),
- Vinci (company): 3.74% (as of July 2019),
- Volkswagen AG: 17.0% ordinary shares (as of August 2019).

In the recent literature on the Gulf countries, the works of the London-based Arab Marxist scholar Hanieh (2011, 2013, 2015, 2016a, 2016b, 2018a, 2018b, 2019a,

[1] https://de.wikipedia.org/wiki/Qatar_Holding.

2019b, 2020)[2] received a considerable global citation presence,[3] library presence,[4] and also class-room presence,[5] which all suggest that a good part of Western scholarship, especially in the academic fields of political science and sociology, follow his argumentation.

His central thesis, in the framework of post-Marxian political economy, is that in the Arab Gulf, a pan-Gulf elite, "*Khaleeji capital*"[6] has emerged. This capital grew "*from state-supported and family-based trading groups in the 1960s and 1970s to the domination of a few massive conglomerates in the contemporary period*" dominated by a "*hierarchical structure around a Saudi-UAE axis.*"

This *Khaleeji* capital, according to Hanieh (2011), plays a pivotal role in the US-dominated transformation of global capitalism in the neoliberal era. "*The entire Middle East is increasingly moving to a single beat—inextricably linked to the rhythms of accumulation and class formation in the Gulf*." He predicts that the importance of the six Gulf Cooperation Council (GCC) states for the supply of hydrocarbons, petrochemicals, and financial flows to the world market will continue to grow, making the Gulf "*a decisive factor for the future trajectory of global capitalism.*"

Pfeifer (2012) correctly summarized the contribution of Hanieh (2011), to the theories of "*global capitalism*" as implying for social science theory about the Gulf countries that

> […] the advent of neoliberalism in the 1980s and the demise of the Soviet Union and the opening of China to the global market in the 1990s enhanced both the role of the United States as the sole "hyperpower" and that of the Gulf as the linchpin of US-led global economic policy in its roles as provider of hydrocarbons to the world (production), major purchaser of capital and consumer goods (exchange), and key supplier of petrodollars to the US-dominated financial system (finance). (Pfeifer, 2012)

According to Hanieh (2018a), the overwhelming majority of foreign investment in the financial sector in the nine non-GCC Arab countries (Algeria, Egypt, Jordan, Iraq, Tunisia, Libya, Syria, and Yemen), now originate in the Gulf. In Lebanon, over 50% of the country's bank assets are held by GCC-related banks, and in the West Bank and Gaza, this figure is 63%, and in Jordan it is 86%. Hanieh (2018a), also shows the importance of Gulf investment into existing Arab banks. Many non-GCC Arab firms have chosen to list their companies on Gulf stock markets. Advani

[2] The work of Hanieh also received a wide reception at the widely disseminated, radical left-wing website https://merip.org/; see especially Pfeifer (2012) and Advani (2019), and their reviews of Hanieh's books. Although we disagree with their political conclusions, our presentation of Hanieh's work is greatly indebted to these excellent reviews, written with a considerable political economic clarity. In general terms, while not sharing the perspective of https://merip.org/ and its virulent "Anti-Americanism" and "Anti-Zionism" at all, it often contains many concrete informations, which are relevant for a strategic analysis of the region, unfortunately packed in a radical language and anti-Western and anti-Israeli thinking.

[3] https://www.scopus.com/freelookup/form/author.uri.

[4] http://classify.oclc.org/classify2/.

[5] https://opensyllabus.org/.

[6] People of the Arabian Peninsula, especially those associated with the Gulf Cooperation Council states.

(2019), captures the importance of these processes, analyzed by Hanieh (2018a), for the theories of "*global capitalism*" with the following words:

> Thus, the liberalization of the financial sector in the Arab World has opened the door to the primacy of Gulf capital. Not only does this allow for the internationalization of Gulf capital, but it also harmonizes domestic accumulation structures across the Arab World to the rhythms and tempos of accumulation in the GCC. (Advani, 2019)

Some of the main conclusions from Hanieh's analysis 2018 are:

- Across the Arab World, there is a high degree of private and public debt—often a result of increasing financialization, expanding credit markets, volatile commodity prices, and food insecurity. 40% of the income of the active population in Tunisia goes towards paying mortgage debt. In Lebanon, 38% of the budget is spent on debt servicing. In Egypt, interest repayments on domestic debt account for 27% of government expenditure. With much of this debt flowing to banks and other financial institutions, Hanieh argues that "*the Arab state thus increasingly mediates the transfer of national wealth to large Gulf-related banks.*"
- Indebted Arab governments are compelled to intensify a politics of austerity, further trapping these societies in cycles of debt.
- Gulf capital constitutes a significant portion of international finance capital.
- Gulf capital has come to strengthen US debt and capital markets, sustaining high US current account deficits through the continual purchases of US treasury bonds. It also plays a pivotal role in enabling the UK banking sector to function as a key node within the global web of financial markets and the offshore banking system.
- Companies such as Credit Suisse, Barclays, Volkswagen, and British Airways all witnessed major investments from Gulf-based investors.
- The region is also an important consumer of commercial services, military imports, machinery and transport equipment, and luxury goods.
- It is estimated that the average Qatari citizen spends $4000 per month on luxury products.
- The GCC has become the largest market for weapons in the world.
- The Gulf has made significant financial investments in the global logistics sector. The Gulf's financial energies accelerate "*the turn time of commodity trade, enabling the Gulf's deeper integration into the global accumulation circuits that stretch from East to West*" (Hanieh, 2018a).

### 4.2.2 *The Gulf and Energy Cycles*

In a somewhat countervailing perspective, Cammett (2018) maintained that

> Oil can bring spectacular wealth; however, it is also central to some explanations for MENA underdevelopment. High oil dependency is negatively associated with industrialization and, as a result, may impede economic development in the longer term [...] None of the region's top exporters of manufactured goods, including Turkey, Tunisia, Jordan, Morocco, and Egypt, have significant oil endowments, while the major oil exporters, such as the Gulf

> states, Libya, and Algeria, have lower levels of industrialization […] Although it may deter industrialization, oil wealth - the primary historical path to development - can buy relatively high quality of life, as captured by intraregional differences in human development levels […] All MENA countries that achieve "high" or "very high" ratings on the UNDP Human Development Index (HDI) are low-population oil exporters […] A cross-regional perspective, however, highlights the fact that wealth does not automatically translate into superior scores on all dimensions of human development. For example, the UAE and Portugal have comparable HDI scores (.818 and .816, respectively), yet the former's GDP per capita is nearly twice that of the latter country. Kuwait's HDI value of .790 is slightly lower than that of Uruguay, yet its GDP per capita ($52,793) is nearly four times larger […]. Within the Middle East, some countries, such as the West Bank and Gaza, have higher literacy rates than their income levels would predict, in part thanks to social practices that value education as a key source of upward mobility. To some degree, oil has shaped the economies of all countries in the region, irrespective of natural resource endowments, by fueling regional flows of labor remittances, aid, and, to a lesser degree, private investment. (Cammett, 2018)

The fluctuations in economic fortunes, brought about rising or falling energy prices, are simply enormous. Looking at the evolution of the price of crude oil since Second World War, one is immediately struck by the clearly visible impression of the oil price hikes since 1973, culminating in May 1980, and the second sharp rise in oil prices from the late 1990s onwards, culminating in the price peaks of June 2008 and March 2011.[7] The dawning of the age of "*Fordism,*" i.e., the societal order around cheap oil, the automobile industry, and mass consumption, was aptly described by Volland (1987): long-wave cycles are caused, for the most part, by the rapid growth of key technologies which ultimately strain the supply of natural resources upon which they are based. This phenomenon is transitory in human history and is analogous to the natural growth mechanism of living organisms. Volland (1987) showed these technological feedback loops associated with coal and steel in the post-Civil War period and petroleum, natural gas, and metal alloys after the Great Depression.

Freeman and Louçã (2018) argue in this context that the key inventions that led to the internal combustion engine, to the automobile, the truck, the tractor, and the tank, were made in Europe, but the innovations that led to the assembly line, to mass production and consumption, were made at the Ford plant in Detroit in the early years of the twentieth century. These technical and organizational innovations gave such an impetus to productivity, profitability, output, and exports of the US automobile industry that the plant became a center of attraction for visitors from all over the world and "Fordism" became a dominant management philosophy as well as an economic and cultural ideology, believed to herald the dawn of a "new economy." Freeman and Louçã (2018) think that with the advent of the "Great Depression" of the 1930s, as with previous clusters of innovations, the assimilation of a major new technology into the social system gave rise to enormous political and social tensions and to structural crises of adjustment, both at national and international levels. These tensions and conflicts led ultimately to a Second World War, a motorized war in which tanks and aircraft proved decisive weapons and oil was the essential fuel for the motorized armed forces. The world economy after the war was no less dependent on abundant cheap oil and became a system dominated by the mass production of cars

[7] https://www.macrotrends.net/1369/crude-oil-price-history-chart.

and consumer durables, as well as motor highways, the mass media, mass tourism, mass education, mass culture, and the political and military supremacy of the USA.

In their re-analysis of the evidence of long, Kondratieff cycles in the global economy, Grinin (2016) came to the conclusion that we can observe the following long half a century economic cycles, closely interrelated with leading technologies and modes of energy production and consumption:

- The first K-cycle 1780–1840s: The textile industry,
- The second K-cycle 1840–1890s: Railway lines, coal, steel,
- The third K-cycle 1890–1940s: Electricity, chemical industry, and heavy engineering,
- The fourth K-cycle 1940–early 1980s: Automobile manufacturing, manmade materials, electronics,
- The fifth K-cycle 1980s–~2020: Micro-electronics, personal computers,
- The sixth K-cycle 2020/2030s–2050/2060s: MANBRIC technologies (med-bionano-roboinfo-cognitive).

Quickly reviewing the geo-strategic consequences of these cycles for the Gulf countries, one can conclude with Grinin et al. (2016) the following:

- Firstly, the USA cannot be replaced by any new hegemon that would have the same set of leadership roles that the USA has today.
- Secondly, the weakening of the leadership capabilities of the USA is inevitable and will become more and more noticeable in the forthcoming years.
- Thirdly, the world is to some extent still interested in a sort of soft US leadership, but not in an USA dictatorship, whose mission is to undermine the power of the opponents by any possible means.
- Fourthly, for the transition to the new world order the global community will have to search for new principles and conditions, to create precedents and desired combinations.
- Fifthly, the movement to the new world order is likely to involve a temporary increase of turbulence and conflict, as well as competition between different versions of the new world order.

The growth of disorder at the transition phase, according to Grinin et al. (2016) is very likely. A radical change in the balance of economic power in the world creates objective conditions for a revision of the world order. However, it does not entail the automatic change in the military-political balance. Grinin et al. (2016) argue that the main vectors of the reconfiguration of the world system are the weakening of the old center of the World System (the USA and the West), the simultaneous strengthening of the positions of a number of peripheral countries, and the increasing role of developing countries in general. Grinin et al. (2016) consider the crises and turmoil in the Middle East precisely as such a geopolitical "reconfiguration crisis" that demands changes in the world order. At the same time, emergence of powerful and probably sudden crises in different societies or regions becomes more probable. For Grinin et al. (2016), their unpredictability may be akin to the one of a major earthquake.

In a similar vein, the world systems scholar Podobnik (1999) analyzed the global cycles of wood, coal, oil, nuclear energy, and hydro-energy since 1800.

For Podobnik (1999), the examination of energy production and consumption trends suggests that the global petroleum regime will become increasingly vulnerable to a variety of disruptions in the coming decades. Global energy consumption[8] is concentrated in a few and powerful countries: China (23.8%), USA (16.8%), Russia (5.6%) India (5.2%), and Japan (3.3%) holding a lion's share, followed by Canada, Germany, South Korea, Brazil, and Iran.

For Podobnik, the most problematic trend is the renewed global reliance on oil produced by a small group of Middle Eastern/Central Asian nations. The latest data from the year 2021 corroborate Podobnik's analysis.[9] The 1.65 trillion barrels of proven oil reserves imply that these reserves correspond to just a little more than four decades of current levels of global oil consumption. The biggest reserves are located in Venezuela (18.2%), Saudi Arabia (16.2%), Canada (10.4%), Iran (9.5%), Iraq (8.7%), Kuwait (6.1%), followed by the United Arab Emirates (5.9%), Russia (4.8%), Libya (2.9%), and Nigeria (2.2%).

For Podobnik (1999), natural gas reserves appeared to be similarly concentrated. Again, latest data from the year 2021 correspond to this diagnosis.[10] Russia is in the lead with 24.3%, followed by Iran (17.3%), Qatar (12.5%), USA (5.3%), Saudi Arabia (4.2%), Turkmenistan (3.8%), United Arab Emirates (3.1%), Venezuela (2.8%), Nigeria (2.6%), and China (2.4%).

Podobnik, already in 1999 also highlighted that with OECD countries are currently consuming over half of the world's commercial energy resources, though they contain less than a quarter of the world's population, and thus, there is a material inequality embedded in the contemporary world energy system that is expected to generate increasing conflict in coming decades. Developing countries, Podobnik argues, will turn increasingly to the Middle East and Central Asia for the oil and gas they need to fuel economic growth to the same resources that many Americans, Europeans, and Japanese tacitly assume are reserved for them.

In accordance with this analysis and also with Devezas (2008), we can discern what Devezas (2008), called "*the perplexing result that a very simple logistic model*" that can fit the data. Based on the works of the Italian physicist Cesare Marchetti (*1927), Devezas speaks of a "*fifty-year pulsation in human affairs.*" Primary energies compete much as species in a biological niche. Not only energy systems, but technologies as well, behave like species competing for the same niche.[11]

Finally, Grinin and Korotayev (2012) highlighted that globalization in general (including a fast diffusion of modern information-communication technologies as well as ideas of norms, relationships, and lifestyles that should be regarded as acceptable and desired) played an immense role in the genesis of the Arab Spring events.

---

[8] https://www.worldometers.info/energy/.

[9] https://www.worldometers.info/oil/#oil-reserves.

[10] https://www.worldometers.info/gas/gas-reserves-by-country/.

[11] https://cesaremarchetti.org/bio.php.

> If we look further back in recent history, we will find one synchronization impulse in 1973 - this was a sharp increase in the world oil prices. A rain of petrodollars pored over the Arab World. Naturally, their primary receivers were the oil-exporting countries, but finally every Arab country received some piece of the "oil pie." [...] For example, both Yemen's received some parts of this pie - through the channels of the Arab aid, but also through massive remittances sent back to their countries by Yemeni labor migrants from the Arab oil-exporting countries. Naturally, this contributed in a very significant way to the acceleration of the modernization of the Arab countries. One of the main components of the Arab countries' modernization was constituted just by the development of modern health care systems. Petrodollars helped to construct networks of hospitals, maternity wards, clinics, and so on. As a result, in the 1970s and 1980s in all the Arab countries we observe a precipitous decline of the death rate in general (see Fig. 8), and infant and child mortality in particular, which against the background of still high birth rates led in the 2000s to the explosive growth of the young population (aged especially between 20 and 24) in the Arab Spring countries". (Grinin & Korotayev, 2012)

For Grinin and Korotayev (2012), a power vacuum may happen in virtually any moment; however, major global crises are more typical for downswing phases of the long Kondratieff waves (K-waves). The characteristic period of those Kondratieff waves is, as we already stated above, 50–60 years, accordingly each of those phases lasts for 20–30 years. It is a common denominator of all K-cycle approaches (see also Grinin et al., 2016) that upswing phases are characterized by faster economic growth rates and less pronounced crises; downswings are characterized by a slower economic growth and more prolonged crises. Grinin and Korotayev conclude that in 1830 and 1848, revolutions in Europe began in the middle of the first K-wave downswing and at its end. After 2008, they argue it appears possible to speak about the start of the downswing phase of the fifth K-wave. Grinin and Korotayev (2012) believe that it is not coincidental that this was accompanied by such a large-scale political event as the Arab Spring.

> This phase is likely to continue up to the 2020s. It cannot be excluded that around that time, as a result of the acceleration of economic growth in a number of Tropical African countries (taking into account their "youth bulges") a wave of protests similar to the Arab Spring will roll over some of them [...] Finally, it appears necessary to stress that each of the aforementioned revolution waves marked large-scale changes in the World System. Those waves gave stimuli to large-scale changes (e.g., after the 1848 revolutions one could observe a strong acceleration of capitalist development and emergence of modern political systems in Europe, which strongly affected the World System periphery as well). That is why there are grounds to suggest that the Arab Spring marks the start of a new global reconfiguration. (Grinin & Korotayev, 2012)

Grinin and Korotayev (2012) do not exclude that (as a result of the weakening of the USA and the West in general) one may expect the emergence of a spectrum of influences (including India, China, et cetera).

They also predicted that

> [...] The United States and Europe have virtually betrayed their allies in whom they heavily invested. As regards the Middle East one has an impression that, first, the Israeli–Palestinian conflict loses its primary importance, and, second, that we observe the deepening of the split in the Islamic Middle East. Since quite a long time one could see here a certain antagonism between Shiite Iran (that supported Shiite movements in Lebanon, Iraq, and some other

countries of the region) and the Sunni monarchies of the Gulf led by Saudi Arabia. (Grinin & Korotayev, 2012)

### 4.2.3 *The Gulf and Economic Cycles*

In this context, Cammett (2018) maintained the following hypothesis:

> A prominent explanation for the relative underdevelopment of MENA countries focuses on the "curse" of oil wealth. This argument emphasizes the correlation between resource abundance and outcomes such as poor economic performance and unbalanced growth, as well as weak state institutions and authoritarianism, among other ills [...]. In its economic dimensions, the resource curse centers on the concept of the "Dutch disease," or the theory that an increase in revenues from natural resources will lead to a decline in a country's industrial sector by raising the exchange rate, which makes the manufacturing sector less competitive. Similarly, states that rely on oil or other forms of windfall profits for a large portion of their revenues are deemed "rentier states," which derive their income from non-productive enterprise. These states concentrate their efforts on distributing wealth to the population, often to buy social peace and pre-empt greater societal demands for accountability, rather than fostering the conditions for the productive generation of wealth in their societies. (Cammett, 2018)

and added:

> Beyond oil, recent analyses emphasize the pervasiveness of corruption and "crony capitalism" in MENA political economies, pointing to poor governance as an explanation for suboptimal economic performance. Corruption and wasta, or influence and personal connections, are prevalent features of economic and social exchange in the region, particularly in comparison with OECD countries [...] Inclusive, accountable governance is assumed to produce positive developmental outcomes by increasing popular influence on policy-making, thereby increasing the probability that policies serve the welfare of the people. With growing emphasis on private sector-led development, good governance has attained increased importance. Respect for the rule of law is critical for firms, which require assurances that their assets will not be expropriated and have a chance of reaping good returns before they will invest. Numerous studies document the alleged "governance gap," or the mismatch between governance and income levels, in the Middle East and North Africa. (Cammett, 2018)

Our following figures show the impressive evidence for the existence of shorter and strong 10–15-year cycles in the Gulf region, casting serious doubt on Hanieh's financial and economic "powerhouse" vision of the Gulf countries and rather supporting the more cautious perspective of Cammett (2018), well-informed by the experiences of decades and centuries of periphery capitalism (Figs. 4.1, 4.2, 4.3, 4.4, 4.5, 4.6, 4.7, 4.8, 4.9, 4.10, 4.11 and 4.12).

## 4.3 Are We Ready for "Another Cuban Missile Crisis?"

At the time of the fall of the Berlin Wall in 1989 and the end of the Soviet Union in 1991, it would have been unthinkable that three decades later there is a "*new*

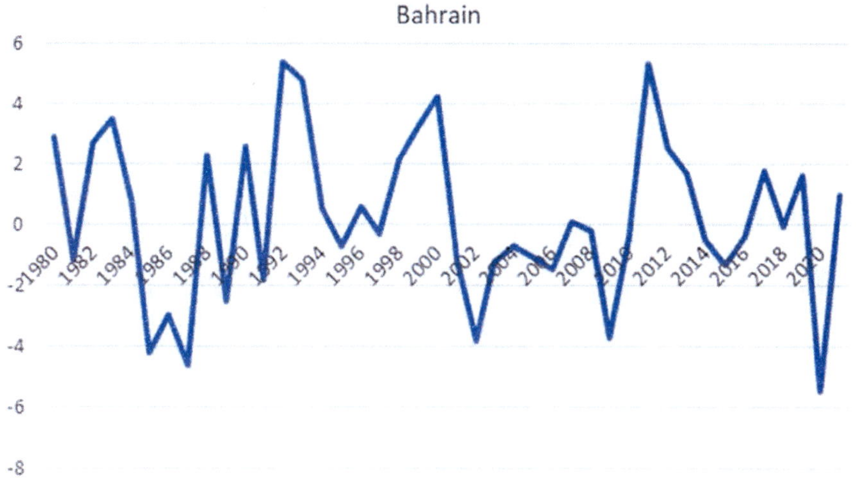


**Fig. 4.1** Bahrain: IMF cycles of economic growth per capita (Gross domestic product per capita, constant prices. Purchasing power parity; percent change)

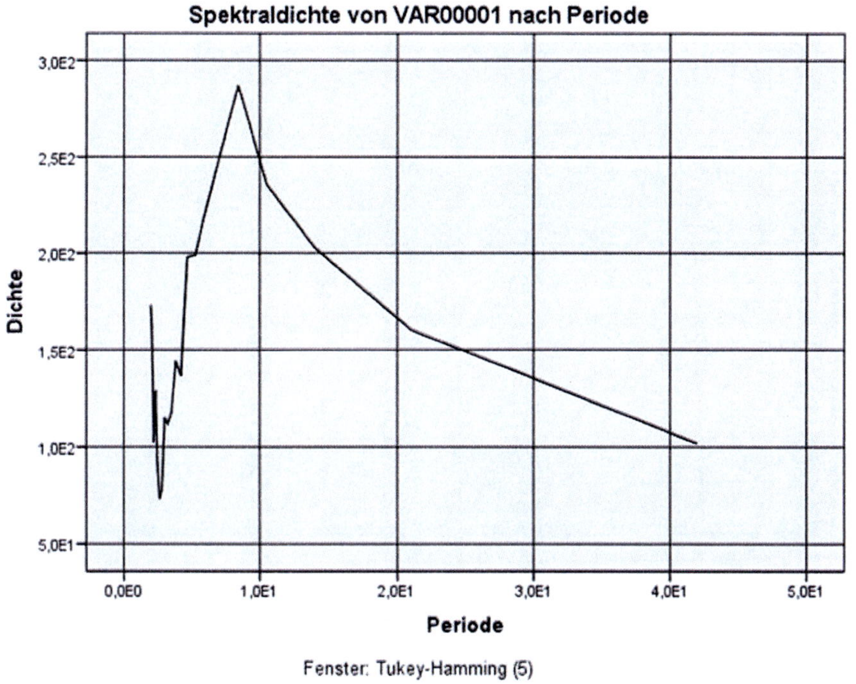


**Fig. 4.2** Bahrain: spectral density analysis of the IMF cycles of economic growth per capita

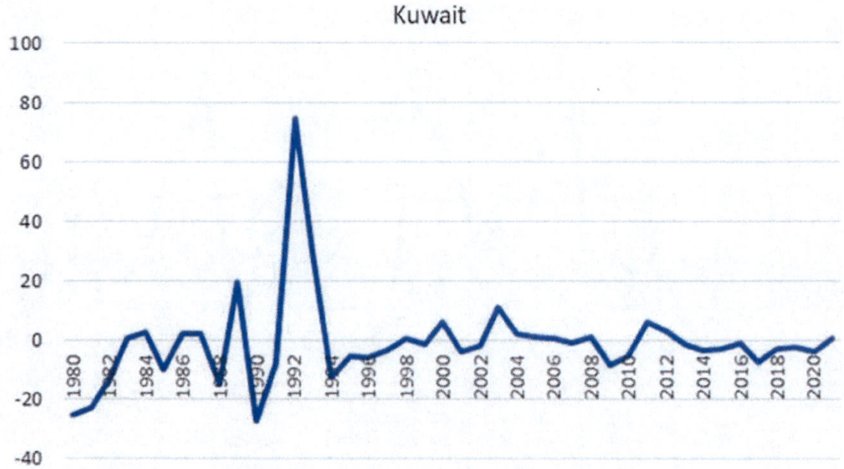


**Fig. 4.3** Kuwait: IMF cycles of economic growth per capita (Gross domestic product per capita, constant prices. Purchasing power parity; percent change)

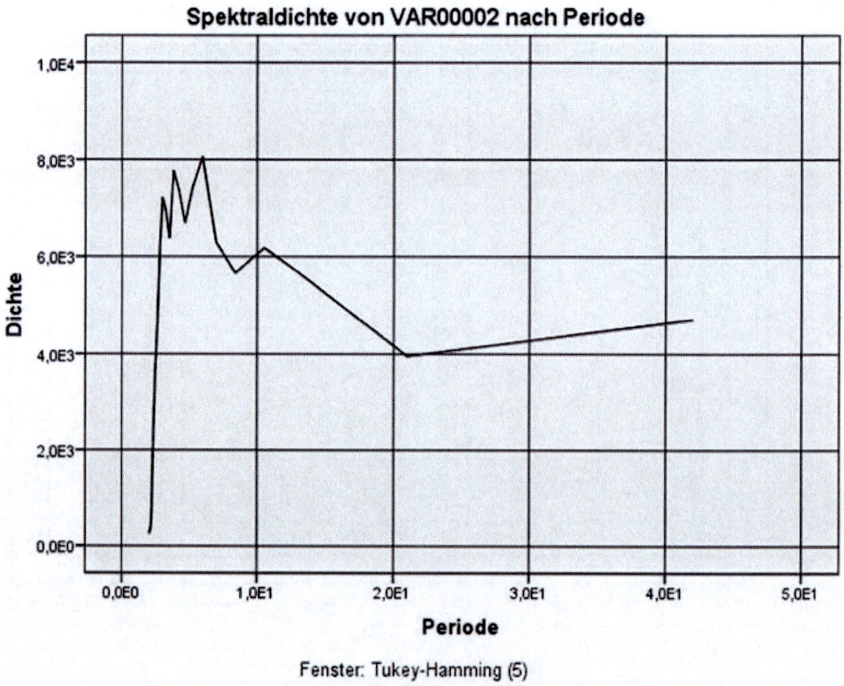


**Fig. 4.4** Kuwait: spectral density analysis of the IMF cycles of economic growth per capita

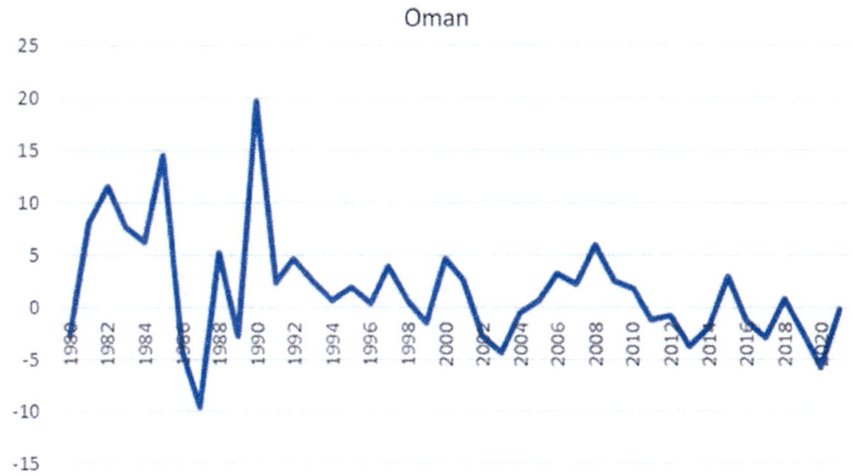


**Fig. 4.5** Oman: IMF cycles of economic growth per capita (Gross domestic product per capita, constant prices. Purchasing power parity; percent change)

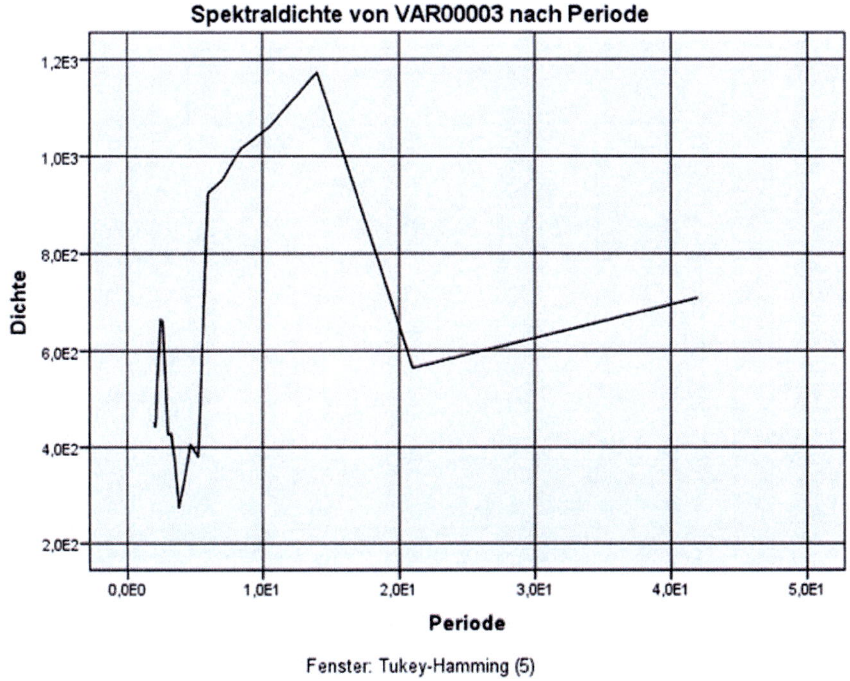


**Fig. 4.6** Oman: spectral density analysis of the IMF cycles of economic growth per capita

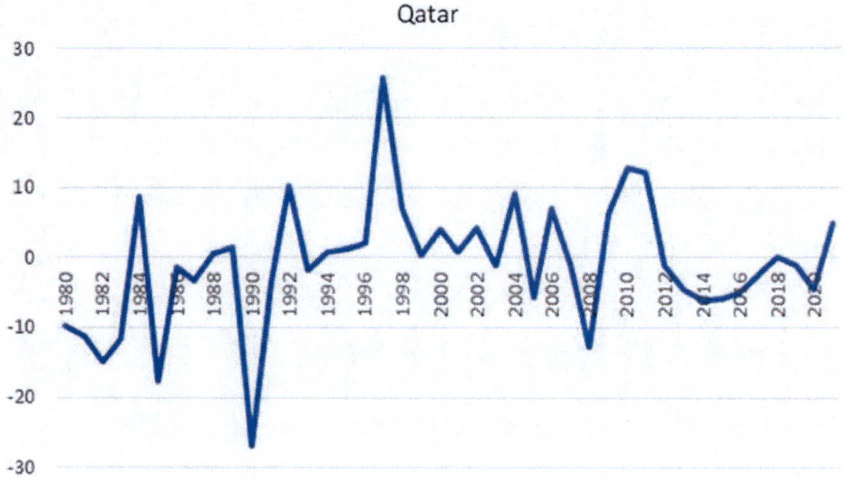


**Fig. 4.7** Qatar: IMF cycles of economic growth per capita (Gross domestic product per capita, constant prices. Purchasing power parity; percent change)

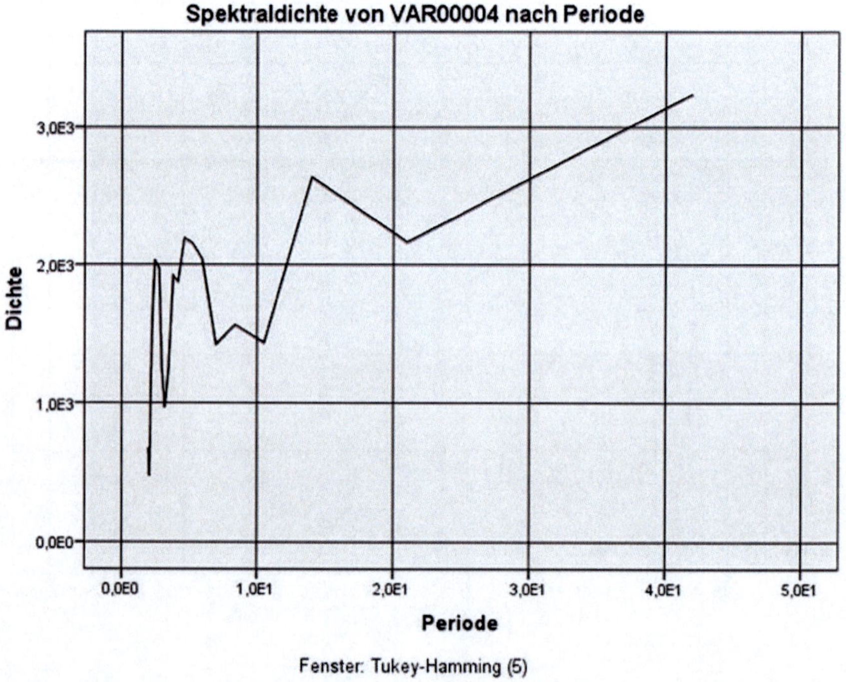


**Fig. 4.8** Qatar: spectral density analysis of the IMF cycles of economic growth per capita

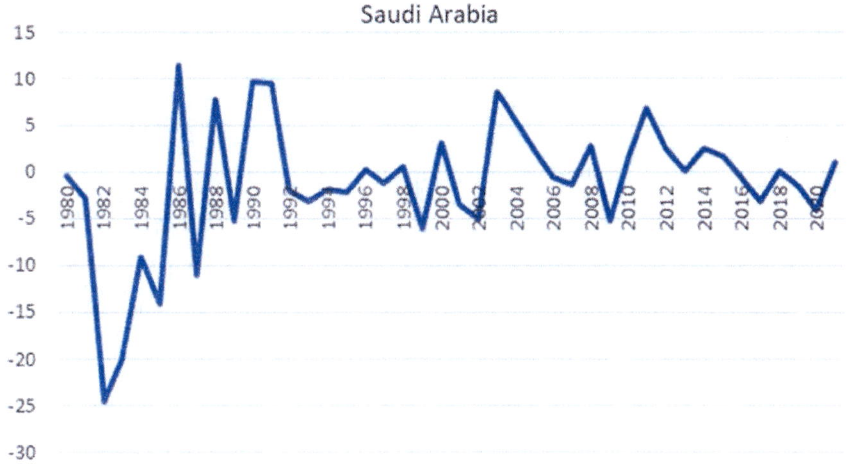


**Fig. 4.9** Saudi Arabia: IMF cycles of economic growth per capita (Gross domestic product per capita, constant prices. Purchasing power parity; percent change)

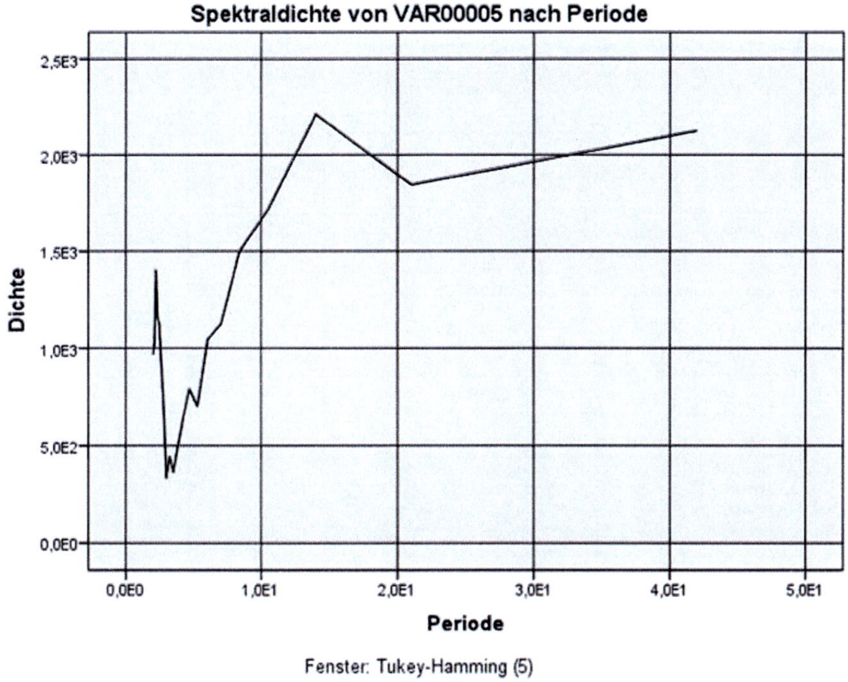


**Fig. 4.10** Saudi Arabia: spectral density analysis of the IMF cycles of economic growth per capita

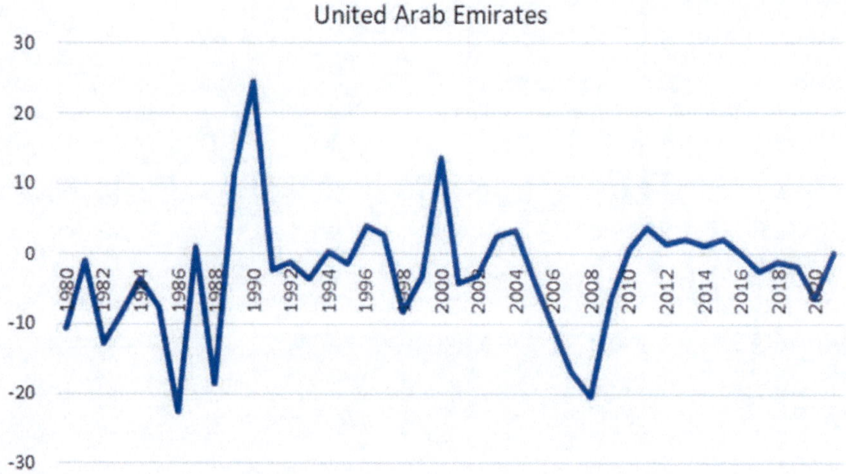


**Fig. 4.11** United Arab Emirates: IMF cycles of economic growth per capita (Gross domestic product per capita, constant prices. Purchasing power parity; percent change)

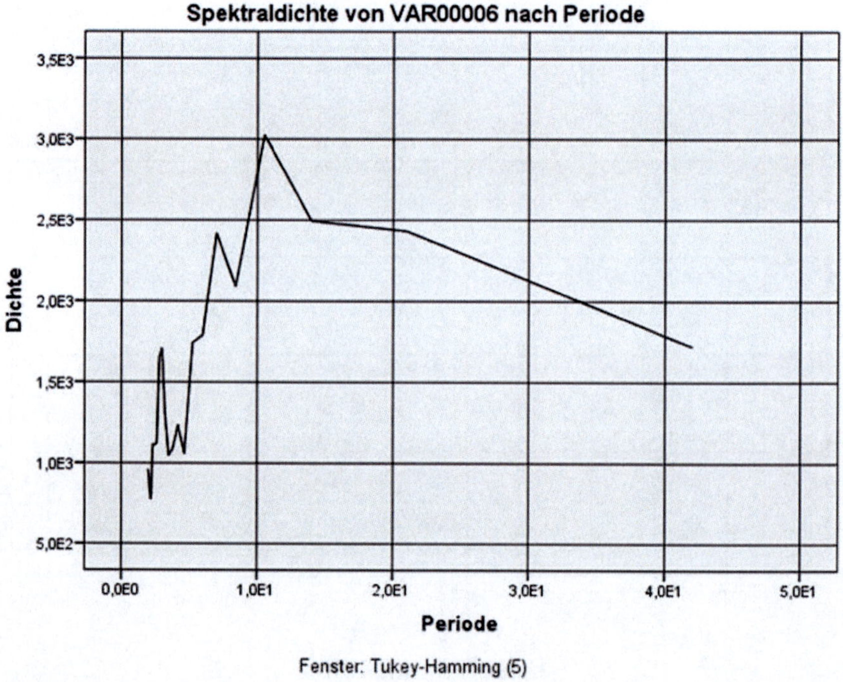


**Fig. 4.12** United Arab Emirates: spectral density analysis of the IMF cycles of economic growth per capita

*Cold War*" between Russia and the West.[12] Not too distantly ago, Russian President Vladimir Putin defiantly even said that he is "*ready for another Cuban Missile Crisis,*" which everybody knows was the worst moment of Soviet-American relations since 1945 and brought the world to the brink of all-out thermo-nuclear war.[13] *Al Jazeera,* on February 21, 2019, reported that

> Russia's military is ready for its own "Cuban Missile Crisis" if the United States is foolish enough to want one and Moscow has the edge when it comes to a nuclear-first attack, President Vladimir Putin said. Putin's comments, made to Russian media late on Wednesday, follow his warning that Moscow would match any US move to deploy new missiles closer to Russia by targeting Western capitals with faster missiles of its own.[14]

And March 20, 2018,

> Chinese president Xi Jinping has delivered a nationalistic speech in which he vowed the nation would "take our due place in the world" and was ready "to fight bloody battles against our enemies".[15]

Not enough that these words were pronounced. The Chairman-for-life Xi Jinping even repeated these warnings in 2020 and told a gathering of politburo officials again that they must "*fight the bloody battle against our enemies.*" He also told a parade of marine troops to: "*focus your minds and energy on preparing to go to war and stay highly vigilant.*"[16]

So, July 1914, all over again? An increasing number of international relations scholars would underline the fact that the current level and bellicose rhetoric of international tensions would have been unthinkable three decades ago (Browning, 2018; Conradi, 2017; Hahn, 2018; Hauser, 2018; Karaganov, 2018; Zhao, 2019).

Were there scholars who predicted all this to happen, even though the majority thought that wars fade away, and the end of history is near? In one of the most disturbing books of the waning twentieth century, the American political scientist Joshua S. Goldstein, in his Ph.D. thesis at Yale University, analyzed the relationship between global economics and global wars with methods of statistics, based on the estimates which were already available at that time of the economic cycles long back into the eighteenth century and wars between the major powers of the world system (the current members of the United Nations Security Council plus Germany) as far back as 1495 (Goldstein, 1988). Goldstein's thesis—that there are not only recurrent long cycles of global economics of around five decades duration, forever associated with the name of the Russian economist Nicolai Kondratieff, but also that there are even more disturbing longer and apparently inescapable cycles of global wars, which

[12] https://www.bbc.com/news/av/world-europe-55356699.

[13] https://www.aljazeera.com/news/2019/2/21/new-cuban-missile-crisis-bring-it-on-says-russias-putin; on the Cuban Missile and the vast literature written on the subject, see Allison (1969).

[14] https://www.aljazeera.com/news/2019/2/21/new-cuban-missile-crisis-bring-it-on-says-russias-putin.

[15] https://www.theguardian.com/world/2018/mar/20/xi-jinping-warns-fight-bloody-battles-chinas-enemies.

[16] https://www.qt.com.au/news/ready-to-fight-china-tensions-rise/4139957/.

already culminated in the conflagrations of the Thirty Years War, 1618–1648, the French Revolutionary Wars, 1793–1815, the German challenge to the global order in the First and the Second World War, 1914–1918, and 1939–1945?

Some global decision-makers, it seems, took these warnings very seriously. In a rare event of cooperation between the NATO Institute for Advanced Studies and the Russian Academy of Sciences (Devezas, 2006), these gloomy predictions were further analyzed in-depth in one of the most memorable books for our turbulent times, now available freely on the Internet. Independently from this unique volume, a number of other world systems scholars reiterated these findings as well and always came back with pretty much the same disturbing predictions.

Goldstein (2005) re-iterated his "*conflict clock*" published in 1988 (Goldstein, 1988: 259). Goldstein was clear about his prediction of the 2020s (i.e., the current era!) as the next real danger zone of world politics: "*In my 1988 book I pointed to the period around the 2020s as a potential danger zone, and it still worries me*" (Goldstein, 2005: 8).

Boswell (1999), in turn, presented a list of ten system evolution steps, starting by the relative decline of the USA, followed by imperial rivalry and global war which the late Terry Boswell expected then between 2010 and 2020, to be followed by another world system sequence, ending by 2080 or 2090 in yet another cycle of relative decline of the hegemon who won the last World War by 2010/2020 ... Maybe his cycles will happen later than expected, but what a frightening scenario.

Work by Attinà and Modelski also suggested that we most probably will not escape the fatal cycle of global leaderships and global contenders. Since the mid-1960s, the defense pact aggregation index which measures the percentage share of defense pact members in the total number of states in the international system, i.e., the control that existing, established mechanisms of world political leadership exercise over global politics, has declined, suggesting that the era of global power by the USA, which was established in 1945, definitely comes to an end and that our era is pretty similar to the era 1850–1878, which was characterized by the de-legitimation of the then British leadership, followed by the de-concentration of the international system and the era of coalition-building between 1878–1914, which ended, as we all too well know, in the catastrophe of 1914.

In the present chapter, we take up these issues anew and re-analyze latest conflict data (great power battle fatalities from all wars, Goldstein (1988) and Peace Research Institute PRIO Data (2020) from 1495 to 2008[17]; and data about the growth of world industrial production 1740–2020, which we collected especially for this volume (see Chap. 3). The database for this Chapter and its predecessor versions have been debated in the literature at length, among others in Lacina and Gleditsch (2005), Lacina et al. (2006), Peace Research Institute, Oslo (PRIO) (2021), Sarkees (2000). Thus, we can assure our readers, who might be perhaps not too familiar with debates in the field of global history that our data or rather estimates as they reach far back in history are considered to be relatively "watertight" by the expert community.

[17] Later data are not available.

It can be shown that the long Kuznets and Kondratieff swings and cycles of capitalist world development that play such an important role in the analysis of global war since 1495 (see Grinin et al., 2016) have indeed not ended after the end of Communism in Europe, and that instability, and not stability, characterizes the world economy, and that there is an "W"-shaped pattern of global conflict since 1495.

As we all know, there were voices that predicted the "*end of history*" in the 1990s (Fukuyama, 1991). The world systems approach always took a more cautious line and did never preclude—at least in principle—a repetition of the insane economic cycles and major power rivalries that characterized the course of history since 1450 and which produced the three mentioned devastating global wars, 1618–1648; 1793–1815; and 1914–1945 (see the by now classic contribution by Goldstein, 1988).

In view of the methodological critique by several authors, including Beck (1991) and Silverberg (2005a, b), against this "*long wave school*" it would seem appropriate rather to talk about "tendencies" or "waves" and not about "cycles. Silverberg's research paper, 2005 applied spectral density analysis to two famous world system time series, Modelski's and Thompson's sea power index and the Levy–Goldstein great power fatalities data series. However, we try to show by using Kondratieff's classic techniques of polynomial regressions as well as by applying spectral analysis and autocorrelation analysis that there are indeed several cycles, and not just one cycle of global economics and politics.

### 4.3.1 The Gulf and the Waning US Hegemony in the World System

Of crucial importance for the GCC countries is the strategic partnerships with the USA. The Israeli Gulf states expert, Yoel Guzansky, whose analyses feature prominently in this book, says:

> Enhancing cooperation with its allies in the Gulf has always been a stated goal of the United States, which played a major role in the formation of the GCC. Paradoxically, however, the important US security role in the Gulf, especially since the early 1980s, has detracted from the six Arab Gulf states' ability to cooperate thoroughly and effectively in the GCC framework: Washington prefers to work directly with the respective capitals on substantive issues. The Arab Gulf states realize that at the present time, there is no attractive substitute for the US security support, but they have doubts about the reliability of the US political commitment. They are therefore also trying simultaneously to step up cooperation with Russia and China, while at the same time taking into account US sensitivities. Better ties with the Gulf states [… are …] also an important goal for Russia. Moscow especially regards Saudi Arabia and the UAE as key Middle East countries, and believes that ties with them will help it advance its political and economic interests. As the political and security weight of Beijing and Moscow in the Gulf grows, the Arab countries there will find it increasingly difficult to maintain this delicate balance.

According to Reuters Factbox (2020), we currently have the following presence of US forces in the Gulf region.

***Bahrain***: US personnel: 5000, The US base, known as Naval Support Activity Bahrain, houses the headquarters of the US Fifth Fleet and US Naval Forces Central Command (USNAVCENT), which reports to US Central Command (CENTCOM). The Fifth Fleet is responsible for the Gulf, the Gulf of Oman, Gulf of Aden, Red Sea, and the Arabian Sea. The region includes three critical chokepoints for shipping—the Strait of Hormuz at the southern end of the Gulf, and the Suez Canal and the Strait of Bab al Mandeb at the northern and southern ends of the Red Sea, respectively.

***Kuwait***: US personnel: 14,500. Kuwait has a large US military presence. The USA uses bases in Kuwait as staging hubs, training ranges and logistical support for regional operations. Bases used by the USA include Camp Arifjan, Camp Buehring, Ali Al Salem Air Field, and the naval base Camp Patriot.

***Oman***: US personnel: a few hundred. The USA clinched a strategic port deal with Oman in March 2019 that would allow the US military better access the Gulf region without having to send ships through the Strait of Hormuz, a maritime chokepoint off Iran that Tehran has threatened to mine. Oman has close defense and security ties with both Britain and the USA. A joint training base with Britain was established there in 2018. Under a 2010 security pact, the USA was allowed to retain a small military presence. It also granted US forces access, with advanced notice and for specialized purposes, to military facilities in Masirah, Muscat, and Thumrait, as well as allowing the USA to pre-stage munitions there.

***Qatar***: US personnel: 10,000. Qatar is home to the Al Udeid Air Base, which includes the forward headquarters of US Air Forces Central Command (AFCENT) as well as the 379th Air Expeditionary Wing, which supports operations throughout the US Central Command area of responsibility and supports the Joint Task Force-Horn of Africa.

***Saudi Arabia***: US personnel: 2500. In October 2019, the US deployed about 3000 troops to Saudi Arabia, including fighter squadrons, an air expeditionary wing, and air defense personnel, at a time of heightened tensions with Iran. The main base is Prince Sultan Air Base, south of Riyadh. US troops also use other Saudi bases. Before the US–Iranian confrontation worsened, US Central Command had 500 people in the Kingdom.

***United Arab Emirates***: US personnel: 5500. More US Navy ships visit the port at Jebel Ali, which can handle vessels up to the size of nuclear-powered aircraft carriers, than any other port outside the USA. Al Dhafra Air Base hosts several US fighter, attack, and reconnaissance aircraft of the US 380th Air Expeditionary Wing.

In recent world systems research literature, Denemark (2018) took up the debates in the wake of Goldstein's work in the 1980s (Goldstein, 1985, 1988) anew and arrived at the conclusion that particular attention has to be paid to the way in which world-systems developments that increase the likelihood of major war have persisted, while those that retard the chances for major war have diminished. These dangers suggest that it may be time to turn some of our attention anew to the dynamics of systemic war and nuclear weapons. Already back in 1996, Pollins, predicted that conflict, likely led by core powers, would reach its zenith in a relatively narrow band of twenty years between 2030 and 2050. Wallerstein (1989) warned of a high probability of conflict around the year 2050. Denemark (2018) consequently underlines Wallerstein's emphasis that war would likely emerge as a result of the decline of the USA and a subsequent struggle for power among competitors. In his synthesis of the debates between major world system scholars on the issue, Denemark (2018) also underlines Arrighi's (1994) argument that major power wars are most likely to emerge in the context of an over-accumulation crisis when declining returns to production and trade drive funds into financial markets.

> Hence, war is associated with downturn and the decline of each major system leader, which will generate increasing interstate competition, tension, and war. (Denemark, 2018).

Denemark interprets the works of Arrighi (1994) and Modelski and Thompson (1996) as suggesting that conditions conducive to global war will emerge about 2030–2040. Pollins (2008) again reviewed the abundant literature in international relations theory on the subject and upheld the empirical relevance of the view, based on Goldstein (1985, 1988) that rising competition in economic and security domains coupled with rising military capabilities culminated, in a systemwide war in which all major powers participated. Such conflagrations, Pollins (2008) in accordance with Goldstein (1985, 1988) argues, clearly affected the global economy and the occurrence of these great wars coincides with the price peak in the Kondratieff long wave. There is an important reciprocal relationship between security and economy at the global level. Economic growth engenders conflict while large wars spur production, innovation, and price inflation. In 1996, Pollins came to the conclusion that

> While the existence of cycles is likely to remain a point of scholarly debate (as it should), the association between these purported cycles and important events in the life of the international system finds increasing empirical support. This study's findings are two. First, based on theoretical arguments found in the core works and new empirical evidence presented here, I have shown that the explanatory domain of hegemony, world leadership cycle, and long wave studies may be fruitfully extended beyond the realm of the great powers to include all states in their investigations of armed conflict. The arguments and knowledge claims of each of the core frameworks were shown to have measurable promise in explaining various levels of conflict for minor and middle as well as great powers. Second, a new model that builds on these frameworks was shown to be superior empirically. The coevolving systems model takes separate account of the contributions to armed conflict made by the long wave and the world leadership cycle, and its success (despite its crude simplicity) points to a new and promising research direction. The improvement in empirical performance shown by this model against existing frameworks was not marginal but significant and occurred under a variety of circumstances and assumptions pertaining to the level of conflict and the nations included. (Pollins, 1996)

Analyzing the linkages between economic cycles and US international conflict involvement, Pollins and Schweller (1999) concluded that global stagnation is a time of little economic energy (both in terms of the level of activity and the rate of growth). It is an inwardly oriented period; foreign engagement is low; and social unrest is high. This picture, Pollins and Schweller (1999) contend, changes in many ways during global "*Rebirth.*" Economic growth rates rise (though the level of such activity remains relatively low since levels are the integral of rates, and the economy is still recovering from Stagnation). Growth in technology and resource use peak at this time. This, according to Pollins and Schweller (1999) is also an inwardly oriented time for the USA. The frequency and probability of US entry into armed disputes is low, and the probability that US foreign policy is

> [...] introverted is strongest during this phase. Global Expansion brings with it a heady atmosphere. US economic growth rates, though a shade lower than the previous period, remain clearly positive. The level of economic activity is higher, reflecting the accumulation of years of good growth. Foreign-policy orientation now makes its largest shift in the entire cycle: from an 82 percent probability of "introversion" [...], we now observe an 86 percent probability of "extroversion." Naval expenditures as well as the probability and frequency of US entry into armed disputes also rise measurably during this time and continue into the next Long Wave phase. Global Stagflation brings significant shifts in the US economic, political, and social environments. Though the economy has expanded to much higher levels than observed at the beginning of the cycle, rates of growth are now lowest, and price inflation peaks-disturbing news to those accustomed to the boom times of recent memory. Export growth peaks, perhaps in an attempt to compensate for slower growth at home. Foreign-policy orientation is likely to remain extroverted, though measurably less so than during the Expansion phase, and the frequency and probability of U.S. entry into armed disputes continues to be high. Linked, possibly, to this soured economic picture and to foreign involvement, social unrest rises measurably. (Pollins & Schweller, 1999)

The recurrence of major power wars in the capitalist world economy from 1495 to the present is one of the most intriguing features of the existing international system. The *x*-axis in our Figs. 4.13, 4.14, 4.15, and 4.16 presents the periods before the major power wars, i.e., 1648, 1816, and 1945. Each world political cycle up to now corresponded to an indented "W"-pattern of untransformed annual battle fatalities from major power wars in thousands of human beings (Fig. 4.17).

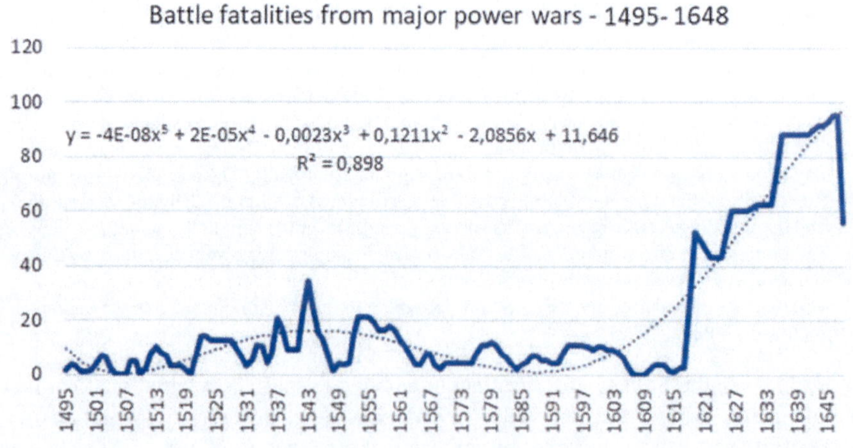


**Fig. 4.13** Battle fatalities from major power wars, 1495–1648

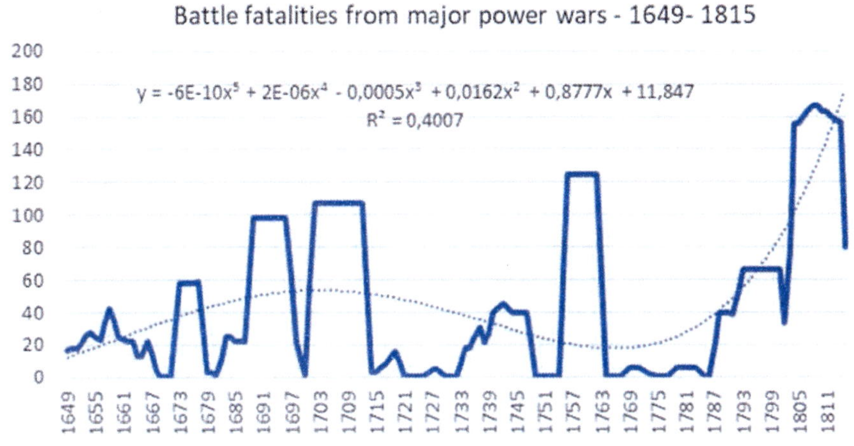


**Fig. 4.14** Battle fatalities from major power wars, 1649–1815

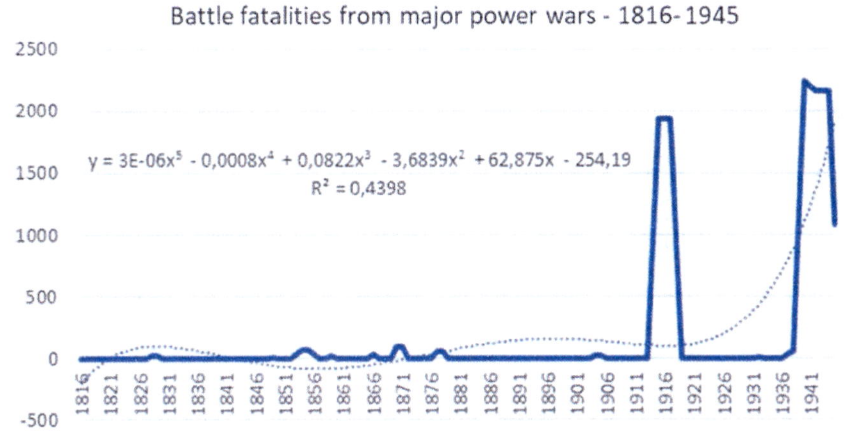


**Fig. 4.15** Battle fatalities from major power wars, 1816–1945

## 4.4 Global War Cycles, 1495–2008

The untransformed data for the war cycles 1495–2008 are statistically well-explained by a polynomial expression of the fifth order over time:

1495–1648: 89.9%
1649–1815: 40.0%
1816–1945: 44.0%
1945–2008: 21.6%; tenth root: 52.2%.

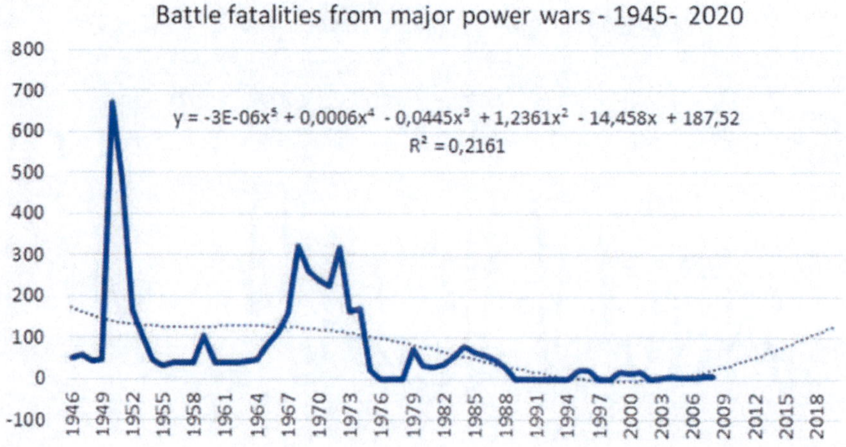


**Fig. 4.16** Battle fatalities from major power wars, 1945–2020

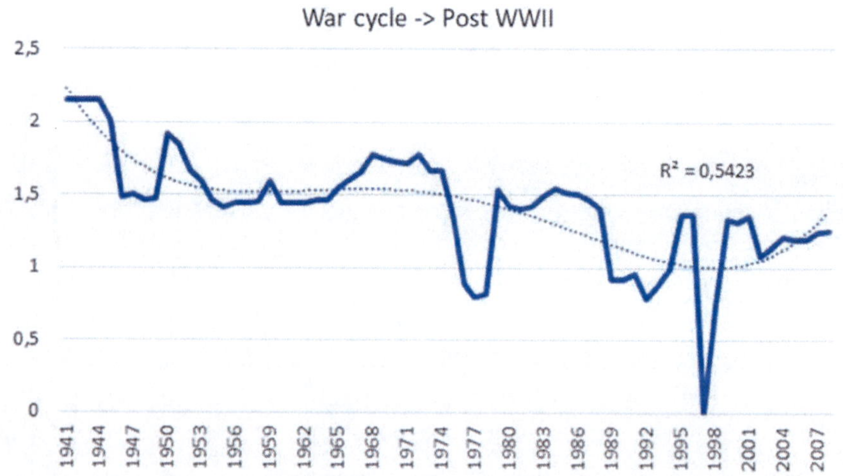


**Fig. 4.17** Battle fatalities from major power wars, 1945–2020 (tenth root of battle fatalities)

The $R^2$ increases considerably, if we take the tenth root of battle fatalities from major power wars:

The international system seems indeed to be characterized according to Goldstein by the following sequence of cycles:

- Global war,
- World hegemony of the dominant power,
- De-legitimization of the international order,
- De-concentration of the global system,

- Global war, et cetera.

The duration of these sub-phases of the international order is approximately one Kondratieff cycle each (an economic cycle of 50–60 years duration, see below), so the unit of time of the international system can be symbolized by the expression—one Kondratieff cycle—1K. An entire hegemonic cycle lasts three Kondratieff cycles.

Different sub-schools, like Goldstein, Russett, Weede or the present author thought it unlikely that a major confrontation between the centers of "world capitalism" might characterize the twenty-first century, and that there are other, more likely candidates for such a terrible scenario. It is important to remember that there are pre-conditions for the "democratic peace" between the major centers of global power, mainly the maintenance of democracy in the countries concerned. The late Immanuel Maurice Wallerstein thought it entirely likely that in the end we will be confronted with a major warfare between Europe and Asia in the twenty-first century (Wallerstein, 2000).

Although this might be termed an absolute exaggeration of projective social scientific writing, a more careful re-reading of the world systems approach—especially the writings of Giovanni Arrighi, George Modelski, Andre Gunder Frank, and Immanuel Wallerstein—does suggest that the twenty-first century will be characterized by a further shift of the world economic centers of gravity towards Asia (see Chap. 3, above), and that the maintenance of the peace of enlarged Europe to the outside world will depend on European Union democracy, technical innovation, and the avoidance of what Arrighi has called—in allusion to the concept of power of historical Venice—a "*terra ferma*" constellation, i.e., the territorial concept of power and the exploitation of the hinterland—*the terra ferma*. In the history of the world economy, major geographical shifts of economic activities and of military power—like in 1340, 1560, 1750, and 1930—were always associated with major wars and with a very deep economic depression.

We agree with Modelski (1999) that global challengers in the world system were always characterized by the interaction of (a) a large army, (b) a large economy, (c) a closed, controlled society, and (d) weak, ethnocentric media. Russett's hypothesis (1994) about the great probability of peace between democracies is an all-important, further element in this debate.

At any rate, the future of the open society in Europe, the exact fulfillment of the Copenhagen criteria of a functioning democracy and market economy in the EU enlargement process, and the future democratization of the Union—also in face of the darker sides of the European heritage and the very idea of an economically united Europe under authoritarian premises under Nazi rule before 1945 (Laughland, 1998)—become decisive whether or not Europe will become a global challenger in Modelski's sense and will determine whether Europe is a new challenger of the new evolving global leadership along the Pacific axis that could fit into the pattern, described by Modelski, as including (a) an oceanic navy, (b) lead industries and fiscal strength, (c) a democratic potential, party system, and (d) a strong active media.

A relatively closed society, combined with high customs, a reliance on military land power, exploitative relations with the internal and immediate external peripheries (the hinterland of Northern Italy before the great Italian wars of the thirteenth and fourteenth centuries, the Spanish colonies and internal peripheries in Hapsburg Europe before the global war 1618, the French internal peripheries and colonies before the Napoleonic Wars, and Germany's "Drang nach Osten" before 1914) and a relatively weak technological home-base characterized the attempts at world power, while the successful world hegemons (the United Provinces of the Netherlands, the UK, and the USA) were each time the practical opposites on all or most of these dimensions: a naval, knowledge-driven, and world-market concept of power, a society open to migration, low customs, a large sea power and a smaller, but mobile and disciplined and readily deployable land-army, and a strong technological home-base.

Significant for hegemonic success each time was also the strong social role of the urban merchant class that was in stark contrast to the ties between the land and capitalism, characteristic of the "rentier" political economy of Hapsburg Europe, France, and Germany, the historic challengers that lost each world leadership contests from 1618 to 1945.

World hegemonies that characterized the workings of world capitalism arise, and they also end. As it is well-known in world system research, especially from the works of Arrighi and Silver, there are signal crises of world capitalism (the usual Kondratieff depressions), and there are terminal crises of the world system, when hegemonies end. Peaceful transitions from one hegemony to the other are among the most intricate questions of peace research and peace policy of our time.

These moments of world hegemony transformation, as Arrighi and Silver (1999) and as Boswell (1999) have shown, are especially dangerous culminations of world depressions and are characterized by a subsequent fight for world hegemony, like during the great crash of the early 1340s, which marked the beginning of the Genoese age (Arrighi) or Portuguese and Genoese age (Modelski), the crash of the 1560s, which marked the beginning of the Dutch era, the depression of the 1750s and 1760s, which marked the beginning of the British era, and the Great Depression in the 1930s, which was the terminal crisis of British world capitalist dominance (Arrighi, 1995). Regulation can be successful, like after 1560 and 1930, and deregulation can be successful, like after 1340, 1760, and—most probably—the 1980s (compiled from Arrighi, 1995).

The following historical data are equally important in this context: 1648: Treaty of Westphalia, the consolidation of Dutch hegemony; 1797: The British fleet defeats the Dutch fleet at Battle of Camperdown. Britain establishes its hegemony. 1919: The British "century" came to an end only when the World War exhausted its energies and drained its reserves. 1945: the cycle of the US hegemony begins. A world hegemony evolves and declines during at least two Kondratieff cycles of each leadership cycle. The risk for such a crisis rapidly increases after 2020 or 2030.

There seems to be ample evidence, reproduced, among others, in Tausch (2003) that the terminal crises of capitalism, like tsunami waves of world politics, have devastating consequences for the well-being of the great majority of humanity. Evidence based on the works of Andre Gunder Frank and Rudolph Rummel suggested that as a consequence of the terrible world depressions of the 1350s (coinciding with the Black Death) and 1750s, Europe lost significantly its share in total world population, while the earthquake of world politics 1900–1950 had an estimated 187.7 million victims in terms of political repression, genocide and democide, i.e., a good part of all the estimated total victims (482 million human beings) of democide, wars, and repression in total human history. Among the major world systems researchers of our time, the late Andre Gunder Frank was most active in championing the idea that there are major economic cycles dating back 5000 years in human history. This idea was put forward by him, for example, in 1994, in an important paper which he presented to the New England Historical Association. A more final word by Frank on these large cycles is to be found, among others, in his "ReOrient" (1998): 248ff. So, where are we now? 1870? 1913? 1938? World systems theory is full of speculation about the future, and much of world systems research writing projects a major global war by around 2020 or 2030.

Attinà interpreted the global political cycles of the world system in accordance with Modelski as:

- Global war macro decision,
- Global power execution,
- De-legitimization and agenda-setting,
- De-concentration and coalition building,
- Global war macro-decision et cetera.

The defense pact aggregation index neatly measures, Attinà argues, these cycles. In Tausch (2007), there is a comparison of these cycles with the long swing of world economic development, Tausch (2007) came to the conclusion that these two cyclical movements are roughly in parallel. The tsunami crisis of world politics precedes the tsunami economic crisis? Tausch (2007) discussed the evidence about global political leadership and the long cycles of world development: declining growth rates and world political de-concentration. This was shown for the period after the Congress of Vienna 1814–1815 and also by analyzing the world political de-concentration and economic stagnation (B-phases of 100-year logistic cycles in the tradition of Wallerstein)—for the period after the Congress of Vienna 1814–1815.

Partially contradicting the liberal globalist optimism inherent in the analysis by de Soysa and Gleditsch (2002), it could be shown already in Tausch (2007, 2010), that the retreat of democracy and human rights on a global scale will most probably go hand in hand with a more conflict-oriented foreign policy in the years to come. Under the weight of globalization and the openings of markets, the weak economies and societies of the former communist world and several democracies of the south witnessed a long-run retreat of civil liberties. Recently, researchers have dealt with the specific aspects of the long-term relative decline of the USA in global politics (Chase-Dunn et al., 2019). But at this stage, it is also important to look still more profoundly into the cycles of reform and regime tightening in Eastern Europe. Just as during the world depression of the 1930s democracy could not survive in the region (Polanyi, 1944), today the danger arises that instability and not democratization will triumph in the end, especially in countries like those of the former USSR, and even in the new member countries of the European Union in Eastern Europe. The turning points in the long waves between the ascents and decline phases (B-phases) were always the beginnings of political decay in the region, while the ascent phases were associated with authoritarian modernization; and the time-lags between the Western cycle and the Eastern semi-periphery and periphery have to be taken into account. The decisive-Kairos-years in earlier periods were (dates before 1756 from Goldstein, 1988): 1509, 1539, 1575, 1621, 1689, 1756, 1832, 1885, 1932, and 1975.

The former hegemonic contenders from earlier global wars slowly slide into an acceptance of their status in the international system. The real power struggle erupts already soon after the great hegemonic war, and through the ups and downs of the history of the system evolve slowly into the hegemonic challenge. Seen in such a way, not 1989, but Korea and Vietnam could become rather the benchmarks of the future W-structure of conflict in the international arena (see Tausch, 2007, there Table 3: Projected Logic of World Power Competition 1495–2002). The triumphalism of America being the unrivaled military hegemonic power of our world notwithstanding, we believe that a serious strategic and political science analysis quickly reveals the direct manpower and other military constraints against continued US military and political dominance in world affairs (already predicted by Adams, 2005).

Our final synopsis about world system tendencies towards peace or towards conflict do not support the generalized hypothesis that with global capitalism peace is "around the corner"; our Figs. 4.18 and 4.19 show the income convergence trajectory of the hegemonic challengers from 1700 to 2020.

Serious quantitative analysis cannot overlook the tragic W-pattern of global war that was already established by Joshua Goldstein in his doctoral thesis in 1988.

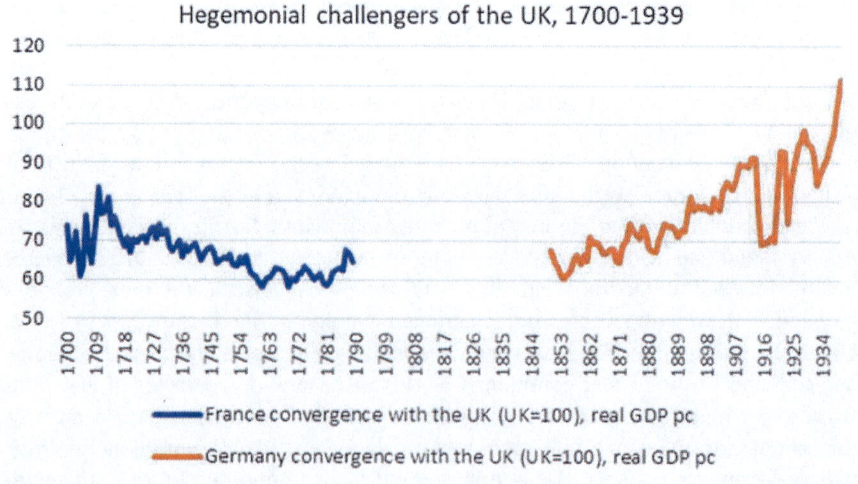


**Fig. 4.18** Hegemonial challengers of the UK, 1700–1939—income convergence/divergence with the UK

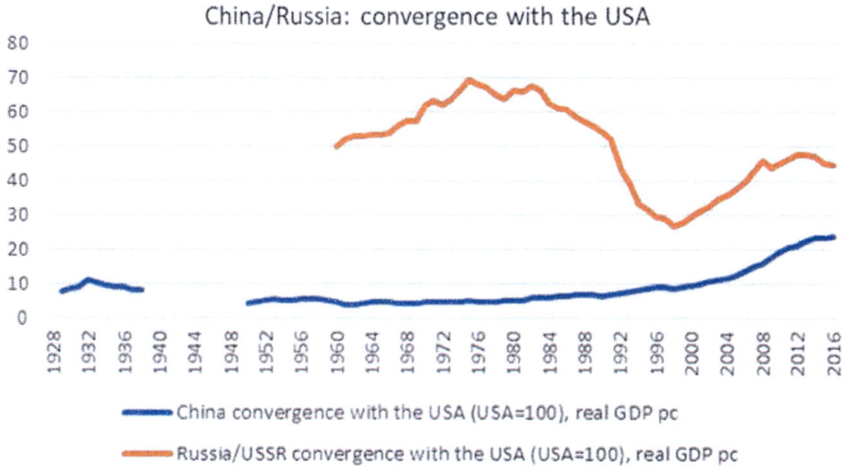


**Fig. 4.19** Hegemonial challengers of the USA, 1928–2020—income convergence/divergence with the USA

New evidence seems to suggest that this cycle has not ended with the end of the cold war, and that preventive social action against "the globalization of warfare" is needed to assure for us, for our children and our habitat a peaceful co-existence in the twenty-first century.

## 4.5 Time Series Analysis Results—The Cycles of Growth and Death

A first glimpse of the grim realities ahead of us is to be seen, when we correlate the data about relative GDP per capita in PPP (UK, 1913 = 100) of the countries of the world system, contained in Barro and Ursúa (2011) and Grinin (2016) with the country death rates of the Great Influenza Epidemic, 1918, documented in Barro et al. (2020). The suggestion is that poorer countries were hit harder by the influenza epidemic of 1918 than countries of the center, although the effect is nonlinear due to the higher death rates in some richer countries. The IBM-SPSS XXIV calculation of the relationship between the natural logarithm of GDP per capita in the last peace year before the First World War in 1913, UK = 100 and its square and the influenza death rate per country in 1918–20 in the 31 countries was significant at the 1% level, the $F$ ratio is 8.46; the adjusted $R^2$ is 33.2%, and the predicted influenza death rate was 138,860 − 61,790 * ln (GDP per capita PPP, UK, 1913 = 100) + 0.7160 * ln (GDP per capita PPP, UK, 1913 = 100)$^2$. The human cost of each pandemic and each depression in the world system is tremendous.

As the world is now in the midst of the COVID-19 emergency, it is also time to remember that already after the economic crisis of 2008, we could witness increases

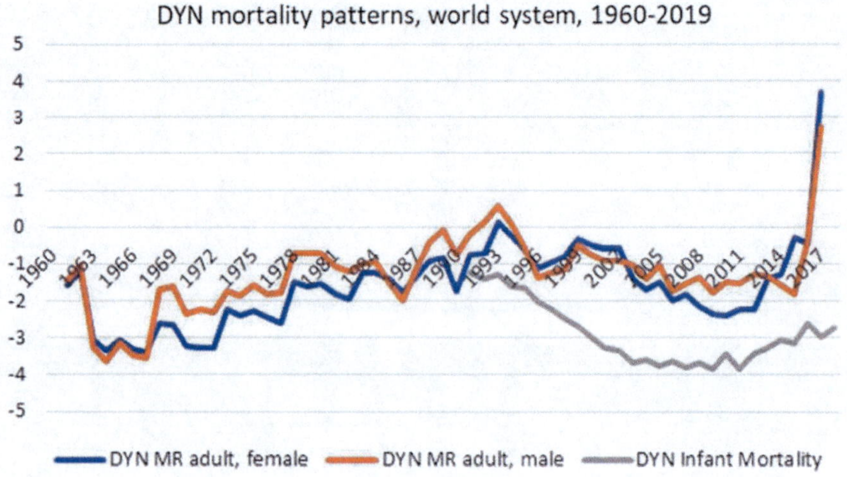


**Fig. 4.20** Mortality increases/decreases in the world system, 1960–2019

in female and male mortality rates and infant mortality rates, as evidenced by the Open Data series of the World Bank. Similarly, rising global female and male mortality rates and global infant mortality rates are to be expected for the coming years as a direct result not only of the COVID-19 pandemic, but also of the economic losses connected with the pandemic and all the lockdowns which happened since early 2020 (Fig. 4.20).

### *4.5.1 Patterns of Global Cycles, 1740–2020*

Sayrs (1993), in one of the most sophisticated articles ever written on the question of global cycles, published in the "*International Studies Quarterly*" came to the conclusion, based on the mathematical model of Markov chains to the empirical data of the conflict intensity and economic cycles, discussed at length in Chap. 2 of this book:

> The theoretical appeal of a periodic cycle is strong, but there is greater cause to believe that whereas war behavior is repetitive, it is not periodic. The existence of a war cycle does not in any way diminish the influence of effects that may serve to alter it. In other words, there is nothing deterministic or mechanistic to my notion of a war cycle or my specification of it as segmented trends. Rather, understanding the war cycle means uncovering a cycle in the presence of the multiplicity of effects that serve to stretch and collapse the intervals of time between cyclical episodes. (Sayrs, 1993)

Sayrs also concluded, based on his analysis of the data with the aid of Markov chains:

**Table 4.1** World depressions (DYN world industrial production growth, WIP, in %), 1770–2020

| | World depressions due to pandemics (DYN WIP) | Pandemic |
|---|---|---|
| 1770 | −3.1 | Russian Plague |
| 1773 | −5.9 | Russian Plague |
| 1774 | −6.2 | Russian Plague |
| 1793 | −13.6 | Philadelphia Yellow Fever |
| 1819 | −8.3 | Cholera |
| 1857 | −2.1 | 3rd Cholera, 3rd Plague |
| 1875 | −2.5 | Fiji Measles |
| 1893 | −3.3 | Russian Flu |
| 1918 | −5.9 | Spanish Flu |
| 1919 | −8.3 | Spanish Flu |
| 1921 | −14.1 | Spanish Flu |
| 1958 | −2.6 | Asian Flu |
| 1982 | −1.0 | HIV Aids |
| 1992 | −0.3 | HIV Aids |
| 1993 | −0.4 | HIV Aids |
| 2001 | −0.1 | HIV Aids |
| 2020 | −8.8 | COVID-19 |

> But, despite my efforts which show that a Kondratieff is not at work for war, prices, or war and prices jointly, much work remains to understand the shorter cycles I do find. Future efforts should apply a fuller specification of associated processes. Specifically, related Kuznets processes in innovation, production, urban development, population migration, and family formation might prove to be quite valuable in understanding the war cycle. Two important outcomes of this research enterprise have been to dispel the view that long waves are Kondratieff-shaped and to suggest that future research may want to focus on shorter, Kuznets-shaped cycles. (Sayrs, 1993)

Linking our debates in Chap. 3 with the data analysis of the present chapter, it emerges from Table 4.1 that each major pandemic in human history was also marked by decline in world industrial production from 1770 to 2020:

In the following, we will quickly present the results of our new encompassing analysis of an old debate (see above).

## 4.6 Are There Cycles of War, Production, and Defense Pacts? A Time Series Analysis

To start, we present in Fig. 4.9 the main parameters of world system research about the evolution of the international system since 1495. For any reader of the study written by Goldstein (1988), this data series will be a "*locus classicus,*" now re-drawn with the data ranging up to 2020. It is the synthesis of all the data, on which the "war cycle" models are based (Senghaas, 1983). The argument by Goldstein (1988) was that there are closely interrelated long cycles of global economics and global warfare (Figs. 4.21 and 4.22).

The Italian international relations scholar Fulvio Attinà added to this the all-important role of defence pacts (Attinà, 2005, 2011, 2021). Attina's Defense Pact

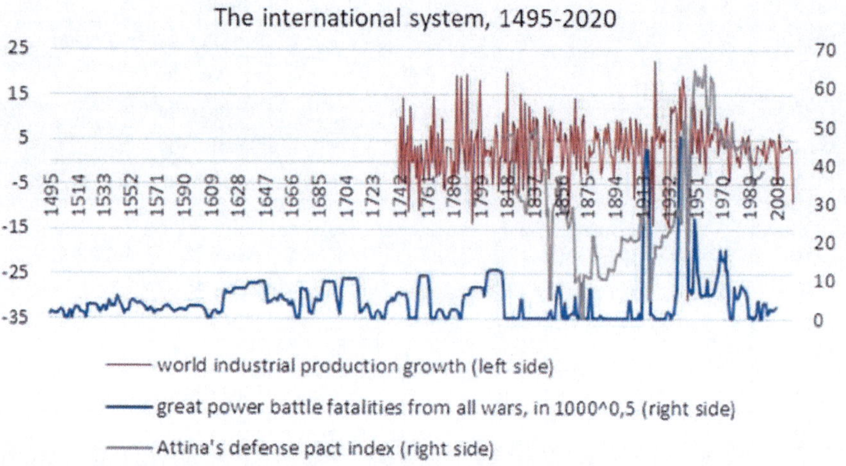


**Fig. 4.21** International system, 1495–2020. World systems research main indicators of the evolution of the international system

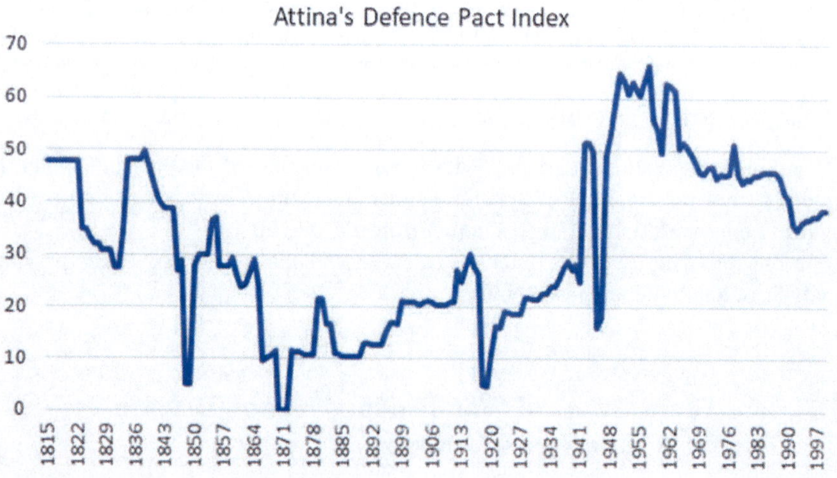


**Fig. 4.22** International system, 1815–2000. Attina's Defense Pact Index

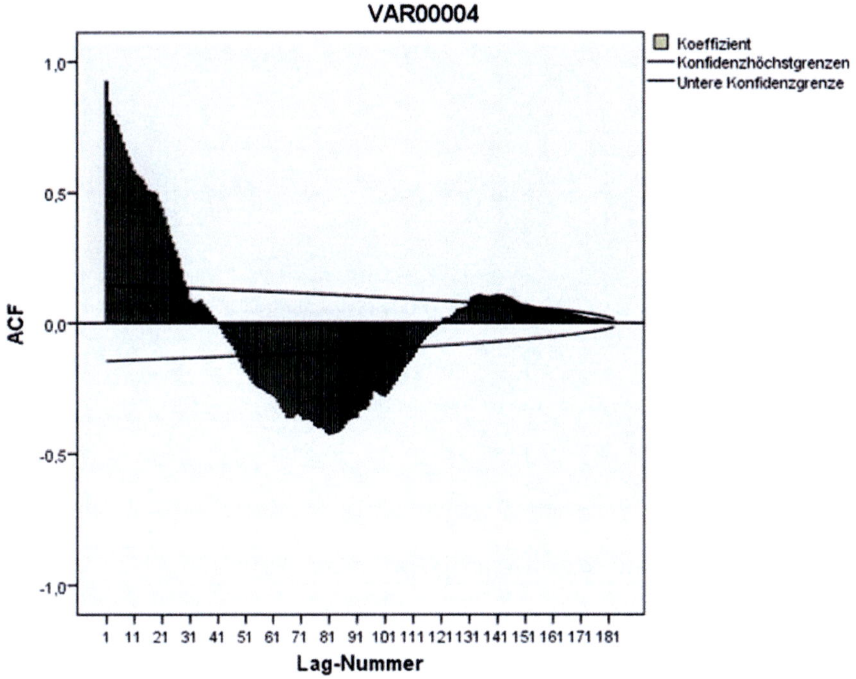


**Fig. 4.23** International system, 1815–2000. Autocorrelation analysis of Attina's Defense Pact Index

Index, 1815–2000 shows the culmination of alliances in the global system since 1815 (Fig. 4.10).

Figure 4.11 applies the econometric and politometric times series technique of autocorrelation analysis, with a clear evidence for an 80-year cycle of the decomposition of global defense pacts since the time of the end of the Napoleonic Wars and the Congress of Vienna, boding very ill for the future of NATO (Fig. 4.23).

Figure 4.12 applies the technique of rolling correlation analysis (70-year window) to the battle fatalities time series about major power conflicts from 1495 to 2008. The troughs of major power conflict waves were around 1615, 1685, 1755, 1875, and in the 2010s, but each time, war rebounds. In accordance with Goldstein (1988), we also can clearly discern the overwhelming importance of the Thirty Years War (1618–1648), the War of the Spanish Succession (1701–1713), the French Revolutionary Wars (1792–1802), the Napoleonic Wars (1803–1815), the First World War (1914–1918), and the Second World War (1939–1945) (Fig. 4.24).

Figures 4.13, 4.14, and 4.15 apply the econometric and politometric time series analysis techniques of spectral analysis (periodogram and spectral density analysis with the IBM SPSS default option for the Window in spectral density analysis) as well as autocorrelation analysis. The evidence suggests to talk about a shorter 30-year cycle of warfare as well as a longer 130-year cycle (Figs. 4.25, 4.26 and 4.27).

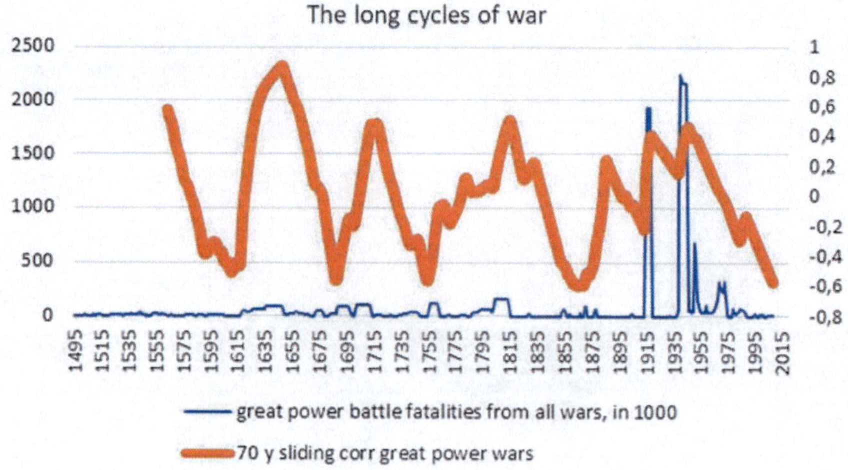


**Fig. 4.24** Cycles of global war 1495–2008—70-year rolling correlation analysis

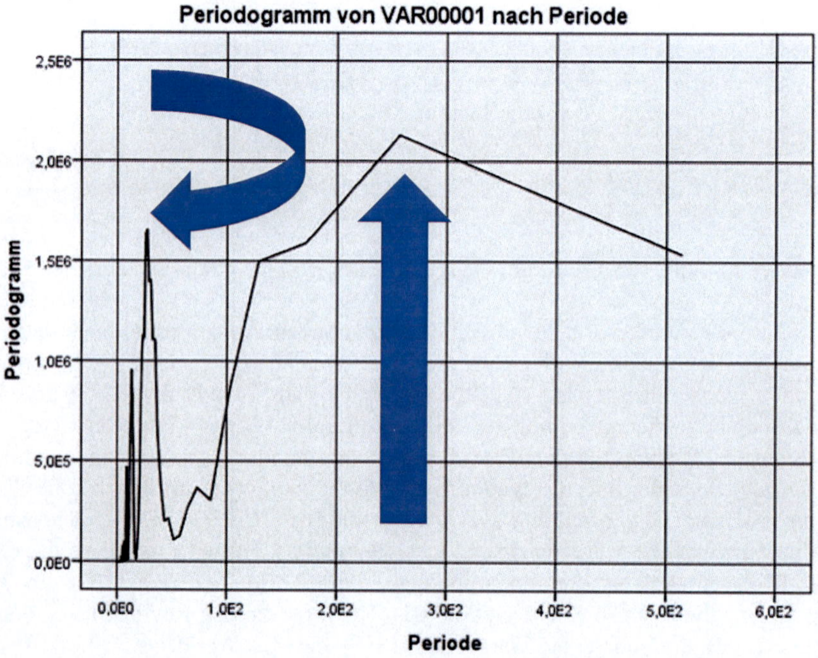


**Fig. 4.25** Periodogram (spectral analysis) of the cycles of global war 1495–2008

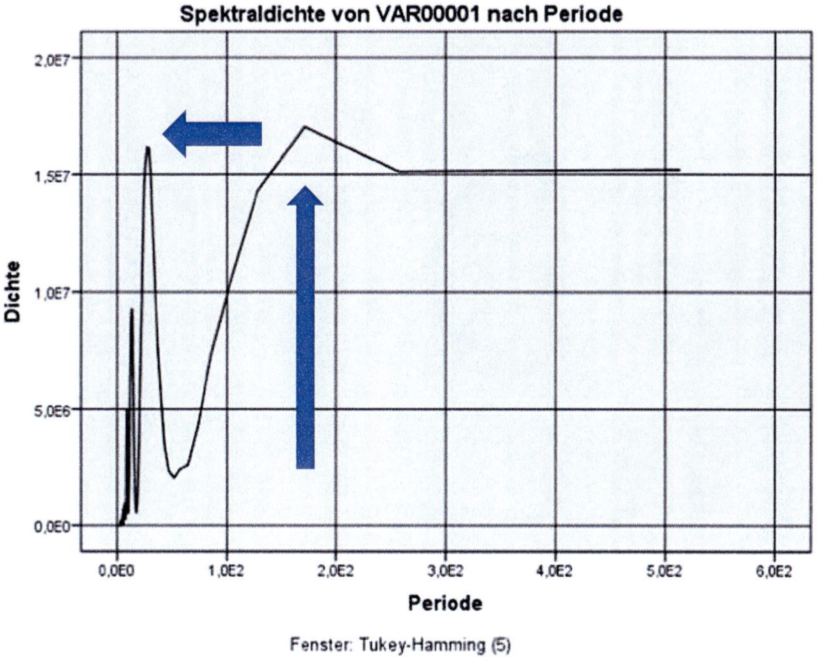


**Fig. 4.26** Spectral density analysis of the cycles of global war 1495–2008

Figure 4.16 now depicts the updated time series of world industrial production growth, 1740–2020, and refers to Chap. 2 for the debate about the data used in this analysis. We highlight the Spanish Flu, 1918–1920 as a precursor of the Great Depression from 1929 onwards. Our Graph shows the original data and the 5-year moving average. Also, this time series, in its aspects before 1975, will be well-known to scholars of international relations, who intensely studied Goldstein's classic, 1988, to this day.[18] We of course cannot debate here the vast econometric literature, written on the subject of the long 50 years economic cycles (Kondratieff waves), and refer our readers to Devezas (2006, 2010, 2012), Grinin et al. (2016), Marchetti (1980, 1986), Marchetti and Ausubel (2013), and Louçã (2019). Our analyses, re-confirming the cyclical nature of wars and economics, could be also a motive for the social science profession to reconsider all the evidence on long, Kondratieff-type fluctuations, from economics to energy production, which the Italian physicist and systems researcher, Cesare Marchetti[19] (*1927) compiled throughout his scientific lifetime, especially

[18] As of March, 5, 2021, Google Scholar recorded 1183 titles, quoting Goldstein (1988).

[19] http://cesaremarchetti.altervista.org/. "*By the rules of the Kondratieff cycles, a new number 1 challenger should have appeared in 1914 + 54 years, or in 1968, and in fact Russia presented itself as a pretender for the throne. The presence of tens of thousands of atomic weapons on each side discouraged direct confrontation, but a bitter war that actually produced millions of casualties was fought by proxy. Assuming that the Kondratieff cycle remains valid, the next competition should*

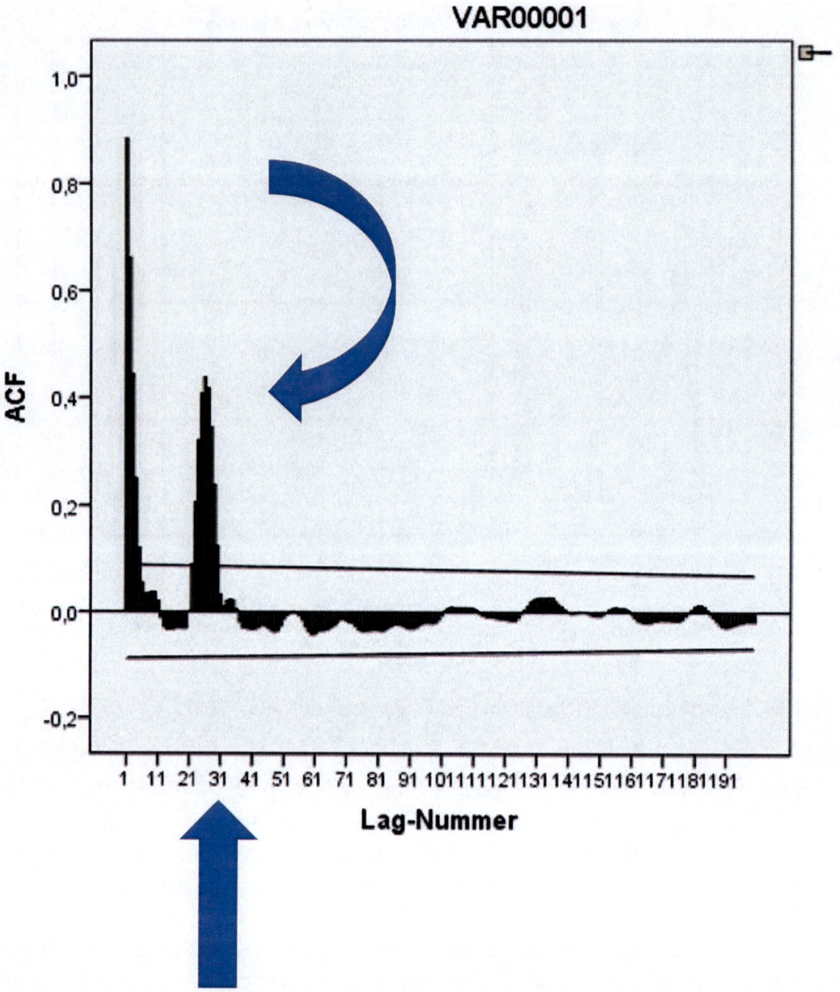


**Fig. 4.27** Analysis of autocorrelation of the cycles of global war 1495–2008

during his time as a researcher at the IIASA Institute in Laxenburg near Vienna,

---

*take place around 1968 + 54 years, or in 2022. This date is a general reference position; a stronger conviction about timing requires specific indicators. One such indicator may come from the fact that world power number 1 often ends by going broke. Being number 1 cures pride and brings benefits of various kinds, but ultimately it is extremely expensive. The previous number 1, Great Britain, toward the end used much of the taxes raised to service its debt. The USA may find itself in a similar position in the next 20 years. Another sign is reinvigorated interest and loud grumbling of nations on, or near, the bases of the number 1 power. If bases start being evacuated, the end is usually in sight*".

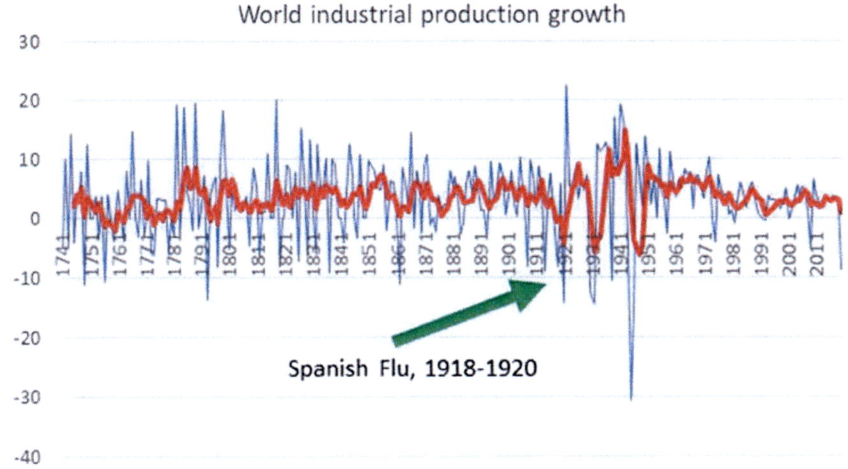


**Fig. 4.28** World industrial production growth, 1740–2020: original data and 5-year moving average

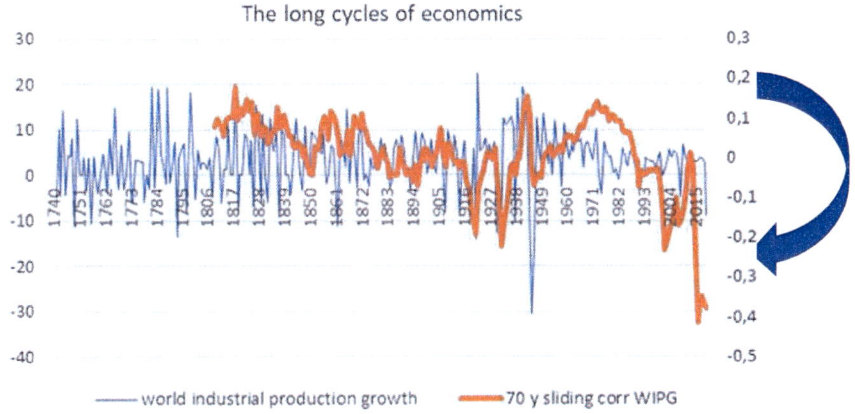


**Fig. 4.29** 70-year rolling time series correlation analysis of world industrial production growth, 1740–2020. Original data and 70-year rolling correlation analysis

Austria, from 1974 onwards until his retirement.[20] Marchetti's breathtaking fifty-five-year cycles (especially Marchetti, 1986) were calculated for infrastructures, energy systems and social behavior over a span of 200 years (Fig. 4.28).

Figure 4.17 applies the technique of a 70-year rolling time series correlation to the data, showing the terminal crisis of capitalism in the 1930s and the terminal crisis of our current period for which we are heading (Fig. 4.29).

[20] https://iiasa.ac.at/.

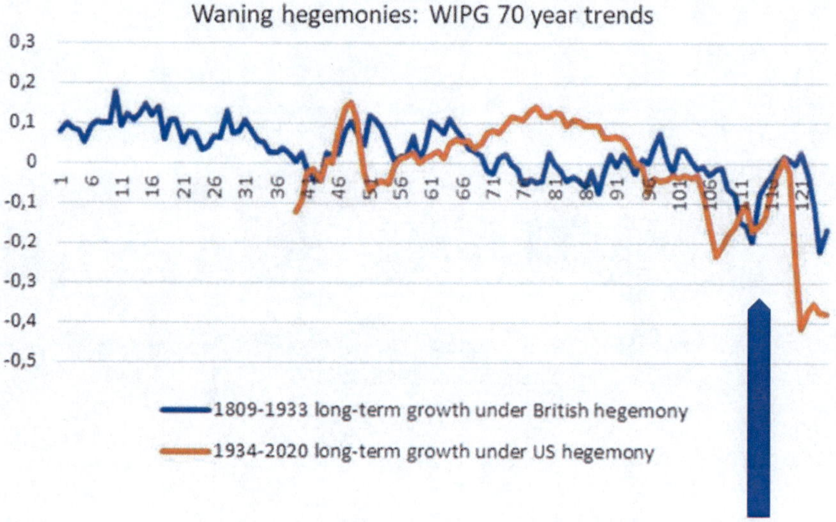


**Fig. 4.30** 70-year rolling time series correlation analysis of world industrial production growth, 1740–2020 shows striking parallels between the waning British and American hegemonies

Figure 4.18 applies the technique of projecting the cycle of the waning British hegemony and the waning American hegemony onto a single graph. The cycle of a waning hegemony took around 130 years in the British case; most probably, the crash of the American hegemony already has set in with full force (Fig. 4.30).

Figure 4.19 summarizes the available evidence from the spectral analysis of the world industrial production growth data, 1740–2020, concentrating here on the periodogram. Other tests, which at least suggest that we can maintain our interpretation, are printed in the Appendix to this Chapter. Our analysis clearly confirms the existence of the Kuznets cycles (22 years); Barro cycles (36 years); Kondratieff cycles (54 years); and Wallerstein cycles (140 years). Our readers should be reminded of the fact that these results were achieved without any prior transformation of the data and without any omission of time periods of measurement between 1740 and 2020 (Fig. 4.31).

## 4.7 The Tsunami of the 1920s and 1930s to Be Repeated Again?

Considering the major power conflict intensity and economic crisis intensity of the global system since 1495 and 1740, respectively, which culminated in the First World War, followed by the Spanish Flu Depression, the Great Depression, and the Second World War, the thought that there might be again major cataclysmic crises on the horizon, which hit the globe with tsunami force, is depressing and hair-raising at the

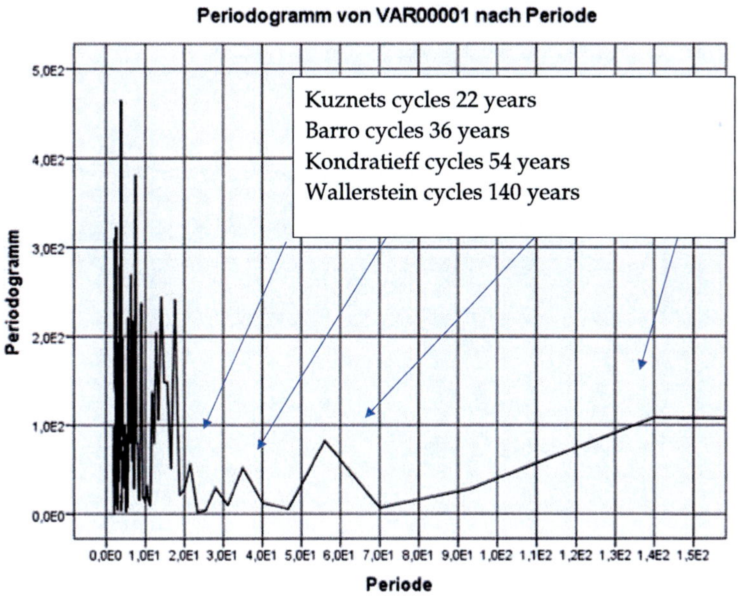


**Fig. 4.31** Periodogram (spectral analysis) of world industrial production growth, 1740–2020 showing the relevance of the Kuznets, Barro, Kondratieff, and Wallerstein cycles

same time. In Fig. 4.20, we analyse the Kondratieff cycle and the Wallerstein cycle in the postwar period, based on rolling correlations for the 54 years Kondratieff cycle and the 140 years Wallerstein cycle (Fig. 4.32).

Figures 4.21 and 4.22 compare the spring tide waves of the global crisis in the aftermath of the Influenza Pandemic 1918–1920 and the current COVID-19 depression: in the Great Depression, all cyclical downswings (16 y, 22 y, 36 y, 54 y, 140 y) coincided. The tidal waves of global crises in the aftermath of the COVID-19 Pandemic 2020 suggest that in the current COVID-19 Depression, again, all cyclical downswings (16 y, 22 y, 36 y, 54 y, 140 y) coincide, which will produce a devastating economic, social, and political crisis which will sweep across our globe in the coming years (Figs. 4.33 and 4.34).

Figures 4.23 and 4.24 show the results of the econometric and politometric cross-correlation analysis CCF of industrial production growth and annual battle fatalities from major power wars in the global system since 1740 onwards. The analysis suggests that a global depression, with a time-lag of about four years, produces a significant upsurge of major power wars (Figs. 4.35 and 4.36).

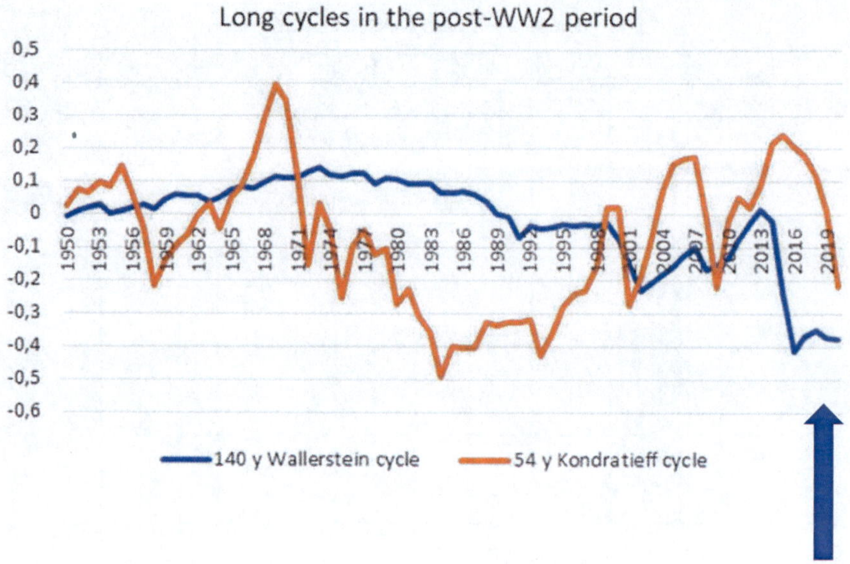


**Fig. 4.32** Kondratieff cycle and the Wallerstein cycle in the postwar period, based on rolling correlations for the 54 years Kondratieff cycle and the 140 years Wallerstein cycle

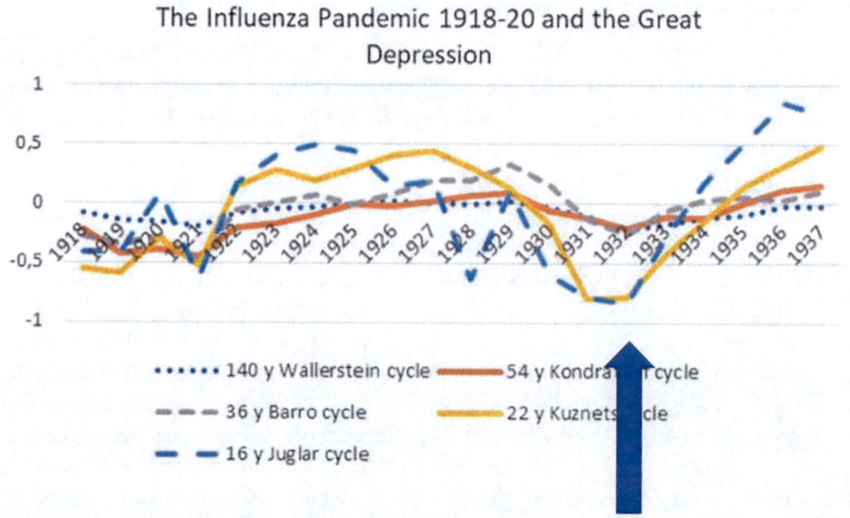


**Fig. 4.33** Spring tide waves of the global crisis in the aftermath of the Influenza pandemic 1918–1920: in the great depression, all cyclical downswings (16 y, 22 y, 36 y, 54 y, 140 y) coincided

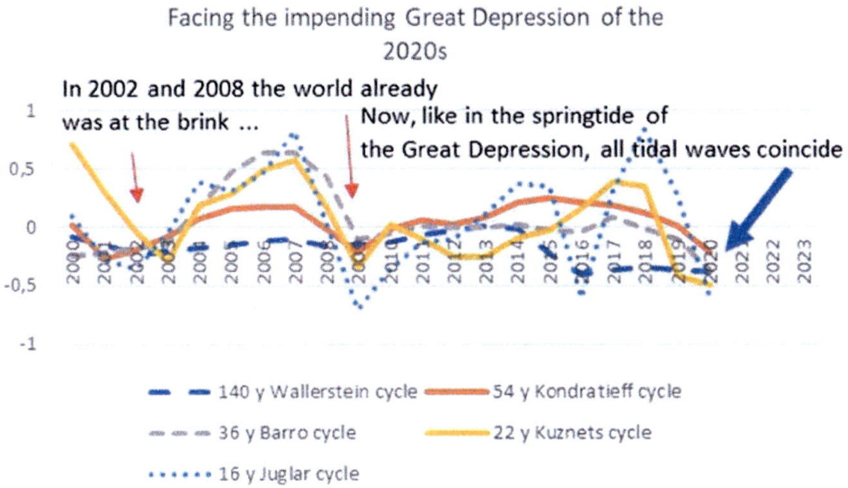


**Fig. 4.34** Tidal waves of global crises in the aftermath of the COVID-19 pandemic 2020: in the current COVID-19 depression, again, all cyclical downswings (16 y, 22 y, 36 y, 54 y, 140 y) coincide

## 4.8 Conclusions and Prospects for the Long Cycles of Global Politics and Economics in the 2020s and 2030s

This chapter is to be considered as the dire synthesis of the predictions and lessons to be drawn from the time series analysis of major power wars since 1495 and global economic cycles since 1740. Our analysis clearly confirms the existence of the Kuznets cycles (22 years); Barro cycles (36 years); Kondratieff cycles (54 years) and Wallerstein cycles (140 years). Our readers should be reminded of the fact that these results were achieved without any prior transformation of the data and without any omission of time periods of measurement between 1740 and 2020.

How gladly we would have pronounced here instead a sentence like this, saying

> our analysis based on the latest empirical data has shown that the critics of earlier world systems predictions of global economic, long-term crises and major power warfare are right. Our article has shown, in particular, that the optimistic predictions by Gat (2017)and Goldstein (2011), could be largely confirmed.

But unfortunately, the absolute opposite is true; and the most dismal of the dismal world systems literature has received a strong empirical support. For the Gulf countries, this chapter implies a grave warning of an impending world political tsunami. The recurrence of major power wars in the capitalist world economy from 1495 to the present is one of the most intriguing features of the existing international system, again confirmed in our data analysis. Each world political cycle up to now corresponded to an indented "W"-pattern of untransformed annual battle fatalities from major power wars in thousands of human beings over time, and the untransformed data for the war cycles 1495–2008 are statistically well explained by a polynomial

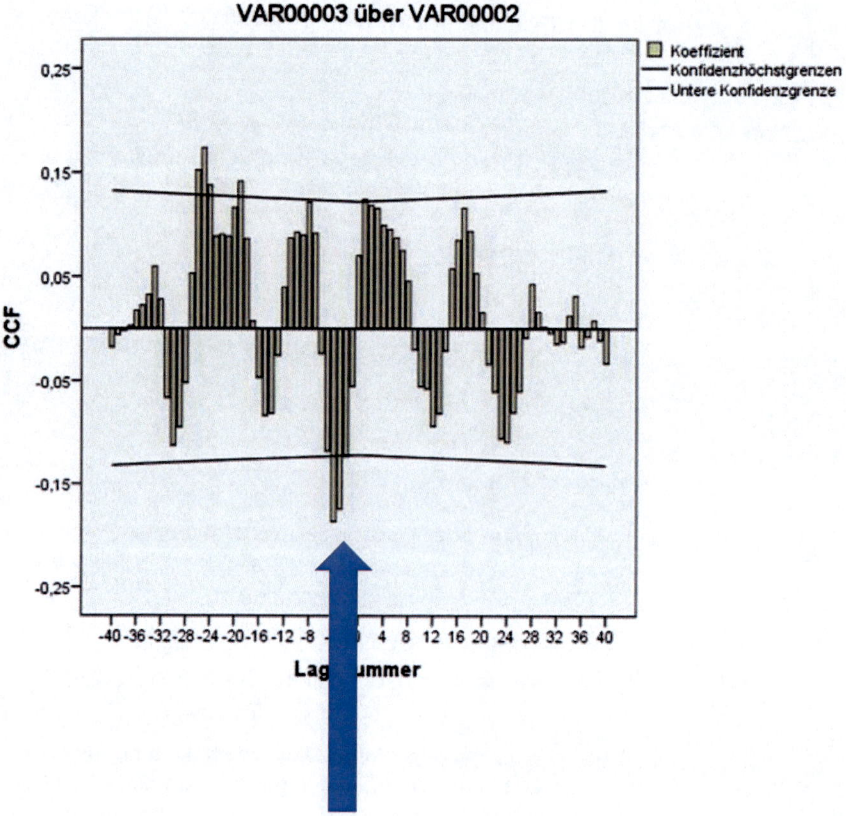


**Fig. 4.35** Cross-correlation analysis CCF: industrial production growth and annual battle fatalities from major power wars

expression of the fifth order over time, with the $R^2$ ranging from 40 to 90%. The international system seems indeed to be characterized by the following sequence of cycles:

- Global war,
- World hegemony of the dominant power,
- De-legitimization of the international order,
- De-concentration of the global system,
- Global war, et cetera.

The duration of these sub-phases of the international order is approximately one Kondratieff cycle each (an economic cycle of 50–60 years duration), so the unit of time of the international system can be symbolized by the expression—one Kondratieff cycle—1K. An entire hegemonic cycle lasts three Kondratieff cycles.

The troughs of major power conflict waves were around 1615, 1685, 1755, 1875, and in the 2010s, but each time, war rebounds. The evidence now suggests to talk

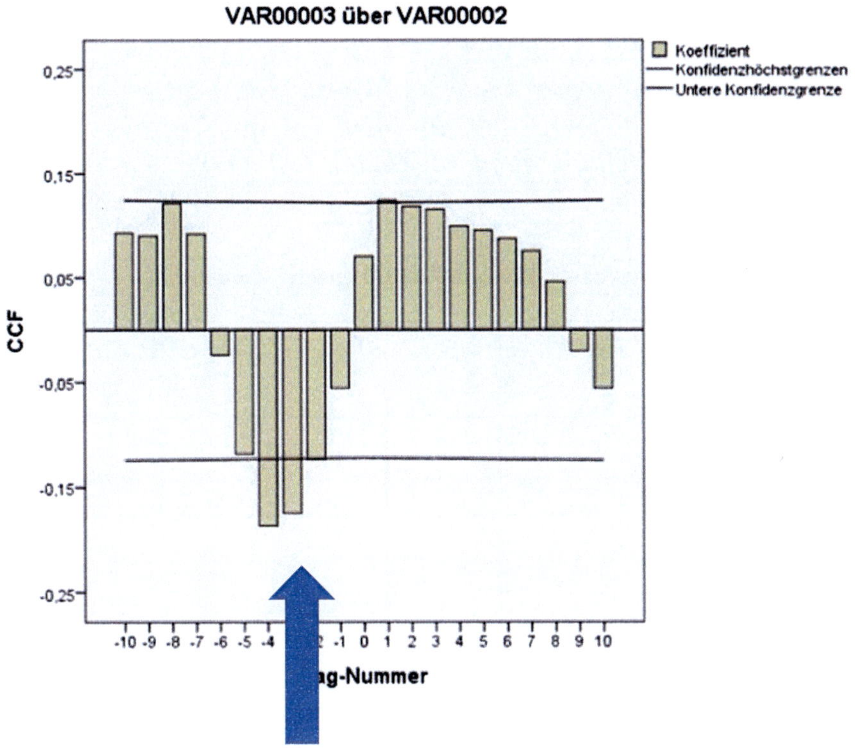


**Fig. 4.36** Cross-correlation analysis CCF: industrial production growth and annual battle fatalities from major power wars—the fatal 4-year distance from a depression to an increase in global power wars (close-up)

about a shorter 30-year cycle of warfare as well as a longer 130-year cycle. The cycle of a waning hegemony took around 130 years in the British case; most probably, the crash of the American hegemony already has set in with full force. The 70-year rolling time series correlation analysis of world industrial production growth, 1740–2020 shows striking parallels between the waning British and American hegemonies.

Considering the major power conflict intensity and economic crisis intensity of the global system since 1495 and 1740, respectively, which culminated in the First World War, followed by the Spanish Flu Depression, the Great Depression, and the Second World War, there is reason to believe that there might be again major cataclysmic crises on the horizon. Our analysis also relies on the results of the econometric and politometric cross-correlation analysis CCF of industrial production growth and annual battle fatalities from major power wars in the global system since 1740 onwards. The analysis suggests that a global depression, with a time-lag of about four years, produces a significant upsurge of major power wars.

## Appendix

### *Econometric Time Series Analysis of World Industrial Production Growth (WIPG), 1740–2020: Spectral Density Analysis with a Default Standard Window (5 Years)*

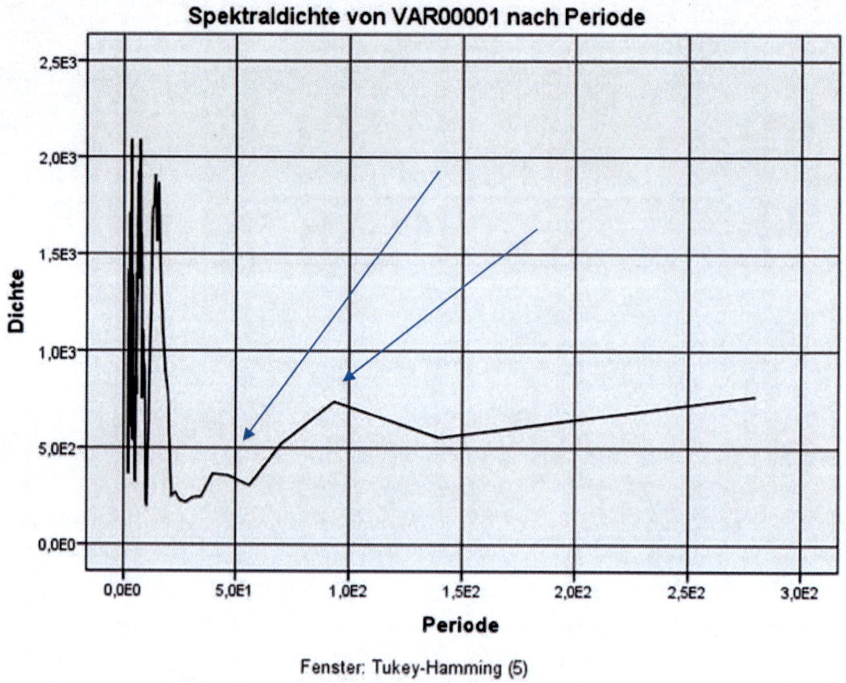

### *Econometric Time Series Analysis of World Industrial Production Growth (WIPG), 1740–2020: Autocorrelation Analysis*

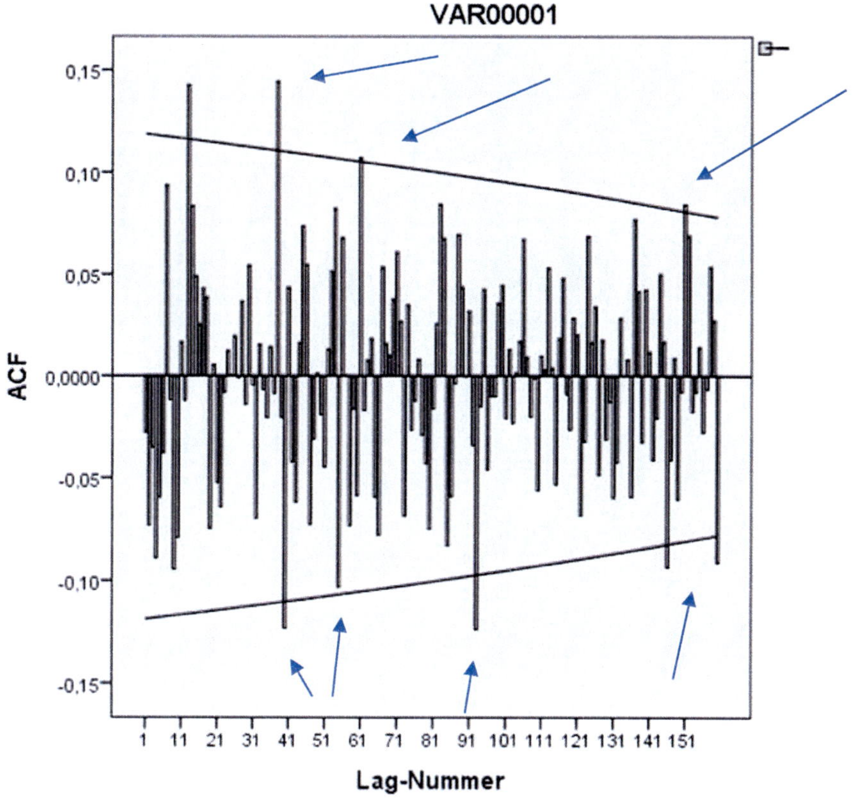


| | |
|---|---|
| 27 y rolling correlation world industrial production growth | 54 y Kondratieff cycle |
| 18 y rolling correlation world industrial production growth | 36 y Barro cycle |
| 11 y rolling correlation world industrial production growth | 22 y Kuznets cycle |
| 8 y rolling correlation world industrial production growth | 16 y Juglar cycle |

### *Econometric Time Series Analysis of World Industrial Production Growth (WIPG), 1740–2020: Rolling Correlation, Kuznets Cycle (22 Years)*

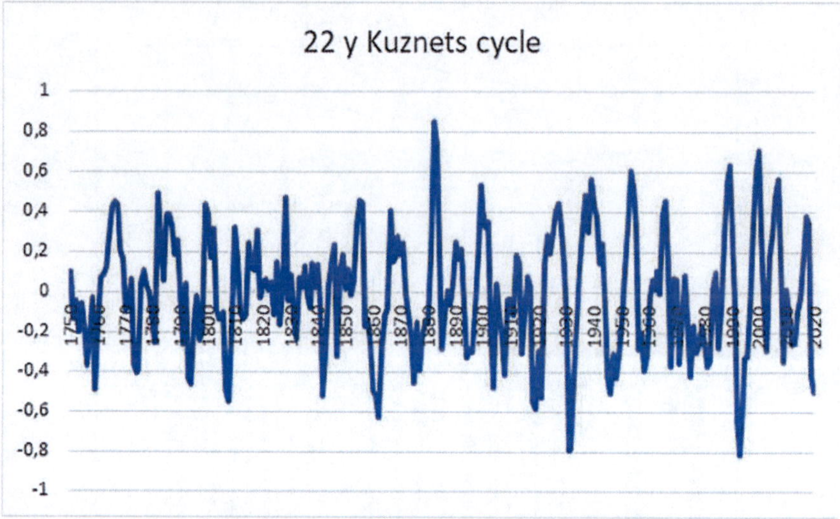

***Econometric Time Series Analysis of World Industrial Production Growth (WIPG), 1740–2020: Rolling Correlation, Barro Cycle (36 Years)***

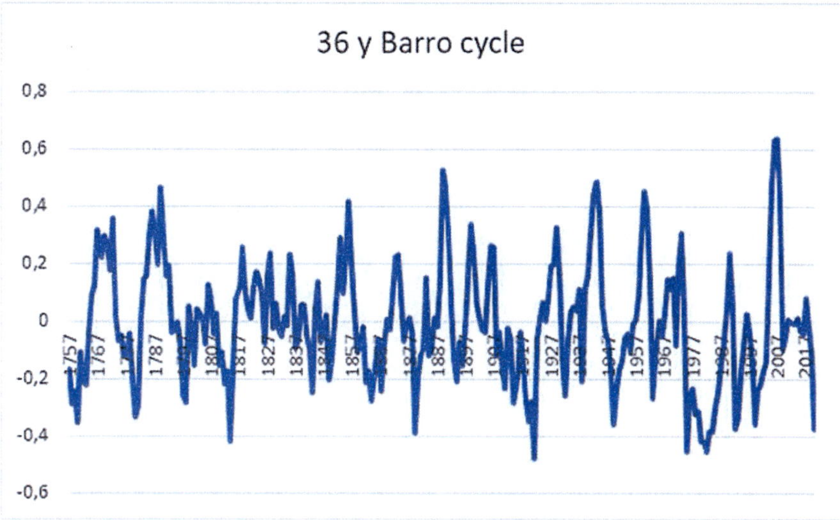

## *Econometric Time Series Analysis of World Industrial Production Growth (WIPG), 1740–2020: Rolling Correlation, Kondratieff Cycle (54 Years)*

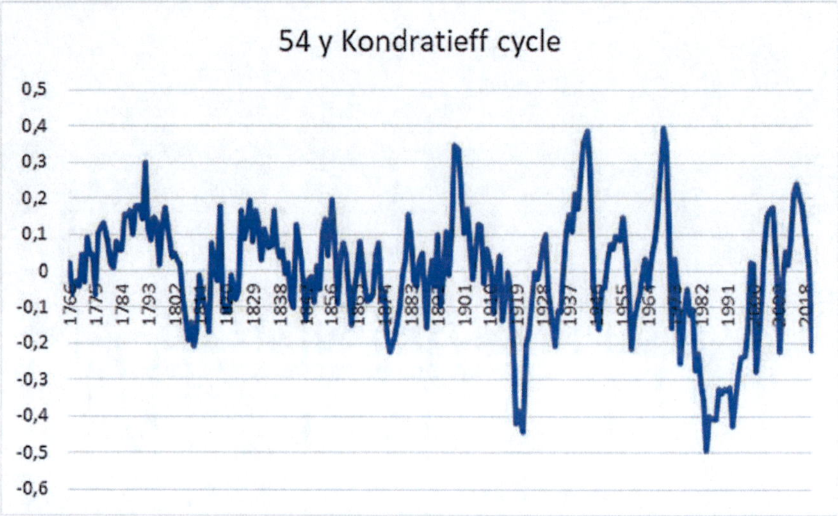


## References


Adams, K. R. (2005). New great powers: Who will they be, and how will they rise? Paper, prepared for presentation at the 2005 Annual Meeting of the International Studies Association, Honolulu, Hawaii, March 2–5, 2005. Available at: http://www.umt.edu/polsci/faculty/adams/greatpower.rtf

Advani, R. (2019). *The imperious rise of gulf capitalism.* Middle East Report Online, August 28, 2019. Available at https://merip.org/2019/08/the-imperious-rise-of-gulf-capital/

Allison, G. T. (1969). Conceptual models and the Cuban missile crisis. *The American Political Science Review, 63*(3), 689–718.

Arrighi, G. (1994). *The long twentieth century: Money, power, and the origins of our times*. Verso.

Arrighi, G. (1995). *The long 20th century. Money, power, and the origins of our times*. Verso.

Arrighi, G., & Silver, B. J. (1999). *Chaos and governance in the modern world system*. University of Minnesota Press.

Attinà, F. (2005). *State aggregation in defense pacts: Systematic explanations*. Jean Monnet Working Papers in Comparative and International Politics, Jean Monnet Centre EuroMed, Department of Political Studies, University of Catania. Available at: http://www.fscpo.unict.it/EuroMed/jmwp56.pdf

Attinà, F. (2011). *The global political system*. Macmillan International Higher Education.

Attinà, F. (2021) The world order lifecycle and world power competition. In F. Attinà (Ed.), *World order transition and the Atlantic area. Global power shift*. Springer. http://doi.org/10.1007/978-3-030-63038-6_2

Beck, N. (1991). The illusion of cycles in international relations. *International Studies Quarterly, 35*, 455–476.

Boswell, T. (1999). In Bornschier & Chase-Dunn (Eds.), *Hegemony and bifurcation points in world history* (pp. 263–284).

Browning, C. S. (2018). Geostrategies, geopolitics and ontological security in the Eastern neighbourhood: The European Union and the 'new cold war.' *Political Geography, 62*, 106–115.

Cammett, M. (2018). Popular grievances in the Arab region: Evaluating explanations for discontent in the lead-up to the uprisings. *Middle East Development Journal, 10*(1), 64–96.

Chase-Dunn, Ch. K., Grimes, P., & Anderson, E. N. (2019). Cyclical evolution of the global right. *Canadian Revue of Sociology/Revue Canadienne de Sociologie, 56*(4), 529–555.

Conradi, P. (2017). *Who lost Russia?* Simon and Schuster.

de Soysa, I., & Gleditsch, N. P. (2002). The liberal globalist case. In B. Hettne & B. Oden (Eds.), *Global governance in the 21st century: Alternative perspectives on world order* (pp. 26–23). Almkvist & Wiksell. Available also at: http://www.egdi.gov.se/pdf/study/study2002_2.pdf

Denemark, R. A. (2018). Nuclear war in the rivalry phase of the modern world-system. *Journal of World-Systems Research, 24*(2), 348–371. http://doi.org/10.5195/jwsr.2018.749. Available at https://jwsr.pitt.edu/ojs/jwsr/article/view/749/1154

Devezas, T. (2008). Energy scenarios: Toward a new energy paradigm. *Futures, 40*(1), 1–16.

Devezas T. C. (Ed.). (2006). Kondratieff waves, warfare and world security. In *Proceedings of the NATO Advanced Research Workshop on the Influence of Chance Events and Socioeconomic Long Waves in the New Arena of Asymmetric Warfare*, Covilhã, Portugal, 14–18 February 2005. IOS Press.

Devezas, T. C. (2010). Crises, depressions, and expansions: Global analysis and secular trends. *Technological Forecasting and Social Change, 77*(5), 739–761.

Devezas, T. C. (2012). The recent crisis under the light of the long wave theory. In L. E. Grinin, T. C. Devezas, & A. V. Korotayev (Eds.), *Kondratieff waves. Dimensions and prospects at the dawn of the 21st century* (pp. 138–175). Uchitel.

Frank, A. G. (1998). *ReOrient: Global economy in the Asian age*. University of California Press.

Freeman, C., & Louçã, F. (2018). *The fourth kondratiev wave: The great depression and the age of oil, automobiles, motorization, and mass production.* https://www.semanticscholar.org/paper/The-Fourth-Kondratiev-Wave%3A-The-Great-Depression-of-Freeman-Lou%C3%A7%C3%A3/1bb37b2adaf05746dcc242bd7900c1f1b21f15bb?p2df

Fukuyama, F. (1991). Liberal democracy as a global phenomenon. *PS: Political Science and Politics (Washington D. C.), 24*(4), 659–664.

Gat, A. (2017). *The causes of war and the spread of peace but will war rebound?* Oxford University Press.

Goldstein, J. S. (1985). Kondratieff waves as war cycles. *International Studies Quarterly, 29*(4), 411–444.

Goldstein, J. S. (1988). *Long cycles. Prosperity and war in the modern age.* Yale University Press.

Goldstein, J. S. (2005). The predictive power of long wave theory, 1989–2004. Prepared for NATO Conference on Kondratieff Waves and Warfare, Covilha, Portugal, February 2005. Available at: http://www.joshuagoldstein.com/jgkond.htm

Goldstein, J. S. (2011). *Winning the war on war: The decline of armed conflict worldwide.* Dutton.

Grinin, L., & Korotayev, A. (2012). Does Arab spring mean the beginning of world system reconfiguration? *World Futures, 68*(7), 471–505.

Grinin, L., Korotayev, A., & Tausch, A. (2016). *Economic cycles, crises, and the global periphery.* Springer.

Hahn, G. M. (2018). *Ukraine over the edge: Russia, the west and the "new cold war"*. McFarland.

Hanieh, A. (2011). *Capitalism and class in the Gulf Arab States.* Palgrave Macmillan.

Hanieh, A. (2013). *Lineages of Revolt: Issues of contemporary capitalism in the Middle East.* Haymarket Books.

Hanieh, A. (2015). *Capitalism and class in the Gulf Arab States.* Palgrave Macmillan.

Hanieh, A. (2016a). Absent regions: Spaces of financialisation in the Arab world. *Antipode, 48*(5), 1228–1248.

Hanieh, A. (2016b). Beyond authoritarianism: Rethinking Egypts long revolution. *Development and Change, 47*(5), 1171–1179.

Hanieh, A. (2018a). *Money, markets, and monarchies: The Gulf Cooperation Council and the political economy of the contemporary Middle East*. Cambridge University Press.

Hanieh, A. (2018b). The contradictions of global migration. *Socialist Register, 55*.

Hanieh, A. (2019a). New geographies of financial power: Global Islamic finance and the Gulf. *Third World Quarterly, 41*(3), 525–546.

Hanieh, A. (2019b). Variegated finance capital and the political economy of Islamic banking in the Gulf. *New Political Economy, 25*(4), 572–589.

Hanieh, A. (2020). New geographies of financial power: Global Islamic finance and the Gulf. *Third World Quarterly, 41*(3), 525–546.

Hauser, G. (2018). *China–eine asiatische Großmacht auf dem Weg zur Weltmacht: Aktuelle Entwicklungen der Außen-und Sicherheitspolitik der Volksrepublik China.* Verlag Barbara Budrich.

Karaganov, S. (2018). The new cold war and the emerging greater Eurasia. *Journal of Eurasian Studies, 9*(2), 85–93.

Lacina, B., & Gleditsch, N. P. (2005). Monitoring trends in global combat: A new dataset of battle deaths. *European Journal of Population, 21*(2–3), 145–166.

Lacina, B., Gleditsch, N. P., & Russett, B. (2006). The declining risk of death in battle. *International Studies Quarterly, 50*(3), 673–680.

Laughland, J. (1998). *The tainted source. The undemocratic origins of the European idea.* Warner Books, Little, Brown and Company.

Louçã, F. (2019). *As time went by—Long waves in the light of evolving evolutionary economics.* SWPS 2019-05. Available at SSRN: https://ssrn.com/abstract=3333183 or http://doi.org/10.2139/ssrn.3333183

Marchetti, C. (1980). Society as a learning system: Discovery, invention, and innovation cycles revisited. *Technological Forecasting and Social Change, 18*(4), 267–282.

Marchetti, C. (1986). Fifty-year pulsation in human affairs: Analysis of some physical indicators. *Futures, 18*(3), 376–388. Available at http://www.cesaremarchetti.org/archive/scan/MARCHETTI-069.pdf

Marchetti, C. & Ausubel, J. H. (2013). Quantitative dynamics of human empires. Color booklet, adapted from: Marchetti and Ausubel (2012). *International Journal of Anthropology, 27*(1–2), 1–62. Available at https://phe.rockefeller.edu/docs/empires_booklet.pdf

Modelski, G. (1999). In Bornschier and Chase-Dunn (Eds.), *From leadership to organization: The evolution of global politics* (pp. 11–39).

Modelski, G., & Thompson, W. R. (1996). *Leading sectors and world powers: The co-evolution of global politics and economics.* University of South Carolina Press.

Peace Research Institute, Oslo (PRIO). (2021). The battle deaths dataset version 3.0 and 3.1. Available at https://www.prio.org/Data/Armed-Conflict/Battle-Deaths/

Pfeifer, K. (2012). *Gulf Juggernaut: Review of Adam Hanieh, capitalism and class in the Gulf Arab States.* Economics: Faculty Publications, Smith College, Northampton, MA. https://scholarworks.smith.edu/eco_facpubs/18

Podobnik, B. (1999). Toward a sustainable energy regime: A long-wave interpretation of global energy shifts. *Technological Forecasting and Social Change an International Journal, 62*(3), 155–172.

Polanyi, K. (1944/1957). *The great transformation.* Beacon.

Pollins, B. (1996). Global political order, economic change, and armed conflict: Coevolving systems and the use of force. *American Political Science Review, 90*(1), 103–117.

Pollins, B. (2008). Globalization and armed conflict among nations. *Conflict Management and Peace Science, 25*(3), 191–205.

Pollins, B., & Schweller, R. L. (1999). Linking the levels: The long wave and shifts in U.S. Foreign Policy, 1790-1993. *American Journal of Political Science, 43*(2), 431–464.

Reuters Factbox. (2020). *U.S. forces in Gulf region and Iraq*. Available at https://www.reuters.com/article/us-iraq-security-usa-presence-factbox-idUSKBN1Z72GF

Russett, B. (1994). The democratic peace. In V. Bornschier & P. Lengyel (Eds.), *Conflicts and new departures in world society* (pp. 21–43). Transaction Publishers.

Sarkees, M. R. (2000). The correlates of war data on war: An update to 1997. *Conflict Management and Peace Science, 18*(1), 123–144.

Sayrs, L. W. (1993). The long cycle in international relations: A Markov specification. *International Studies Quarterly, 37*(2), 215–237.

Senghaas, D. (1983). The cycles of war and peace. *Bulletin of Peace Proposals, 14*(2), 119–124. https://doi.org/10.1177/096701068301400202

Silverberg, G. (2005a). *When is a wave a wave? Long waves as empirical and theoretical constructs from a complex systems perspective*. Maastricht Economic Research Institute on Innovation and Technology, MERIT Infonomics Research Memorandum series, 016. Available at: http://edata.ub.unimaas.nl/www-edocs/loader/file.asp?id=1057

Silverberg, G. (2005b). *When is a wave a wave? Long waves as empirical and theoretical constructs from a complex systems perspective*. Available at: http://collections.unu.edu/eserv/UNU:1187/rm2005-016.pdf

Tausch, A. (2003). The European Union: Global challenge or global governance? 14 world system hypotheses and two scenarios on the future of the union. In G. Kohler & E. J. Chaves (Eds.), *Globalization: Critical perspectives* (pp. 93–197). Nova Science Publishers.

Tausch, A. (2007). War cycles. *Social Evolution and History (Moscow), 6*(2), 39–74.

Tausch, A. (2010). Paul Boccara's analysis of global capitalism. The return of the Bourbons, and the breakdown of the Brussels/Paris neo-liberal consensus. *Entelequia. Revista Interdisciplinar, 12*, 105–147.

Volland, C. (1987). A comprehensive theory of long wave cycles. *Technological Forecasting and Social Change an International Journal, 32*(2), 123–145.

Wallerstein, I. (1989). The capitalist world-economy: Middle-run prospects. In I. Wallerstein (Ed.), *Geopolitics and geoculture: Essays on the modern world-system* (pp. 123–136). Cambridge University Press.

Wallerstein, I. (2000). *The essential Wallerstein*. The New Press.

Zhao, M. (2019). Is a new cold war inevitable? Chinese perspectives on US–China strategic competition. *The Chinese Journal of International Politics, 12*(3), 371–394.

# Chapter 5
# Learning from Classical Arab Writers: G-Zero and the Cultural Aspects of Changing Global Leadership: Rediscovering Ibn Khaldun's "*Asabiyya*" with the Latest *PEW Global Survey* and *World Values Survey* Data

**Abstract** This chapter analyzes the current vacuum in global leadership with global opinion survey data from the latest freely available PEW Global Survey, 2018 and World Values Survey, 2017–2020 data files, using multivariate factor analysis of the opinions of the global citizenry. On a population-weighted basis, the Chinese leadership model is accepted by only 27.63% of the global population. The dismal United States Leadership under President Donald Trump was backed by only 50.17% of the citizens of our globe, while 52.51% of global citizens "voted" for Russian President Vladimir Putin, and 57.16% were expressing support for French President Emmanuel Macron, and 60.35% were in favor of German Chancellor Angela Merkel. The inroads, Chinese or Russian leadership concepts could make among Western publics are considerable. For the Gulf countries, this global leadership vacuum, now often being referred to as "G-Zero" has considerable implications. We find that secular orientations have an influence on the growing distance to America and test the analytical merits of two approaches to deal with the role of religion in that context: Inglehart's model of religion's sudden decline (Inglehart, in Religion's sudden decline. What's causing it, and what comes next? Oxford University Press, 2021) and the model adapted from classical Arab historiography, based on Ibn Khaldun (1332–1406). We present empirical tests for both approaches, and find that the corrosive effects of social decay in the wake of secularization in many countries of the world have been underestimated. We finally present some estimates on how global values are distributed among the world's generations.

**Keywords** PEW global survey · World values survey · Chinese leadership · President Donald Trump · Russian President Vladimir Putin · French President Emmanuel Macron · German Chancellor Angela Merkel · G-zero · Ronald F. Inglehart · Religion · Arab historiography · Ibn Khaldun (1332–1406) · Secularization



A. Tausch, *The Future of the Gulf Region*, Gulf Studies 2,
https://doi.org/10.1007/978-3-030-78299-3_5

## 5.1 Introduction to G-Zero and the Cultural Aspects of Changing Global Leadership

In the literature on global governance (see Grigoryev & Pabst, 2019), it has become fashionable to talk about a leadership vacuum in the international system, now also being called a "*G Zero*" world as a reflection of failing global governance by the G-20 (see Bremmer & Roubini, , just to mention a single very influential article on the issue). The *G-Zero* approach, i.e., the hypothesis about such a dangerous vacuum and a growing failure of global governance, has become all the more relevant due to the current COVID-19 pandemic.

Bremmer and Roubini (2011) diagnosed that the expanded group of leading economies, the G-20, has moved away from a would-be concert of nations to a cacophony of competing voices as the diversity of political and economic values within the group has asserted itself. For Bremmer and Roubini (2011), there is also no viable G-2—a US–Chinese alternative for the pressing transnational problems on the horizon, a diagnosis which was made all the more dramatically relevant by COVID-19 from January 2020 onwards. Bremmer and Roubini (2011) already correctly foresaw that there is no G-3 alternative either, that is to say, a grouping of the USA, Europe, and Japan: For Bremmer and Roubini (2011), the USA lacks the resources to continue as the primary provider of global public goods. Europe was and still is fully occupied with saving the Eurozone, and COVID-19 compounded the European North–South dilemma even further (see Chap. 3 of this book). For Bremmer and Roubini (2011), Japan was likewise tied down with complex political and economic problems at home, made all the more dramatic by the Fukushima Daiichi nuclear disaster on Friday, March 11, 2011. Early on, world systems scholars already foresaw the dangers of such a vacuum evolving in the international system (Kohler & Chaves, 2003).

In geostrategic terms, the western democracies, which still were the dominant political, military, and economic powers at the turn of the millennium, are still centered around the North Atlantic Treaty Organization (NATO) with its current 30 members. In 1949, there were just twelve founding members of the Alliance: Belgium, Canada, Denmark, France, Iceland, Italy, Luxembourg, the Netherlands, Norway, Portugal, the UK and the USA. Today, the other member countries are: Greece and Turkey (1952), Germany (1955), Spain (1982), the Czech Republic, Hungary and Poland (1999), Bulgaria, Estonia, Latvia, Lithuania, Romania, Slovakia and Slovenia (2004), Albania and Croatia (2009), Montenegro (2017) and North Macedonia (2020).[1] Apart from NATO, the USA as the still leading military power of the West also maintains so-called *major non-NATO ally (MNNA)* relations with a host of other countries. MNNA is a designation given by the US government to close allies that have strategic working relationships with the US Armed Forces outside NATO (Hendrickson, 2011). Currently, these countries are the following eighteen

[1] https://www.nato.int/cps/en/natolive/topics_52044.htm.

nations,[2] and there are none the less than six Arab countries among them, two of them, Bahrain and Kuwait, are countries in the Gulf region:

Afghanistan (2012)
Argentina (1998)
Australia (1987)
Bahrain (2002)
Brazil (2019)
Egypt (1987)
Israel (1987)
Japan (1987)
Jordan (1996)
Kuwait (2004)
Morocco (2004)
New Zealand (1997)
Pakistan (2004)
Philippines (2003)
South Korea (1987)
Taiwan (2003)
Thailand (2003)
Tunisia (2015).

Apart from the Gulf countries Bahrain and Kuwait, Qatar has become a realistic candidate to become the country number nineteen on this list. The USA Department of State, on its official Web site, praises the pro-Western stance of Qatar and underlines that the USA and Qatar are working collectively toward the common goal of a stable, secure, and prosperous Middle East. The State Department emphasizes that Qatar is also a vital US partner on a wide range of regional security issues, a member of the *Global Coalition to Defeat ISIS*, and provides critical basing access supporting US military operations throughout the Middle East. The USA, the State Department underlines, works with Qatar and other members of the Gulf Cooperation Council to increase cooperation on border security, maritime security, military preparedness, arms transfers, cybersecurity, and counterterrorism. The access, basing, and overflight privileges granted by Qatar facilitate US operations against al-Qa'ida and are a unique geostrategic asset for the USA in the region.[3]

But how strong is the West, especially its social and value cohesion after four years of the disastrous Trump Presidency which ended by the Storming of Capitol Hill on January 6, 2021? And what is the standing of the USA and the West in general among the global publics? Has the global public support for America declined sharply? Are the Gulf countries then banking their long-term security on the wrong candidates, or is there a dimension, about which the secularized West can start to learn from the Gulf countries? And what role is played by religion in all this? And is the West in a

[2] https://www.state.gov/major-non-nato-ally-status/.

[3] https://www.state.gov/u-s-security-cooperation-with-qatar.

process of decline from—so to say—within, caused by the erosion of fundamental values?

In international political science, the decline of America's leadership is meanwhile an omnipresent theme of debates for more than two decades now. For Kristensen (2017), it is clear that with the election of Donald Trump in 2016 and Brexit on January 1, 2021 and the resulting uncertainties over the future of NATO, the EU, free trade, and climate agreements, there is now a more urgent sense that we are not only witnessing the rise of new powers but the decline of the so-called *liberal international order* installed after the Second World War. Second, and most remarkably, Kristensen says that the order is collapsing by way of abdication rather than a hegemonic clash between rising and declining powers. Third, and as a consequence, the discourse on a global leadership vacuum, turmoil and world disorder is gaining prominence. A number of high-profile studies, among them (Beeson, 2018; Katzenstein, 2013; Steinbock, 2014; Sussman, 2017; Wæver, 2018); reiterated this perspective.

Do religion and the values held by a society affect the chances for global leadership in the twenty-first century? The American sociologist Ronald F. Inglehart, in his most recent book publication (Inglehart, ) which bears the provocative title "*Religion's Sudden Decline. Why It's Happening and What Comes Next*" and which was released just in these days of the crisis in America and the assault on Capitol Hill on January 6, 2021, provides a far-reaching perspective on the crisis of American (and Western) society which cannot be overlooked by global publics and especially by publics in the Middle East (see below).

Inglehart's starting point is that as the twenty-first century began, religion was spreading rapidly. The collapse of Communism, according to Inglehart (2021), had left a psychological vacuum that was being filled with resurgent religion; fundamentalism was a rising political force in the USA, and the 9/11 attacks drew attention to the power of militant Islam. There were claims of a global resurgence of religion.

An analysis of religious trends from 1981 to 2007 in 49 countries containing 60% of the world's population did not, Inglehart (2021) says, find a global resurgence of religion—most high-income countries were becoming less religious—however, it did show that in 33 of the 49 countries studied, people had become more religious. But since 2007, (Inglehart, 2021) maintains, things have changed with surprising speed. From 2007 to 2020, an overwhelming majority (43 out of 49) of these same countries became less religious. This decline in religious belief is strongest in high-income countries but it is evident across most of the world (Inglehart, 2021).

The countermodel to the secularistic sociology of the waning of religion on a global scale, published by Inglehart (2021), which we propose here is based on classical Arab historiography and the idea of social and political cycles of rising and then waning leaderships, going back to the idea of "*Asabiyya*" ("*social cohesion*"), first described by Ibn Khaldun (1332–1406) in his seminal work "*Muqaddimah.*"

This chapter then attempts to analyze these issues with data from the *PEW Spring Survey*, 2018 and the latest edition of the *World Values Survey*, 2017–2020. In the Background section of this chapter, we briefly debate the relevant theoretical concepts and then present the results of our analysis. Conclusions and prospects round up the chapter.

## 5.2 Background on G-Zero and the Cultural Aspects of Changing Global Leadership

The most dramatic shift away from religion took place, (Inglehart, 2021) maintains, among the US American public. For years, (Inglehart, 2021), argues, the USA had been the key case demonstrating that economic modernization need not produce secularization. But recently, the American public has, according to Inglehart (2021), been moving away from religion along with virtually all other high-income countries—in fact, religiosity has been declining more rapidly in the US than in most other countries.

In attempting to explain what had happened, (Inglehart, 2021) says that several forces are driving this trend but the most powerful one is according to Inglehart the waning grip of a set of beliefs closely linked with the imperative of maintaining high birth rates. Inglehart's argument is the following: for many centuries, most societies assigned women the role of producing as many children as possible and discouraged divorce, abortion, homosexuality, contraception, and any other sexual behavior not linked with reproduction. Virtually all major world religions encouraged high fertility because it was necessary, in the world of high infant mortality and low life expectancy that prevailed until recently, for the average woman to produce five to eight children in order to simply replace the population. Religions that did not emphasize these norms gradually disappeared. A growing number of countries have now attained high life expectancies and drastically reduced infant mortality rates, making these traditional cultural norms no longer necessary. According to Inglehart (2021), this process did not happen overnight. The major world religions had presented pro-fertility norms as absolute moral rules and firmly resisted change. According to Inglehart (2021), people only slowly give up the familiar beliefs and societal roles they have known since childhood, concerning gender and sexual behavior. But when a society reaches a sufficiently high level of economic and physical security, younger generations grow up taking that security for granted and the norms around fertility recede. Ideas, practices, and laws concerning gender equality, divorce, abortion, and homosexuality are, Inglehart's argument is, now changing rapidly. Almost all high-income societies have recently reached a tipping point where the balance shifts from pro-fertility norms being dominant, to individual-choice norms being dominant.

Inglehart also thinks that several other factors help explain the waning of religion. In the USA, politics explains part of the decline. Since the 1990s, Inglehart maintains, the Republican Party has sought to win support by adopting conservative Christian positions on same-sex marriage, abortion, and other cultural issues. But this appeal to religious voters has had the corollary effect of pushing other voters, especially young liberal ones, away from religion. The uncritical embrace of President Donald Trump by conservative evangelical leaders has, Inglehart argues, accelerated this trend. And the Roman Catholic Church has lost adherents because of its own crises in connection with abuse by Catholic priests (Inglehart, 2021).

Inglehart's following empirical argument merits the special attention in our chapter. Inglehart (2021) squarely believes that although some religious conservatives warn that the retreat from faith will lead to a collapse of social cohesion and public morality, the evidence, so Inglehart's claim, does not support this claim. Surprising as it may seem, Inglehart (2021) says, countries that are less religious actually tend to be less corrupt and have lower murder rates than religious ones.

Contrary to some other studies, most notably (Mensah, 2014), Inglehart believes that religious countries actually tend to be more corrupt than secular ones. This pattern also, Inglehart argues, applies to other crimes, such as murder. Inglehart concedes that pandemics such as the current one will reduce peoples' sense of existential security. If the pandemic was to endure for decades, or lead to an enduring Great Depression, the theory underlying Inglehart's most recent book, 2021 implies that the cultural changes of recent decades would reverse themselves.

We now introduce a counterconcept, based on classical Arab historiography. The classic Arab historian and philosopher Ibn Khaldun (1332–1406) (Khaldun, 2005; furthermore Alatas, 2006; Amirabedini, 2014; Arnason, 2004; Chapra, 2008; Irwin, 1997; Tibi, 1996) explained in his most important book "*Muqaddimah*" (*Introduction to History*) that historical change and the succession of dynasties are a function of the interactions between nomadic culture and urban civilization. His major contribution in this context consisted in the analysis of the correlation between "*Asabiyya*," i.e., the social cohesion of a society and political power. He argued that the strong social cohesion of tribal peoples enabled them to conquer urbanized regions and build regimes and civilizations, but that these conquests were undone by the tribes' gradual loss of "*Asabiyya*" in the urban setting, leading to new conquests by tribal peoples still strong in desert cohesiveness. Although power was the basis of rulership and royal authority was established through military might, the "*glue*" that held societies together (Tibi, 1996) was "*Asabiyya*," based on kinship and religion and stronger in tribal than in urban society. Conquerors, whose social cohesion was weak, were soon overcome by the civilization of the conquered and gave way to a new conquering group. Rulers would be successful only so long as they remained just; as the rulers' level of luxury increased, so did their level of exploitation, and injustice soon produced division and "*the ruin of civilization*" (Tibi, 1996). Ibn Khaldun also foresaw the tendency toward economic cycles, which he linked, quite in a modern fashion, to this "value change" (see also Chap. 4 of the present book, and Chap. 6).

Ibn Khaldun (pp. 227–229) puts it as follows:

> The first generation retains the desert qualities, desert toughness, and desert savagery. (Its members are used to) privation and to, sharing their glory (with each other); they are brave and rapacious. Therefore, the, strength of group feeling continues to be preserved among them. They are sharp and, greatly feared. People submit to them. Under the influence of royal authority and a life of ease, the second, generation changes from the desert attitude to sedentary culture, from privation to, luxury and plenty, from a state in which everybody shared in the glory to one in, which one man claims all the glory for himself while the others are too lazy to strive, for (glory), and from proud superiority to humble subservience. Thus, the vigor of, group feeling is broken to some extent. People become used to lowliness and, obedience. But many of (the old virtues) remain in them, because they had had, direct personal contact with the first generation and its conditions, and had observed, with their

> own eyes its prowess and striving for glory and its intention to protect and, defend (itself). They cannot give all of it up at once, although a good deal of it may go. They live in hope that the conditions that existed in the first generation may, come back, or they live under the illusion that those conditions still exist.

Ibn Khaldun goes on to say:

> The third generation, then, has (completely) forgotten the period of desert, life and toughness, as if it had never existed. They have lost (the taste for) the, sweetness of fame and (for) group feeling, because they are dominated by force. Luxury reaches its peak among them, because they are so much given to a life of, prosperity and ease. They become dependent on the dynasty and are like women and, children who need to be defended (by someone else). Group feeling disappears, completely. People forget to protect and defend themselves and to press their, claims. With their emblems, apparel, horseback riding, and (fighting) skill, they, deceive people and give them the wrong impression. For the most part, they are, more cowardly than women upon their backs. When someone comes and demands, something from them, they cannot repel him. The ruler, then, has need of other, brave people for his support. He takes many clients and followers. They help the, dynasty to some degree, until God permits it to be destroyed, and it goes with, everything it stands for. As one can see, we have there three generations. In the course of these three, generations, the dynasty grows senile and is worn out.

Final decay arrives in the fourth generation:

> Therefore, it is in the fourth, generation that (ancestral) prestige is destroyed. […] Three generations last one hundred and twenty years, as stated before. As a, rule, dynasties do not last longer than that many years, a few more or a few less, save when, by chance, no one appears to attack (the dynasty). When senility, becomes preponderant (in a dynasty), there may be no claimant (for its power, and, then nothing will happen), but if there should be one, he will encounter no one, capable of repelling him. If the time is up, (the end of the dynasty) cannot be postponed for a single hour, no more than it can be accelerated.

Contemporary quantitative Kondratieff cycle researchers (see Chap. 4), like Rennstich (2007), stumbled onto the issue again, without duly taking into account the theoretical forerunner Ibn Khaldun. Rennstich linked the four-generation cycle to the Nobel-laureate winning four-generation family sage "*Buddenbrooks. The Decline of a family*" by the German novelist Mann (1924), and called the four-generation cycle the *Buddenbrooks cycle*. The novel by Thomas Mann, all about the rise and dramatic decline of a rich merchant family in Germany indeed contains a dire and general message for societal systems and is a model of analysis, also to be applied to the contemporary, post-Second World War world order. Independently of Rennstich, Devezas/Corredine, already in 2001, developed a similar and mathematical model, also based on four generations. Also, Joseph Alois Schumpeter, touched upon this issue in his 1950 "*Capitalism, Socialism and Democracy*" when he stated that capitalism will not disappear from the face of the earth by a socialist revolution, but by the gradual erosion of capitalist individual family values and the rise of a socialist-oriented intellectual elite and bureaucracy. In terms of modern social science theory, the following Ibn Khaldunian scheme can be proposed. Societies do not generally experience a smooth transformation from a traditional and survival-oriented value set toward a secular oriented value set. The "*Asabiyya*"-based countermodel to Inglehart relies on a factor analytical evaluation of variables from the *World Values Survey* and combines theoretical insights of Inglehart with those by Ibn Khaldun. At the very beginning of the modernization process, societies are traditional, religious, and pre-modern; but at the same time, they are also "active" societies with functioning neighborhood structures, and very clear perceptions of what is "good" and "evil," and a high respect of the law. As the modernization process proceeds, traditional religious values seem to be on the losing side of the societal equation. The uneasiness of the "religious right" in the USA, of the Islamist camp in many countries of the Muslim world, and the religious right in many developing countries is best understood by the process, whereby traditional values get lost at the same time with rising secularization. Respect of the law in such vital areas as tax morale, non-acceptability of bribery and government benefits fraud decreases at middle stages of development. Islamists are of course wrong in assuming that countries of the Muslim world are the only ones to suffer from a downward trend in public morale.

*Asabiyya* is defined empirically by the residuals from factor scores on value decay in the world system, as the current economic crisis and the COVID-19 pandemic strike the populations of our planet, and cross-national data. *Asabiyya* in the twenty-first Century, as a way out from the modernization trap of societies, characterized by large-scale social anomaly, is high secularism combined with a low social decay score, thus avoiding the "modernization trap" of an increasingly secular society, which accepts cheating on taxes; accepts government benefits fraud and taking bribes.

According to the *World Values Survey* data, 2017–2020, analyzed in this chapter, the most secular countries today are Japan, New Zealand, Australia, Denmark, Switzerland, Norway, Czech Republic, Germany, Netherlands, and Finland. The least secular countries are Nigeria, Pakistan, Bangladesh, Indonesia, Ethiopia, Zimbabwe, Jordan, Egypt, Iraq, and Burma/Myanmar.

Value decay (empirical analysis, see below), is highest, according to our study, in Peru, Mexico, Bolivia, Tunisia, Guatemala, Nicaragua, Iraq, Lebanon, Egypt, and Colombia. The countries with the lowest value decay (empirical analysis, see below) are Denmark, Norway, Tajikistan, China, Netherlands, Vietnam, Bangladesh, Iceland, Burma/Myanmar, and Finland.

As churches, mosques, and religious temples of all denominations are empty in the West, the East and increasingly also in the global South and since "G'd" is said not to "exist," everything becomes feasible, acceptable, and even becomes practice: cheating on taxes, taking bribes, receiving government benefits even if you are not entitled for them. Only at high levels of development, trends seem to reverse again for the better. *Asabiyya* is defined empirically by the residuals from factor scores on value decay in the world system, as the current economic crisis and the COVID-19 pandemic strike the populations of our planet, and cross-national data. *Asabiyya* in the twenty-first century is high secularism combined with a low social decay score.

Finally, we show on the basis of our *World Values Survey* data on social decay and with cross-national data quoted at length in this volume (see Chap. 3) that resilience in the current world crisis is closely connected with these dimensions. The loss of trust in many leading Western societies is statistically closely connected with CV-19 Deaths per 1 million population, and CV-19 Cases per 1 million population; with

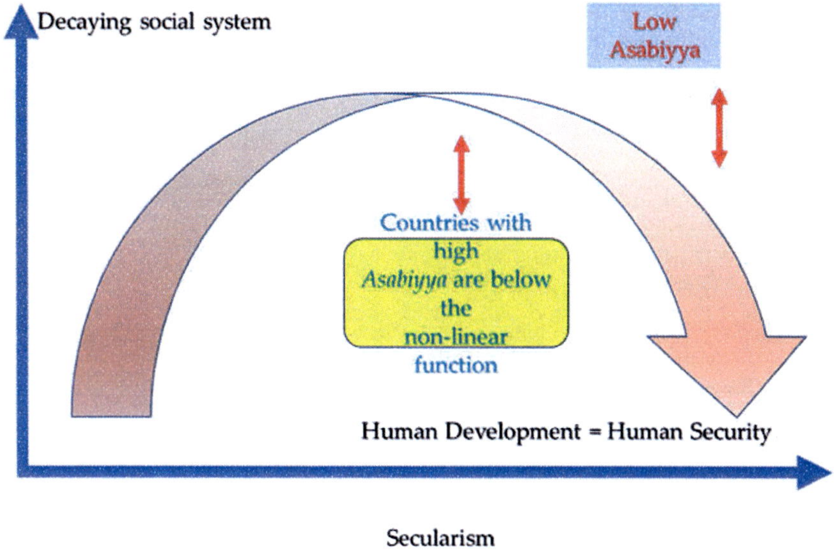


**Fig. 5.1** Asabiyya

unemployment during the present crisis; and with disturbing deficits in the current account balances (Fig. 5.1).

## 5.3 Research Design for the Study of "G-Zero"

In the discipline of quantitative analysis of international relations, it has been long established to use data from cross-national opinion surveys (Alker et al., 1973; Bornschier, 2005; Deutsch, 1979). In this chapter, we will test our propositions with such data, using factor analysis (Alker et al., 1973; Deutsch, 1979). In the first section of our chapter, we will analyze the latest available *PEW Spring Survey* data, 2018. In the current empirical political science literature, multivariate analyses, based on PEW data are now commonplace, and it suffices here to quote the recent article by Arıkan (2021) on public attitudes toward climate change. Our PEW sample covers 3.52 billion people (7483 global population, i.e., 47% global coverage) and the countries of the sample are mentioned in Chap. 2 of this work.

The variables for the multivariate analysis include several great power dimensions and a religion variable and were, in alphabetical order:

1. age
2. agree: China's government disrespects personal freedom
3. agree: China's power no threat
4. agree: climate change no threat

5. dissatisfied with democracy in the country
6. elected officials do not care what people think
7. financial situation worse than 20 years ago or no change
8. gender female
9. agree: ISIS no threat
10. most people do not live in dangerous areas
11. most politicians are not corrupt
12. no confidence in President Xi Jinping
13. agree: North Korean nuke no threat
14. **religion not important in life**
15. unfavorable toward China
16. unfavorable toward Russia
17. unfavorable toward US
18. agree: US government disrespects personal freedom.

The total *N* was 16,043 persons. In presenting the research strategy for this chapter, we should not miss drawing our reader's attention to one more recent exemplary work, based, *inter alia*, on PEW data, (Ciftci,2016), on soft power, religion, and Anti-Americanism in the Middle East. This study used PEW Research Center's Global Attitudes Project (GAP) data from the 2012 GAP surveys and included responses from Egypt, Tunisia, Jordan, and Lebanon.

In the second part of our chapter, we further test our propositions about the current vacuum in global leadership with an analysis of the *World Values Survey* data, here the latest WVS, 2017–2020 comprising 73 variables from 79 countries (see also our publicly available electronic appendix).[4,5] This analysis is based on promax factor analysis (see Chaps. 2, 6, and 8) and on plotting average human values by date of birth to gain an understanding of the support for various values across the generations. The 73 variables of the multivariate analysis can be seen in the publicly available electronic appendix.[6]

## 5.4 Research Results from PEW Spring Survey 2018, Covering 47% of the Population of Our Globe

Having presented the theoretical concepts (see above) and the methodology (see Chap. 2 of the present book) at length, our presentation of the results can now be rather rapid. On a population-weighted basis (see Table 5.1), the Chinese leadership model,

[4] https://www.academia.edu/45321329/Factors_affecting_the_Gulfs_future_Global_cycles_and_value_change_Electronic_Appendix.

[5] Freely available from https://www.academia.edu/44295690/GULF_2020_Covid_19_PUBLIC_ACCESS_Data_base_for_the_book_Factors_affecting_the_Gulfs_future_Global_cycles_and_value_change_Springer_Publishing_Series_https_www_springer_com_series_16417_Gulf_Studies_edited_by_Prof_Mizanur_Rahman_Qatar_University.

[6] See Footnote 4.

**Table 5.1** Global confidence, 2018, in the leaders of China, France, Germany, Russia and the USA according to the PEW Spring Survey, 2018

| Country | Confidence in Xi | Confidence in Trump | Confidence in Putin | Confidence in Macron | Confidence in Merkel |
|---|---|---|---|---|---|
| Russia | 69.26 | 22.36 | 85.41 | 35.54 | 34.78 |
| Philippines | 61.00 | 82.15 | 69.13 | 73.62 | 72,36 |
| Tunisia | 66.91 | 14.77 | 61.50 | 67.09 | 68.44 |
| Nigeria | 67.39 | 68.52 | 53.45 | 58.63 | 58.59 |
| Kenya | 67.85 | 60.06 | 51.69 | 59.77 | 59.38 |
| India | 34.08 | 75.09 | 77.69 | 60.83 | 65.01 |
| Indonesia | 54.66 | 34.67 | 53.62 | 59.49 | 55.72 |
| South Africa | 60.51 | 43.26 | 46.89 | 52.99 | 56.53 |
| Israel | 42.79 | 57.53 | 37.31 | 42.76 | 54.54 |
| Australia | 47.04 | 32.42 | 24.71 | 77.71 | 78.06 |
| Germany | 35.87 | 8.65 | 35.66 | 80.64 | 70.75 |
| Korea, South | 37.50 | 45.85 | 33.09 | 70.28 | 82.32 |
| Canada | 46.21 | 20.90 | 22.82 | 75.21 | 80.25 |
| Argentina | 30.40 | 13.37 | 34.92 | 31.37 | 31.03 |
| United States | 44.00 | 48.72 | 20.74 | 69.95 | 68.21 |
| United Kingdom | 43.38 | 25.68 | 20.14 | 63.40 | 69.42 |
| Greece | 15.67 | 16.52 | 46.98 | 33.67 | 17.35 |
| Hungary | 28.29 | 34.47 | 32.94 | 42.61 | 34.45 |
| Italy | 19.79 | 28.34 | 34.37 | 32.30 | 40.70 |
| Netherlands | 38.47 | 17.30 | 13.47 | 79.28 | 89.90 |
| Sweden | 32.75 | 17.60 | 16.26 | 81.01 | 89.54 |
| France | 28.97 | 8.39 | 19.68 | 63.64 | 78.66 |
| Mexico | 24.34 | 5.43 | 23.71 | 26.00 | 33.13 |
| Japan | 16.67 | 32.65 | 25.16 | 70.01 | 82.11 |
| Brazil | 17.81 | 16.04 | 16.67 | 22.82 | 33.88 |
| Spain | 15.38 | 6.39 | 8.69 | 50.00 | 57.00 |
| Poland | 9.96 | 35.68 | 5.94 | 32.17 | 37.35 |
| **Total global population** | **27.63** | **50.17** | **52.51** | **57.16** | **60.35** |

whose support we measure by support expressed for President Xi Xinping, is accepted by only 27.63% of the global population. The dismal United States Leadership under President Donald Trump was backed by only 50.17% of the citizens of our globe, while 52.51% of global citizens "voted" for Russian President Vladimir Putin, while 57.16% were expressing support for French President Emmanuel Macron, and finally 60.35% were in favor of German Chancellor Angela Merkel. The inroads, Chinese or Russian leadership concepts could make among Western publics are considerable. Our choropleth maps in Figs. 5.2, 5.3, 5.4, 5.5, and 5.6 further underline our argument about a weakening West in a time of global crisis.

It is to be noted, that each Western leadership—Trump, Macron, and Merkel, also appears to have severe favourability problems in some neighboring countries, and, as is the case with Macron and Merkel, in their own home countries as well.

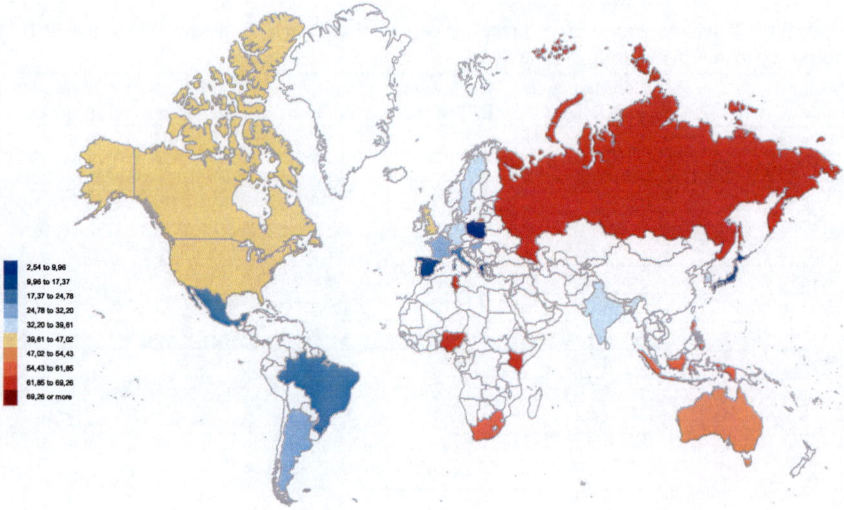


**Fig. 5.2** Confidence in Xi (27.63% of the global population)

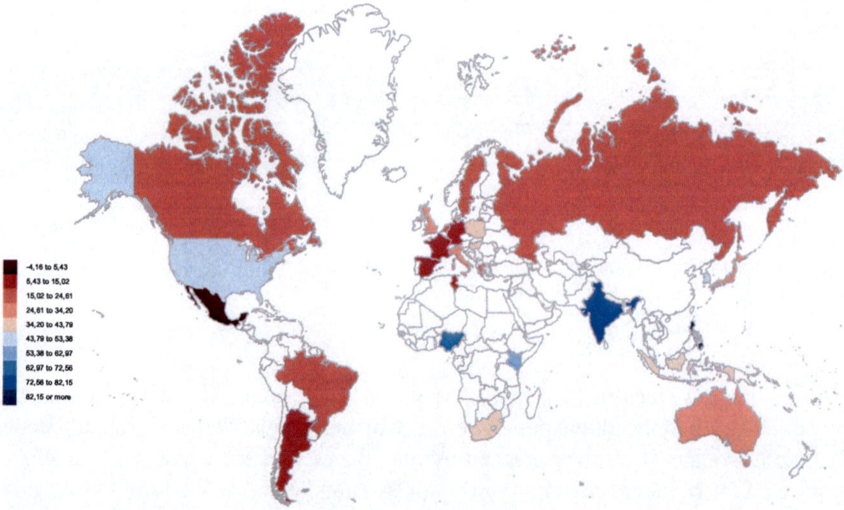


**Fig. 5.3** Confidence in Trump (50.17% of the global population)

The promax factor analysis of the PEW data reveals interesting patterns. There are six factors, explaining almost 53% of the total variance of the opinions of the citizens of our globe. Distance to China and Russia still goes hand in hand; individuals, skeptical of China, tend to be skeptical of Russia, and vice versa, and of course, support for China coincides with support for Russia. The second factor shows that

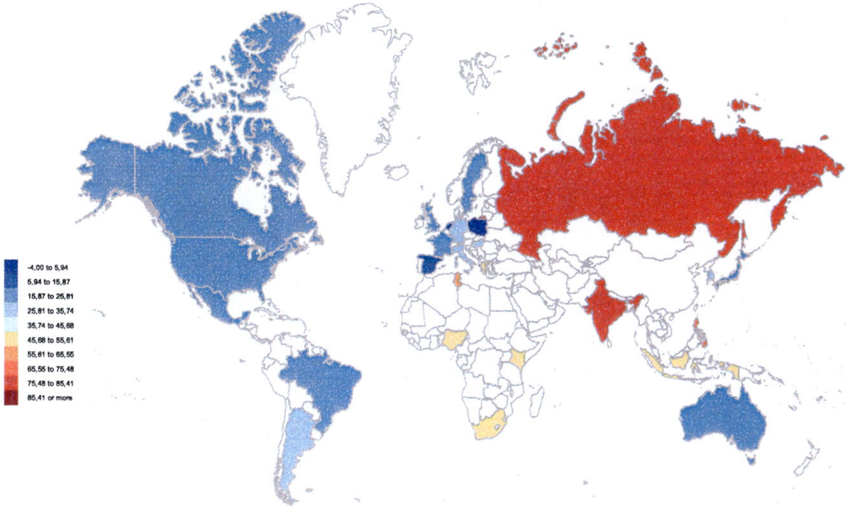


**Fig. 5.4** Confidence in Putin (52.51% of the global population)

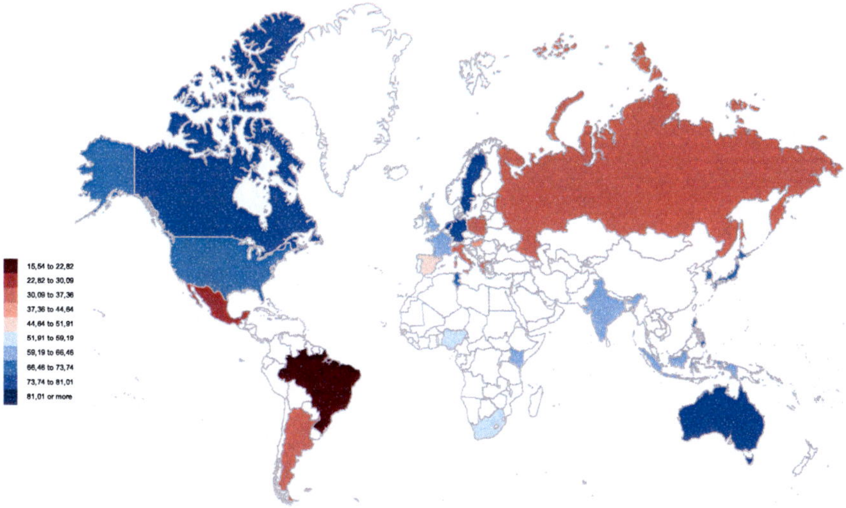


**Fig. 5.5** Confidence in Macron (57.16% of the global population)

a low level of threat perception from the Islamist terror organization ISIS goes hand in hand with a low level of threat perception from North Korea, climate change, and Chinese power, reflecting a strongly isolationist value orientation. Anti-Americanism tends to be combined with the rejection of other powers; global publics, which perceive no corruption around them, tend to enjoy security; and furthermore, the

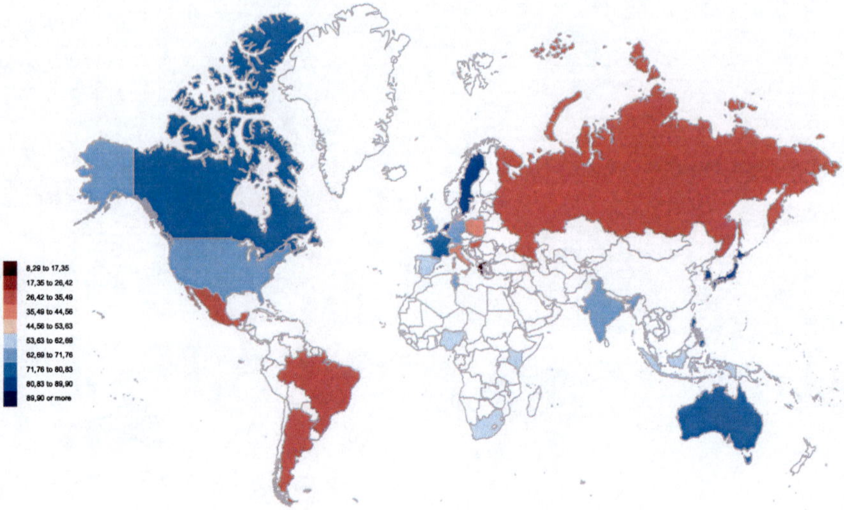


**Fig. 5.6** Confidence in Merkel (60.35% of the global population)

promax factor analysis with the PEW data also shows that there is a considerable alienation of the younger generations, and that secularism is supported above all by people of male gender. Our model explains 52,802% of the total variance. Figures 5.7 and 5.8 as well as Table 5.2 and the data reproduced in our freely available electronic

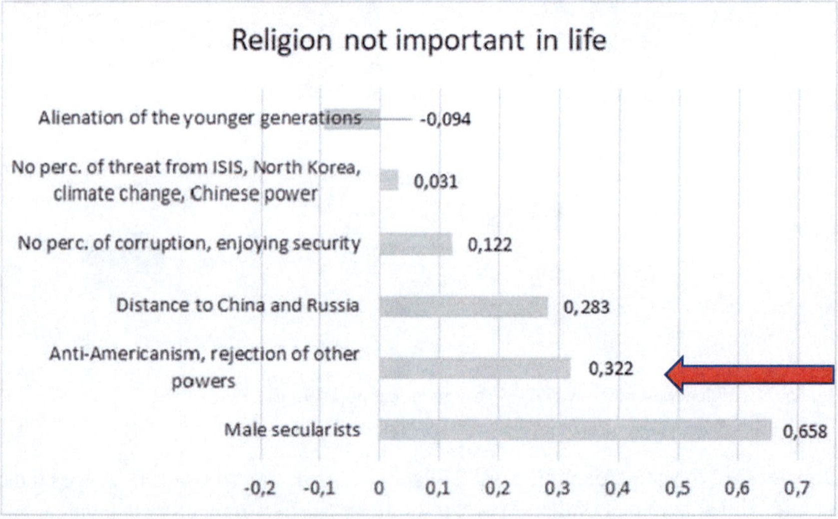


**Fig. 5.7** Loadings of the variable "*Religion not important in life*" with the six promax factors of our model

**Fig. 5.8** The G-Zero world—loadings of the variables "*financial situation worse than 20 years ago*" and "*dissatisfied with democracy in the country*" with the six promax factors of our model

appendix[7,8] further highlight the results of our factor analysis of the PEW data.

The

- least **distance to China and Russia** was to be observed in Tunisia, Kenya, Russia, Nigeria, Indonesia, South Africa, Philippines, India, Israel, and Argentina, while the **strongest reserves against Chinaand Russia** were felt in Japan, Poland, USA, Spain, France, Sweden, Australia, Italy, Netherlands, and Canada.

For the following factors, we again mention the ten countries with the highest factor scores and the ten countries with the lowest factor scores.

- **No perception of threat from ISIS, North Korea, climate change, Chinese power**: Highest: Israel, Russia, Nigeria, Tunisia, South Africa, Sweden, Canada, Hungary, UK, and Netherlands. Lowest: in South Korea, France, Japan, Greece, Spain, Italy, Indonesia, India, Philippines, and Kenya.
- **Anti-Americanism, rejection of other powers**: Highest: France, Netherlands, Sweden, Spain, Canada, Greece, Germany, UK, Mexico, and Australia and Lowest: in India, Philippines, Nigeria, Kenya, Indonesia, South Korea, Israel, South Africa, USA, and Poland.
- **No perception of corruption, enjoying security**: Highest in Indonesia, Mexico, Philippines, Netherlands, Sweden, Germany, Canada, Australia, Japan, and

[7] See Footnote 4.

[8] See Footnote 5.

**Table 5.2** Factor correlations of the promax factor analysis PEW September 2018 Global Survey

| Component correlation | Distance to China and Russia | No perc. of threat from ISIS, North Korea, climate change, Chinese power | Anti-Americanism, rejection of other powers | No perc. of corruption, enjoying security | Alienation of the younger generations | Male secularists |
|---|---|---|---|---|---|---|
| Distance to China and Russia | 1.000 | − 0.197 | 0.275 | − 0.007 | 0.001 | 0.230 |
| No perception of threat from ISIS, North Korea, climate change, Chinese power | | 1.000 | − 0.058 | 0.066 | − 0.055 | − 0.022 |
| Anti-Americanism, rejection of other powers | 0.275 | − 0.058 | 1.000 | − 0.075 | 0.008 | 0.282 |
| No perception of corruption, enjoying security | − 0.007 | 0.066 | − 0.075 | 1.000 | − 0.195 | − 0.044 |
| Alienation of the younger generations | 0.001 | − 0.055 | 0.008 | − 0.195 | 1.000 | 0.081 |
| Male secularists | 0.230 | − 0.022 | 0.282 | − 0.044 | 0.081 | 1.000 |

Kenya; Lowest: in Greece, Tunisia, Russia, Italy, South Africa, Nigeria, India, USA, Argentina, and Hungary.

- **Alienation of the younger generations**: Highest: in Mexico, Brazil, Argentina, Greece, Tunisia, South Africa, Spain, Kenya, Nigeria, and Italy; Lowest: in Sweden, Netherlands, Poland, Australia, Canada, Germany, Indonesia, India, UK, and Israel.
- **Male secularists**: Highest: in Spain, South Korea, Japan, Sweden, France, UK, Netherlands, Germany, Australia, and Hungary; Lowest: in Indonesia, Philippines, Kenya, Nigeria, South Africa, Tunisia, India, Russia, Mexico, and Israel.

Figure 5.7 shows the promax factor loadings of the variable "religion is not important in life" with the six factors of our model, and the clear result that secularism goes hand in hand with Anti-Americanism and the rejection of other powers, so typical for countries like France, Netherlands, Sweden, Spain, Canada, Greece, Germany, UK, Mexico, and Australia. In the twenty-first century, America's friends typically are the publics in India, Philippines, Nigeria, Kenya, Indonesia, South Korea, Israel, South Africa, and Poland, most of them characterized by lower rates of secularism. Figure 5.8 and Table 5.2 show the further important relationships of our factor analytical model.

## 5.5 Research Results from the Latest Wave of the *World Values Survey*, 2017–2020: Promax Factor Analysis

Our following analysis was based on 73 variables from the latest edition of the *World Values Survey*. We again applied promax factor analysis, and due to the structure of the *eigenvalues* of the model (see Fig. 5.9), we just interpreted eight factors, which explained 35.26% of the total variance.

Due to the loadings of the factors, we interpreted them as

- Secularism
- No confidence in societal institutions
- Feminism
- Distrust
- Democracy and redistribution
- Corruption and shadow economy
- Marginalization
- Elderly people.

In order not to overburden the presentation with statistical results, we mention the factor loadings in the publicly available electronic appendix[9,10] and concentrate

[9] See Footnote 4.

[10] See Footnote 5.

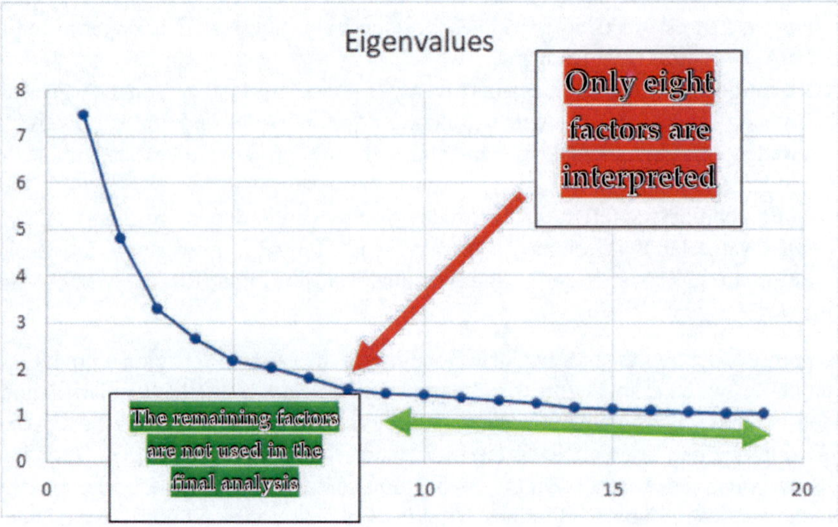


**Fig. 5.9** Eigenvalues of the factor analysis, *World Values Survey*, 2017–2020

here rather on the promax factor correlations (see Table 5.3 and Fig. 5.10). Contrary to Inglehart's diagnosis, secularism is indeed connected to an erosion of trust and an acceptancy of corruption and the shadow economy.

In accordance with the advanced parametric indicator methodology presented in Chap. 2, relying on factor analysis, we multiplied the factor scores of all the countries (see publicly available electronic appendix)[11,12] with the eigenvalues of the following factors from our model to arrive at a parametric indicator of decaying social systems:

- No confidence in societal institutions 4.796
- Distrust 2.634
- Corruption and shadow economy 2.019
- Marginalization 1.827.

The results of this operation are to be seen in Table 5.4 and Fig. 5.11:

The global map of social decay is seen in Fig. 5.11. Truly vast areas of Latin America, Latin Europe, Eastern and Southeastern Europe, the mainly Orthodox regions of the former USSR, and the Arab World are affected by strong social decay, while only a few fortunate European countries (Scandinavia, Estonia, Belarus, UK, Belgium, Portugal, Macedonia, Montenegro, the Slovak Republic, Turkey, and the mainly German speaking countries Austria, Germany, and Switzerland) seem to be able to escape from social decay. Our choropleth map shows as well, how inexorably the center of gravitation not only of the world economy, but also of value development already has shifted toward East and Southeast Asia and the Pacific

[11] See Footnote 4.

[12] See Footnote 5.

**Table 5.3** Promax factor correlations *World Values Survey*, 2017–2020

| Correlations | Secularism | No confidence in societal institutions | Feminism | Distrust | Democracy and redistribution | Corruption and shadow economy | Marginalization | Elderly people |
|---|---|---|---|---|---|---|---|---|
| Secularism | 1.000 | 0.157 | 0.528 | −0.337 | 0.161 | 0.160 | −0.133 | 0.138 |
| No confidence in societal institutions | 0.157 | 1.000 | 0.199 | 0.181 | −0.092 | −0.019 | 0.009 | 0.081 |
| Feminism | 0.528 | 0.199 | 1.000 | −0.337 | 0.166 | 0.054 | −0.216 | 0.126 |
| Distrust | −0.337 | 0.181 | −0.337 | 1.000 | −0.173 | −0.040 | 0.162 | −0.115 |
| Democracy and redistribution | 0.161 | −0.092 | 0.166 | −0.173 | 1.000 | −0.057 | −0.152 | 0.104 |
| **Corruption and shadow economy** | **0.160** | **−0.019** | **0.054** | **−0.040** | **−0.057** | **1.000** | **0.116** | **−0.129** |
| Marginalization | −0.133 | 0.009 | −0.216 | 0.162 | −0.152 | 0.116 | 1.000 | −0.068 |
| Elderly people | 0.138 | 0.081 | 0.126 | −0.115 | 0.104 | −0.129 | −0.068 | 1.000 |

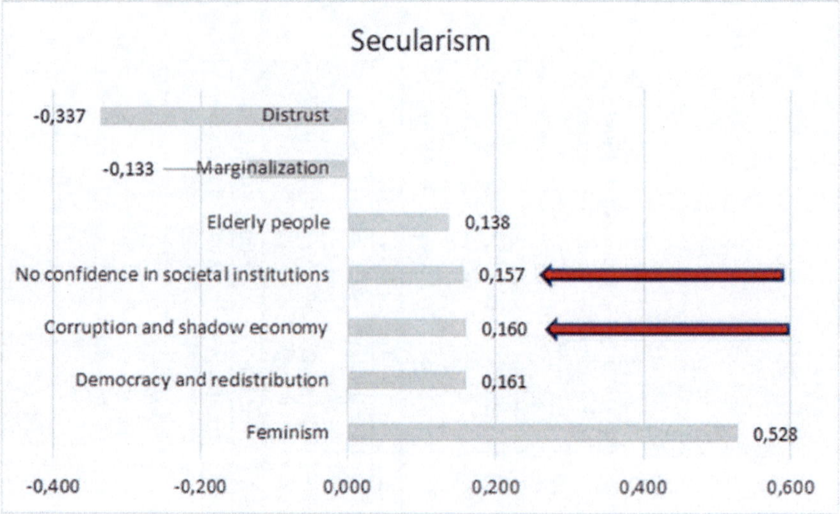


**Fig. 5.10** Contradictions of secularism: increasing low confidence in societal institutions, inclination toward accepting corruption and the shadow economy

region, with its dynamism affecting also already the Indian Ocean region and the countries in Sub-Saharan Africa. We already mentioned that value decay is highest according to our study, in Peru, Mexico, Bolivia, Tunisia, Guatemala, Nicaragua, Iraq, Lebanon, Egypt, and Colombia. The countries with the lowest value decay are Denmark, Norway, Tajikistan, China, Netherlands, Vietnam, Bangladesh, Iceland, Burma/Myanmar, and Finland.

Figure 5.12 now adapts the logic put forward in Fig. 5.1 to the world of the real data as they emerge from this chapter and our cross-national database.[13]

The scatterplot is in the expected shape from Fig. 5.1, and the $R^2$ is 24.2%. *Asabiyya* 1 is defined by the residuals below (positive) or above (negative) the nonlinear regression line.

Figure 5.13 projects these residuals onto a choropleth world map. Countries with a high *Asabiyya* manage to have low social decay, even if their Human Development Index is high, while countries with a negative *Asabiyya* are characterized by higher social decay irrespective of their Human Development level.

Table 5.5 further elaborates the entire drama of social decay irrespective of the level of the Human Development Index, which now affects large parts of the mainly Roman Catholic and also the Protestant West and the mainly Orthodox East, and also the Arab countries with available data, while countries in the religious traditions of Buddhism and Hinduism, as well as China with its mainly Confucian heritage, as well as some notable countries in the Muslim world, like Indonesia, Pakistan, former Soviet Central Asia, Iran and Turkey seem to escape from the fatal societal logic of

---

[13] See Footnote 5.

**Table 5.4** Decaying social system and the G-Zero

| | Decaying social system |
|---|---|
| Denmark | − 5.905 |
| Norway | − 5.888 |
| Tajikistan | − 5.007 |
| China | − 4.845 |
| Netherlands | − 4.269 |
| Vietnam | − 4.121 |
| Bangladesh | − 3.786 |
| Iceland | − 3.669 |
| Myanmar | − 3.388 |
| Finland | − 3.266 |
| Thailand | − 2.855 |
| New Zealand | − 2.853 |
| Macau SAR | − 2.591 |
| Switzerland | − 2.571 |
| Ethiopia | − 2.549 |
| Kyrgyzstan | − 2.522 |
| Pakistan | − 2.521 |
| Germany | − 2.131 |
| Austria | − 2.104 |
| Kazakhstan | − 1.981 |
| Indonesia | − 1.859 |
| Turkey | − 1.807 |
| United Kingdom | − 1.737 |
| Nigeria | − 1.685 |
| Azerbaijan | − 1.543 |
| Philippines | − 1.531 |
| Iran | − 0.994 |
| Portugal | − 0.876 |
| Estonia | − 0.844 |
| Taiwan ROC | − 0.839 |
| Australia | − 0.730 |
| Montenegro | − 0.581 |
| Japan | − 0.441 |
| United States | − 0.226 |
| Hungary | 0.002 |
| Macedonia | 0.017 |
| Malaysia | 0.068 |
| Georgia | 0.117 |
| Andorra | 0.213 |
| Belarus | 0.269 |
| Zimbabwe | 0.331 |
| Spain | 0.464 |
| Poland | 0.586 |
| Cyprus | 0.694 |

(continued)

**Table 5.4** (continued)

| | |
|---|---|
| Korea, South | 0.809 |
| Italy | 0.870 |
| Lithuania | 0.955 |
| France | 0.974 |
| Hong Kong SAR | 0.976 |
| Slovakia | 1.050 |
| Puerto Rico | 1.179 |
| Armenia | 1.277 |
| Jordan | 1.969 |
| Argentina | 2.144 |
| Slovenia | 2.145 |
| Bulgaria | 2.287 |
| Bosnia and Herzegovina | 2.374 |
| Romania | 2.519 |
| Russia | 2.522 |
| Greece | 2.528 |
| Czech Republic | 2.716 |
| Chile | 2.722 |
| Brazil | 2.739 |
| Croatia | 3.429 |
| Albania | 3.752 |
| Ukraine | 3.760 |
| Serbia | 4.191 |
| Ecuador | 4.241 |
| Colombia | 4.301 |
| Egypt | 4.766 |
| Lebanon | 4.805 |
| Iraq | 4.811 |
| Nicaragua | 4.965 |
| Guatemala | 5.088 |
| Tunisia | 5.259 |
| Bolivia | 5.776 |
| Mexico | 6.119 |
| Peru | 6.774 |

Fig. 5.1. It emerges from Fig. 5.13 and from Table 5.5, just how deeply the West—here especially the Western countries in the Roman Catholic tradition—are already affected by a deep moral and social crisis, stemming from the inability to avoid social decay in an era of secularization. This inability, which also affects the Arab World, also negatively affects the ability to come to terms with the current COVID-19 pandemic, and it further undermines the fabric of trust which is so important for future economic growth and socio-economic development. Of all social scientists, the late Harvard economist Alberto Alesina (1957–2020) deeply understood how distrust undermines economic growth (Alesina & Ferrara, 2000, 2002; Alesina &

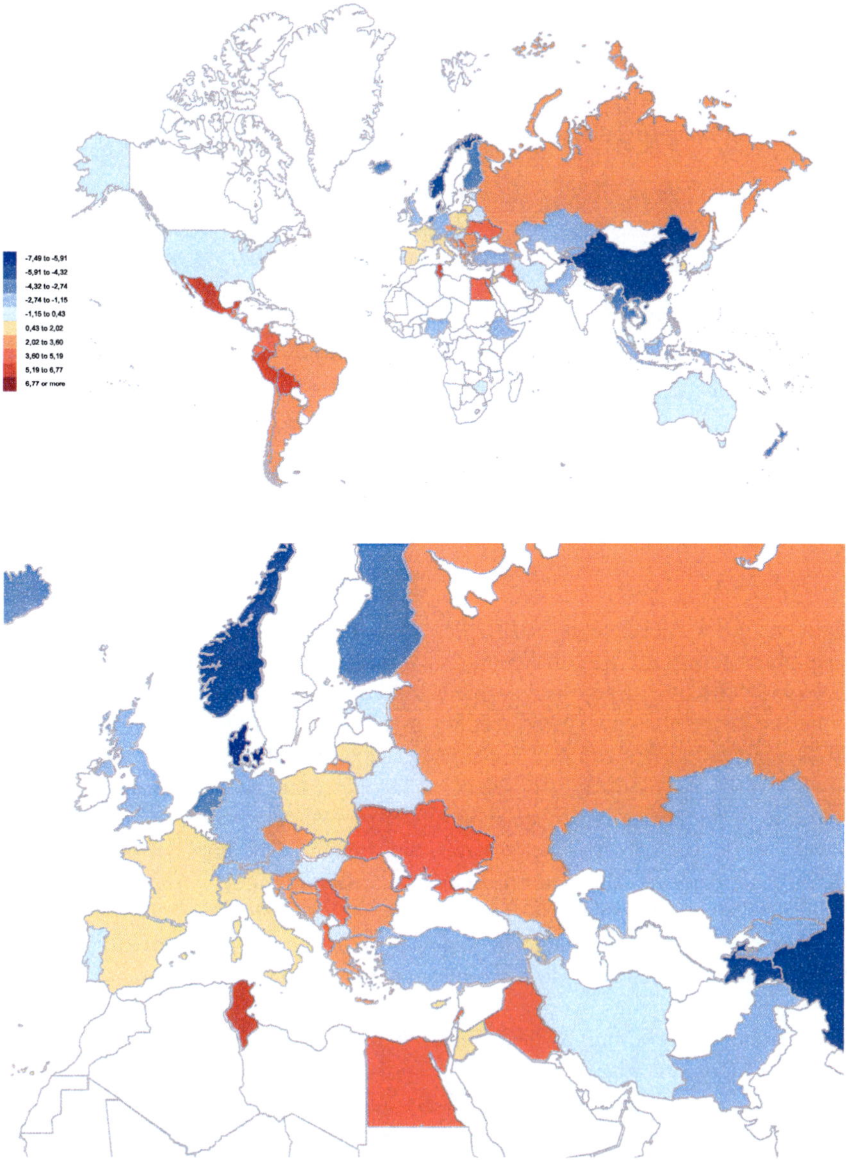


**Fig. 5.11** Decaying social systems

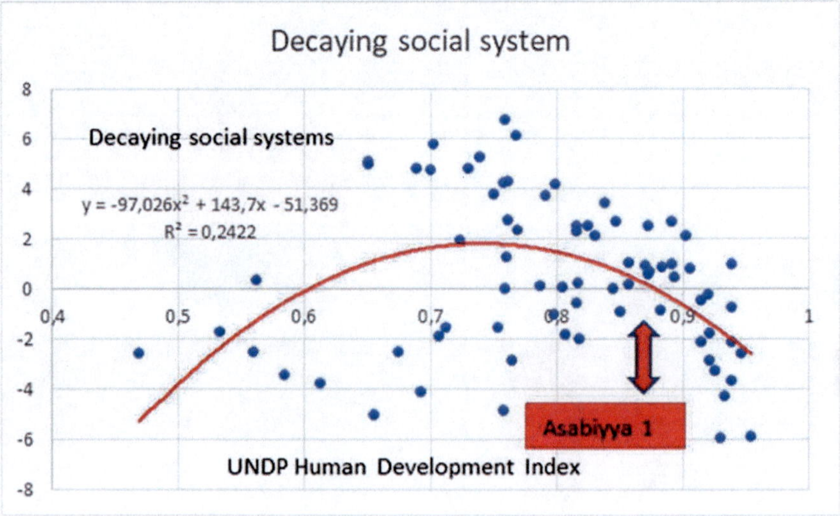


**Fig. 5.12** Human Development Index and decaying social systems

Guiliano, 2015; Alesina et al., 2015). The Pearson Bravais correlations of *Asabiyya* 1 clearly bear testimony to this reading of actual global developments.

Figure 5.14 now draws a choropleth map of global secularism, as it emerges from our factor analytical interpretation of the *World Values Survey* data, 2017–2020, analyzed in this chapter. We already mentioned above that according to the *World Values Survey* data, 2017–2020, analyzed in this chapter, the most secular countries were Japan, New Zealand, Australia, Denmark, Switzerland, Norway, Czech Republic, Germany, Netherlands, and Finland. The least secular countries are Nigeria, Pakistan, Bangladesh, Indonesia, Ethiopia, Zimbabwe, Jordan, Egypt, Iraq, and Burma/Myanmar.

Figure 5.15 now again adapts the logic put forward in Fig. 5.1 above to the world of the real data as they emerge from this chapter and our cross-national database.[14] The scatterplot is in the expected shape of Fig. 5.1, and the $R^2$ is 24.2%. *Asabiyya* 2 is again defined, as before, by the residuals below (positive) or above (negative) the nonlinear regression line (Fig. 5.15).

Figure 5.16 again projects these residuals onto a choropleth world map. Countries with a high *Asabiyya* 2 manage to have low social decay, even if their Secularization is advanced, while countries with a negative *Asabiyya* 2 are characterized by higher social decay irrespective of their Secularization level. *Asabiyya,* 2 is highest in Tajikistan, Vietnam, Thailand, China, Kyrgyzstan, Burma/Myanmar, Kazakhstan, Turkey, Norway, and Denmark. *Asabiyya,* 2 is lowest in Czech Republic, Peru, Egypt, Iraq, Bolivia, Mexico, Tunisia, Nicaragua, Guatemala, and Slovenia.

Table 5.6 further elaborates again the entire drama of social decay irrespective of the level of Secularization which now affects large parts of the West and the mainly

[14] See Footnote 5.

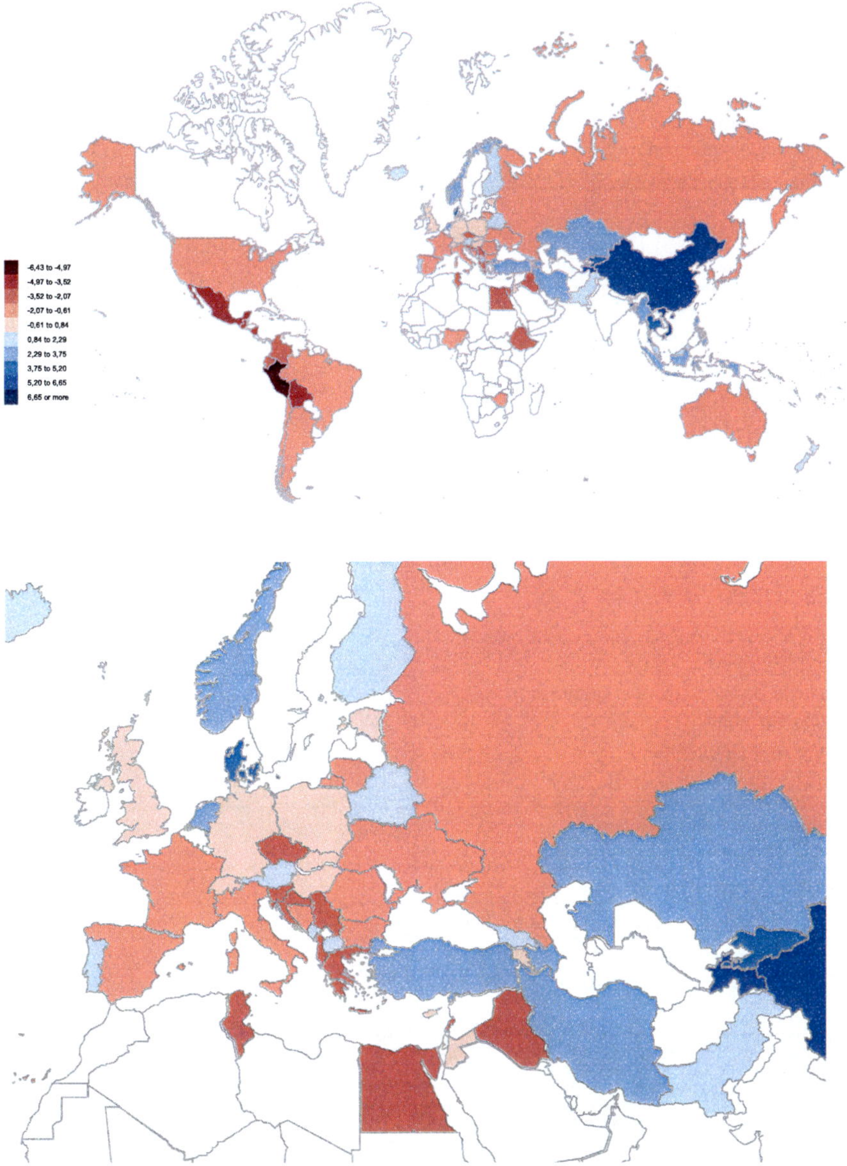


**Fig. 5.13** *Asabiyya* 1—little social decay compared to the level of the Human Development Index

**Table 5.5** The bi-variate correlations of *Asabiyya* 1

| | Pearson correlation with *Asabiyya* 1 | Error *p* | *N* |
|---|---|---|---|
| Trust in institutions | 0.672 | 0.000 | 41 |
| Gallup poll about satisfaction: Trust in national government | 0.568 | 0.000 | 70 |
| Gallup poll about satisfaction: Safety | 0.500 | 0.000 | 72 |
| Gallup poll about satisfaction: Trust in other people | 0.472 | 0.000 | 66 |
| Gallup poll about satisfaction: Education quality | 0.414 | 0.000 | 72 |
| Gallup poll about satisfaction: Efforts to deal with the poor | 0.372 | 0.001 | 72 |
| Gallup poll about satisfaction: Health care quality | 0.330 | 0.005 | 72 |
| DYN GDP pc PPP 2020 | 0.317 | 0.006 | 74 |
| Total population in millions | 0.301 | 0.011 | 70 |
| % world population | 0.300 | 0.012 | 69 |
| Share of Buddhists per total population | 0.297 | 0.011 | 72 |
| GSN Economic | 0.294 | 0.033 | 53 |
| Share of people without religion per total population | 0.271 | 0.021 | 72 |
| DYN HDI 1990–2018 | 0.266 | 0.032 | 65 |
| DYN GDP pc PPP 2021 | 0.266 | 0.023 | 73 |
| Gallup poll about satisfaction: Local labor market | 0.250 | 0.034 | 72 |
| Closing economic gender gap | 0.248 | 0.043 | 67 |
| Share of Hindus per total population | 0.248 | 0.036 | 72 |
| CV-19 Cases per 1 million population, end September 2020 | – 0.237 | 0.040 | 75 |
| Arab League | – 0.259 | 0.025 | 75 |
| Military expenditures per GDP | – 0.260 | 0.045 | 60 |
| GINI Income | – 0.282 | 0.029 | 60 |
| Democracy measure | – 0.298 | 0.014 | 68 |
| CV-19 Deaths per 1 million population, end September 2020 | – 0.326 | 0.004 | 75 |
| Homicide rate | – 0.349 | 0.003 | 72 |
| Quintile share income difference between richest and poorest 20% | – 0.354 | 0.004 | 65 |
| Accepting the market economy | – 0.426 | 0.005 | 41 |
| Share of Roman Catholics per total population | – 0.459 | 0.000 | 72 |

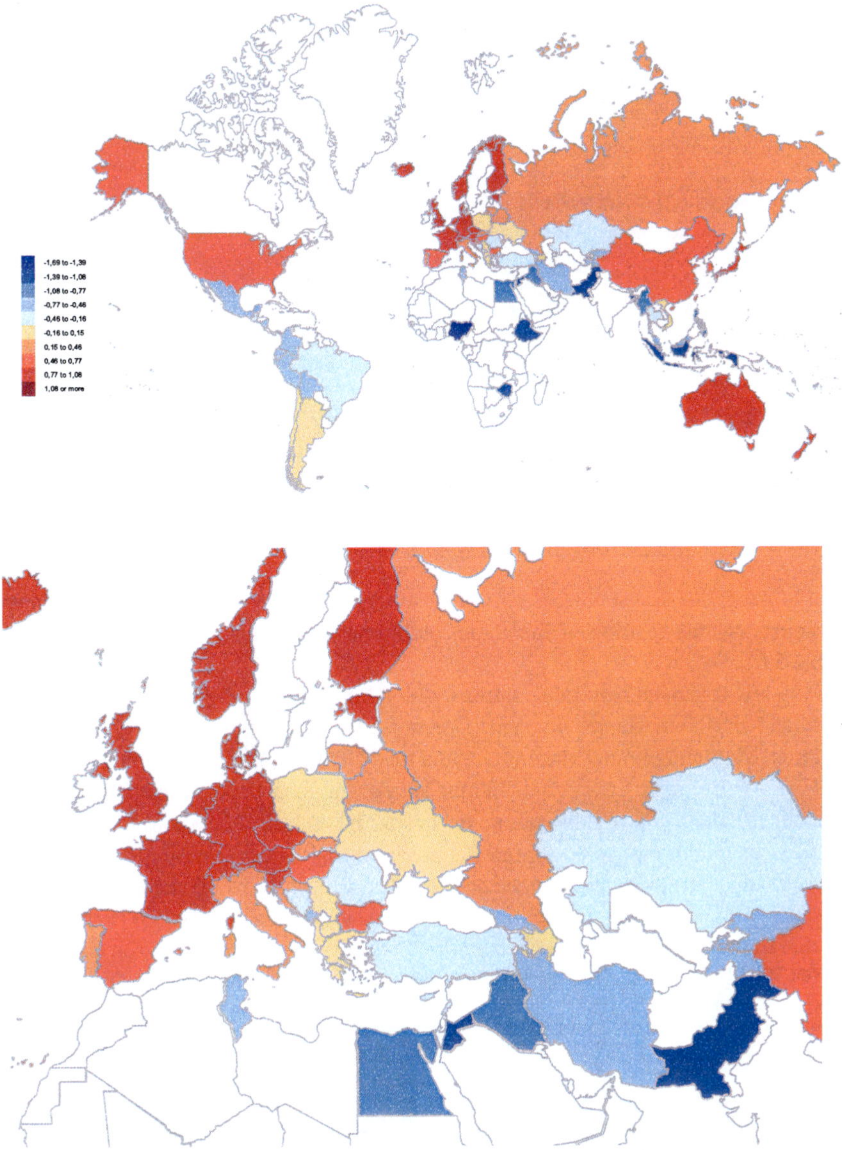


**Fig. 5.14** Secularism

Orthodox East, and also the Arab countries with available data. Again, countries in the religious traditions of Buddhism and Hinduism, as well as China with its mainly Confucian heritage, and some fortunate Muslim countries again seem to escape from this fatal societal logic. Also, the correlations of *Asabiyya* 2 again clearly bear

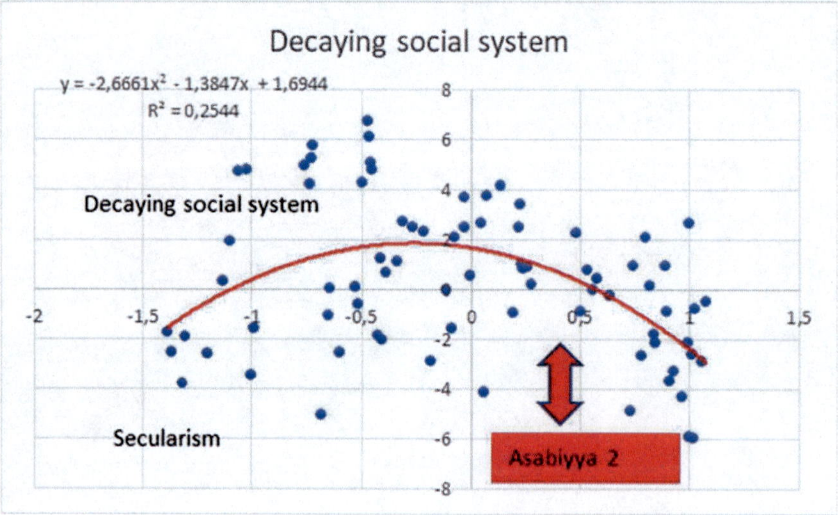


Fig. 5.15 Secularism and decaying social systems

testimony to this reading of the actual global developments (Figs. 5.15, 5.16, and Table 5.6).

The few fortunate European nations with a higher *Asabiyya* 2 are—clockwise—Portugal, the Netherlands, Iceland, Denmark, Norway, Austria, Finland, Poland, Belarus, Montenegro and Macedonia, and the Muslim majority country Turkey.

Figures 5.17 and 5.18 provisionally answer Inglehart's question (2021) about the direct effects of secularization on economic morality (claiming government benefits to which one is not entitled, and accepting bribes). While in both cases, distrust and marginalization increase the disposition toward claiming government benefits to which one is not entitled, and accepting bribes, secularism is negatively related to these phenomena. This could be regarded as a positive test for Inglehart's theory, 2021. But in the following paragraph, we will show that the picture is more complicated (see below).

## 5.6 Research Results from the Latest Wave of the *World Values Survey*, 2017–2020: The Buddenbrooks Cycle (Generational Aspects of Human Values)

In the first round of tests for Inglehart's hypotheses about value change, 2021, we attempted to find out whether or not there is a difference in the distribution of value orientations across the generations between global citizens and global Muslim citizens. Are the adherents of Islam and the global citizenry. Are the adherents of global

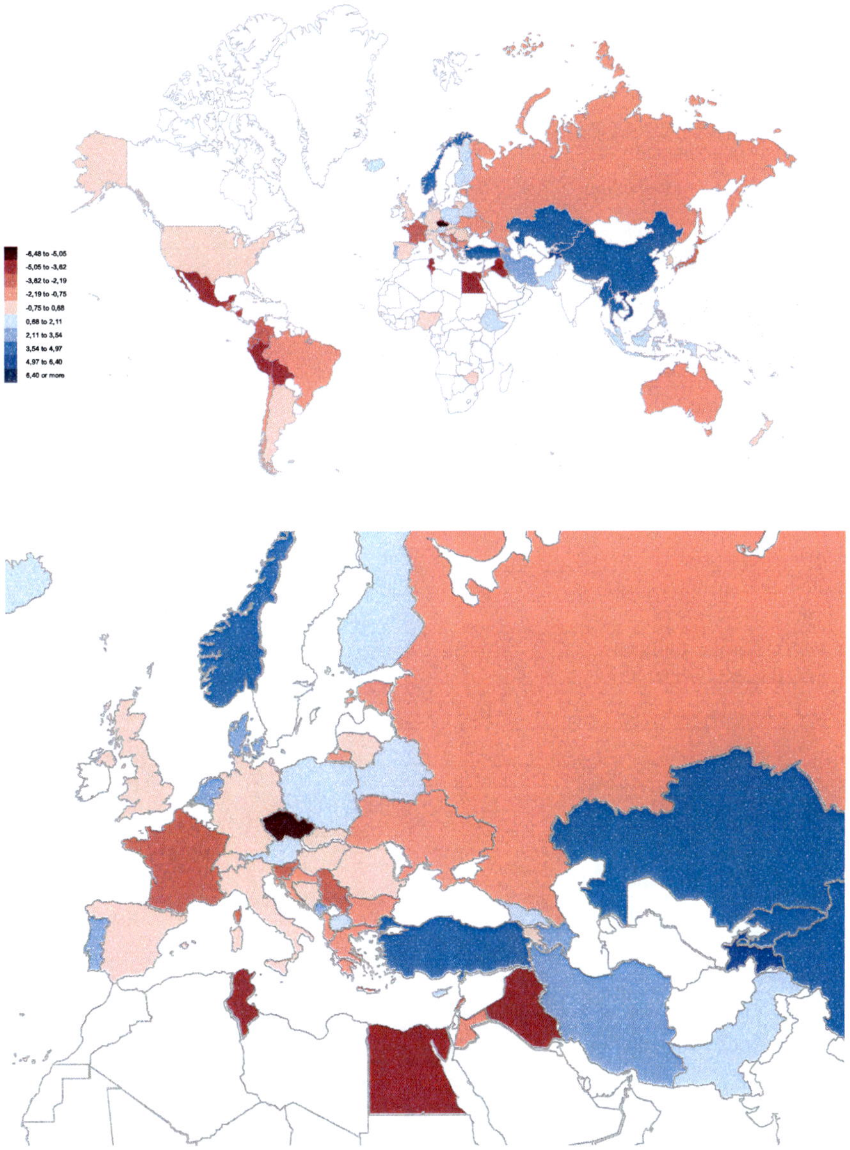


**Fig. 5.16** *Asabiyya* 2—little social decay compared to the level of secularism

Islam more immune from the corrosive effects, which secularization seems to have, as it is often alleged in the public debate, especially among the younger generations, or is the picture much more complex? Based on the *World Values Survey*, WVS Longitudinal 1981–2016, SPSS Version 20180912, we exactly can get the following results based on the sufficient number of respondents when we plot average values

**Table 5.6** The bi-variate correlations of *Asabiyya* 2

| | Pearson correlation with Asabiyya 2 | Error *p* | *N* |
|---|---|---|---|
| Trust in institutions | 0.615 | 0.000 | 42 |
| Gallup poll about satisfaction: Trust in national government | 0.468 | 0.000 | 70 |
| Gallup poll about satisfaction: Safety | 0.427 | 0.000 | 72 |
| Gallup poll about satisfaction: Trust in other people | 0.408 | 0.001 | 66 |
| Gallup poll about satisfaction: Education quality | 0.378 | 0.001 | 72 |
| DYN HDI 2000–2010 | 0.337 | 0.004 | 72 |
| Gallup poll about satisfaction: Efforts to deal with the poor | 0.302 | 0.010 | 72 |
| Average annual HDI growth 2000–2013 | 0.298 | 0.015 | 66 |
| Share of Buddhists per total population | 0.296 | 0.011 | 73 |
| Gallup poll about satisfaction: Health care quality | 0.251 | 0.034 | 72 |
| CV-19 Cases per 1 million population. end September 2020 | – 0.238 | 0.037 | 77 |
| Deaths per total cases | – 0.241 | 0.036 | 76 |
| GINI Income | – 0.280 | 0.030 | 60 |
| Democracy measure | – 0.281 | 0.020 | 68 |
| MENA country | – 0.288 | 0.011 | 78 |
| CV-19 Deaths per 1 million population, end September 2020 | – 0.322 | 0.005 | 76 |
| Quintile share income difference between richest and poorest 20% | – 0.340 | 0.006 | 65 |
| Arab League | – 0.362 | 0.001 | 78 |
| Distrust of the army and the press | – 0.373 | 0.011 | 46 |
| Share of Roman Catholics per total population | – 0.430 | 0.000 | 73 |
| Accepting the market economy | – 0.439 | 0.004 | 42 |

per each birth year of the global citizenry and the global Muslim citizenry from 1926 to 1996, i.e., investigating value change for a very long haul of seven decades among the generations. All figures which we can present, could be important inputs for the further debates about G-Zero in the world system and the rise of global populism. Figure 5.19 clearly shows that there is the stark reality of declining support for gender equality as a goal of democracy among the global younger generations among the

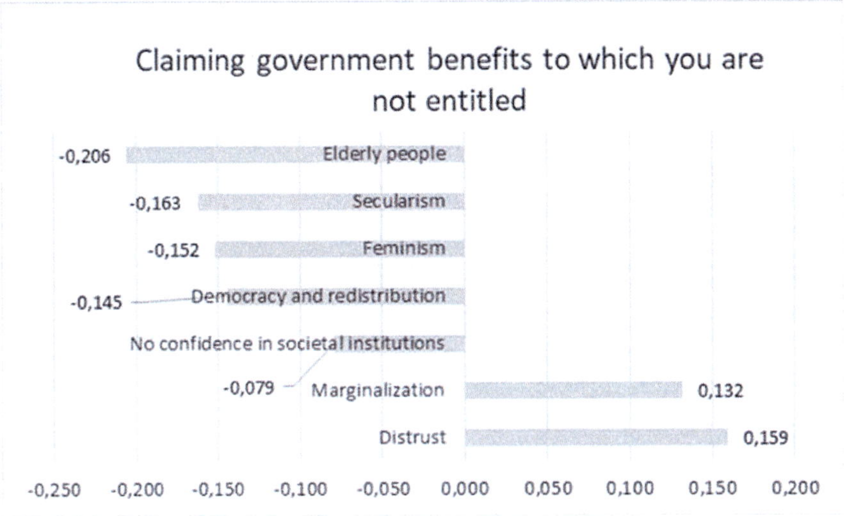


**Fig. 5.17** Factor loadings of claiming government benefits to which you are not entitles with the eight promax factors of our model

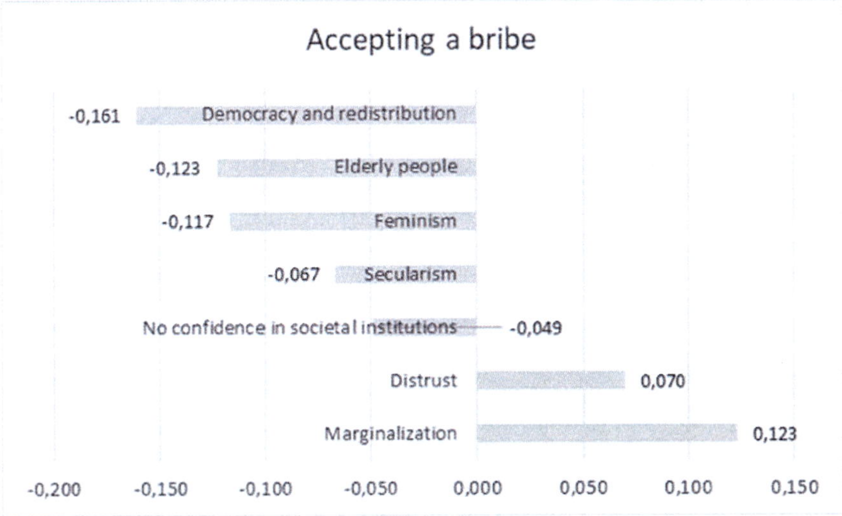


**Fig. 5.18** Factor loadings of accepting a bribe with the eight promax factors of our model

global citizenry, while among the global Muslims, their initially lower support for gender equality as a goal of democracy now converges with that of the global citizens.

Figure 5.20 is as surprising as Fig. 5.19. There is a rising support for the role of religious authorities in the interpretation of the laws in a democracy among the

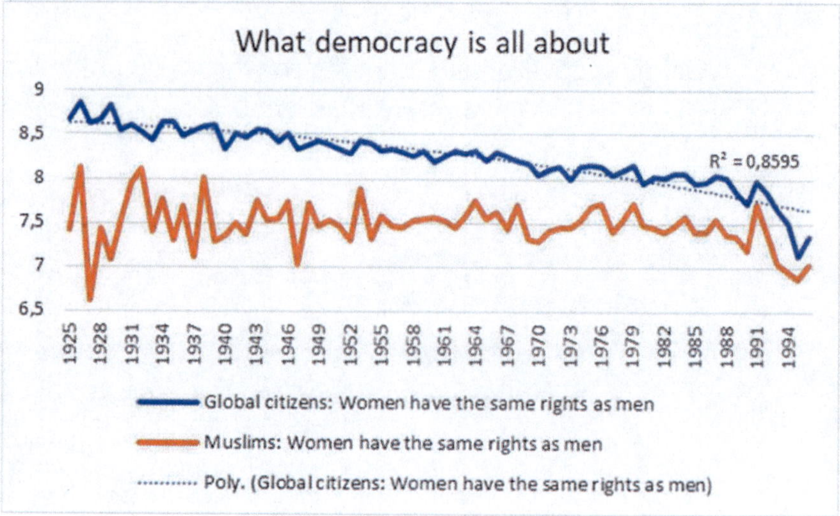


**Fig. 5.19** Declining support for gender equality as a goal of democracy among the global younger generations

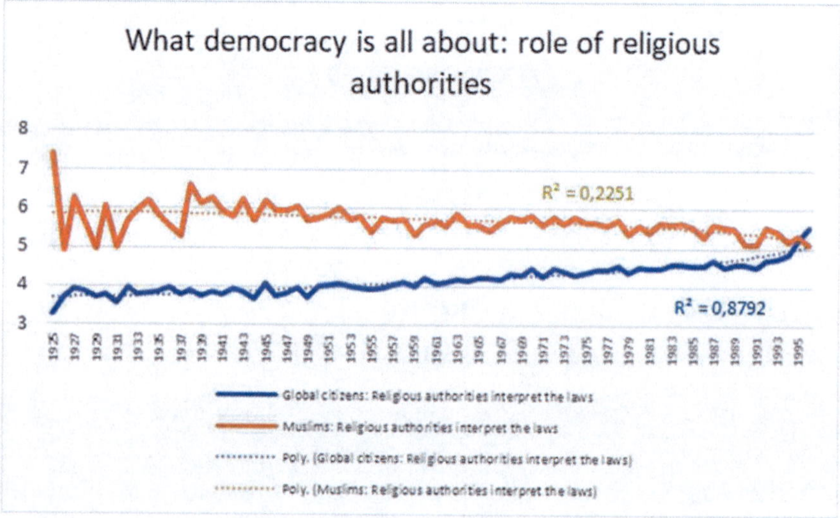


**Fig. 5.20** Rising support for the role of religious authorities in the interpretation of the laws in a democracy among the global younger generations

global younger generations, while this tendency is originally higher among the older generations of Muslims, but now converged with the global citizenry, to be overtaken by the new religious zeal of the younger generations of the citizens of our globe, born in the 1990s.

Based on the very latest release of the *World Values Survey* data, the WVSEVS Joint release, SPSS Version 2021, we cannot distinguish well enough between the Muslim global citizens and the global citizenry in general due to the smaller samples of Muslims now included in that latest version of the *World Values Survey*. But what we can do now empirically with these data is not only to plot the distribution of core global values by birth year, but also to show how contested a terrain these values have become, suggesting that the deep divisions which now characterize America (the one six storm on Capitol Hill, 2021) are to be found in many countries around the world, and that the divisions, measured by the standard deviations of human values, increase among the younger generations. Figure 5.21 clearly shows that there is less support for the goal of democracy that people choose leaders in free elections among the younger global generations on a global level, and that this goal is more contested among the younger generations.

Figure 5.22 is as devastating in its implications as the previous images. There is more and more support for the idea that it is bad to have a democracy among the younger global generations and also the standard deviation of this item increases markedly with the birth year of the global citizenry.

Figure 5.23 is proof for the worst fears to be concluded from an impartial reading of Ibn Khaldun or the Buddenbrooks novel by the German poet and Nobel Laureate for Literature, Thomas Mann. There is a clear and growing acceptancy of corruption

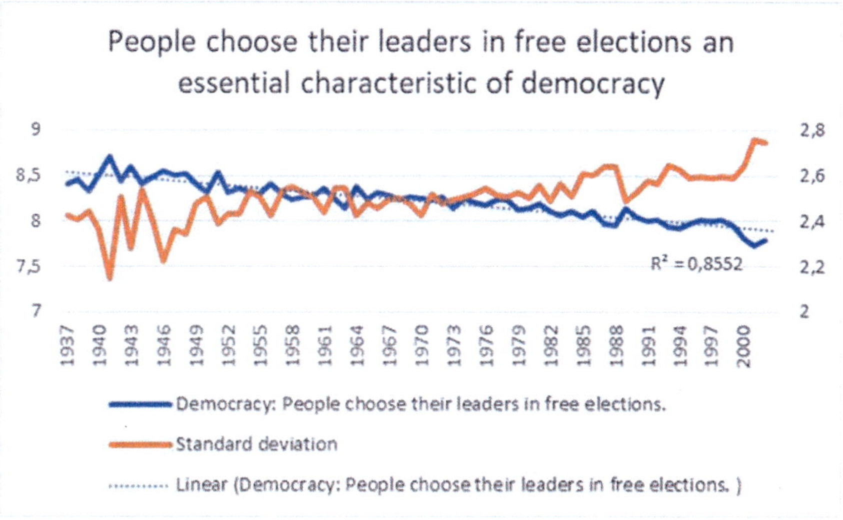


**Fig. 5.21** Support for the goal of democracy that people choose leaders in free elections—global tendency by birth year and standard deviation by birth year

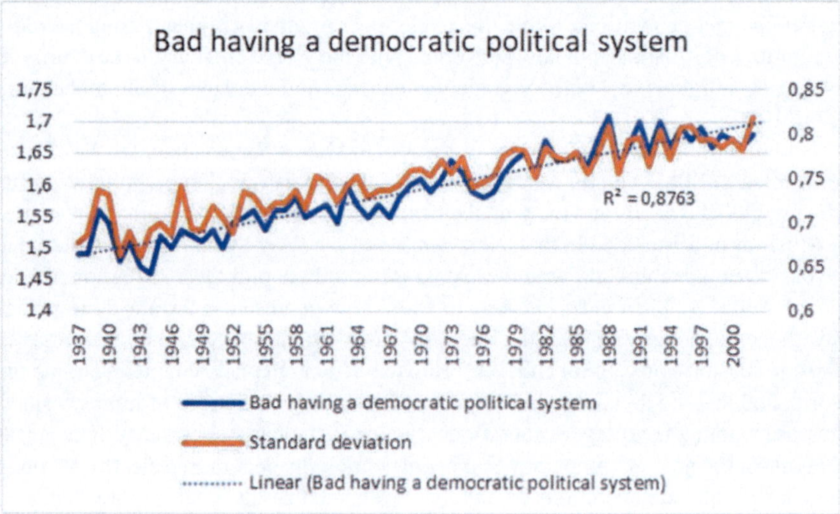


**Fig. 5.22** More and more support for the idea that it is bad to have a democracy among the younger global generations

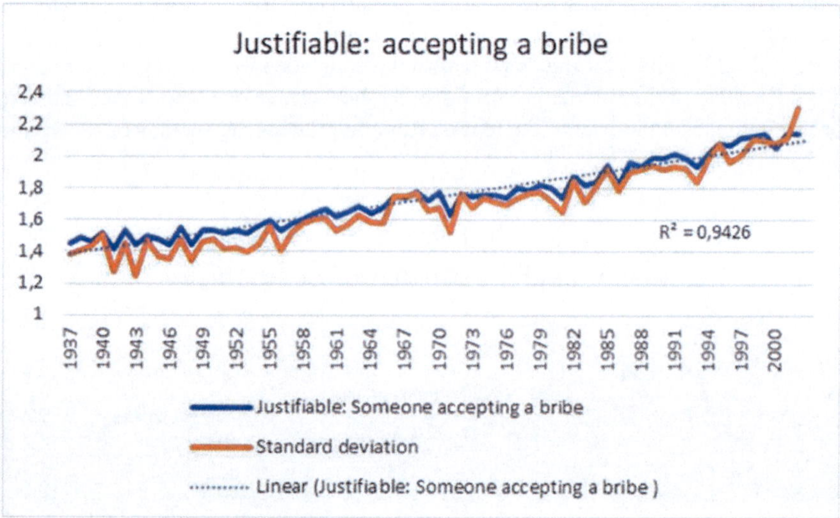


**Fig. 5.23** The Ibn Khaldun/Buddenbrooks cycle: growing acceptancy of corruption by the global younger generations

by the global younger generations of our globe, and also the standard deviation on this item increases over the generations.

## 5.7 Perspectives and Conclusions About the "G-Zero" World

In this chapter, we attempted to estimate the global leadership vacuum in the international system, now also being called a "G-Zero" world from the perspective of the countries of the Gulf. While at first sight, the western democracies are now united in the 30 nation North Atlantic Treaty Organization (NATO) and apart from NATO, the USA as the still leading military power of the West also maintains so-called major non-NATO ally (MNNA) relations with eighteen nations, and there are none the less than six Arab countries among them, two of them, Bahrain and Kuwait, are countries in the Gulf region, the foundations of this architecture seem rather weak.

We analyzed data from the PEW Spring Survey, 2018, and from the latest edition of the *World Values Survey*, 2017–2020. On a population-weighted basis, the Chinese leadership model, competing with the West, whose support we measure by support expressed for President Xi Xinping, is accepted by only 27.63% of the global population. The dismal United States Leadership under President Donald Trump was backed by only 50,17% of the citizens of our globe, while 52.51% of global citizens "voted" for Russian President Vladimir Putin, while 57.16% were expressing support for French President Emmanuel Macron, and finally 60.35% were in favor of German Chancellor Angela Merkel. The inroads, Chinese or Russian leadership concepts could make among Western publics are considerable.

The promax factor analysis of the PEW data reveals interesting patterns. There are six factors, explaining almost 53% of the total variance of the opinions of the citizens of our globe. Distance to China and Russia still goes hand in hand; individuals, skeptical of China, tend to be skeptical of Russia, and vice versa, and of course, support for China coincides with support for Russia. The second factor shows that a low level of threat perception from the Islamist terror organization ISIS goes hand in hand with a low level of threat perception from North Korea, climate change, and Chinese power, reflecting a strongly isolationist value orientation. Anti-Americanism tends to be combined with the rejection of other powers; global publics, which perceive no corruption around them, tend to enjoy security; and furthermore, the promax factor analysis with the PEW data also shows that there is a considerable alienation of the younger generations, and that secularism is supported above all by people of male gender. Secularism goes hand in hand with Anti-Americanism and the rejection of other powers, so typical for countries like France, Netherlands, Sweden, Spain, Canada, Greece, Germany, UK, Mexico, and Australia. In the twenty-first century, America's friends typically are the publics in India, Philippines, Nigeria, Kenya, Indonesia, South Korea, Israel, South Africa, and Poland, most of them characterized by lower rates of secularism.

As a result of our factor analysis of global values from the *World Values Survey*, edition 2017–2020, we found that truly vast areas of Latin America, Latin Europe, Eastern and Southeastern Europe, the mainly Orthodox regions of the former USSR, and the Arab World are affected by strong social decay, while only a few fortunate European countries (Scandinavia, Estonia, Belarus, UK, Belgium, Portugal, Macedonia, Montenegro, the Slovak Republic, Turkey, and the mainly German speaking countries Austria, Germany, and Switzerland) seem to be able to escape from social decay. It emerges just how deeply the West—here especially the Western countries in the Roman Catholic tradition—are already affected by a deep moral and social crisis, stemming from the inability to avoid social decay in an era of secularization. This inability, which also affects the Arab World, and it also negatively affects the ability to come to terms with the current COVID-19 pandemic, and it further undermines the fabric of trust which is so important for future economic growth and socio-economic development.

Of all social scientists, the late Harvard economist Alberto Alesina (1957–2020) deeply understood how distrust undermines economic growth (Alesina & Ferrara, 2000, 2002; Alesina & Guiliano, 2015; Alesina et al., 2015).

Analyzing generational differences of human values, we found that there is the stark reality of declining support for gender equality as a goal of democracy among the global younger generations among the global citizenry. There is a rising support for the role of religious authorities in the interpretation of the laws in a democracy among the global younger generations, while this tendency is originally higher among the older generations of Muslims, but now converged with the global citizenry, to be overtaken by the new religious zeal of the younger generations of the citizens of our globe, born in the 1990s. There is less support for the goal of democracy that people choose leaders in free elections among the younger global generations on a global level, and this goal is more contested among the younger generations. There is more and more support for the idea that it is bad to have a democracy among the younger global generations, and also the standard deviation of this item increases markedly with the birth year of the global citizenry. There is a clear and growing acceptancy of corruption by the global younger generations of our globe, and also the standard deviation on this item increases over the generations.

The devastations which the current Corona COVID-19 pandemic will create will coincide with these long-term structural downward shifts in global value development.

For the countries of the Gulf, these developments will mean that global uncertainties and insecurities will continue, and that its political leadership will have to find answers for the challenges of social decay. Classic Arab historiography (Ibn Khaldun) already warned the world clearly enough of such possible events to come.

# References


Alatas, S. (2006). Ibn Khaldun and contemporary sociology. *International Sociology Review of Books; Journal of the International Sociological Association, 21*(6), 782–795.

Alesina, A., & Ferrara, E. L. (2000). *The determinants of trust (No. w7621)*. National Bureau of Economic Research.

Alesina, A., & Ferrara, E. L. (2002). Who trusts others? *Journal of Public Economics, 85*(2), 207–234.

Alesina, A., & Giuliano, P. (2015). Culture and institutions. *Journal of Economic Literature, 53*(4), 898–944.

Alesina, A., Algan, Y., Cahuc, P., & Giuliano, P. (2015). Family values and the regulation of labor. *Journal of the European Economic Association, 13*(4), 599–630.

Alker, H. R., Deutsch, K. W., & Stoetzel, A. H. (Eds.). (1973). *Mathematical approaches to politics*. Elsevier Science and Technology.

Amirabedini, A. (2014). Two development theories: Ibn-i-Khaldun and Wallerstein. *Campus-Wide Information Systems, 31*(1), 63–74.

Arıkan, G. (2021). Public attitudes towards climate change: A cross-country analysis. *British Journal of Politics and International Relations., 23*(1), 158–174.

Arnason, J. (2004). Civilization and state formation in the Islamic context: Re-reading Ibn Khaldūn. *Thesis Eleven, 76*(1), 29–48.

Beeson, M. (2018). Asia's leadership deficit and regional crises. *Asia Europe Journal, 16*(2), 141–154.

Bornschier, V. (2005). *Culture and politics in economic development*. Routledge.

Bremmer, I., & Roubini, N. (2011). AG-zero world-the new economic club will produce conflict, not cooperation. *Foreign Affairs, 90*, 2.

Chapra, M. U. (2008). Ibn Khaldun's theory of development: Does it help explain the low performance of the present-day Muslim world? *Journal of Socio-Economics, 37*(2), 836–863.

Ciftci, S. (2016). Soft power, religion, and anti-Americanism in the Middle East. *Foreign Policy Analysis, 12*(3), 374–394.

Deutsch, K. W. (1979). *Tides among nations*. Free Press.

Devezas, T. C., & Corredine, J. T. (2001). The biological determinants of longwave behavior in socioeconomic growth and development. *Technological Forecasting and Social Change, 68*(1), 1–57.

Grigoryev, L., & Pabst, A. (Eds.). (2019). *Global governance in transformation: Challenges for international cooperation*. Springer Nature.

Hendrickson, R. C. (2011). Potential NATO partners—Political and military utility for NATO. In *NATO: The power of partnerships* (pp. 163–180). Palgrave Macmillan.

Inglehart, R, F. (2021). *Religion's sudden decline. What's causing it, and what comes next?* Oxford University Press.

Irwin, R. (1997). Toynbee and Ibn Khaldun. *Middle Eastern Studies, 33*(3), 461–479.

Katzenstein, P. J. (2013). Reading the right signals and reading the signals right: IPE and the financial crisis of 2008. *Review of International Political Economy, 20*(5), 1101–1131.

Khaldun, I. (1332–1406) (2005). The Muqaddimah: An introduction to history. Ibn Khaldun; translated and introduced by Franz Rosenthal; abridged and edited by N. J. Dawood, with a new introduction by Bruce B. Lawrence'. Princeton University Press. Available at https://asadullahali.files.wordpress.com/2012/10/ibn_khaldun-al_muqaddimah.pdf

Kohler, G., & Chaves, E. J. (Eds.). (2003). *Globalization: Critical perspectives*. Nova Science Publishers.

Kristensen, P. M. (2017). After abdication: America debates the future of global leadership. *Chinese Political Science Review, 2*(4), 550–566.

Mann, Th. (1924). Buddenbrooks, translated from the German of Thomas Mann, by H.T. Lowe-Porter. A.A. Knopf.

Mensah, Y. (2014). An analysis of the effect of culture and religion on perceived corruption in a global context. *Journal of Business Ethics., 121*(2), 255–282.

Rennstich, J. (2007). Is globalization self-organizing?' In G. Modelski, T. C., Devezas, & W. R. Thompson (Eds.), *Globalization as evolutionary process. Modeling global change* (pp. 74–107). Routledge.

Schumpeter J. A. (1950). *Capitalism, socialism and democracy.* Harper & Row.

Steinbock, D. (2014). The erosion of America's defense innovation. *American Foreign Policy Interests, 36*(6), 366–374.

Sussman, G. (2017). Reheating the cold war: Us, Russia, and the revival of rollback. *Perspectives on Global Development and Technology, 16*(6), 736–755.

Tibi, B. (1996). Der Wahre Imam: der Islam von Mohammed bis zur Gegenwart. Piper.

Wæver, O. (2018). A post-western Europe: Strange identities in a less liberal world order. *Ethics and International Affairs, 32*(1), 75–88.

# Chapter 6
# Value Change and Patterns of Development in the Rich Arab Countries

**Abstract** This chapter starts from the assumption that even advanced and very internationally recognized Gulf studies do not deliver us any more systematic clues about the social value patterns, held by the publics at large in the region. First of all, we debate here the Gulf country implications of research in the context of the GLOBE Project at Simon Fraser University in Canada and its hypothesis about an "Arabic cluster" of business leadership culture. We then review the research by Geert Hofstede and associates, followed by our discussion of the Gulf country perspectives in the value research by Shalom Schwartz and Ronald F. Inglehart. We round up our theoretical surveys by a debate about the Islamicity Index, developed by the American researcher Hossein Askari. In our own empirical work, which attempts to answer questions left over from current approaches, we first present an Index of Civil Society and an Index of Religious Tolerance, comparing Gulf countries with all the countries of the world with comparable World Values Survey data. We then develop our own multivariate scales, based on a comprehensive promax factor analysis with World Values Survey data. Combining the **World Values Survey** data about the acceptancy of

1. The non-violent and law-abiding society
2. Democracy movement
3. Climate of personal non-violence
4. Trust in institutions
5. Happiness, good health
6. No redistributive religious fundamentalism
7. Accepting the market economy
8. Feminism
9. Involvement in politics
10. Optimism and engagement
11. No welfare mentality, acceptancy of the Calvinist work ethics

in a parametric Index construction from factor analytical data, it emerges that Qatar performs relatively well in global society, while Kuwait and Bahrain are in the lowest third global performance "group." It would be wrong to assume that there is an "Arabic Cluster" of human and social values. An analysis of the development patterns of rich Arab countries, compatible with the comparative analysis of societal values of



A. Tausch, *The Future of the Gulf Region*, Gulf Studies 2,
https://doi.org/10.1007/978-3-030-78299-3_6

this chapter, finalizes our essay and shows the relevance of overcoming still existing restrictive gender norms and attitudes favoring inequality.

**Keywords** Globe project · Geert Hofstede · Power distance · Uncertainty avoidance · Inglehart · Gulf states · Global values · Gulf development patterns

## 6.1 Introduction: Value Change and Patterns of Development in the Rich Arab Countries

Even advanced and very internationally recognized Gulf studies do not deliver us any more systematic clues about the social value patterns, held by the publics at large in the Gulf region (Guzansky, 2015; Guzansky & Berti, 2013; Guzansky & Even, 2020; Guzansky & Segal, 2020; Guzansky & Winter, 2020). With all the world economic and strategic importance of the Gulf region, this is even more surprising: not too many (to say the least) comparative studies of the patterns of value development in the Gulf. Based on the data from the ***World Values Survey*** (see Chap. 2), one particular essay (Gengler et al., 2013) looked at civic life and democratic citizenship in Qatar. This study explored the political orientations of ordinary Qatari citizens. Qatar citizen involvement in societal organizations is not associated with higher appreciation for democracy, nor again with those values and behaviors thought to be essential to it. Rather, associational life in Qatar is simply an extension of traditional society and the prevailing regime, with those most involved being those who derive the most benefit and who would thus stand to lose most from any revision of the political status quo.

Another important study (Kabasakal & Bodur, 2002) attempted to study leadership patterns among business elites, based on data from the ***GLOBE Project*** (see below) in Egypt, Morocco, Turkey, Kuwait and Qatar, and found that societies in the "Arabic cluster" all tend to be highly group-oriented, hierarchical, masculine, and low on future orientation. But outstanding business leaders from the cluster were found to be team-oriented and charismatic.

With all the attention that global science has paid to the analysis of Islamism, Salafism, and Terrorism after 9/11 (just to mention one particular study, featuring on the region, Wiktorowicz, 2006), the dearth of quantitative, opinion-survey supported evidence on the Gulf is indeed surprising. As we shall see in Chap. 8 of this study, we quote from Moaddel, 2006, who showed that fifteen of the nineteen terrorists who attacked the World Trade Center and the Pentagon on 9/11, 2001 in America were Saudi citizens. In view of the strategic importance of the region this lack of systematic, comparative, survey-driven studies on the values of publics in the Gulf are even more surprising.

A few salient available studies were rather written in the tradition of transnational business management and tertiary education management, given the fact that a number of high-profile Western Universities now opened their affiliates in the region. How to do profitable and efficient business and university education in a cultural environment, perceived to be different from the West, was the overriding concern of these studies (just to mention a few, Al-Alawi et al., 2007; Eickelman, 1992; Kabasakal & Bodur, 2002; Rettab et al., 2009).

Given these rather unsatisfactory conditions of current Gulf research (see also Chap. 1), we attempt to achieve here with our chapter two aims:

- an encompassing and representative analysis of the societal values of the Gulf population, as far as current data (see Chap. 2) allow us to do so,
- an analysis of the development patterns of rich Arab countries, compatible with the comparative analysis of societal values.

## 6.2 Background: The State of the Art of Global Value Research (the Globe Project, Hofstede, Schwartz, Inglehart, and Their Indicators for the Gulf Region)

Amidst the current global crisis, how well-prepared are the Gulf countries to assume a more active role in global leadership along partners in the G-20, in the United Nations, and, hopefully, one day also in the OECD? In this context, the systematic study of values held by the populations of the Gulf countries is of paramount importance.

Is there, as was maintained in international business research (Kabasakal & Bodur, 2002) really an "Arabic cluster" of values influencing the way people are doing business? How does this "cluster" affect the countries of the Gulf?

The discipline of global value research made enormous methodological developments over the last decades and is now an integral part of global sociology (Davidov et al., 2008, 2011; Gonzales, 2011). As it is well-known, value research and the sociology of religion are an integral part of sociology (Morel, 1972, 1977, 1986, 1997, 1998, 2003). Richard F. Inglehart initiated the repeated and constant standard **national** population surveys over time in the *Eurobarometer*[1] project and later in the ***World Values Survey*** (Inglehart and Norris and associates in Inglehart, 1988, 1990, 2000, 2006, 2008; Inglehart & Baker, 2000; Inglehart & Norris, 2003, 2009, 2012, 2016; Inglehart & Welzel, 2003, 2009, 2010; Norris & Inglehart, 2002, 2011, 2012, 2015; Norris et al., 2003)[2]. The regional coverage of the ***World Values Survey*** project quickly transcended developed Western democracies to include in wave 1, 1981–1984 already South Korea and Mexico; and in wave 2, 1990–1994 already a number of former communist and also developing countries. ***World Values Survey*** data are now available from some 100 countries which contain some 90% of the world's population, still using a largely common and stable questionnaire with almost 400,000 representative respondents.[3] There was also a growing inclusion of representative Muslim publics in these surveys (Tausch & Moaddel, 2009; Tausch et al., 2014; Yesilada & Noordijk, 2010). Never before in human history we have known as much about the different values and cultures across the globe, and never before have

[1] http://ec.europa.eu/public_opinion/index_en.htm.

[2] http://lcsr.hse.ru/en/Inglehart.

[3] http://www.worldvaluessurvey.org/WVSContents.jsp?CMSID=WhatWeDo.

we known as much about the changes of values and cultures over time. It should be stressed at the outset that here we are dealing with the subjective views of representative populations around the globe regarding core values of an "*Open Society*". We are not saying that, say, country *x* or *y* are more "*open societies*" than, say, country *z*. But what we are saying is that the ***World Values Survey*** data suggest that according to the combined results for the values under scrutiny here the populations in several transition countries and developing countries today are already more characterized by a higher commitment to a *"Civil Society"* than several Western countries and the undisputed successful cases of transition and democratization in Eastern Europe, South and East Asia and Latin America.

Thus, the rankings, presented in this chapter, surely are based on legitimate factor analytical procedures, applied to the ***World Values Survey*** data, reflecting opinions of the respective populations. We are well aware that some of the results, presented here, could be regarded by many as paradoxical, counter-intuitive or even misleading.

But apart from the fact that any analyst with access to the statistical software used here—or in fact to any version of alternative advanced statistical software programs—will come to the same or very similar conclusions: that the openly accessible ***World Values Survey*** data, reflecting opinions of the respective populations, suggest exactly or something very similar to the results structure, achieved here by applying promax factor analysis procedures to the ***World Values Survey*** data.

All the evidence suggests that there is a certain room for optimism regarding the leadership culture and open society orientation in the countries of the Gulf region.

Basically, four social scientific traditions gained an overwhelming prominence in the study of global values:

- the ***GLOBE Project*** of Simon Fraser University in Canada, headed by A. Dastmalchian,
- the research, initiated by the late G. Hofstede from Maastricht University,
- S. Schwartz from the Hebrew University in Jerusalem, and
- R. F. Inglehart from the University of Michigan.

In Tausch et al. (2014), a massive attempt was made to analyze the statistical relationships between the three approaches to global value research which caught the imagination of the majority of the political science and sociology profession dealing with the subject, Hofstede, Schwartz, and Inglehart. Interestingly enough, the ***GLOBE Project*** had a massive reception in international business studies, but hardly ever was mentioned in political science and sociology literature.

Our survey of the existing leadership and cross-cultural studies is based on the original contributions of these authors (especially Hofstede, 1980, 1984, 1988; Schwartz, 1992, 1994) and the surveys offered by Tausch et al. (2014), and we refer interested readers to the extensive bibliography, offered at the end of this chapter.

So, the ***GLOBE Project*** (Global Leadership and Organizational Behavior Effectiveness), just like the competing approach by G. Hofstede, was focused primarily on business leadership culture. The ***GLOBE Project*** was and still is focused on the values of business executives, while the late G. Hofstede began his studies by surveying the cultural values of employees of the transnational US company IBM.

S. Schwartz, in turn, began his international value studies by surveying the values of international teachers. Thus, comparative value studies in the field of international business studies and international business ethics (the research efforts by Hofstede, the subsequent ***GLOBE project***, and Schwartz) basically started by comparing global cultures, but used limited samples across nations—both in terms of the social inclusiveness of the strata surveyed and in terms of the number of persons interviewed. Although we also find in the research programs inspired by Dastmalchian, Hofstede, and Schwartz several attempts to use additional country-level data from the ***World Values Survey*** or the *European Social Survey,* we feel that the time has come to re-analyze the entire question of the underlying dimensions of leadership cultures, using national samples from the truly encompassing with nation-wide representative samples from the ***World Values Survey*** data.

In the 1990s, the ***GLOBE project*** examined the relationships among societal culture, leadership, and organizational practices in collaboration with more than 200 researchers from 62 societies. ***GLOBE 2020*** focuses on four research questions:

1. To what extent have the cultures of the countries in the original ***GLOBE project*** study changed?
2. And what are the major societal drivers of culture change?
3. What is the relationship between national culture and antecedents of interpersonal trust in different countries?
4. What insights can be gained by expanding the number of countries mapped by the ***GLOBE project*** to include relatively unstudied countries in the Middle East and North Africa region?

> **GLOBE 2020** will collect cultural information from a large number of the countries in Africa and the Middle East that were not part of the original **GLOBE** study, enabling more nuanced cross-comparative research for years to come. We are interested in assessing and examining **GLOBE's** cultural and leadership dimensions and the relationships between them with particular interest in societies that are experiencing a rapid cultural transition. (MENA region)[4]

The 2014 survey focused on the following four questions:

1. How does national culture influence the kinds of leadership behaviors expected in a society?
2. What CEO behaviors generally lead to success?
3. What are some distinctions between the high-performing CEOs (i.e., superior), and underperforming CEOs (i.e., inferior)?
4. How important is it that CEO leadership behaviors match the leadership expectations within a society?

Almost 70 researchers collected data from over 1000 CEOs and over 5000 senior executives in corporations in a variety of industries in 24 countries.[5] The project claims that this sample of countries represents cultures in almost all cultural regions

[4] https://globeproject.com/about.

[5] https://globeproject.com/study_2014.

of the world. The goal was to have a minimum of 40 CEOs from each country which were selected from a wide variety of industries.

The ***GLOBE project*** found that cultural values indirectly predict behavior through the manifestation of culturally endorsed leadership expectations—values themselves do not directly predict behaviors. Leaders learn what is expected in a culture and operate in a socially desirable fashion. For instance, cultures that value performance orientation (a major cultural dimension) such as the USA and Germany expect leaders to be participative, and it was these leaders who showed some of the highest levels of participative leadership.

The second ***GLOBE*** objective was to determine the extent to which CEO leadership behavior leads to effective TMT (technology, media and telecommunications) teams and organizational success across the participating societies. GLOBE's major hypothesis was that the fit between a CEO's behavior and his/her society's leadership expectations will predict the CEO's effectiveness. To produce extraordinary results, global leaders need to understand and exceed the leadership expectations in the cultures they are interacting with.[6]

According to Hofstede, who investigated above all the value systems of the global employees of a particular multinational US company (IBM), there are four to six basic clusters of international value systems, and they are all defined along the scales of how different national societies handle ways of coping with inequality, ways of coping with uncertainty, the relationship of the individual with her or his primary group, and the emotional implications of having been born as a girl or as a boy (Kang & Kwon, 2018). The massive global research on Hofstede's theory did not avoid two important questions in the context of our research: Are Hofstede's categories compatible with the openly available global general population survey data from the *World Values Survey* (see below and Chap. 2 as well as Chaps. 6, 7, and 8 of this book), and which conclusions are to be drawn for the Arab countries?

Almutairi et al. (2020) addressed the clustering of seven Arab States within Hofstede's national culture dimensions, and its implications for international firms targeting the MENA region. Almutairi et al. (2020) argues that re-examining the Arab States' national cultures can provide interested firms with a better understanding of the differences and similarities between them. Their paper utilized Hofstede's so-called VSM 13 questionnaires to measure the seven Arab States on all of Hofstede's six dimensions. The research's translation and methods were overseen by Hofstede and approved before subsequent analysis. In total, 1400 surveys were disseminated to the Arab States, but only 200 for each country, so the sample size was very low indeed. Almutairi et al. (2020) found evidence that national cultural differences do exist across the Arab States.

Hofstede's power distance scale (Kang & Kwon, 2018) is generally lowest in Scandinavian and in the Anglo-Saxon democracies, and also in the social welfare state Austria, and it is, generally speaking, highest in the ex-USSR, in India and Southeast Asia and in China, in Mexico and Venezuela and in some countries of Central and Eastern Europe. According to Almutairi et al., the power distance dimension ranks

[6] See Footnote 5.

Iraq, Libya, and Egypt as countries with very high power-distance, while the Gulf states and Lebanon are ranked around average on the index.

The most collectivistic countries are to be found in the Andean region of Latin America, in China and in countries of South and Southeast Asia, while the highest individualism is to be found in the USA and in the Anglo-Saxon democracies, with high scores also to be found in the Netherlands, Denmark, and also in Italy and in Hungary. Almutairi et al. (2020) remark that the MENA region is known to embrace family cohabitation, in which children are encouraged and expected to live with their families even after reaching their adult age. Hofstede labeled the Arab States as collective, but Almutairi et al. (2020) observed a disparity between their rankings, with Saudi Arabia and Lebanon as being the highest on the individualism rank.

The highest masculinity scores are to be found in some countries of Central and Eastern Europe, in Japan and in Venezuela, while the countries of Scandinavia and the Baltic region, the Netherlands and, interestingly enough, Chile and Portugal are characterized by high femininity as well.

Uncertainty avoidance is highest in Greece, Portugal and Uruguay and is generally highest in Roman Catholic and Orthodox cultures, and also in Japan and South Korea, while it is lowest in Protestant cultures, but also in China and in Iran, in South Asia and in Southeast Asia.

Hofstede's long-term orientation (LTO) is highest in South Korea, Japan, China, in Russia and the Baltic Republics, and in Germany and some other European countries, while it is especially absent in Latin America, in the Anglo-Saxon democracies and Scandinavia, and in Morocco and Iran, with most of the other European countries and the countries of South and Southeast Asia classified in the same ranks.

Hofstede's Indulgence Factor is highest in Venezuela, in countries of the Caribbean and in Mexico, and, interestingly enough, also in Sweden, New Zealand, Australia, the UK, the US, and Canada and in some European countries (Denmark, Netherlands, Switzerland and Austria) and in Chile, while Restraint is highest in Pakistan, in most former communist countries, and in China. Most continental European countries seem to be characterized by low indulgence and higher restraint.

Table 6.1 lists the admittedly preliminary results by Almutairi et al. (2020).

For Hofstede and his associates, there are very clear **underlying dimensions of these six dimensions** and social and **cultural history**. Some of these conclusions, not necessarily fully shared by us, are nevertheless really breathtaking theoretical perspectives over the last two millennia of global history:

- Power distance scores are high for Latin, Asian and African countries and smaller for Anglo and Germanic countries. **Uncertainty avoidance** scores are higher in Latin countries, in Japan, and in German speaking countries, lower in Anglo, Nordic, and Chinese culture countries. **Individualism** prevails in developed and Western countries, while collectivism prevails in less developed and Eastern countries; Japan takes a middle position on this dimension. **Masculinity** is high in Japan, in some European countries like Germany, Austria *and* Switzerland, and moderately high in Anglo countries; it is low in Nordic countries and in the Netherlands and moderately low in some […] countries like France, Spain and Thailand. **Long-term orientation** scores are highest in East Asia, moderate in Eastern and Western Europe, and low in the Anglo world, the Muslim world, Latin America and Africa. **Indulgence** scores are highest in Latin America, parts of

**Table 6.1** Geert Hofstede's value categories in the Arab World according to Almutairi et al. (2020) and Tausch et al. (2014)

| | Power distance | Individualism | Masculinity | Uncertainty avoidance | Long-term orientation | Indulgence versus restraint |
|---|---|---|---|---|---|---|
| Kuwait | 73 | 39 | 45 | 70 | 19 | 29 |
| Saudi Arabia | 72 | 48 | 43 | 64 | 27 | 14 |
| United Arab Emirates | 74 | 36 | 52 | 66 | 22 | 22 |
| Lebanon | 62 | 43 | 48 | 57 | 22 | 10 |
| Iraq | 97 | 31 | 53 | 96 | 12 | 23 |
| Egypt | 80 | 37 | 55 | 55 | 42 | −2 |
| Libya | 103 | 35 | 66 | 67 | 15 | 74 |
| World average (62 countries) | **59** | **46** | **49** | **67** | **50** | **48** |

Africa, the Anglo world and Nordic Europe; restraint is mostly found in East Asia, Eastern Europe and the Muslim world.

- The grouping of country scores points to some of the **roots of cultural differences**. These should be sought in the common history of similarly scoring countries. All **Latin countries**, for example, **score relatively high on both power distance and uncertainty avoidance**. Latin countries (those today speaking a Romance language, i.e., Spanish, Portuguese, French, Italian or Romanian) have inherited at least part of their civilization from the Roman Empire. The Roman Empire in its days was characterized by the existence of a central authority in Rome, and a system of law applicable to citizens anywhere. This established in its citizens' minds the value complex which we still recognize today: centralization fostered large power distance and a stress on laws fostered strong uncertainty avoidance. The Chinese empire also knew centralization, but it lacked a fixed system of laws: it was governed by men rather than by laws. In the present-day countries once under Chinese rule, the mindset fostered by the empire is reflected in large power distance but medium to weak uncertainty avoidance. The Germanic part of Europe, including Great Britain, never succeeded in establishing an enduring common central authority and countries which inherited its civilizations show smaller power distance. Assumptions about historical roots of cultural differences always remain speculative but in the given examples they are plausible. In other cases, they remain hidden in the course of history.
- Power distance is correlated with the use of violence in domestic politics and with income inequality in a country. Uncertainty avoidance is associated with Roman Catholicism and with the legal obligation in developed countries for citizens to carry identity cards. Individualism is correlated with national wealth and with mobility between social classes from one generation to the next. Masculinity is correlated negatively with the percent of women in democratically elected governments. Long-term orientation is correlated with school results in international comparisons. Indulgence is correlated with sexual freedom and a call for human rights like free expression of opinions.[7]

Figures 6.1 and 6.2 summarize our results in graphical fashion.

Just as Hofstede, the works of the Israeli psychologist Shalom Schwartz received an enormous international reception, especially in the expanding field of international

[7] http://www.geerthofstede.nl/dimensions-of-national-cultures.

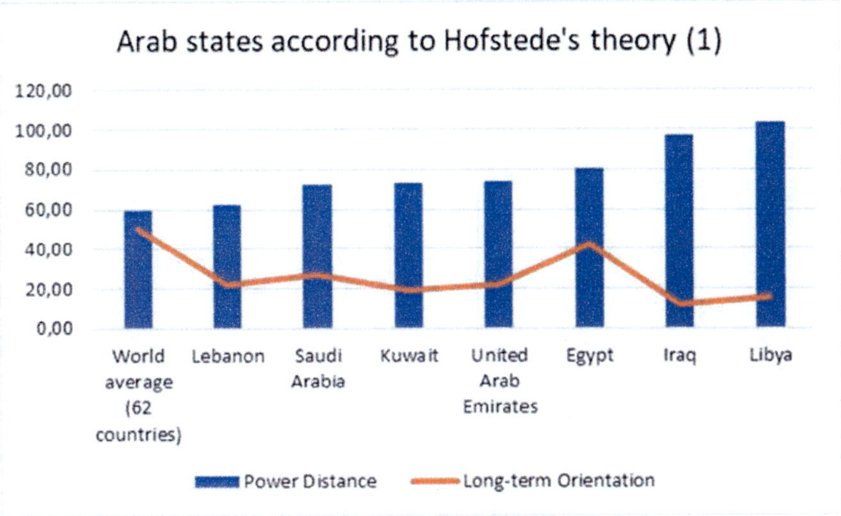


**Fig. 6.1** Arab states according to Hofstede's theory (1)

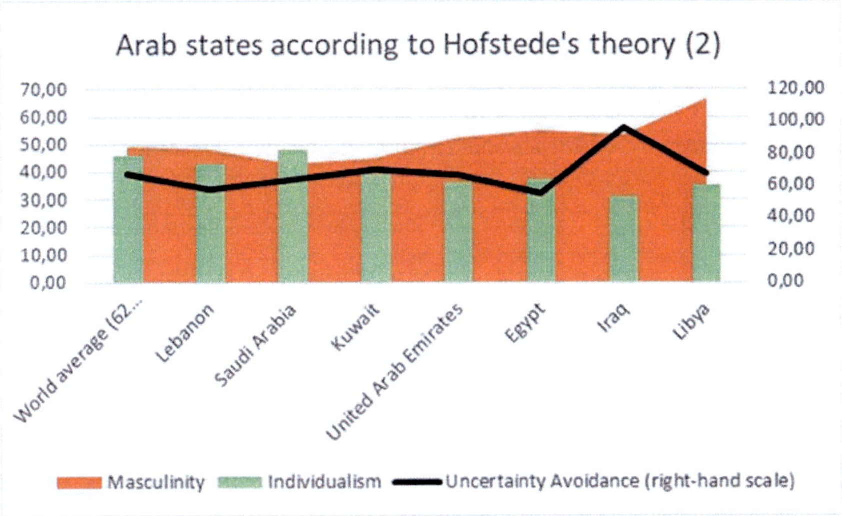


**Fig. 6.2** Arab states according to Hofstede's theory (2)

business studies. Schwartz (2006a, b, 2009) highlights a famous map of global values. But there is a problem of limited country samples. Schwartz began his investigations not on generalized surveys of the total population, but on relatively small samples of global schoolteachers and college students. Most samples included only between 180 and 280 respondents. Schwartz identifies seven basic cultural orientations and the structure of interrelations among them: West European, English-speaking, Latin American, East European, and South Asian, Confucian influenced, and African and Middle Eastern.

His seven country-level cultural dimensions are

1. Embeddedness,
2. Hierarchy,
3. Mastery,
4. Affective autonomy,
5. Intellectual autonomy,
6. Egalitarianism,
7. Harmony.

Correlations between the sample means were used in a multidimensional scaling analysis to assess the presence of seven cultural orientations and the relations among them. Muslim societies rank very high on values of embeddedness. Embeddedness combines such values as social order, respect tradition, forgiving, moderateness, obedience, politeness, cleanness, national security, devoutness, wisdom, self-discipline, family security, honoring elders, reciprocation of favors, protecting the public image. In cultures with an emphasis on embeddedness, people are viewed as entities embedded in the collectivity. Meaning in life comes largely through social relationships, through identifying with the group, participating in its shared way of life, and striving toward its shared goals. "*Embedded cultures emphasize maintaining the status quo and restraining actions that might disrupt in-group solidarity or the traditional order. Important values in such cultures are social order, respect for tradition, security, obedience, and wisdom*" (Schwartz, 2006a, b).

The *World Values Survey* runs several items from the Schwartz categories.

1. It is important to this person to think up new ideas and be creative.
2. It is important to this person to be rich.
3. It is important to this person living in secure surroundings.
4. It is important to this person to have a good time.
5. It is important to this person to help the people nearby.
6. It is important to this person being very successful.
7. It is important to this person adventure and taking risks.
8. It is important to this person to always behave properly.
9. It is important to this person looking after the environment.
10. It is important to this person tradition.
11. It is important to this person to do something for the good of society.

Each time, respondents were being asked to scale themselves whether such a person is "*very much like me*" (1) to "*such a person is not at all like me*" (6). Direct *World Values Survey* data about this are available from 60 countries.

A careful reading of the Schwartz categories and projections of his scales on real and concrete choropleth maps would suggest that there is a considerable discrepancy of his research results with the interpretation of Muslim civilization by Huntington (see also, Tausch et al., 2014). In the work of this great Israeli scholar, Muslim societies are seen in a much more positive way than as they are seen by Huntington (see again, for an extensive debate, Tausch et al., 2014). Although the country results for embeddedness, hierarchy and egalitarianism still fit in with the Huntington's predictions for the Muslim majority countries, but the country results for mastery, affective autonomy, intellectual autonomy and longing for harmony suggest a completely different picture of Muslim societies: in contrast to what one would have to expect from Huntington's works, key Muslim countries are even at the international forefront of the values of affective autonomy, intellectual autonomy, and the quest for harmony.

Figure 6.3 summarizes the results of Table 6.2 in a graphical fashion.

Eldad Davidov pays great attention to the statistical-mathematical foundations of comparative value analysis, including the detection of contradictory statements in survey research to reduce the measurement error. He further developed the Schwartz approach. Among his major works one finds Davidov et al. (2008, 2011), where Davidov and associates base their analysis on a cascade of human values.

In his 2008 study on the contradictions of the European integration process, Davidov also highlighted the close interrelationship between gender prejudice, homophobia, anti-Semitism, xenophobia, social out-groups (as for example homeless people), Islamophobia and the overall devaluation of newcomers.

Inglehart, in some of his main publications, developed an interpretation of global value change (Inglehart, 1988, 1990, 2000, 2006, 2008; Inglehart & Baker, 2000; Inglehart & Norris, 2003, 2009, 2012, 2016; Inglehart & Welzel, 2003, 2009, 2010; Norris & Inglehart, 2002, 2011, 2012, 2015; Norris et al., 2003), which rests on a

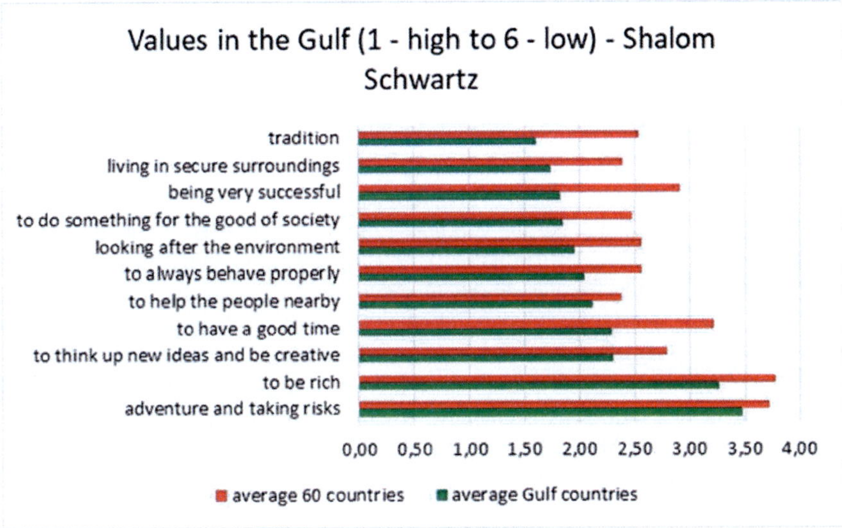


**Fig. 6.3** Values in the Gulf by international comparison

**Table 6.2** Values in the Gulf according to Shalom Schwartz

| Country/region | Kuwait | Qatar | Average Gulf countries | Average Arab countries | Average 60 countries | Difference—Gulf—World |
|---|---|---|---|---|---|---|
| Adventure and taking risks | 3.27 | 3.66 | 3.47 | 3.66 | 3.71 | −0.25 |
| To help the people nearby | 2.12 | xx | 2.12 | 2.19 | 2.37 | −0.25 |
| To think up new ideas and be creative | 2.48 | 2.12 | 2.30 | 2.69 | 2.79 | −0.49 |
| To be rich | 3.17 | 3.36 | 3.27 | 3.46 | 3.77 | −0.51 |
| To always behave properly | 2.26 | 1.82 | 2.04 | 2.29 | 2.55 | −0.51 |
| Looking after the environment | 2.40 | 1.50 | 1.95 | 2.38 | 2.55 | −0.60 |
| To do something for the good of society | 2.25 | 1.43 | 1.84 | 2.26 | 2.47 | −0.63 |
| Living in secure surroundings | 2.03 | 1.45 | 1.74 | 1.97 | 2.38 | −0.64 |
| Tradition | 1.93 | 1.28 | 1.61 | 1.96 | 2.53 | −0.93 |
| To have a good time | 2.52 | 2.04 | 2.28 | 2.74 | 3.21 | −0.93 |
| Being very successful | 2.07 | 1.57 | 1.82 | 2.49 | 2.91 | −1.09 |

well-known two-dimensional scale of global values and global value change. It is based on the statistical technique of factor analysis of up to over twenty key ***World Values Survey*** variables from the originally more than 900 ***World Values Survey*** items. These 900 items cover practically all major areas of human concern, from religion to politics, and from economic to social life.

The two Inglehart dimensions are: (1) the traditional/secular-rational dimension and (2) the survival/self-expression dimension. In a factor analysis of ten indicators, these two dimensions explain more than 70% of the cross-national variance. Each of these dimensions is strongly correlated with scores of other important variables. For Inglehart and Baker (2000), all of the preindustrial societies show relatively low levels of tolerance for abortion, divorce, and homosexuality and tend to emphasize male dominance in economic and political life. There is a deference to parental authority, and the importance of family life is of primary concern, and these societies are relatively authoritarian. Most of them place strong emphasis on religion. Advanced industrial societies tend to have the opposite characteristics. When survival is uncertain, cultural diversity seems threatening. When there is not "enough to go around," foreigners are seen as dangerous outsiders who may take away one's sustenance. People cling to traditional gender roles and sexual norms and emphasize absolute rules and family norms in an attempt to maximize predictability in an uncertain world. Conversely, when survival begins to be taken for granted, ethnic and cultural diversity become increasingly acceptable. Beyond a certain point, diversity is not only tolerated and it may be positively valued because it is interesting and stimulating. In advanced industrial societies, people seek out foreign restaurants to taste new cuisine; they pay large sums of money and travel long distances to experience exotic cultures. Changing gender roles and sexual norms no longer seem threatening.

In Table 6.4 and Fig. 6.4, we document value preferences in the education of children in the Gulf region according to the *World Values Survey*. The values in the education of children play an important part in Inglehart's sociology.

Figure 6.4 summarizes the results of Table 6.3 in graphic fashion.

In Inglehart (2017), the theory of materialism/postmaterialism is further specified and applied to the Muslim world. According to Inglehart, gender norms and democracy strongly interact:

> In every stable democracy, a majority of the public disagrees with the statement: Men make better political leaders than women. None of the societies in which less than 30% of the public rejects this statement has a democratic government. In China, one of the world's most autocratic countries, a majority of the public agrees that men make better political leaders than women, despite a party line that has long emphasized gender equality. (Inglehart, 2017)

With a comparative focus on the Gulf region, in this book, we analyze the weight of such factors as the loss of religion and the rise of the shadow economy, including

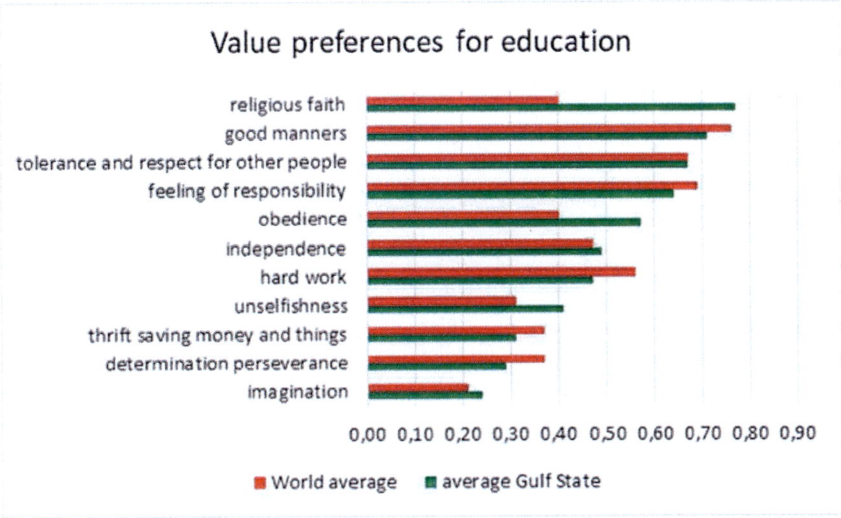


**Fig. 6.4** Value preferences for education

**Table 6.3** Values for raising children according to the *World Values Survey*

| | Kuwait | Qatar | Saudi Arabia | Average Gulf State | Average 10 other Arab states | World average (100 countries and territories) | Difference Gulf-world average |
|---|---|---|---|---|---|---|---|
| Religious faith | 0.74 | 0.85 | 0.71 | 0.77 | 0.76 | 0.40 | 0.37 |
| Good manners | xx | xx | 0.71 | 0.71 | 0.87 | 0.76 | −0.05 |
| Tolerance and respect for other people | 0.68 | 0.77 | 0.56 | 0.67 | 0.67 | 0.67 | 0.00 |
| Feeling of responsibility | 0.60 | 0.76 | 0.57 | 0.64 | 0.67 | 0.69 | −0.05 |
| Obedience | 0.56 | 0.47 | 0.67 | 0.57 | 0.52 | 0.40 | 0.17 |
| Independence | 0.42 | 0.34 | 0.70 | 0.49 | 0.39 | 0.47 | 0.02 |
| Hard work | 0.51 | 0.49 | 0.42 | 0.47 | 0.51 | 0.56 | −0.09 |
| Unselfishness | 0.39 | 0.33 | 0.52 | 0.41 | 0.30 | 0.31 | 0.10 |
| Thrift saving money and things | 0.36 | 0.25 | 0.31 | 0.31 | 0.29 | 0.37 | −0.06 |
| Determination perseverance | 0.28 | 0.18 | 0.40 | 0.29 | 0.25 | 0.37 | −0.08 |
| Imagination | 0.28 | 0.12 | 0.31 | 0.24 | 0.15 | 0.21 | 0.03 |

**Table 6.4** Toward a UNDP type of index of the global democratic civil society

| Country/region | Maximum | Minimum | Worst value | Best value |
|---|---|---|---|---|
| Important child qualities: tolerance and respect for other people | 0.900 | 0.360 | 0.360 | 0.900 |
| Neighbors: people of a different religion | 0.570 | 0.010 | 0.570 | 0.010 |
| Men make better political leaders than women do | 3.430 | 1.500 | 1.500 | 3.430 |
| Reject: University is more important for a boy than for a girl | 3.700 | 2.300 | 2.300 | 3.700 |
| Democracy: religious authorities should interpret the laws | 7.830 | 1.800 | 7.830 | 1.800 |
| Justifiable: for a man to beat his wife | 4.820 | 1.160 | 4.820 | 1.160 |

in leading Western countries, and we start out from a frame of reference which was provided by the American economist Robert Barro. He believes that the networks and interactions fostered by churches and religious denominations are important elements of social capital (Barro & McCleary, 2003).

For Inglehart, such phenomena as bribery, corruption, tax evasion, cheating the state to get government benefits for which one would not be entitled, but also the healthy activism of citizens in volunteer organizations, already described by Etzioni (1968), hardly exist, while the rich data base of the ***World Values Survey*** provides ample evidence about these phenomena and their occurrence in world society.

Our statistics and choropleth maps of global value adherence underline a preoccupation about the loss of spiritual values, which also unites leaders of the major world religious denominations.[8] Among the major political scientists in Europe, Wilfried Röhrich (Röhrich, 2004, 2010) very clearly focused the attention of the discipline on the systematic study of the interaction between religion and politics on a global level. As the late Rabbi Lord Jonathan Sacks in his address at the Pontifical Gregorian University on December 12, 2012, in Rome correctly highlighted, the religious roots of the market economy and democratic capitalism can never be forgotten (Sacks, 2012).[9]

And Rabbi Lord Sacks went on to underline the following aspect, which achieves highest significance in our own empirical results:

> But trust is not a dispensable luxury. It is the very basis of our social life. (Sacks, 2012)[10]

Lord Sacks argues also that affluence makes you complacent. You no longer have the moral and mental energy to make the sacrifices necessary for the defense of freedom. Inequalities grow. The rich become self-indulgent. The poor feel excluded. There are social divisions, resentments, injustices. Society no longer coheres. People do not feel bound to one another by a bond of collective responsibility. Individualism prevails. Trust declines. Social capital wanes.

> Those who believe that liberal democracy and the free market can be defended by the force of law and regulation alone, without an internalised sense of duty and morality, are tragically mistaken. (Ibid.)

Secularization is part and parcel of the decay of family values and economic values, which threatens to affect negatively the very existence of the capitalist market economy. This simple and compelling message from one of the greatest economists of all times, Joseph Alois Schumpeter, was already published in (1950). For Schumpeter, the loss of family values destroys the fabric of capitalist family-owned businesses. Today, we can add: If overall values decline, then, also tax morale and the human work ethic, considered to be so vital for economic progress since the days of the German sociologist Max Weber, will decline as well.

The economics profession, that is, mathematical, quantitative economics, already began to make large-scale use of the ***World Values Survey*** data, integrating the ***World Values Survey*** country-level results into international economic growth accounting.

---

[8] http://www.peace.ca/concludingstatement.htm and http://www.catholiceducation.org/en/religion-and-philosophy/social-justice/has-europe-lost-its-soul.html Invalid reference.

[9] See Footnote 9.

[10] http://www.catholiceducation.org/en/religion-and-philosophy/social-justice/has-europe-lost-its-soul.html.

## 6.3 *Arab Opinion Index*—A Snapshot of Arab Opinion Today

A survey of the present state of global value research would be absolutely incomplete, if it were not to mention the results of the most encompassing survey of Arab opinion today, the *Arab Opinion Index,* organized by the *Arab Center for Research and Policy Studies* (*ACRPS*) in Qatar, and unfortunately largely overlooked in Western literature.

- Only 27% of all Arab respondents, mostly in the Gulf region, reported that their household income was sufficient. 26% of respondents reported that they lived "in a state of need."
  - Forty eight percentage evaluate the political situation in their countries as positive, compared to 47% who evaluate it as negative. Most respondents in the Gulf states evaluated the political situation as positive.
  - More than half of all Arabs said that unemployment, high prices, poor economic conditions, and poverty are the most important challenges facing their country.
  - Twenty-two percentage of all Arabs want to emigrate.
  - Thirty-nine percentage of all Arabs define democracy as a guarantee of political and civil freedoms. A majority of public opinion (76%) supports democracy while 17% oppose it. Seventy-four percentage believe that democracy is the most appropriate system of governance for their countries.
  - The *ACRPS* also asked the following question: „*Back in 2011, several Arab countries witnessed revolutions and popular protests, in which people took to the streets in demonstrations. What is your assessment of that?*" 58% consider the uprisings to have been positive, and 28% consider them to have been negative.

- Most of the Arab public considers itself as "religious to some extent" (63%). Twelve percentage define themselves as "not religious," and 23% as "very religious." 49% are supporting the principle of the separation of religion from political life while 44% of respondents opposes this.

## 6.4 Research Design for the Study of Value Change and Development Patterns in the Gulf

Given the fact, that in the literature we find hardly any guiding posts for a comparative analysis of value patterns in the Gulf, we decided to run our own comparisons with the relevant data, described in Chap. 2 of this work. We will first evaluate indicators of civil society and religious tolerance, move on to a general factor analytical evaluation of the value systems of Gulf countries by global comparison, and then move on to look into the development patterns of richer Arab countries, including the promising development of tertiary education.

Results about value change and development patterns in the Gulf.

### 6.4.1 Democratic Civil Society in the Gulf?

During the dark days of the Second World War, Sir Karl Popper wrote in faraway New Zealand the text of his monumental *"Open Society and Its Enemies*.. With a stark turn to the worse in many regions of the world in 2014/2015 (sharp Russia/West tensions over the Ukraine, destabilization of many countries of the Middle East due to Islamist terrorism, *et cetera),* it is time to test the global support for the two principle aims of the West, democracy and a market economy. To this end, we analyzed the global support for Open Society in the tradition of Popper (2012) and Bergson (1935), based on the freely available data of the ***World Values Survey,*** latest release, 2010–2014.[11]

We constructed a kind of "UNDP Index" of the civic culture of Arab countries by international comparison with the existing data,[12] selecting the ***World Values Survey*** items on the civic culture of tolerance, accepting gender equality, secularization and non-violence. Our index, which can be interpreted as a parameter for the maturity of civil society for democratic governance, weights the variables equally. According to the UNDP Index methodology, for each country the worst value of a given variable is subtracted from the observed country value of a given variable. This resulting number is then divided by the difference between the best and the worst value among the entire group of countries of the variable in question, yielding component indices ranging from 0 (worst performance) to 1 (best performance). The resulting overall index—in our case the index of a democratic civil society—is nothing but the average of the six chosen components[13]:

- Important child qualities: tolerance and respect for other people.
- Accepting neighbors: People of a different religion.
- Men do not make better political leaders than women do.
- University is equally important for a boy and for a girl.
- Democracy: religious authorities should not interpret the laws.
- Unjustifiable: For a man to beat his wife.

In constructing the index, we took particular care to select variables with a maximum country coverage in the vast ***World Values Survey*** data base in order to achieve not only a substantial depth of the index, but also a very wide geographical coverage across nations and cultures. The original ***World Values Survey*** items were used in the following fashion in the construction of the UNDP type of indicator:

The country results for this procedure are the following: Sweden, Norway, and Andorra are the countries best combining the civic culture of tolerance, accepting gender equality, secularization, and non-violence, while the three worst placed nations on earth are Mali, Bahrein, and Yemen. All Muslim countries in our 77 countries and territories with full data which were under investigation here are below the global average; and the best placed Muslim country is post-Soviet Kazakhstan; and the best placed Arab country is Qatar. While some Arab countries might perform, here

[11] http://www.worldvaluessurvey.org/wvs.jsp.

[12] WVS_Longitudinal_1981_2014_MULTIVAR_spss_v2015_04_20.sav.

[13] http://hdr.undp.org/en/content/human-development-index-hdi.

**Table 6.5** An index of a democratic civil society

| Global rank (77 countries) | Country/region | UNDP-type index of a democratic civil society |
|---|---|---|
| 50 | Qatar | 0.514 |
| 76 | Bahrain | 0.328 |

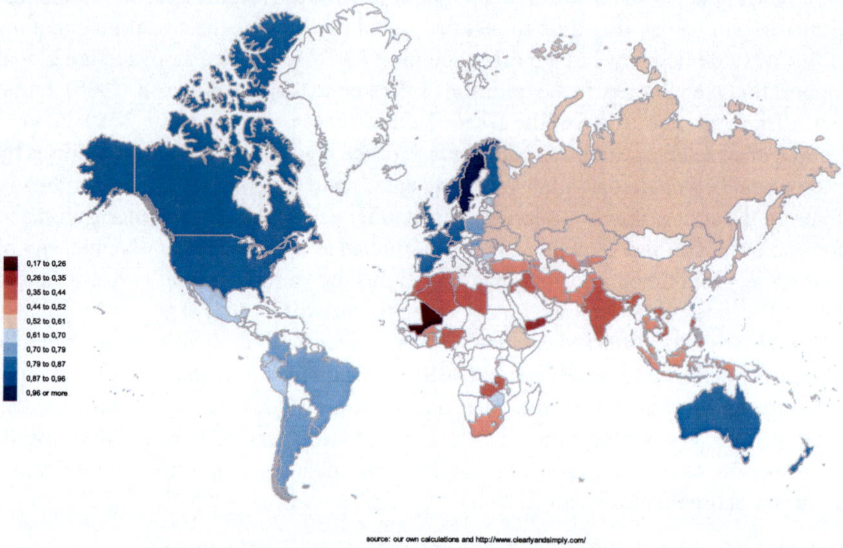


**Fig. 6.5** Index of a democratic civil society (UNDP type of index)

and there, in a rather surprising and positive fashion, it is especially the combination of the six dimensions, where the Arab World really fails (Table 6.5).

Figure 6.5 portrays the values of these calculations onto a choropleth global map. Figure 6.6 shows the global distribution of the country values of the "*Effective Democracy Index*" developed by Alexander et al. (2012), combining human rights and the freedom from corruption. Arguably, the Effective Democracy Index is the most reliable single democracy indicator today.

## 6.4.2 *The Global Tolerance Index in the Gulf*

For the calculation of the Global Religious Tolerance Indicator, we again relied on the well-established methodology of the United Nations Human Development Programme and its UNDP Human Development Index (UNDP, 2014, and Chap. 3). The *World Values Survey* offers fairly encompassing and comparable data on tolerance items. The chosen *World Values Survey* tolerance indicators were:

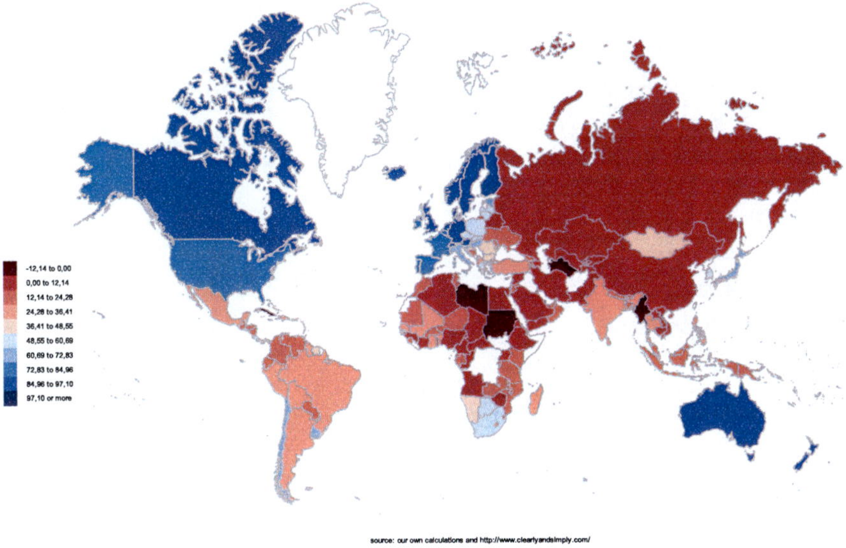


**Fig. 6.6** Index of effective democracy

1. Disagree or strongly disagree: the only acceptable religion is my religion (mean) F203.
2. Agree or strongly agree: all religions should be taught in public schools (mean) F204.
3. Agree or strongly agree: people who belong to different religions are probably just as moral as those who belong to mine (mean) F205.
4. Trust completely or somewhat: people of another religion (mean) G007_35B.
5. Meaning of religion: do good to other people (%-percentages) F200.

In Chap. 2, we already highlighted that since the 1990s, the United Nations Development Program (UNDP, 2014) calculates the internationally recognized "*Human Development Index*," which equally weights life expectancy, education, and real income. Life expectancy, education, and real incomes are projected on a scale from 0 (worst value) to 1 (best value). Although UNDP calculation methods have become somewhat more complicated in recent years, the simple rationale remains: Our Index of Religious Tolerance is the average of the five components, other religions are acceptable; all religions should be taught in public schools; people who belong to different religions are probably just as moral as those who belong to mine; trust people of another religion; and the meaning of religion is to do good to other people.

For 59 states of the world there are complete data. While, e.g., in Sweden and the USA 30% or less of the population have no confidence in people with a religious denomination other than their own, these percentages in Algeria; Armenia; Yemen; Kyrgyzstan; Libya; Morocco; Mexico; Palestinian Territories; Peru; Romania; Tunisia; and Uzbekistan are over 70% each. Unfortunately, the clear North–South divide of religious tolerance on our globe also corresponds to a clear denominational

**Table 6.6** Religious tolerance index—total populations

| Country/region | Bahrain | Kuwait | Qatar |
|---|---|---|---|
| Not only my religion is acceptable | 0.307 | 0.201 | 0.000 |
| All religions should be taught in public schools | 0.665 | 0.223 | 0.069 |
| People who belong to different religions are probably just as moral as those who belong to mine | 0.569 | 0.204 | 0.321 |
| Distrust: People of another religion (B) | 0.570 | 0.530 | 0.584 |
| Do good to other people | 0.457 | 0.346 | 0.219 |
| Religious Tolerance Index | 0.514 | 0.301 | 0.239 |
| Global rank (59 countries) | 34 | 50 | 55 |

gap. Among the ten states with the lowest general religious tolerance, based on our five indicators, there are nine predominantly Muslim states.

It is being mentioned that there are substantial ranges of examples of best and worst practice within the Muslim communities of the world. According to our data, flagship models of a liberal and tolerant Islam can be encountered in Trinidad and Tobago, and also in Georgia, India, Ghana, and South Africa.

Religious tolerance is also a matter for urgent consideration by the global South, and not just for the global North and the global migration recipient countries.

The disappointing ranks of Bahrain, Kuwait, and Qatar are at the end of Table 6.6.

So, the best performers were Sweden, Trinidad and Tobago, Brazil, New Zealand, Australia, and the worst performers were Algeria, Yemen, Morocco, Palestinian Territories, and Qatar. Figure 6.7 shows the basic wisdom of classical modernization theories in political science. As correctly predicted by these theories (see Tausch et al., 2014), there is a very close causal and statistical relationship between "modern," open-society oriented values, democracy and development. In his 2012 analysis on the subject, Barro found that in an 80-country panel since the 1960s, the convergence rate for per capita GDP is around 1.7% per year. In the post-1960s panel, Barro's analyses found support for modernization theories, highlighting the positive effects of per capita GDP and schooling on democracy and the maintenance of law and order. Barro (2012) also found positive effects of per capita GDP and schooling on the Polity indicator for democracy. Our own modest contribution to this debate in the context of the future of democracy in the Arab World is to be seen in the proof of the combination of the low degree of the development of a democratic civil society and low effective democracy in the Arab World. The development of a democratic civil society explains none the less than 67% of the variance of the effective democracy indicator. If the Arab World wants democracy, it must devise ways to become a more democratic civil society, combining a commitment to educate children in a spirit of tolerance and respect for other people, accepting neighbors, including people of a different religion, a commitment to gender equality in politics and education, a separation of law making and religious authorities, and a commitment against male domestic violence.

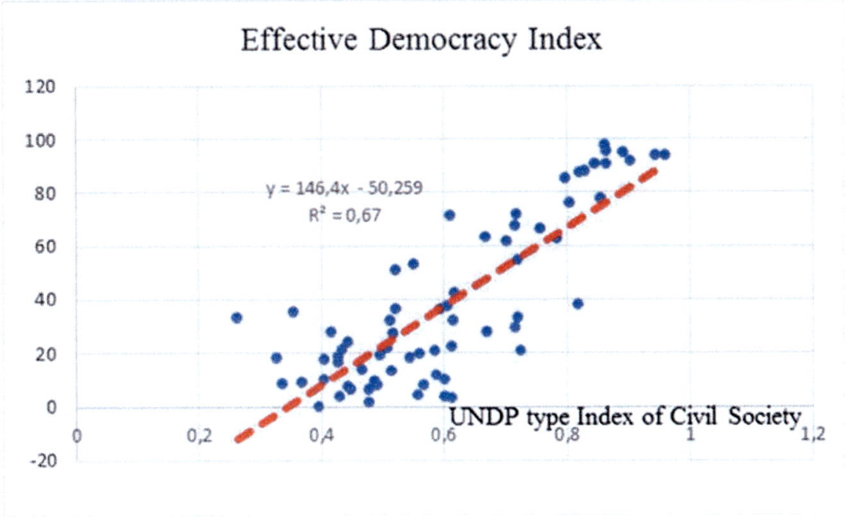


**Fig. 6.7** Index of a democratic civil society and effective democracy

For sure, our predictions for the development of democracy in the Arab World are thus rather pessimistic.

### 6.4.3 *Comparative Aspects of the Development of Arab Civil SocietyEvidence from the Islamicity Index*

The data about the democratic civil society in the Arab World must be confronted by highlighting the fact that the religion of Islam can be reconciled with an Open Society. At this stage, we attempt to introduce a unique, novel and forward-looking new approach to reconcile Popper's Open Society and the theology of Islam, namely the writings of the Shia Muslim US scholar Hossein Askari (Rehman & Askari, 2010a; Askari et al., 2015).[14] His *Index of Islamicity* is the instrument how a society fulfills what Askari[15] calls Islam as a rules based religion. Askari says:

> Islam is a rules-basedreligion. The Holy Quran provides the foundational rules. These rules were interpreted and put into practice by the Prophet Mohammad (sawa) in the first Muslim community in Medina. The Prophet Mohammad (sawa) shepherded and served the first Muslim community in Medina subject to the support and concurrence of the community. Under his guidance the Muslim community flourished with justice as its hallmark. Muslims are asked to study the Quran and the life of the Prophet Mohammad (sawa), to internalize the rules prescribed by The Almighty, and to follow them to develop just and flourishing communities. […] When we simply used our eyes to look at the Muslim countries, we failed

[14] http://islamicity-index.org/wp/islamicity-rankings/.

[15] http://hossein-askari.com/islamicity/.

to see a high degree of rule compliance. We could not see the socio-economic outcomes that we would expect from rule-compliant Muslim countries. We realized that a benchmark or index was needed to assess the degree of rule compliance or "Islamicity," to serve as an indicator of needed political, social, and economic reforms.

Askari, in his research, attempts to answer the following questions:

- What should be the characteristics of a truly rule-compliant Muslim community?
- Where could the data be found on these characteristics?
- How could diverse data be summed up (or what importance or weight should be assigned to each characteristic) in a single number or index?

The Islamicity Index (see above) was further developed in Rehman and Askari (2010b), and in Askari and Mohammadkhan (2015). Cross-correlating the Index of Islamicity with standard indicators of socio-economic development, reported in Tausch and Heshmati (2013), we get (Table 6.7).

Thus, we can say that the indicator is a valid instrument to measure socio-economic progress. Figure 6.8 now shows the additional vital link between our democratic civil society indicator, developed above, and the Islamicity Index. Following Askari's argument, the real precondition to develop a truly Islamic society, guided by

**Table 6.7** Correlates of the Islamicity Index

| Variable | Pearson correlation with Islamicity Index | $R^2$ |
|---|---|---|
| Overall, 35 development index | 0.884 | 78,193 |
| Corruption avoidance measure | 0.880 | 77,445 |
| Human Development Index (HDI) value 2004 | 0.835 | 69,792 |
| Rule of law | 0.831 | 69,011 |
| UNDP-type index of a democratic civil society | 0.830 | 68,881 |
| Gender empowerment index value | 0.827 | 68,443 |
| Happy life years | 0.791 | 62,614 |
| Environmental Performance Index (EPI) | 0.755 | 57,035 |
| Global tolerance index | 0.752 | 56,614 |
| Tertiary enrollment | 0.744 | 55,355 |
| Life satisfaction (0–10) | 0.738 | 54,442 |
| No redistributive religious fundamentalism | 0.730 | 53,360 |
| Life expectancy (years) | 0.729 | 53,143 |
| UNDP education index | 0.719 | 51,645 |
| 2000 Economic Freedom Score | 0.718 | 51,591 |
| Comparative price levels (US = 1.00) | 0.708 | 50,092 |
| Civil and Political Liberties violations | −0.742 | 55,100 |
| Combined Failed States Index | −0.921 | 84,886 |

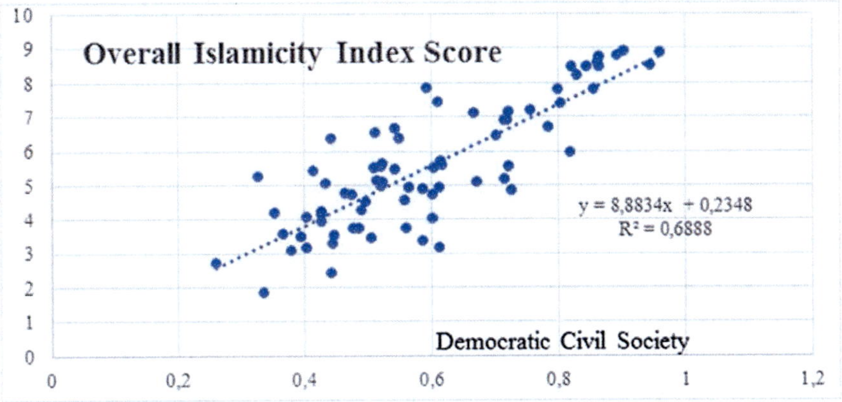


**Fig. 6.8** Democratic civil society and Islamicity Index

truly Quranic principles and directed toward an Open Society, would be to develop precisely the patterns of societal tolerance, captured by our democratic civil society index. The development of tolerance, measured by the democratic civil society index, explains almost 69% of the variance of Askari's Islamicity Index, highlighted also in choropleth map in Fig. 6.9. We agree with Askari's conclusions based on the classical Muslim nineteenth-century philosopher Muḥammad 'Abduh (1849–1905) who said:

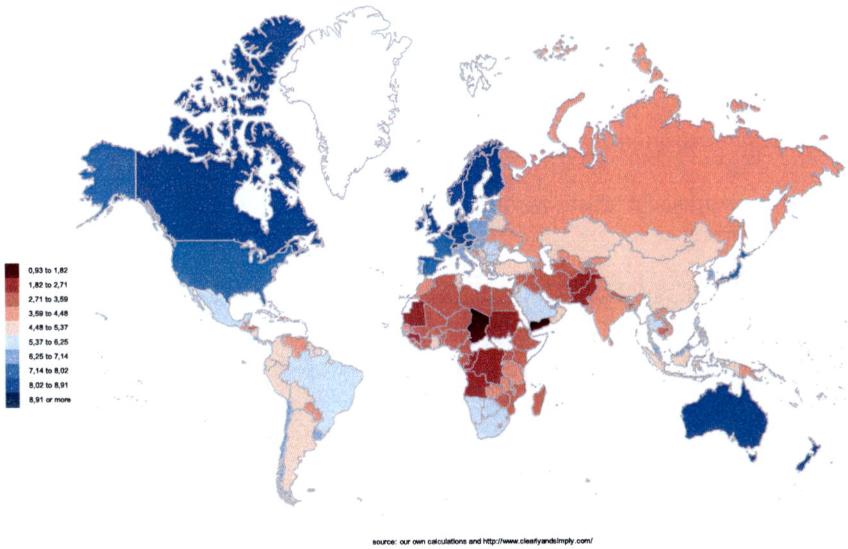


**Fig. 6.9** Islamicity Index, 2015

I went to the West and saw Islam, but no Muslims; I got back to the East and saw Muslims, but no Islam.[16]

The development of a democratic civil society explains none the less than 68.9% of the variance of the Islamicity indicator (see Fig. 6.8).

It is to be underlined that Islamicity—as defined by Askari and his associates is highest in the most highly developed Western nations, and is lowest in several African states, and is generally very low also in the Arab World. Only the Gulf countries are an exception to this and are on a same level as, say, the BRICS countries Brazil and South Africa.

### *6.4.4 Comparative Aspects of Value Development in the Gulf*

In the following, we use ***World Values Survey*** data first to classify the identification of global publics with pivotal aspects of democracy and the market economy, and we then analyze how the country values for the general public depend on per capita incomes. These calculations are necessary in order to risk some predictions about the future trajectories of democracy in different cultural systems, as per capita incomes rise.

Our analysis of the ***World Values Survey*** data derived the following factor analytical scales, well compatible with a large social scientific literature, discussed at length in Solomon and Tausch (2020, 2021).

1. The non-violent and law-abiding society,
2. Democracy movement,
3. Climate of personal non-violence,
4. Trust in institutions,
5. Happiness, good health,
6. No redistributive religious fundamentalism,
7. Accepting the market economy,
8. Feminism,
9. Involvement in politics,
10. Optimism and engagement,
11. No welfare mentality, acceptancy of the Calvinist work ethics.

We mention here briefly the salient factor loadings, explaining 10% or more of a variable:

**The Violent and Lawless Society**

0.796 Justifiable: Avoiding a fare on public transport
0.765 Justifiable: Stealing property
0.760 Justifiable: Claiming government benefits

[16] See Footnote 15.

| | |
|---|---|
| 0.732 | Justifiable: Someone accepting a bribe |
| 0.560 | Justifiable: Violence against other people |
| 0.451 | Justifiable: For a man to beat his wife. |

**Democracy Movement**

| | |
|---|---|
| Democracy: Civil rights protect people's liberty against oppression | 0.753 |
| Democracy: People choose their leaders in free elections | 0.738 |
| Democracy: Women have the same rights as men | 0.704 |
| Democracy: Governments tax the rich and subsidize the poor | 0.493 |
| Importance of democracy | 0.493 |
| Democracy: The state makes people's incomes equal | 0.448. |

**Climate of Personal Violence**

| | |
|---|---|
| Justifiable: For a man to beat his wife | 0.846 |
| Justifiable: Parents beating children | 0.795 |
| Justifiable: Violence against other people | 0.786 |
| Justifiable: Someone accepting a bribe | 0.604 |
| Justifiable: Stealing property | 0.587. |

**Lack of Trust in Institutions**

| | |
|---|---|
| No confidence: The Government | 0.776 |
| No confidence: The Police | 0.717 |
| No confidence: The Press | 0.715 |
| No confidence: The United Nations | 0.637. |

**Unhappiness, Poor Health**

| | |
|---|---|
| State of health (bad) (subjective) | 0.771 |
| Feeling of unhappiness | 0.716 |
| Age | 0.440 |

| | |
|---|---|
| I don't see myself as a world citizen | 0.405 |
| Insecurity in neighborhood | 0.364. |

**Redistributive Religious Fundamentalism**

| | |
|---|---|
| Democracy: Religious authorities interpret the laws | 0.687 |
| not important in life: Religion | −0.596 |
| Democracy: The state makes people's incomes equal | 0.460 |
| Democracy: Governments tax the rich and subsidize the poor | 0.389. |

**Rejecting the Market Economy**

| | |
|---|---|
| Competition [good or] harmful | 0.760 |
| Hard work does not bring success | 0.733 |
| [Private vs.] state ownership of business | 0.353. |

**Feminism**

| | |
|---|---|
| Reject: men make better political leaders than women do | 0.717 |
| University is not more important for a boy than for a girl | 0.682. |

**Distance to Politics**

| | |
|---|---|
| No interest in politics | 0.849 |
| Not important in life: Politics | 0.837. |

**Nihilism**

| | |
|---|---|
| Not important in life: Friends | 0.690 |
| Not important in life: Leisure time | 0.669 |
| Not important in life: Work | 0.495 |
| Not important in life: Family | 0.478. |

**Welfare Mentality, Rejection of the Calvinist Work Ethics**

| | |
|---|---|
| Supporting larger income differences | −0.677 |

| | |
|---|---|
| Not important in life: Work | 0.467 |
| Not important in life: Religion | 0.400 |
| Democracy: The state makes people's incomes equal | 0.395. |

Our Index construction is described in Table 6.8. This is applied in the following Table 6.9.

In Table 6.9, we see the results of this parametric index construction and also an information about the percentile performance of the Gulf countries. The (parametric) rank (achieved by multivariate analysis) of the Gulf countries among 59 nations of the world is as follows:

Qatar 9

**Table 6.8** Weighting of the factors for construction of a parametric indicator of the development of civil society

| | Factor | Eigenvalue | % of variance explained | Cumulative % | Multiplication factor | Weight |
|---|---|---|---|---|---|---|
| The non-violent and law-abiding society | 1 | 4.263 | 10,931 | 10,931 | −1 | −4.263 |
| Democracy movement | 2 | 2.574 | 6601 | 17,532 | +1 | 2.574 |
| Climate of personal non-violence | 3 | 2.26 | 5794 | 23,326 | −1 | −2.26 |
| Trust in institutions | 4 | 1.929 | 4947 | 28,273 | −1 | −1.929 |
| Happiness, good health | 5 | 1.864 | 4779 | 33,052 | −1 | −1.864 |
| No redistributive religious fundamentalism | 6 | 1.554 | 3986 | 37,037 | −1 | −1.554 |
| Accepting the market economy | 7 | 1.434 | 3676 | 40,714 | −1 | −1.434 |
| Feminism | 8 | 1.245 | 3193 | 43,907 | +1 | 1.245 |
| Involvement in politics | 9 | 1.197 | 3.07 | 46,977 | −1 | −1.197 |
| Optimism and engagement | 10 | 1.141 | 2926 | 49,904 | −1 | −1.141 |
| No welfare mentality, acceptancy of the Calvinist work ethics | 11 | 1.075 | 2756 | 52.66 | −1 | −1.075 |

**Table 6.9** Percentile performance (Arab civil societies are among the top percent of global society)

| | Overall civil society index | Global Rank | Global percentile performance (country among top ... %) |
|---|---|---|---|
| Sweden | 7.047 | 1 | 1695 |
| Trinidad and Tobago | 5.751 | 2 | 3390 |
| Australia | 5.487 | 3 | 5085 |
| Japan | 5.466 | 4 | 6780 |
| Netherlands | 5.216 | 5 | 8475 |
| Ghana | 4.760 | 6 | 10,169 |
| Germany | 4.274 | 7 | 11,864 |
| Uzbekistan | 4.250 | 8 | 13,559 |
| Qatar | 3.749 | 9 | 15,254 |
| Cyprus | 3.500 | 10 | 16,949 |
| Uruguay | 3.496 | 11 | 18,644 |
| Spain | 3.197 | 12 | 20,339 |
| United States | 3.197 | 13 | 22,034 |
| Romania | 2.920 | 14 | 23,729 |
| Poland | 2.802 | 15 | 25,424 |
| Taiwan | 2.745 | 16 | 27,119 |
| Georgia | 2.562 | 17 | 28,814 |
| Thailand | 2.523 | 18 | 30,508 |
| Turkey | 2.121 | 19 | 32,203 |
| South Korea | 1.906 | 20 | 33,898 |
| Armenia | 1.852 | 21 | 35,593 |
| Zimbabwe | 1.789 | 22 | 37,288 |
| Brazil | 1.752 | 23 | 38,983 |
| Tunisia | 1.656 | 24 | 40,678 |
| China | 1.514 | 25 | 42,373 |
| Chile | 1.312 | 26 | 44,068 |
| Estonia | 1.157 | 27 | 45,763 |
| Malaysia | 1.029 | 28 | 47,458 |
| Ecuador | 0.945 | 29 | 49,153 |
| Slovenia | 0.730 | 30 | 50,847 |
| Colombia | 0.631 | 31 | 52,542 |
| Rwanda | 0.402 | 32 | 54,237 |
| Argentina | 0.342 | 33 | 55,932 |
| Morocco | 0.249 | 34 | 57,627 |
| Jordan | 0.199 | 35 | 59,322 |
| Libya | 0.079 | 36 | 61,017 |
| Nigeria | 0.042 | 37 | 62,712 |
| Yemen | −0.205 | 38 | 64,407 |
| Azerbaijan | −0.301 | 39 | 66,102 |
| Kazakhstan | −0.367 | 40 | 67,797 |
| Kuwait | −0.840 | 41 | 69,492 |
| Peru | −0.931 | 42 | 71,186 |
| Mexico | −0.947 | 43 | 72,881 |
| Kyrgyzstan | −0.958 | 44 | 74,576 |
| Pakistan | −1.223 | 45 | 76,271 |
| Singapore | −1.482 | 46 | 77,966 |
| Hong Kong | −1.876 | 47 | 79,661 |
| Belarus | −2.711 | 48 | 81,356 |
| Palestinian Occupied Territories | −2.997 | 49 | 83,051 |
| Ukraine | −3.060 | 50 | 84,746 |
| Iraq | −3.306 | 51 | 86,441 |
| Egypt | −3.878 | 52 | 88,136 |
| Algeria | −4.422 | 53 | 89,831 |
| Bahrain | −4.426 | 54 | 91,525 |
| Russia | −4.609 | 55 | 93,220 |
| Lebanon | −5.183 | 56 | 94,915 |
| Philippines | −5.774 | 57 | 96,610 |
| South Africa | −9.691 | 58 | 98,305 |
| India | −10.498 | 59 | 100,000 |

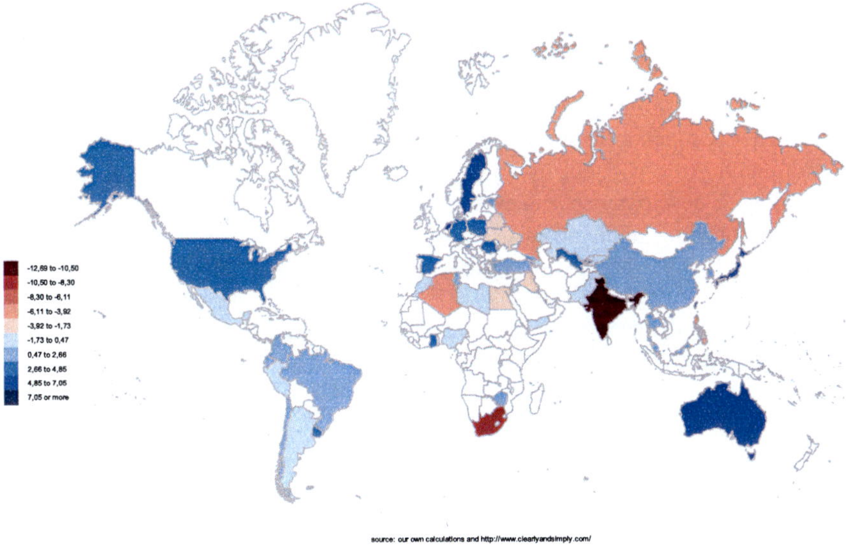


**Fig. 6.10** Overall Civil Society Index, based on the size of the factor analytical eigenvalues

| | |
|---|---|
| Kuwait | 41 |
| Bahrain | 54. |

Figure 6.10 projects the above Table 6.9 onto a choropleth map of the world.

### *6.4.5 The Chances for democracy in the Gulf Region and the Kuznets Trap*

In the following analysis, we plot ***World Values Survey*** data about the acceptancy of

1. The non-violent and law-abiding society,
2. Democracy movement,
3. Climate of personal non-violence,
4. Trust in institutions,
5. Happiness, good health,
6. No redistributive religious fundamentalism,
7. Accepting the market economy,
8. Feminism,
9. Involvement in politics,
10. Optimism and engagement,
11. No welfare mentality, acceptancy of the Calvinist work ethics.

among overall society, Muslim communities and among active Roman Catholics against the respective natural logarithm of GDP per capita. What are the results of this endeavor? We find strong empirical support for the idea first proposed by Nobel Laureate Simon Kuznets (1976) that social crisis phenomena, such as inequality, rise with higher per capita incomes, only to level off at higher stages of development. Our research gives evidence to the fact that not only macro-economic phenomena, but also value developments and even patterns of prejudice are U-shaped or inverted U-shaped functions of the natural logarithm of GDP per capita. We call this process the Kuznets trap: rising crises phenomena in the transition phase from the rural to the urban life style, accompanied by an implosion of basic values, necessary for the functioning of a democratic society and a market economy.

Muslim communities at all stages of development still show currently an empirically existing deficit of accepting the fundamental values of democracy and the market economy in comparison to the development paths of overall society (x-axis is the n log of GDP per capita) (Fig. 6.11).

To proceed with our analysis, we reproduce in Table 6.10 the correlation matrix of the components of our factor analysis at the global level.

Correlations greater than or equal to ±0.100.

In the following, we will reflect on religion, work ethics and economic growth.

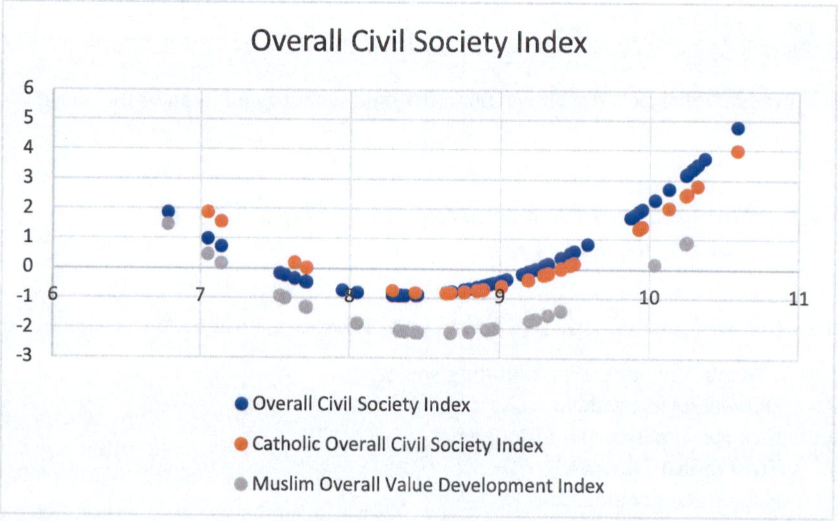


**Fig. 6.11** Value development in global society, as predicted by our data and GDP per capita[17]

[17] For the GDP per capita data, see Tausch and Heshmati (2013).

**Table 6.10** Correlation matrix of components at the global level

| Component | The violent and lawless society | Democracy movement | Climate of personal violence | Lack of trust in institutions | Unhappiness, poor health | Redistributive religious fundamentalism | Rejecting the market economy | Feminism | Distance to politics | Nihilism |
|---|---|---|---|---|---|---|---|---|---|---|
| Democracy movement | −0.139 | | | | | | | | | |
| Climate of personal violence | 0.405 | −0.225 | | | | | | | | |
| Lack of trust in institutions | | | | | | | | | | |
| Unhappiness, poor health | | | | 0.138 | | | | | | |
| Redistributive religious fundamentalism | | | | | | | | | | |
| Rejecting the market economy | 0.236 | | 0.125 | | | | | | | |
| Feminism | | | −0.201 | | | −0.120 | | | | |
| Distance to politics | | | | 0.161 | | | | | | |
| Nihilism | 0.105 | | | | 0.198 | −0.100 | 0.261 | | 0.101 | |
| Welfare mentality, rejection of the calvinist work ethics | | 0.120 | | | 0.103 | | 0.324 | −0.101 | 0.108 | 0.295 |

## 6.5 Religion, Work Ethics, and Economic Growthan Optimistic Message for the Gulf Countries

Following Hayek (1960), we think that values like hard work which brings success, competition, and private ownership of business play an overwhelming role in the twenty-first century capitalism, and simply cannot be overlooked in empirical global value research.

Starting with the usual World Bank economic growth data, we immediately see that the centers of economic growth since the crisis of 2008 shifted inexorably toward the countries of the Pacific and the Indian Ocean arena, and away from the secularized West. Even in their wildest anti-Western dreams, the opponents of the West would not have been able to imagine what has come true today—the tremendous reduction of Western economic power. What radical Islamist movements could hitherto not achieve—the undermining of Western military power—the economic crisis and the rise of new economic global players accomplished within the timespan of a few years. Not the bombs of Mr. Osama Ben Laden, but the relentless economic crisis starting in 2007 and now COVID-19 achieved this decline.

Barro speaks decidedly in favor of the importance of religion for sound economic growth and long-run economic well-being. Barro and McCleary instead of viewing "religious beliefs" as an "impediment" of economic growth, tend to see them as requirements of a resilient society today:

> Our central perspective is that religion affects economic outcomes mainly by fostering religious beliefs that influence individual traits such as thrift, work ethic, honesty, and openness to strangers. For example, beliefs in heaven and hell might affect these traits by creating perceived rewards and punishments that relate to 'good' and 'bad' lifetime behavior. (Barro & McCleary, 2003: 779)

Religious beliefs stimulate growth because they help to sustain aspects of individual behavior that enhance productivity. Respect of parents is related in a clear-cut positive manner to economic growth, as well as the belief in hell.

*The decline of "economic morality" has many visible results—the shadow economy, bribery, corruption*, etc. The economic profession already invested lots of energy over recent years to investigate these phenomena (see Tausch et al., 2014). Hofstede, Schwartz, Davidov, and Inglehart, the three major existing sociological and psychological theories about global values do not talk about the shadow economy at all.

The author of the present book, of course, is well aware of the vast debate on religion and globalization in the context of the theories, debated elsewhere (Tausch et al., 2014). Eisenstadt (1968) already underlined the pivotal role of Protestantism in the rise of the Enlightenment traditions in the West. In accordance with Guiso et al. (2003), we think that Eisenstadt's theory deserves a more than passing mentioning here, because he moved away from an analysis of a direct causal link between Protestantism and capitalism to focus on the "transformative potential" of religions.

The transformative potential is defined by Eisenstadt as the *"capacity to legitimize, in religious or ideological terms, the development of new motivations, activities, and institutions which were not encompassed by their original impulses and views"* (Eisenstadt, 1968). As Guiso *et al.* correctly emphasized, Eisenstadt's main contribution in the context of the debate was to show that Protestantism redefined political and social institutions and had influence on the reformulation of roles within the economic sphere.

In view of the global quest for an ethics of tolerance and understanding, one might also argue that starting from Montesquieu (1989) in the Western Christian tradition, the Enlightenment would be indeed inseparable from the development of a culture of tolerance and the market economy (Allen, 2008; Holmes, 2006; just to mention a few). Important traditions of Enlightenment in other world religions—to name here only Judaism and Islam, without neglecting the other global religions—must also be named in this context (Feiner & Naor, 2011; Morgan, 2007; Sacks, 1998, 2003, 2005, 2014). Europe's dominant denomination, Roman Catholicism, only joined the traditions of Enlightenment in the Second Vatican Council. And while the Roman Catholic Church now forcefully condemns Anti-Semitism, it is still ambivalent at best on the "second pillar" of the Anti-Enlightenment prejudice,[18] which targeted Free Masonry.

But for empirical economists, who are not primarily in sacred scriptures but in economic behavior of human beings, the evidence published on the relationship between religion, denominations, societal ethics and economic growth is far from clear and conclusive. There are, if one wishes to say so, different shades of very contradicting evidence, and among the dozens and dozens of articles and books published on the subject, we refer our readers to Tausch et al. (2014) for further debates. Guiso et al. (2003) is an especially noteworthy source in this context, because the article is based on a very comprehensive analysis of the ***World Values Survey*** data, which were available to the authors at the time of the writing of their article. Guiso et al. (2003) found that on average religion is good for the development of stronger institutions. *Religious people trust others more, trust the government more, are less willing to break the law, and believe more in the fairness of the market. Active churchgoers are not more intolerant toward immigrants than the rest of the population.* Finally, both a religious upbringing and a active religious participation increase trust toward government institutions. Religious upbringing and affiliation are associated with a reduced willingness to break any sort of legal rule. People attending religious services on a more regular basis are more willing to trade off equality for incentives, and in particular, they favor more private ownership.

Inglehart is right in one respect: Human development leads to value change: increasing support for democracy, against redistributive religious fundamentalism, and feminism, but, as has been already shown in Tausch et al. (2014), there is an erosion of support for the market economy and the Calvinist work ethic.

---

[18] http://www.jewishvirtuallibrary.org/jsource/judaica/ejud_0002_0007_0_06772.html.

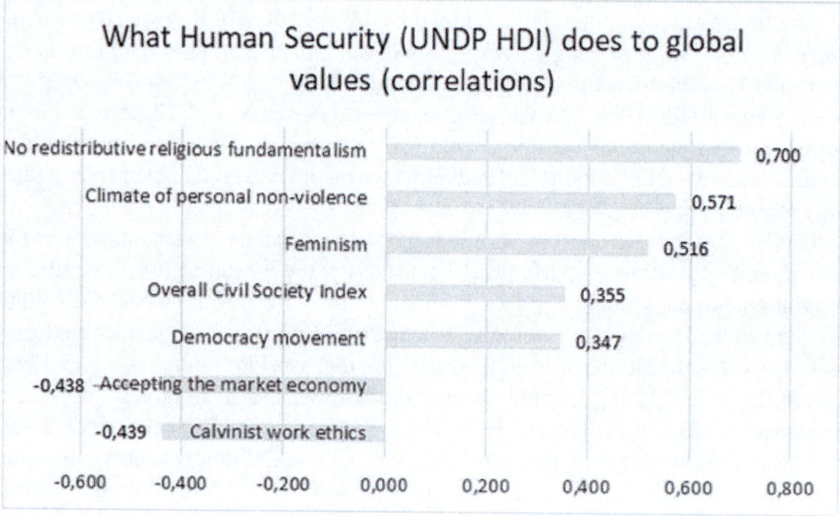


**Fig. 6.12** What Human security does to global values

In Fig. 6.12, we show the results of our analysis on the trade-off between human security and our promax factor analytical values. While redistributive fundamentalism decreases with human security, and the climate of personal non-violence, feminism, and the democracy movement increase with human security, accepting the market economy and the Calvinist work ethics decrease with the level of human security. This must be regarded as a major societal problem for the advanced countries.

In a dynamic perspective, increases in human security over time (1990–2018) are clearly connected with the intergenerational increase of the overall civil society index and the intergenerational increase of feminism according to our measurement of intergenerational value change, introduced in Chap. 2. That is to say, growing human security over time indeed affects the spread of "modern" civil society and feminist values among the younger generations (Fig. 6.13).

Figures 6.14 and 6.15 answer the question, where the level (Fig. 6.14) or the dynamics of human security(Fig. 6.15) hardly affect global values (Fig. 6.14) or the spread of values among the younger generations (Fig. 6.15). The factors, where there is hardly any linear effect of human security on human values, are trust, the law-abiding society, involvement in politics, optimism and engagement, and happiness and good health (Fig. 6.14).

Increases in human security over time hardly affect the intergenerational change of the democracy movement, optimism and engagement, the climate of personal non-violence, the Calvinist work ethics, the involvement in politics, the law-abiding

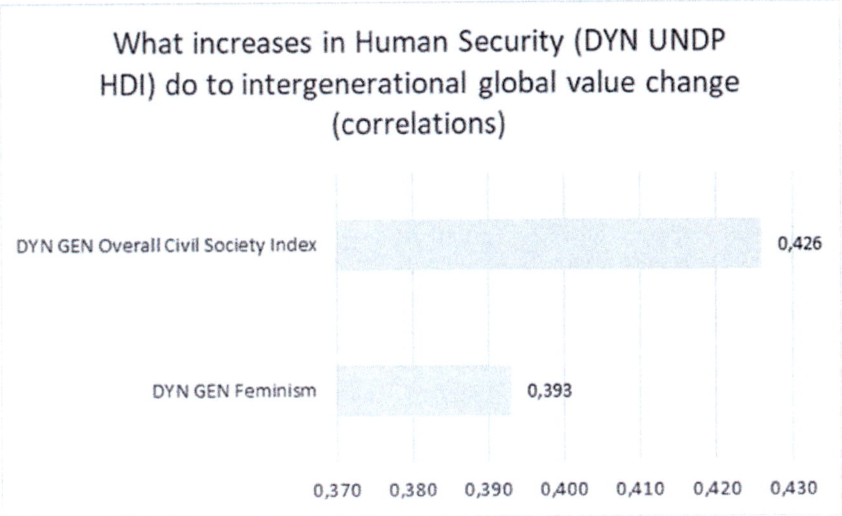


**Fig. 6.13** What increases in human security, 1990–2018, do to intergenerational value change

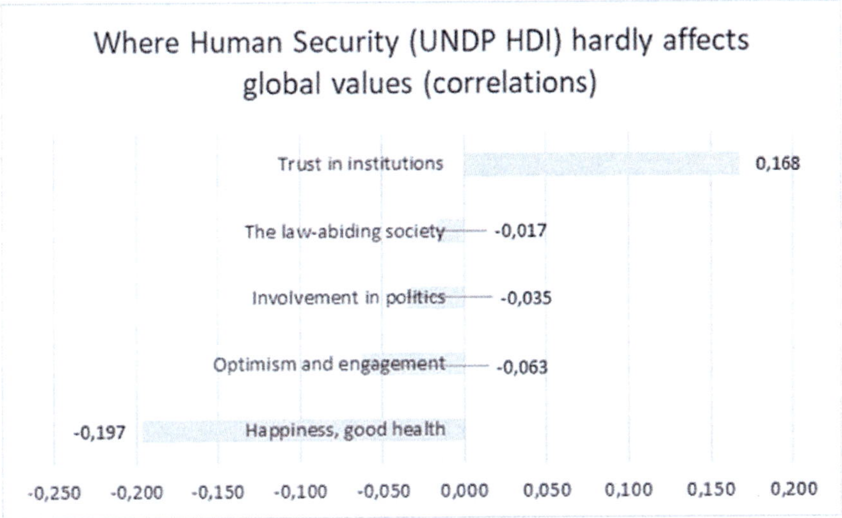


**Fig. 6.14** Where human security hardly affects global values

society, happiness and good health, accepting the market economy, no redistributive religious fundamentalism, and trust in institutions.

Table 6.11 summarizes these informations, contained in the above images, in the form of a statistical table.

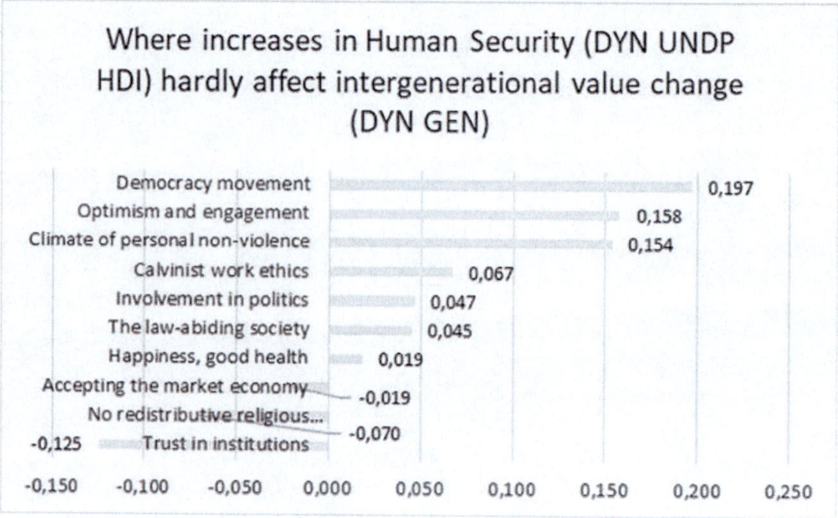


**Fig. 6.15** Where increases in human security, 1990–2018, hardly affect intergenerational value change

### 6.5.1 *Development Outcomes of the Gulf Countries in Light of the Analysis of Global Values*

In the following, we compare the value systems of the Gulf countries, analyzed above, with the "surpluses" and deficits of the development process of the region. To this end, we calculated the partial correlations of Gulf Cooperation Council membership in the world system with a host of developmental indicators, prepared for this study (https://www.academia.edu/44295690/GULF_2020_Covid_19_PUBLIC_ACCESS_Data_base_for_the_book_Factors_affecting_the_Gulfs_future_Global_cycles_and_value_change_Springer_Publishing_Series_https_www_springer_com_series_16417_Gulf_Studies_edited_by_Prof_Mizanur_Rahman_Qatar_University).

As to the data sources and definitions, interested readers are invited to consult this website. We kept the Human Development Index and its square as well as membership in the Arab League and the Muslim population share constant (HDI, $HDI^2$, Arab League and Muslim population share). Table 6.12 lists the results of this exercise.

It is evident that the countries of the Gulf are characterized by an intensive inward migration. The Gulf development model has the following pluses to its favor. We list these achievements in descending order, independent from the Human Development Index and its square as well as independent from the membership of a country in the Arab League and the Muslim population share.

- Trust in institutions
- Gallup poll about satisfaction: Local labor market

**Table 6.11** The dynamics of global value change—intergenerational value differences and human development

| xx | Correlation with DYN HDI, 1990–2018 | Error *p* | N | xx | Correlation with HDI, 2018 | Error *p* | N |
|---|---|---|---|---|---|---|---|
| DYN (based on inverted ranks) GEN the law-abiding society | 0.045 | 0.756 | 51 | The law-abiding society | −0.017 | 0.901 | 58 |
| DYN (based on inverted ranks) GEN democracy movement | 0.197 | 0.165 | 51 | Democracy movement | 0.347 | 0.008 | 58 |
| DYN (based on inverted ranks) GEN climate of personal non-violence | 0.154 | 0.281 | 51 | Climate of personal non-violence | 0.571 | 0.000 | 58 |
| DYN (based on inverted ranks) GEN trust in institutions | −0.125 | 0.380 | 51 | Trust in institutions | 0.168 | 0.209 | 58 |
| DYN (based on inverted ranks) GEN happiness, good health | 0.019 | 0.893 | 51 | Happiness, good health | −0.197 | 0.138 | 58 |
| DYN (based on inverted ranks) GEN no redistributive religious fundamentalism | −0.070 | 0.624 | 51 | No redistributive religious fundamentalism | 0.700 | 0.000 | 58 |
| DYN (based on inverted ranks) GEN accepting the market economy | −0.019 | 0.894 | 51 | Accepting the market economy | −0.438 | 0.001 | 58 |
| DYN (based on inverted ranks) GEN feminism | 0.393 | 0.004 | 51 | Feminism | 0.516 | 0.000 | 58 |
| DYN (based on inverted ranks) GEN involvement in politics | 0.047 | 0.743 | 51 | Involvement in politics | −0.035 | 0.794 | 58 |
| DYN (based on inverted ranks) GEN optimism and engagement | 0.158 | 0.268 | 51 | Optimism and engagement | −0.063 | 0.640 | 58 |
| DYN (based on inverted ranks) GEN No welfare mentality, acceptancy of the Calvinist work ethics | 0.067 | 0.638 | 51 | No welfare mentality, acceptancy of the Calvinist work ethics | −0.439 | 0.001 | 58 |
| DYN (based on inverted ranks) GEN Overall Civil Society Index | 0.426 | 0.002 | 51 | Overall Civil Society Index | 0.355 | 0.006 | 58 |

**Table 6.12** Partial correlations of GCC membership (HDI, HDI$^2$, Arab League and Muslim population share constant)

| | GCC membership | Error *p* | df. |
|---|---|---|---|
| Immigration—share of population 2005 (%) | 0.621 | 0.000 | 163 |
| Carbon emissions per capita | 0.560 | 0.000 | 159 |
| Annual population growth rate, 1975–2005 (%) | 0.526 | 0.000 | 159 |
| Trust in institutions | 0.511 | 0.000 | 51 |
| Net migration rate | 0.503 | 0.000 | 168 |
| military expenditures per GDP | 0.501 | 0.000 | 121 |
| Restrictive gender norms, UNDP (2019) GSN Physical integrity | 0.396 | 0.001 | 67 |
| Gallup poll about satisfaction: Local labour market | 0.389 | 0.000 | 145 |
| CV-19 Cases per 1 million population, end September 2020 | 0.382 | 0.000 | 167 |
| Richest 1% | 0.368 | 0.003 | 61 |
| Happiness, good health | 0.366 | 0.007 | 51 |
| Annual rate of growth international migration, 1960–2005 | 0.354 | 0.000 | 144 |
| Net international migration rate, 2005–2010 | 0.351 | 0.000 | 160 |
| Gallup poll about satisfaction: efforts to deal with the poor | 0.344 | 0.000 | 141 |
| Restrictive gender norms, UNDP (2019) GSNI2 | 0.338 | 0.005 | 67 |
| Restrictive gender norms, UNDP (2019) GSNI | 0.310 | 0.010 | 67 |
| Gallup poll about satisfaction: education quality | 0.304 | 0.000 | 126 |
| Involvement in politics | 0.304 | 0.027 | 51 |
| Migrants 2017 percentage of total population | 0.294 | 0.000 | 171 |
| FPZ (free production zones) employment as % of total population | 0.289 | 0.000 | 163 |
| Ecological footprint (g ha/cap) | 0.286 | 0.001 | 132 |
| Restrictive gender norms, UNDP (2019) GSN political | 0.257 | 0.033 | 67 |
| *DYN GEN accepting the market economy* | *0.252* | *0.069* | *51* |
| Mean economic growth rate crisis years | 0.248 | 0.001 | 162 |
| Asylum recognition rate | 0.247 | 0.005 | 126 |
| Gallup poll about satisfaction: health care quality | 0.228 | 0.006 | 142 |
| Gallup poll about satisfaction: freedom of choice | 0.227 | 0.006 | 145 |
| Social globalization, de facto index | 0.217 | 0.004 | 171 |
| DYN HDI 2010–2018 | 0.213 | 0.005 | 170 |
| comparative price levels (US = 1.00) | 0.200 | 0.012 | 157 |
| Gallup poll about satisfaction: job | 0.199 | 0.017 | 141 |

(continued)

**Table 6.12** (continued)

| | | | |
|---|---|---|---|
| Gallup poll about satisfaction: trust in national government | 0.196 | 0.022 | 134 |
| LFPR 55-59 | 0.183 | 0.020 | 159 |
| Civil and political liberties violations | 0.180 | 0.021 | 162 |
| Migration balance 2013 | 0.168 | 0.027 | 171 |
| Migration balance 2017 | 0.165 | 0.030 | 171 |
| *GINI income* | *0.155* | *0.080* | *127* |
| DYN GDP pc PPP 2020 | 0.154 | 0.046 | 168 |
| *Gallup poll about satisfaction: overall life satisfaction index* | *0.153* | *0.064* | *145* |
| *Economic globalization, de facto index* | *0.144* | *0.060* | *169* |
| *LEX 2013 life expectancy, 2013* | *−0.128* | *0.093* | *171* |
| *Share of international immigrant stock* (%) | *−0.138* | *0.093* | *147* |
| *Female survival probability of surviving to age 65 female* | *−0.143* | *0.071* | *158* |
| *Comparative price level 2013* | *−0.147* | *0.064* | *158* |
| *Gallup poll about satisfaction: trust in other people* | *−0.153* | *0.093* | *119* |
| *ESI-index environment sustainability index (Yale Columbia)* | *−0.157* | *0.069* | *134* |
| *Nurses and midwives (per 1000 people), 2010* | *−0.157* | *0.066* | *137* |
| *Happy planet index, HPI* | *−0.159* | *0.067* | *132* |
| *Global terrorism index* | *−0.161* | *0.053* | *144* |
| *Life expectancy (years) (by 2010)* | *−0.161* | *0.062* | *132* |
| MNC PEN: DYN MNC PEN 1995–2005 | −0.168 | 0.032 | 160 |
| *Closing economic gender gap* | *−0.173* | *0.051* | *126* |
| Deaths per total cases | −0.175 | 0.024 | 164 |
| Foreign savings rate | −0.180 | 0.027 | 148 |
| Life expectancy at birth | −0.181 | 0.017 | 171 |
| *Unemployment 2021* | *−0.181* | *0.089* | *87* |
| Closing political gender gap | −0.182 | 0.040 | 126 |
| DYN GDP pc PPP 2021 | −0.194 | 0.012 | 167 |
| Closing of global gender gap overall score 2009 | −0.202 | 0.022 | 126 |
| CV-19 fatality rate per total cases | −0.203 | 0.008 | 167 |
| Closing health and survival gender gap | −0.208 | 0.018 | 126 |
| EU-recipient of global migration from this country in %, 2017 | −0.211 | 0.005 | 171 |

(continued)

**Table 6.12** (continued)

| | | | |
|---|---|---|---|
| International stock of refugees | −0.214 | 0.008 | 149 |
| Current health expenditure (% of GDP), 2017 | −0.215 | 0.005 | 168 |
| EU-recipient of global migration from this country in %, 2013 | −0.215 | 0.005 | 171 |
| Per capita world class universities | −0.217 | 0.005 | 163 |
| Thousand asylum seekers per total population (in millions) | −0.218 | 0.007 | 149 |
| Social globalization, de jure index | −0.220 | 0.004 | 171 |
| *DYN GEN optimism and engagement* | *−0.244* | *0.078* | *51* |
| Pearson correlation 1992–2014 GDP pc growth | −0.250 | 0.004 | 128 |
| Health expenditure as % of GDP | −0.257 | 0.001 | 169 |
| Overall 35 variable development index, based on 7 dimensions | −0.269 | 0.002 | 123 |
| Effective democracy index | −0.281 | 0.000 | 158 |
| DYN GEN no redistributive religious fundamentalism | −0.282 | 0.041 | 51 |
| Democracy measure | −0.290 | 0.000 | 143 |
| Environmental performance index (EPI) | −0.306 | 0.000 | 140 |
| Overall 35 variable development index | −0.309 | 0.000 | 163 |
| Share of people with no bias | −0.310 | 0.010 | 67 |
| Tertiary enrollment | −0.335 | 0.000 | 138 |
| ESI index component social and institutional capacity | −0.341 | 0.000 | 134 |
| Social protection (ILO) | −0.356 | 0.000 | 156 |
| Accepting the market economy | −0.358 | 0.008 | 51 |
| Overall 30 variable development index | −0.376 | 0.000 | 163 |
| Gender empowerment index value | −0.388 | 0.001 | 68 |
| Slope 2007–2014 GDP pc growth | −0.420 | 0.000 | 128 |
| Democracy movement | −0.435 | 0.001 | 51 |

- Happiness, good health
- Gallup poll about satisfaction: Efforts to deal with the poor
- Gallup poll about satisfaction: Education quality
- Involvement in politics
- DYN GEN accepting the market economy
- mean economic growth rate crisis years
- Gallup poll about satisfaction: Healthcare quality
- Gallup poll about satisfaction: Freedom of choice
- Social globalization, de facto index
- DYN HDI 2010–2018
- comparative price levels (US = 1.00)

- Gallup poll about satisfaction: Job
- Gallup poll about satisfaction: Trust in national government
- LFPR 55-59 (Labor Force Participation Rate of Older Workers)
- DYN GDP pc PPP 2020
- Gallup poll about satisfaction: Overall life satisfaction index.

Equally, the countries of the Gulf could avoid the following negative phenomena, independent from the Human Development Index and its square as well as independent from the membership of a country in the Arab League and the Muslim population share:

- CV-19 Fatality rate per total cases
- unemployment 2021
- foreign savings rate
- CV-19 Deaths per total cases
- Growth of MNC penetration (MNC PEN: DYN MNC PEN 1995–2005)
- Global terrorism index.

However, ceteris paribus, the development model of the Gulf countries is characterized by a high amount of these following negative development tendencies, independent from the Human Development Index and its square as well as independent from the membership of a country in the Arab League and the Muslim population share:

- Carbon emissions per capita,
- Restrictive gender norms, UNDP (2019) GSN Physical integrity,
- CV-19 cases per 1 million population, end September 2020,
- Richest 1%,
- Restrictive gender norms, UNDP (2019) GSN I2,
- Restrictive gender norms, UNDP (2019) GSN I,
- ecological footprint (g ha /cap),
- Restrictive gender norms, UNDP (2019) GSN Political,
- Civil and Political Liberties violations,
- GINI Income Inequality.

Also, the Gulf development model did not achieve to maximize these following beneficial processes and indicators, independent from the Human Development Index and its square as well as independent from the membership of a country in the Arab League and the Muslim population share:

- Democracy movement (based on *World Values Survey*),
- Slope 2007–2014 GDP pc growth (long-term growth),
- Gender empowerment index value,
- Overall 30 variable development index,
- Accepting the market economy (based on *World Values Survey*),
- Social protection (ILO),
- ESI Index Component Social and Institutional Capacity,
- Tertiary enrollment,

- Share of people with no gender norms bias (based on *World Values Survey*), UNDP, 2019,
- Overall 35 variable development index,
- Environmental Performance Index (EPI),
- Democracy measure,
- Change of social norms among the generations (based on the present study) (DYN GEN): No redistributive religious fundamentalism,
- Effective Democracy Index,
- Overall 35 variable development index, based on seven dimensions,
- Health expenditure as % of GDP,
- Pearson correlation 1992–2014 GDP pc growth (long-term growth),
- Change of social norms among the generations (based on the present study) DYN GEN optimism and engagement,
- Social globalization, de jure index,
- Per capita world class universities,
- Current health expenditure (% of GDP), 2017,
- Closing health and survival gender gap,
- Closing of global gender gap overall score 2009,
- DYN GDP pc PPP 2021,
- Closing political gender gap,
- Life expectancy at birth,
- Closing economic gender gap,
- Life expectancy (years) (by 2010),
- Happy Planet Index, HPI,
- ESI Index, Environment Sustainability Index (Yale Columbia),
- Nurses and midwives (per 1,000 people), 2010,
- Gallup poll about satisfaction: trust in other people,
- Comparative price level 2013,
- Female survival probability of surviving to age 65 female,
- LEX 2013 Life Expectancy, 2013.

Thus, gender equality and less restrictive gender norms as well as steps in the direction of democracy are a key to the Gulf's future.

### 6.5.2 Prospects and Conclusions About Value Change and Patterns of Development in the Rich Arab Countries

This chapter started from the assumption that even advanced and very internationally recognized Gulf studies do not deliver us any more systematic clues about the social value patterns, held by the publics at large in the region. First of all, we debated here the Gulf country implications of research in the context of the GLOBE Project at Simon Fraser University in Canada, and its hypothesis about an "Arabic cluster" of business leadership culture. We then reviewed the research by Geert Hofstede and associates, followed by our discussion of the Gulf country perspectives in the value research by Shalom Schwartz and Ronald F. Inglehart. We rounded up our theoretical surveys by a debate about the Islamicity Index, developed by the American researcher Hossein Askari.

In our own empirical work, which attempted to answer questions left over from current approaches, we first presented a (nonparametric) Index of Civil Society and an Index of Religious Tolerance, comparing Gulf countries with all the countries of the world with comparable *World Values Survey* data. The Index of Civil Society relied on the WVS items.

- Important child qualities: tolerance and respect for other people.
- Accepting neighbors: people of a different religion.
- Men do not make better political leaders than women do.
- University is equally important for a boy and for a girl.
- Democracy: religious authorities should not interpret the laws.
- Unjustifiable: For a man to beat his wife.

The three worst placed nations on earth are Mali, Bahrein, and Yemen. All Muslim countries in our 77 countries and territories with full data which were under investigation here are below the global average; and the best placed Muslim country is post-Soviet Kazakhstan; and the best placed Arab country is Qatar. Of the 77 countries ranked by our UNDP-type index of a democratic civil society, Qatar is rank 50 and Bahrein is rank 76.

For the calculation of the Global Religious Tolerance Indicator, we again relied on the well-established methodology of the United Nations Human Development Programme and its UNDP Human Development Index (nonparametric index). The chosen World Values Survey tolerance indicators were:

1. Disagree or strongly disagree: The only acceptable religion is my religion (mean)
2. Agree or strongly agree: All religions should be taught in public schools (mean)
3. Agree or strongly agree: People who belong to different religions are probably just as moral as those who belong to mine (mean)
4. Trust completely or somewhat: people of another religion (mean)
5. Meaning of religion: Do good to other people (%-percentages).

For 59 states of the world, there are complete data. Among the ten states with the lowest general religious tolerance, based on our five indicators, there are nine predominantly Muslim states. Only public opinion in predominantly Christian Armenia is among this laggard group. Bahrein is rank 34 of the 59 nations with complete data, Kuwait is rank 50, and Qatar is rank 55.

We then developed our own multivariate scales, based on a comprehensive promax factor analysis with World Values Survey data. Combining the **World Values Survey** data about the acceptancy of

1. The non-violent and law-abiding society,
2. Democracy movement,
3. Climate of personal non-violence,
4. Trust in institutions,
5. Happiness, good health,
6. No redistributive religious fundamentalism,
7. Accepting the market economy,
8. Feminism,
9. Involvement in politics,
10. Optimism and engagement,
11. No welfare mentality, acceptancy of the Calvinist work ethics.

in a parametric Index construction from factor analytical data, it emerges that Qatar performs relatively well in global society, while Kuwait and Bahrain are in the lowest third global performance "group".

Our analysis also shows the results of our analysis on the trade-off between human security and our promax factor analytical values. While redistributive fundamentalism decreases with human security, and the climate of personal non-violence, feminism, and the democracy movement increase with human security, accepting the market economy and the Calvinist work ethics decrease with the level of human security. In a dynamic perspective, increases in human security over time (1990–2018) are clearly connected with the intergenerational increase of the overall civil society index and the intergenerational increase of feminism according to our measurement of intergenerational value change, introduced in Chap. 2. That is to say, growing human security over time indeed affects the spread of "modern" civil society and feminist values among the younger generations.

An analysis of the development patterns of rich Arab countries, compatible with the comparative analysis of societal values of this chapter, finalized our essay and showed the relevance of overcoming restrictive gender norms and attitudes favoring inequality.

## References


Al-Alawi, A. I., Al-Marzooqi, N. Y., & Mohammed, Y. F. (2007). Organizational culture and knowledge sharing: Critical success factors. *Journal of Knowledge Management, 11*(2), 22–42.

Alexander, A. C., Inglehart, R., & Welzel, C. (2012). Measuring effective democracy: A defense. *International Political Science Review, 33*(1), 41–62.

Allen, M. (2008). *The theological origins of modernity*. University of Chicago Press.

Almutairi, S., Heller, M., & Yen, D. (2020). *Reclaiming the heterogeneity of the Arab states*. Cross Cultural & Strategic Management. Available at: https://bura.brunel.ac.uk/bitstream/2438/21344/2/FullText.pdf

Askari, H., et al. (2015). Introduction to Islamic economics: Theory and application. In *Wiley finance series*. Wiley.

Askari, H., & Mohammadkhan, H. (2015). *Islamicity indices: The seed for change*. Springer.

Barro, R. J. (2012). *Convergence and modernization revisited*. Harvard University.

Barro, R. J., & McCleary, R. M. (2003). Religion and economic growth across countries. *American Sociological Review, 68*(5), 760–781.

Bergson, H. (1935). *The two sources of morality and religion* (R. Audra & C. Brereton, with the Assistance of W. Horsfall Carter, Trans.). Macmillan and Co., Ltd.

Davidov E., Schmidt P., & Schwartz, S. H. (2008). Bringing Values Back in: The adequacy of the European social survey to measure values in 20 Countries. *Public Opinion Quarterly, 72*(3), 420–445.

Davidov E., Schmidt P., & Billiet J. (2011). *Cross-cultural analysis: Methods and applications*. New York

Eickelman, D. F. (1992). Mass higher education and the religious imagination in contemporary Arab societies. *American Ethnologist, 19*(4), 643–655.

Eisenstadt, S. N. (1968). *The protestant ethic and modernization: A comparative view*. Basic Books.

Etzioni A. (1968). *The active society: A theory of societal and political processes*. Collier-Macmillan; Free Press.

Feiner, S., & Naor, C. (2011). *Jewish enlightenment*. University of Pennsylvania Press.

Gengler, J., Tessler, M., Al-Emadi, D., & Diop, A. (2013). Civic life and democratic citizenship in Qatar: Findings from the First Qatar World Values Survey. *Middle East Law and Governance, 5*(3), 258–279.

Gonzales, J. I. (2011). The culture and development index (CDI): Measuring values and attitudes associated with development in selected Asian and Latin American Countries. *Culture, 1*(1), 42–50.

Guiso, L., Sapienza, P., & Zingales, L. (2003). People's opium? Religion and economic attitudes. *Journal of Monetary Economics, 50*(1), 225–282.

Guzansky, Y. (2015). *The Arab Gulf states and reform in the Middle East: Between Iran and the "Arab Spring" (Ser. Palgrave pivot)*. Palgrave Macmillan.

Guzansky, Y., & Berti, B. (2013). *Is the New Middle East Stuck in Its Sectarian past? The Unspoken dimension of the "Arab Spring"* (pp. 57135–57151). Orbis. https://doi.org/10.1016/j.orbis.2012.10.009

Guzansky, Y., & Even, S. (2020). *The economic crisis in the Gulf states: A challenge to the "contract" between rulers and ruled*. INSS Insight No. 1327, June 1, 2020. Available at https://www.INSS.org.il/publication/gulf-states-economy/?offset=7&posts=201&outher=Yoel%20Guzansky

Guzansky, Y., & Segal, E. (2020). *All in the family: Leadership changes in the Gulf*. INSS Insight No. 1378, August 30, 2020. Available at: https://www.INSS.org.il/publication/gulf-royal-families/?offset=1&posts=201&outher=Yoel%20Guzansky

Guzansky, Y., & Winter, O. (2020). *Apolitical Normalization: A New Approach to Jews in Arab States*. INSS Insight No. 1332, June 8, 2020. Available at: https://www.INSS.org.il/publication/judaism-in-the-arab-world/?offset=5&posts=201&outher=Yoel%20Guzansky

Hayek, F. A. (1960). *The constitution of liberty*. University of Chicago Press.

Hofstede, G. (1980). Motivation, leadership, and organization: Do American theories apply abroad? *Organizational Dynamics, 9*(1), 42–63.

Hofstede, G. (1984). Hofstede's culture dimensions: An independent validation using rokeach's value survey. *Journal of Cross-Cultural Psychology, 15*(4), 417–433.

Hofstede, G. (1988). The confucius connection: From cultural roots to economic growth. *Organizational Dynamics., 16*(4), 5–21.

Holmes, D. L. (2006). *The faiths of the founding fathers*. Oxford University Press.

Inglehart, R. (2017). Changing values in the Islamic world and the West. In M. Moaddel & M. J. Gelfand (Eds.), *Values, political action, and change in the Middle East and the Arab Spring* (pp. 3–24). Oxford University Press.

Inglehart, R. F. (1988). The renaissance of political culture. *American Political Science Review, 82*(04), 1203–1230.

Inglehart, R. F. (1990). *Culture shift in advanced industrial countries*. Princeton University Press.

Inglehart, R. F. (2000). Globalization and postmodern values. *Washington Quarterly, 23*(1), 215–228.

Inglehart, R. F. (2006). Mapping global values. *Comparative Sociology, 5*(2), 115–136.

Inglehart, R. F. (2008). Changing values among western publics from 1970 to 2006. *West European Politics, 31*(1–2), 130–146.

Inglehart, R. F., & Baker, W. E. (2000). Modernization, cultural change, and the persistence of traditional values. *American Sociological Review, 65*(1), 19–51.

Inglehart, R. F., & Norris, P. (2003). *Rising tide: Gender equality and cultural change around the world*. Cambridge University Press.

Inglehart, R. F., & Norris, P. (2009). *The true clash of civilizations. Foreign policy*, November 4, 2009. Available at: http://foreignpolicy.com/2009/11/04/the-true-clash-of-civilizations/

Inglehart, R. F., & Norris, P. (2012). The four horsemen of the apocalypse: Understanding human security. *Scandinavian Political Studies, 35*(1), 71–95.

Inglehart, R. F., & Norris, P. (2016). *Trump, brexit, and the rise of populism: Economic have-nots and cultural backlash*. HKS Working Paper No. RWP16-026. Available at SSRN: http://ssrn.com/abstract=2818659

Inglehart, R. F., & Welzel, C. (2003). Political culture and democracy: Analyzing cross-level linkages. *Comparative Politics, 36*(1), 61–79.

Inglehart, R. F., & Welzel C. (2009). *How development leads to democracy. What we know about modernization*. Foreign Affairs, March, April (freely available at: http://www.foreignaffairs.com/articles/64821/ronald-Inglehart-and-christian-welzel/how-development-leads-to-democracy).

Inglehart, R. F., & Welzel, C. (2010). Changing mass priorities: The link between modernization and democracy. *Perspectives on Politics, 8*(02), 551–567.

Kabasakal, H., & Bodur, M. (2002). Arabic cluster: A bridge between east and west. *Journal of World Business, 37*(1), 40–54.

Kang, M. Y., & Kwon, J. W. (2018). Hofstede cultural dimension measuring through world values surveys. *Asia-Pacific Journal of Business, 9*(2), 137–152.

Kuznets, S. (1976). *Modern economic growth: Rate, structure and spread*. Yale University Press.

Montesquieu C.-L. de Secondat (1689–1755). (1989). *The spirit of the laws* (A. M. Cohler, B. C. Miller, & H. S. Stone, Trans.). Cambridge Texts in the History of Political Thought. Cambridge University.

Morel, J. S. (1972). *Glaube und Säkularisierung: Religion im Christentum als Problem*. Tyrolia-Verlag.

Morel, J. S. (1977). *Enthüllung der Ordnung: Grundbegriffe und Funktionen der Soziologie*. Tyrolia-Verlag.

Morel, J. S. (1986). *Ordnung und Freiheit: Die soziologische Perspektive*. Tyrolia-Verlag.

Morel, J. S. (1997). *Soziologische Theorie: Abriss der Ansaetze ihrer Hauptvertreter*. Oldenbourg.

Morel, J. S. (1998). *Glauben hat Zukunft: Persönliche Wege einer neuen Sinnfindung*. Tyrolia-Verlag.

Morel, J. S. (2003). *Radikale Kirchenreform: Für eine mutige Erneuerung: Fakten und religionssoziologische Theorien über die Krise der katholischen Kirche von 1950 bis in die Gegenwart als Entscheidungsgrundlagen notwendiger Reformen.* Tyrolia-Verlag.

Morgan, M. H. (Ed.). (2007). *Lost history: The enduring legacy of muslim scientists, thinkers, and artists. Foreword by King Abdullah II of Jordan.* National Geographic.

Norris, P., & Inglehart, R. F. (2002). Islamic culture and democracy: Testing the 'clash of civilizations' thesis. *Comparative Sociology, 1*(3), 235–263.

Norris, P., & Inglehart, R. F. (2011). *Sacred and secular: Religion and politics worldwide.* Cambridge University Press.

Norris, P., & Inglehart, R. F. (2012). Muslim integration into Western cultures: Between origins and destinations. *Political Studies, 60*(2), 228–251.

Norris, P., & Inglehart, R. F. (2015). Are high levels of existential security conducive to secularization? A response to our critics. In *The changing world religion map* (pp. 3389–3408). Springer Netherlands.

Norris, P., Kern, M., & Just, M. R. (2003). *Framing terrorism: The news media, the government, and the public.* Routledge.

Popper, K. S. (2012). *The open society and its enemies.* Routledge.

Rehman, S. S., & Askari, H. (2010). An Economic Islamicity Index (EI 2). *Global Economy Journal, 10*(3), 1–37.

Rehman, S. S., & Askari, H. (2010). How Islamic are Islamic Countries? *Global Economy Journal, 10*(2), 1–30.

Rettab, B., Brik, A. B., & Mellahi, K. (2009). A study of management perceptions of the impact of corporate social responsibility on organisational performance in emerging economies: The case of Dubai. *Journal of Business Ethics, 89*(3), 371–390.

Röhrich, W. (2004). *Die Macht der Religionen: Glaubenskonflikte in der Weltpolitik.* Beck.

Röhrich, W. (2010). *Rückkehr der Kulturen: die neuen Mächte in der Weltpolitik.* Nomos.

Sacks, J. (1998). *Morals and markets. Seventh annual IEA Hayek memorial lecture given in London. June 2.* Institute of Economic Affairs.

Sacks, J. (2003). *The dignity of difference: How to avoid the clash of civilizations.* Continuum.

Sacks, J. (2005). *To heal a fractured world: The ethics of responsibility.* McGill-Queens University Press.

Sacks, J. (2012). *Has Europe lost its soul?* Address given at the Pontifical Gregorian University. December 12, 2012.

Sacks, J. (2014). *The religious other: Hostility, hospitality, and the hope of human flourishing.* Lexington Books.

Schumpeter J. A. (1950). *Capitalism, socialism and democracy.* Harper & Row.

Schwartz, S. H. (1992). Universals in the content and structure of values: Theoretical advances and empirical tests in 20 countries. *Advances in Experimental Social Psychology, 25*(C), 1–65.

Schwartz, S. H. (1994). Are there universal aspects in the structure and contents of human values? *The Journal of Social Issues a Journal of the Society for the Psychological Studies of Social Issue, 50*(4), 19–45.

Schwartz, S. H. (2006a). A theory of cultural value orientations: Explication and applications. *Comparative Sociology, 5*(2), 137–182.

Schwartz, S. H. (2006b). *Basic human values: An overview.* The Hebrew University of Jerusalem. http://segr-did2.fmag.unict.it/Allegati/convegno%207-8-10-05/Schwartzpaper.pdf

Schwartz, S. H. (2009). *Cultural value orientations: Nature & implications of national differences.* The Hebrew University of Jerusalem, Israel Science Foundation Grant No. 921/02. Available at http://blogs.helsinki.fi/valuesandmorality/files/2009/09/Schwartz-Monograph-Cultural-Value-Orientations.pdf

Solomon, H., & Tausch, A. (2020). Islamism, crisis and democratization: Implications of the World Values Survey for the Muslim world. In *Perspectives on development in the Middle East and North Africa (Mena) region.* Springer.

Solomon, H., & Tausch, A. (2021). Arab MENA countries: Vulnerabilities and constraints against democracy on the eve of the global Covid-19 crisis. In *Perspectives on development in the Middle East and North Africa (Mena) region.* Springer.

Tausch, A., & Heshmati, A. (2013). *Globalization, the human condition, and sustainable development in the twenty-first century: Cross-national perspectives and European implications.* Anthem Press.

Tausch, A., & Moaddel M. (2009). *What 1.3 billion Muslims really think. An answer to a recent Gallup study, based on the 'World Values Survey'.* Nova Science Publishers.

Tausch, A., Heshmati, A., & Karoui, H. (2014). *The political algebra of global value change. General models and implications for the Muslim world.* Nova Science Publishers.

Wiktorowicz, Q. (2006). Anatomy of the Salafi movement. *Studies in Conflict and Terrorism, 29*(3), 207–239.

Yesilada, B., & Noordijk, P. (2010). Changing values in Turkey: Religiosity and tolerance in comparative perspective. *Turkish Studies, 11*(1), 9–27.

# Chapter 7
# Antisemitism—The Second Contemporary Pandemic and What the Gulf Countries Can Do to Overcome It

**Abstract** The Anti-Defamation League (nADL) in its recent update of its empirical research on global Antisemitism estimated that today more than 1 billion people on earth harbor antisemitic attitudes (https://global100.adl.org/map). This chapter takes a new look at the ADL (ADL—Anti-Defamation League, 2014) data on global Antisemitism in 101 countries and the performance of the Gulf countries in overcoming Antisemitism. ADL (ADL—Anti-Defamation League, 2014) estimated the average rate of Antisemitism in the Middle East and North Africa to be 74%. The Gulf countries, including those which recently signed the historic "Abraham accords" with the State of Israel, are among the countries with the highest rates of Antisemitism, world-wide. Such an appalling high rate of Antisemitism could be also a destabilizing element in the peace process, which has begun so recently. Explaining these high rates of Antisemitism is a challenge for comparative social science, which by and large, is only in the beginning phase to assess the cross-national implications of the ADL 100 data. Berggren and Nilsson (2020) recently proposed to rely on the interaction between economic freedom and the rule of law to statistically explain adequately these variations in the international rates of Antisemitism. It is impossible; however, our argument is, to exclude patterns of Islamism and the societal processes which lead to it from the statistical models used to explain the ADL (ADL—Anti-Defamation League, 2014) rates of Antisemitism.

We also look at the World Values Survey—data on Antisemitism (rejection of Jewish neighbors as a proxy of the ADL 100 data)—and establish that the opinions and structural personal background data:

- Important child qualities: religious faith
- Satisfaction with financial situation of household
- Private versus state ownership of business,
- Government responsibility
- How often do you attend religious services (never) (never)?
- Gender (female),
- Highest educational level attained

are all drivers of Antisemitism, while

- Not important in life: Religion
- Satisfaction with your life



A. Tausch, *The Future of the Gulf Region*, Gulf Studies 2,
https://doi.org/10.1007/978-3-030-78299-3_7

- Reject opinion: Jobs scarce: Men should have more right to a job than women
- Reject opinion: Men make better political leaders than women do
- Reject opinion: University is more important for a boy than for a girl,
- Reject opinion: Politicians who do not believe in God are unfit for public office
- Age

block against Antisemitism in the 28 countries and territories with full data. Our analysis is based on 24,858 interviews.

**Keywords** Relation of economics to social values · Index numbers and aggregation · Antisemitism · MENA region · Religious radicalism

## 7.1 Introduction: Antisemitism—The Second Contemporary Pandemic

In the light of the current developments and the "thaw" in the relations between Israel and several Arab States, including the Gulf countries Bahrain and United Arab Emirates, this chapter takes up anew the issue of Antisemitism on a global scale. Evolving the approach, already developed in Solomon and Tausch (2020, 2021), Tausch (2014, 2016a, 2016b, 2017a, 2017b), Tauschs and Obirek (2020), we undertake to present a completely new and expanded analysis on the subject, bringing in especially new and hitherto not tested variables of gender inequality into the focus of analysis of the drivers of Antisemitism. As we highlighted in Chap. 1, the UNDP (2019, current issues) with good reason underlines the role of still existing restrictive gender norms in shaping a good part of the social landscape of global society. Do these still existing restrictive gender norms have an influence on Antisemitism? And is there a hope that—as still existing restrictive gender norms fade away with development, the forces propelling societies toward xenophobia, racism, and Antisemitism will loose in strength as well?

Our book is characterized by the overall assumption that human development and the expansion of world class education in the region brings about an overall value change in the direction of accepting less restrictive gender norms, and of more tolerance. Optimistic as we are, we also assume that value change in these positive directions will also reduce the still lamentably high rates of Antisemitism in the Gulf region, which were documented in ADL (2014) and among others in our past publications, referred to above, on the subject.

The Anti-Defamation League (ADL) in its recent update of its empirical research on global Antisemitism estimated that eight decades after the *Shoah* a staggering 1 billion people on earth harbor antisemitic attitudes (https://global100.adl.org/map). ADL (2014) estimated the average rate of Antisemitism in the Middle East and North Africa to be 74%. The Gulf countries, including those which recently signed the historic "*Abraham accords*" with the State of Israel, are among the countries with the highest rates of Antisemitism, world-wide. Such an appalling high rate of Antisemitism could be a destabilizing element in the peace process, which has begun so recently, and is a societal attitude absolutely in contrast to the spirit and the text of these historic accords. Explaining these high rates of Antisemitism is a challenge for comparative social science, which by and large, is only in the beginning phase to assess this phenomenon in an adequate fashion. Berggren and Nilsson (2020) recently proposed to rely on the interaction between economic freedom and the rule of law to statistically explain adequately these variations in the international rates of Antisemitism. It would be impossible; however, our argument is to exclude patterns of Islamism and the societal processes which lead to it from the statistical models used to explain the ADL (2014) rates of Antisemitism in 101 countries. Building on earlier research by the present author, cross-national variations in the rates of Antisemitism across the globe are explained by

- Low education,
- Deficient employment,
- The modernization crisis of societies, captured by the Kuznets curve (Kuznets, 1955),
- The detrimental effects of patriarchy,
- Adherence to religious denominations in whose history there were significant trends of Antisemitism in the past,
- The lack of societal trust.

We also undertake to test this new approach, sensitive to the role gender norms have in the formation of antisemitic attitudes, with the *World Values Survey* data on Antisemitism (rejection of Jewish neighbors as a proxy, see also Solomon and Tausch (2020), Tausch and Obirek (2020).These data are from 28 countries and territories, are based on 24,858 interviews with full data, and test the influence of

- Age,
- Gender (female),
- Priorities for government sector or private sector in development,
- Highest educational level attained,
- Religious commitment (several variables),
- Role in society for religion,
- Gender norms (several variables),
- Satisfaction with financial situation of household,
- Satisfaction with your life

on the degree of Antisemitism, measured by the proxy of the rejection of Jewish neighbors (see Chap. 2, and also Solomon & Tausch, 2020; Tausch & Obirek, 2020, with data on the calibration of this proxy measure). As we highlighted in Chap. 2, we calibrated our proxy results for 23 countries. The two measurement scales have 56.16% of the variance in common, i.e., our second-best solution using *World Values Survey* data indeed captures more than half of the variance of the far superior measurement scale, the ADL (2014) data series.

So, here, first, we debate the background, discussing recent other published research on these issues, and then successively introduce the cross-national drivers of Antisemitism, as we specify them in our comparative analysis. We then proceed the cross-national results, based on ADL data, and compare these results with the OLS regression, based on *World Values Survey* data. Concluding prospects round up the presentation.

## 7.2 Background on Antisemitism—The Second Contemporary Pandemic and What the Gulf Countries can do to Overcome It

The analysis of Antisemitism with its catastrophic results for the Jewish people can look back on more than a century of fruitful studies with a vast and still growing literature (just to mention a few, Cohen, 2018a, 2018b; Jikeli & Allouche-Benayoun, 2012; Kertzer, 2007; Michael, 2008; Rosenfeld, 2013; von Bieberstein, 1977; Wistrich, 2010; see also Bauer, 1993; Wistrich, 2004, 2007, 2010). The ADL, 2014 study was a hallmark in previous research, because it opened the way for cross-national, empirical comparisons of the rates of Antisemitism.

We already highlighted in Chap. 2 that the data provided by the ADL (2014) are based on survey researched attitudes and opinions toward Jews in more than 100 countries around the world. With a sample size of 53,100 total interviews among citizens aged 18 and over across 101 countries and the Palestinian Territories in the West Bank and Gaza, this vast global survey is without parallels in the world today and is an invitation for cross-national research. ADL analyzed negative stereotypes (11 stereotypes) about Jews, and respondents who said at least 6 out of 11 statements are "probably true" were considered to hold antisemitic attitudes.

The negative stereotypes used in the ADL Survey ADL 100 (see Chap. 2) were about Jews being more loyal to Israel than to the country they live in; having too much power in the business world, in international financial markets, in global affairs, in the US government, the global media; that they do not care about what happens to anyone but their own kind; that they think they are better than other people; talk too much about what happened to them in the Holocaust; that people hate Jews because of the way Jews behave; and that Jews are responsible for most of the world's wars.

The overall ADL GLOBAL 100 Index Score (ADL, 2014) is 26%, that is to say at least 26% of the citizens of our globe—more than 1 billion people—are anti-Semitic according to the ADL definition. This reflects the percentage of global respondents who say that at least six of the eleven negative stereotypes tested in the ADL (2014) study are "*probably true*." In the world regions, the results are as follows (weighted percentages).

**Middle East and North Africa (MENA): 74%**

Eastern Europe: 34%
Western Europe: 24%
Sub-Saharan Africa: 23%
Asia: 22%

**Table 7.1** Antisemitism in the Arab World by international comparison among 101 countries

| Global rank (101 countries) | Country | ADL 100—Antisemitism |
|---|---|---|
| 87 | Saudi Arabia | 74.0 |
| 88 | Egypt | 75.0 |
| 89 | Oman | 76.0 |
| 90 | Lebanon | 78.0 |
| 91 | Morocco | 80.0 |
| 92 | Qatar | 80.0 |
| 93 | United Arab Emirates | 80.0 |
| 94 | Bahrain | 81.0 |
| 95 | Jordan | 81.0 |
| 96 | Kuwait | 82.0 |
| 97 | Tunisia | 86.0 |
| 98 | Algeria | 87.0 |
| 99 | Libya | 87.0 |
| 100 | Yemen | 88.0 |
| 101 | Iraq | 92.0 |

Americas: 19%

Oceania: 14%

In the Arab World, by international comparison, these rates are extremely high (Table 7.1):

In Fig. 7.1, we summarize the results from the ADL (2014) study in a geographical fashion in order to estimate the geographical distribution of the strength of societal Antisemitism.

The systematic social scientific study of global values and opinions, used in this Chapter, has of course a long and fruitful history in the social sciences (for a survey of this vast literature, see Chap. 6 of the present work as well as Tausch et al., 2014a, 2014b).

What do we know from this type of sociology on the phenomenon of Antisemitism? Adorno et al. (2019) in their pathbreaking study on the authoritarian personality[1] already attempted to establish the hypothesis that subjects with some religious affiliations are more prejudiced than those without affiliation, but no significant differences between Protestants and Catholics exist. Adorno et al. (2019) also maintained that there is a low but significant negative relation of intelligence and education to ethnocentrism. Other factors mentioned in that famous study are parental relations, childhood experiences, the conception of the self, and the dynamics and

[1] As of January 21, 2021, Google scholar lists none the less than 19,157 follow-up studies to this work.

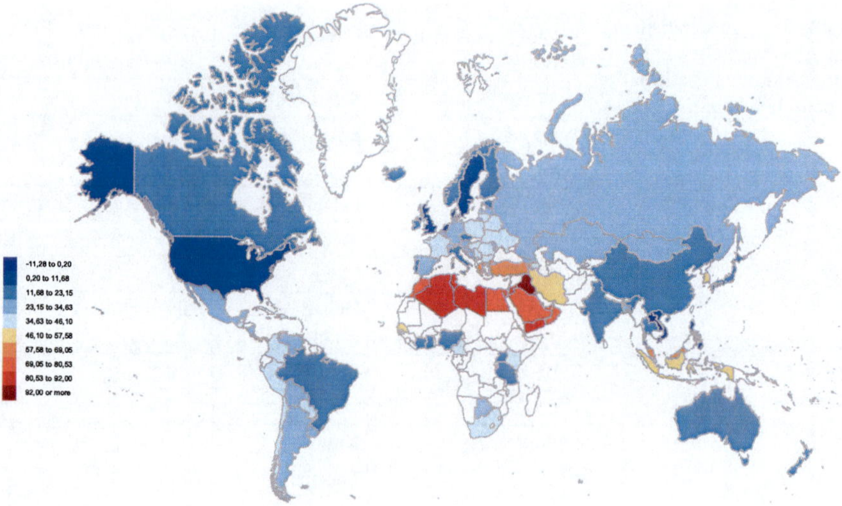


**Fig. 7.1** Antisemitism, ADL 100

the organization of personality. Political and economic ideas, religious ideology, and syndromes all determine the development of the authoritarian personality. A veritable endless number of studies meanwhile has researched the connections between Antisemitism and authoritarianism, and between Antisemitism and sexism (see, just to mention two of them here Auestad, 2015; Young-Bruehl, 1996). This chapter is well within the tradition to study Antisemitism with the help of rigorous quantitative methods and applies a comparative perspective in focus (see the influential study by Glock and Stark 1966 on Antisemitism in Canada; furthermore, Aichinger & Fine, 2017; Brym et al., 1993; Cohen, 2018a, 2018b; Farnen et al., 2005; Klein & Streib, 2018; Klein et al., 2018).

We start from the assumption that still existing restrictive gender norms (UNDP, 2019, current issues), typically to be found among adherents of the opinion that University education is more important for a boy than for a girl, will have a profound effect on the general pattern of prejudice, held by an individual, including Antisemitism (Adams et al., 1997). In the theoretical social science literature, there is a large consensus that sexism, repressive gender norms, and Antisemitism are deeply interconnected, and that still existing restrictive gender norms and their corresponding social structures enhance antisemitic tendencies (Eisenstein, 2014; Guillaumin, 2002; Lutz et al., 1995; Smith, 1988; Williams, 1989). A very large number of empirical studies, but also government reports, have already highlighted this close inter-relationship (for a survey of this literature, see Solomon & Tausch, 2020). While in some advanced industrial countries, like the UK, there is now a gender gap in tertiary education to the detriment of males,[2] in a large number of countries not only enrollment rales in tertiary education are often heavily male-biased; but also professed opinions on the issue consent with this large-scale discrimination (Chzhen & Bruckauf, 2017). To state that "*A university education is more important for a boy than for a girl*" must be regarded as the "*locus classicus*" of the discourse of gender discrimination and as the one-catch-all measurement scale, capturing this phenomenon. In fact, this variable is one of the best documented in the entire *WVS* project and is highly correlated with a host of other indicators (Tausch et al. 2014a, 2014b).

We also start from the assumption that the effects of denomination and commitment to religion on Antisemitism per se are not as important as the effects of a religious fundamentalist world outlook that says that "*politicians who don't believe in G'd are unfit for public office*". The importance of "*religious fundamentalism*" measured here by the mentioned *WVS* item has been also stated frequently in the relevant literature, among others, in Solomon and Tausch (2020). Such an opinion is not restricted to so-called developing countries. Even in the USA, an influential Evangelical Republican, 2016 Presidential Contender Senator Ted Cruz (Texas), who by the way also supported President Trump all the way even after the assault on Capitol Hill on January 6, 2021,[3] made headlines when he exactly declared in 2016 that *an Atheist is not fit to be President of the country.*[4]

What is the relationship between Antisemitism, xenophobia, and "*existential security*"? In view of the contemporary lamentable implosion of societal trust by majority populations in Western democracies in their respective political systems, we are also inclined to mention here as well Inglehart's recent theory of "*cultural evolution*" (Inglehart, 2018) in the explanation of xenophobia, racism, and Antisemitism. Inglehart argues that people's values and behavior are shaped by the degree to which survival is secure; it was precarious for most of history, which encouraged a heavy emphasis on group solidarity, a rejection of outsiders, and obedience to strong leaders. High levels of existential security encourage openness to change, diversity, and new ideas. The unprecedented prosperity and security of the post-war era brought cultural change, the environmentalist movement, and the spread of democracy. But, Inglehart says, in recent decades, diminishing job security and rising inequality have led to authoritarian reactions. In the perspective of Inglehart's theory, growing unease with "*multiculturalism*" and "*migration*" coincides with a rising inequality in many countries of the Western world.[5]

---

[2] Guardian, May 12, 2016, available at https://www.theguardian.com/education/2016/may/12/universitygender-gap-scandal-thinktank-men.

[3] https://www.latimes.com/opinion/story/2021-01-07/ted-cruz-donald-trump-riot-congress-joe-biden.

[4] Huffington Post, Dec. 21, 2016, available at https://www.huffingtonpost.com/entry/ted-cruz-atheists_us_5640b613e4b0411d30719f52. On Antisemtism and Evangelical Christianity, see Kressel, 2007; Rausch, 1993.

[5] For most recent time series data about inequality, see University of Texas Inequality Project, available at https://utip.lbj.utexas.edu/data.html.

We will now look at some of the major studies on the subject of the comparison of global rates of Antisemitism, which are of relevance here.

Of all the approaches under scrutiny here, the work by the Swedish economist Niclas Berggren and associates, who in a series of articles already could successfully linked tolerance to economic freedom, seems to be especially noteworthy, because their writings on tolerance and Antisemitism are linked to a very elaborate socioeconomic theory, i.e., the liberal approach to economic freedom (Berggren & Bjørnskov, 2019; Berggren & Jordahl, 2006; Berggren & Nilsson, 2013, 2014, 2015, 2016, 2020, Berggren et al., 2019). In Berggren and Nilsson (2020), the authors examined how the variation in Antisemitism across countries can be explained by economic freedom. They proposed two mechanisms. First, the more economic freedom, the greater the scope of market activities. Second, a key type of institution undergirding the market is an effective and fair legal system, or the rule of law. The authors used the ADL Global 100 survey of antisemitic attitudes and relate them to the Economic Freedom of the World Index[6] and its five areas. Their empirical findings confirm that the more economic openness, the more Antisemitism; and the stronger the rule of law, the less Antisemitism. The variables used in their main multiple regression model on the rates of global Antisemitism are

- Economic freedom,
- Size of government,
- Legal structure and security of property rights,
- Access to sound money,
- Freedom to trade internationally,
- Regulation of credit, labor, and business.

The indicators of "good governance" all reduce Antisemitism (small size of government, legal structure and security of property rights, access to sound money, freedom to trade internationally, liberal regulation of credit, labor, and business). The control variables are

- Civil liberties violations,
- Dependency ratio,
- Education,
- Ethnic fractionalization,
- Log GDP per capita,
- Political rights violations,
- Religious fractionalization,
- Share of Christians per total population,
- Share of Jews per total population,
- Share of Muslims per total population,
- Urban population.

Berggren and Nilsson (2020) say:

---

[6] https://www.heritage.org/index/about.

> As we argued, the quality of the legal system (EFW2) reduces suspicions and fears that those who are different will engage in opportunistic behaviour, and this effect has also been shown to be hold when it comes to other minorities. [...] Regarding the control variables, a higher level of economic development is associated with less antisemitism, and the same holds for the ratio of those outside of the labour force to those in it, implying that countries in which the working population cares for a large share of young and old dependents, this comes with more encompassing values. While Antisemitism is lower when a relatively large share of the population in a country is Jewish, our measures of adherence to other religions are positive and often significant: A larger fraction of the population being Muslim or Christian is associated with more antisemitic views (although significance is weak in the latter case). Interestingly, this also holds true for more urban populations. These results corroborate the findings of Gouda and Gutmann (2019), who show that discrimination against religious minorities is higher in Muslim countries (especially those with Sharia law), but also the result in Berggren et al. (2019), which shows that tolerance towards gay people is lower among religious second-generation immigrants in Europe stemming from Muslim-dominated countries.

Beyer (2019)presented a global comparison of antisemitic attitudes and their functions. His sample consisted of 18 countries from five continents surveyed in 2011 by the Pew Research Center ($n = 13{,}975$). These data were analyzed using multilevel analyses. For Beyer, 2019 Antisemitism, on the one hand, offers a pseudo-explanation for rapid social change by equating capitalism, "American imperialism," and globalization with Jews, and on the other hand is used for the construction of nationalistic boundaries. Nationalism and anti-globalization views are particularly important predictors in countries with higher GDP per capita. Those countries, however, have comparatively low base rates of anti-Semitism, whereas Muslim countries have rather high ones. Cross-level parameters show that especially Muslims living in countries that are rather democratic but have a low GDP and where Muslims are a minority report negative views of Jews.

Kaplan, already back in 2006, in his analysis of Anti-Israel sentiment and Antisemitism in Europe links the discourse surrounding the Israeli-Palestinian conflict, and extreme criticisms of Israel (e.g., Israel is an apartheid state, the Israeli Defense Forces (IDF) deliberately target Palestinian civilians), and policy proposals (e.g., boycott of Israeli academics and institutions, divest from companies doing business with Israel), with Antisemitism (for only Israel being singled out). Kaplan (2006) reports research results from a survey of 500 citizens in each of 10 European countries, analyzing the relationship of anti-Israel views with antisemitic stereotypes. Even after controlling for numerous potentially confounding factors, Kaplan (2006) finds that anti-Israel sentiment consistently predicts the probability that an individual is anti-Semitic, with the likelihood of measured anti-Semitism increasing with the extent of anti-Israel sentiment observed.

In his Ph.D. thesis, Alexander, 2018, analyzed how anti-Semitism is related to country stability. Alexander (2018), assumes that feelings of instability will drive individuals and groups toward espousing anti-Semitic attitudes. Particular attention in this research project was paid to political instability as manifested in immigration levels, regime duration, conflict, and terrorism. Three hypotheses are tested through the use of OLS regression and a dataset that brings together the 2014 cross-national data on anti-Semitic attitudes with country-level variables that act as indirect indicators of political instability. Alexander (2018) defines as his major findings the statistically significant impacts of (1) a country's presence in the Middle East North Africa (MENA) region, (2) a regime's duration, and (3) terrorism outside of the MENA region on levels of anti-Semitic attitudes.

It is certain that few studies sufficiently consider the harsh realities of the lack of religious freedom and their role in the formation of structures and even state support for Antisemitism in many Muslim societies in the world today. The Israeli researcher Jonathan Fox from Bar Ilan University, in a series of global publications, duly documented these tendencies which culminated in Iran under President Mahmoud Ahmadinejad (Fox, 2007, 2011, 2013, 2014, 2015, 2016, 2019; Fox, & Flores, 2009; Fox et al., 2018). Tausch (2014), in an article, published in Israel, was the first to evaluate the cross-national aspects of the ADL data. The results seemed, Tausch (2014) argued, to indicate that the future trajectory of antisemitism will inexorably shift further toward the axis of the current Middle East conflict and the dual hatred of the State of Israel and of Jews. Tausch pleaded in favor of future waves of the *World Values Survey* project to include in future items from the ADL 100 survey into their own questionnaires. Tausch (2014) highlighted that membership in the Arab League and Muslim population shares wield significant influence on the level of Antisemitism in a country. This theme was further developed in Tausch, 2016. First, Tausch, 2016 analyzed with a promax factor analysis (on the methodology, see Chap. 2) the determinants of what led representative global *World Values Survey (WVS)* global Muslim interview partners to reject to have a Jewish neighbor. Tausch also identified the extent of relationships between Antisemitism, the economic and social situation, religion data, and opinions on terrorism among global Muslim publics based on the global Pew Research Center surveys. Finally, Tausch evaluated *Arab Barometer* survey data on 'moderate Islamism' and its relationship to Antisemitism. All quantitative evidence supported the hypothesis that Islamism is deeply connected to Antisemitism. Tausch also indicated that Muslim dissatisfaction and dissent with society, often mentioned as the drivers of Islamism, are in fact connected to Muslim secularism and a distance from Islamism. Channeling this dissent in secular left- and right-wing protest parties would be an important future task for the stabilization of Arab and Muslim democracies. In Tausch, 2017, the author relied on the analysis of PEW and *World Values Survey* data from Muslim publics in different countries around the globe to analyze by multivariate promax factor analysis and standard OLS multiple regressions which factors contributed to the approval or the rejection of terrorist acts, measured by such variables as the opinion on suicide bombing, the strict application of Shari'a law and the favourability of terrorist organizations such as Hamas and Hezbollah. Tausch also analyzed the drivers of general opinions on Islamic extremist groups and reactions to the ambitions of Iran and its nuclear program.

Tausch, 2017, thought that it would be wrong to define radical Islamism only in terms of the identification with outright support for the immediate "bomb-throwing terror", while neglecting the underlying ideological radicalism and also the ongoing radicalization of such antisemitic organizations as the Muslim Brotherhood or the

Turkish Milli Görüs, which both start, like the most radicalized factions of Islamist terrorism, from the intense hatred of "Jews and Free Masons" and Western civilization as such. These groups appear for many on both sides of the Atlantic as "moderate Islamists" and worthy partners of dialog, while in reality they provide the fertile ground from which the armed terrorist groups can develop. Tausch highlighted the role of the omnipresent hatred of America and the West *("Occidentalism").* Tausch also pointed to the intense competition between Islamist and secular, Marxist terror groups which still exist in the Middle East and the entire Muslim world, and the Sunni/Shia competition as well as regional quests for hegemony. With Tibi (1993, 2007, 2009, 2012, 2013a, 2013b, 2015), he analyzed the close connection between the Islamist sharia ideology and the overall aspects of Islamism. Based on PEW data, Tausch showed that the two main drivers of Muslim opposition against suicide bombing are the rejection of honor killing and the rejection of the death penalty against Muslims who chose to leave the Muslim community altogether. The cross-national econometric and politometric analyses of the drivers of the Global Terrorism Index at the nation state data level support our general contentions.

In Tausch and Obirek (2019), the authors highlight that from Roman Catholicism's early days, Antisemitism was its "Original Sin." With its Second Vatican Council Declaration "*Nostra aetate*," many would have hoped that the Church buried the demons of Catholic Antisemitism for good. While the study of global Antisemitism received important new and global empirical insights from the ADL (Anti-Defamation League) 100 study (ADL, 2014), comparative studies on Antisemitism among "practicing" global Roman Catholics, the authors found, are rather lacking. The Tausch, Obirek, 2019, study focused on indicators of Antisemitism of entire countries in comparison with their practicing Roman Catholic communities, i. e. those Catholics who attend Sunday Mass regularly, the so-called *Dominicantes*, which still make up some 45% of the global 1.3 billion Roman Catholics according to the population weighted data of the study. The proxy results based on *World Values Survey* data on the rejection of Jewish neighbors about the anchoring of "*Nostra aetate*" in the hearts and minds of active global Catholics based on *World Values Survey* suggested the following tendencies: On a global scale, about one in five practicing Roman Catholics still rejected to have a Jewish neighbor, irrespective of all the Church's teaching on Judaism since the Second Vatican Council; and that *ceteris paribus,* adherence to Catholicism, Orthodox Christianity, and Islam, all are still to be considered as lamentable and significant drivers of the rate of societal Antisemitism (ADL data) in standard OLS multiple regression analyses. Tausch and Obirek also found in an inter-religious comparison that only the Catholic *Dominicantes* in Argentina and the USA were among the world's top 10% performers in overcoming Antisemitism, while the Catholic *Dominicantes* in Venezuela, Bosnia, Nigeria, Slovakia, South Africa, and South Korea were among the world's lamentable bottom 2/3 of communities in overcoming Antisemitism.

Finally, Solomon and Tausch (2020) highlighted the role of the intervening variables such as restrictive attitudes on gender and the religion–state relationship in the explanation of Antisemitism. Western Evangelical and Oriental Christianity, Islam, Hinduism, and Buddhism are performing badly on this account, and there is also a clear global North–South divide for these phenomena. Challenging patriarchic gender ideologies and fundamentalist conceptions of the relationship between religion and state, which are important drivers of Anti-Semitism, would be an important task in the future. Multiculturalism must be aware of prejudice, patriarchy and religious fundamentalism in the global South.

## 7.3 Research Design of the Study of Antisemitism as the Second Global Pandemic

So, basically, two approaches are applied here: to explain Antisemitism as reported by the ADL (2014) by the aggregate national data of the 101 countries concerned, and secondly, to look into what happens in the hearts and minds of the citizens of our globe when they say in their *World Values Survey* interviews that they reject to have a Jewish neighbor.

In the following, we first name the variables, which we test in cross-national standard OLS "development accounting" equations for their influence on the variations in the global rates of Antisemitism, and the hypotheses we connect with them in the light of the surveyed literature in alphabetical order:

- **Closing overall gender gap** (relevant studies: Cohen, 2018a, 2018b; Cole, 2017; Hausmann et al., 2012). Highest occurrence of this variable in the global sample: Iceland, Finland, Norway, Sweden, New Zealand. Lowest: Yemen, Chad, Pakistan, Benin, Saudi Arabia. **Hypothesis: Variable will reduce the amount of Antisemitism.**
- **Employment measured especially by the Labor Force Participation Rate of the 55–59-year-old people** (relevant studies: Bauer, 1993; Don, 1986; Gendell, 2008; Legge, 1996; Whiting, 2005; Worrell, 2019). Highest occurrence of this variable in the global sample: Malawi, Madagascar, Tanzania, Rwanda, Mozambique. Lowest: Jordan, Turkey, Iran, Bosnia and Herzegovina, Syria. **Hypothesis: Variable will reduce the amount of Antisemitism.**
- **Income (nat log EU = 100), Income (nat log EU = 100)$^2$** (relevant studies: Acemoglu & Robinson, 2002; Dinda, 2004; Tausch, 2014). Highest occurrence of this variable in the global sample: Luxembourg, Norway, Qatar, Switzerland, Australia. Lowest: Malawi, Burundi, Central African Republic, Congo, Democratic Republic of the, Niger. **Hypothesis: Variable will first increase and then reduce the amount of Antisemitism.**
- **Mean Years of Schooling** (relevant studies Barro & Lee, 1996; Bobo & Licari, 1989; Curtis, 1997; Kurthen et al., 1997). Highest occurrence of this variable in the global sample: Germany, USA, Australia, Norway, Israel. Lowest: Burkina Faso, Niger, Chad, Guinea, Mali. **Hypothesis: Variable will reduce the amount of Antisemitism.**
- **Share of Catholics per total population** (relevant studies: Maritain, 1939; Michael, 2008; Modras, 2005).Highest occurrence of this variable in the global

sample: Malta, Colombia, Ecuador, Poland, Venezuela. Lowest: Bhutan, Yemen, Afghanistan, Nepal, Maldives. **Hypothesis: Variable will increase the amount of Antisemitism.**

- **Share of Muslims per total population** (relevant studies: Jikeli, 2015; Maccoby, 2006). Highest occurrence of this variable in the global sample: Maldives, Mauritania, Tunisia, Yemen, Morocco. Lowest: Haiti, Angola, Paraguay, Guatemala, Papua New Guinea. **Hypothesis: Variable will increase the amount of Antisemitism.**
- **Share of Orthodox Christians per total population** (relevant studies: Verkhovsky, 2002, 2004; Rossman, 2002).Highest occurrence of this variable in the global sample: Greece, Armenia, Romania, Bulgaria, Cyprus. Lowest: Burma/Myanmar, Cuba, Equatorial Guinea, Laos, China. **Hypothesis: Variable will increase the amount of Antisemitism.**
- **Trust** (relevant studies: Alesina, & La Ferrara, 2002). Highest occurrence of this variable in the global sample: Denmark, Morocco, Finland, China, Sweden. Lowest: Lebanon, Albania, Turkey, Ecuador, Botswana. **Hypothesis: Variable will reduce the amount of Antisemitism.**

In the second part of our Chapter, we analyze whether the above hypotheses are compatible with the evidence on the individual level of respondents across the globe in the *World Values Survey* data on the rejection of Jewish neighbors. Our analyses are based on the following variables:

- Age
- Gender (female)
- Government responsibility
- Highest educational level attained
- How often do you attend religious services (never)?
- Important child qualities: religious faith
- Not important in life: Religion
- Private vs state ownership of business
- Reject opinion: Jobs scarce: Men should have more right to a job than women
- Reject opinion: Men make better political leaders than women do
- Reject opinion: Politicians who don´t believe in God are unfit for public office
- Reject opinion: University is more important for a boy than for a girl
- Satisfaction with financial situation of household
- Satisfaction with your life.

## 7.4 Results: The Kuznets Curves of Global Development and Antisemitism

Well, in presenting the results, we have to start from a very robust observation: that many societal crisis phenomena are crises of modernization, and that societies on their path to become less dependent on agriculture, and there is a rapid process of

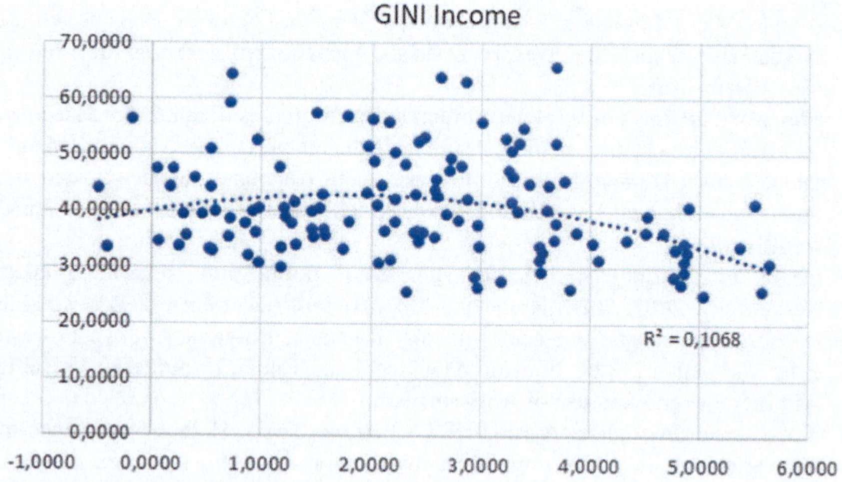


**Fig. 7.2** Original curve showing the relationship between income levels and economic inequality, established by Kuznets (1955) with the empirical data of today. *x-axis* natural logarithm of income per capita. *y-axis* GINI Income Inequality

urbanization, inequality increases, as predicted very correctly by Kuznets, and with it, phenomena like xenophobia and Antisemitism.

In view of the vast literature on the Kuznets curve (see, among others Acemoglu & Robinson, 2002; Dinda, 2004; Tausch, 2014), it is paramount to test the nonlinear relationship of Antisemitism with development[7] with the classical hypothesis by Nobel Laureate Kuznets (1955) in mind: that (originally) inequality—and (later on in the massive literature written on the subject) other social and other phenomena—are a curve-linear function of development. As of January 20, 2021, Google Scholar lists none the less than 15,901 studies quoting the original article by Kuznets (1955),[8] and it could be quite logical that along with development and rising inequality at middle income or development levels, Antisemitism increases as well. Figure 7.2 through to Fig. 7.4 prove the stability of this Kuznets hypothesis for income inequality, ecological footprint, economic growth, and effective democracy, leading up to Fig. 7.4 and its proof that Antisemitism again is a function of development levels.

Cross-national empirical social science and ecological science would assume with Dinda (2004), that there is also an environmental Kuznets curve, an idea which has

[7] Based on the data freely available from https://www.academia.edu/44295690/GULF_2020_Covid_19_PUBLIC_ACCESS_Data_base_for_the_book_Factors_affecting_the_Gulfs_future_Global_cycles_and_value_change_Springer_Publishing_Series_https_www_springer_com_series_16417_Gulf_Studies_edited_by_Prof_Mizanur_Rahman_Qatar_University.

[8] https://scholar.google.com/scholar?hl=en&as_sdt=0%2C5&q=Kuznets%2C+S.+%281955%29.+Economic+Growth+and+Income+Inequality.+American+Economic+Review%2C+45%2C+1%3A+1-28&btnG.

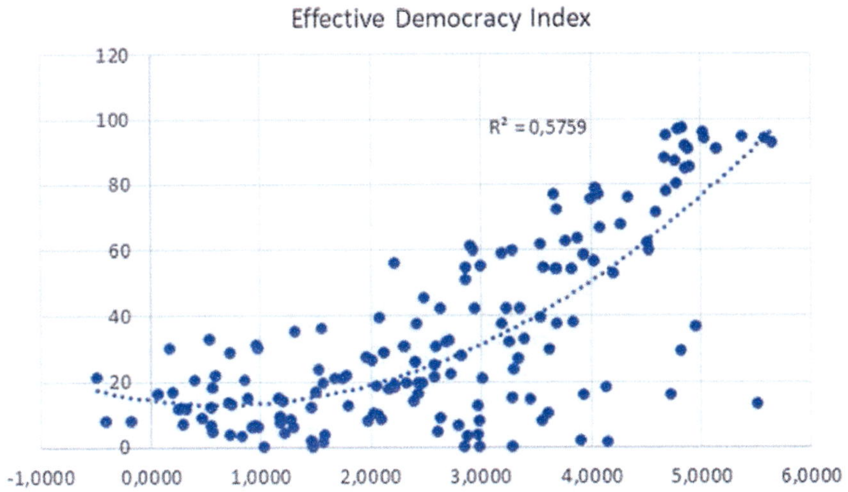


**Fig. 7.3** Relationship between income levels and effective democracy is nonlinear. *x-axis* natural logarithm of income per capita. *y-axis* Effective Democracy Index, combining human rights and freedom from corruption

been very extensively debated in the literature[9]. As Dinda specifies, the environmental Kuznets curve (EKC) hypothesis postulates an inverted-U-shaped relationship between different pollutants and per capita income, i.e., environmental pressure increases up to a certain level as income goes up; after sthat, it decreases. But this has been contested in the literature.

Applying the measurement of effective democracy as a combination of political and civil rights and the absence of corruption, as proposed by Alexander et al. (2012), we can also proceed to plot the scattergram of the relationship between income per capita and effective democracy, again showing the nonlinear relationship between development levels and development outcomes (Fig. 7.3).

In Fig. 7.4, we plot the relationship between income levels and Antisemitism.

Following Inglehart and Norris (2012), one could achieve practically identical results with the Human Development Index on the x-axis. Interestingly enough, we can also mention here that in the countries of the Arab League, Antisemitism is lower at higher levels of Human Development (polynomial expression, EXCEL graph, almost 22% of variance in common), while in the member countries of the Organization of Islamic Cooperation as a whole, Antisemitism is higher at higher levels of Human Development (polynomial expression, EXCEL graph, almost 39% of variance in common) (Figs. 7.5 and 7.6).

[9] Google scholar lists none the less than 3210 articles referring to Dinda's article by January 20, 2021.

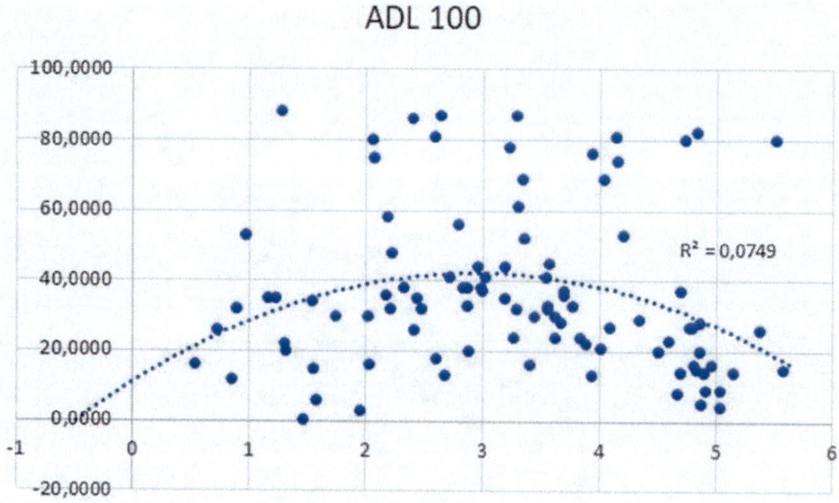


**Fig. 7.4** Relationship between income levels and Antisemitism resembles the U-shaped curve between income levels and inequality, discovered by Kuznets in (1955). *x-axis* natural logarithm of income per capita. *y-axis* Antisemitism, as evidenced by the ADL 100 global survey.

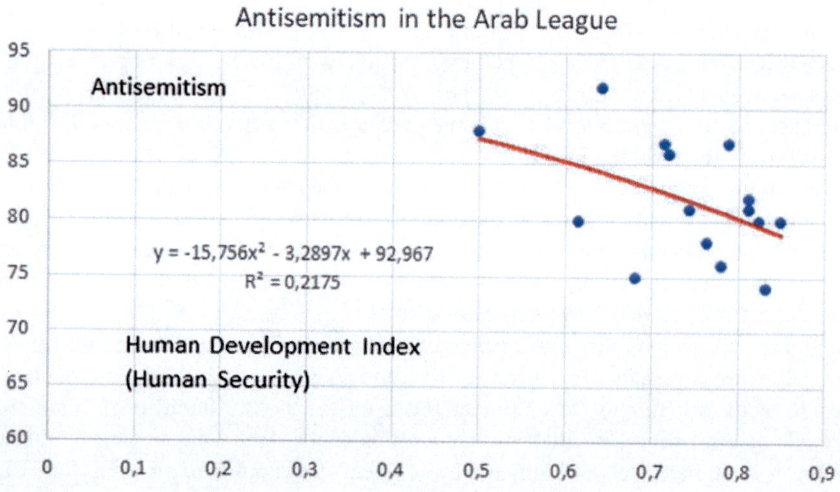


**Fig. 7.5** Human security (Inglehart & Norris, 2012) and antisemitism in the Arab League. *x-axis* UNDP Human Development Index as an indicator of human security. *y-axis* Antisemitism, as evidenced by the ADL 100 global survey.

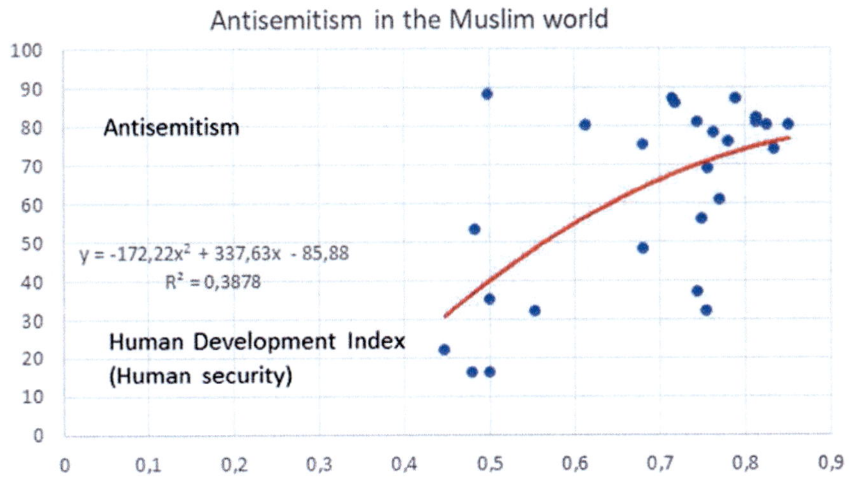


**Fig. 7.6** Human security (Inglehart & Norris, 2012)and antisemitism in the Muslim world (OIC). *x-axis* UNDP Human Development Index as an indicator of human security. *y-axis* Antisemitism, as evidenced by the ADL 100 global survey.

## 7.5 Results: Religion and Antisemitism—The Cross-National Evidence from the ADL Data

First, our partial correlations (Kuznets curve constant)[10]: they very clearly support the view that feminism is one of the most noteworthy blocks against societal Antisemitism, while immigration variables, human rights violations, a high share of military personnel and military expenditure, carbon emissions per capita, inequality, world economic openness, and economic growth after the crisis of 2008, which largely benefitted the global periphery and semi-periphery, population growth rates and above all data referring to a Muslim heritage are unfortunately and clearly associated with Antisemitism independent from the development level of a nation.

The most decisive blocks against Antisemitism appear to be, according to this patrial correlation analysis, Feminism, and gender equality, a favorable opinion on Israel, a secular worldview that rejects religious authorities interpreting the laws, overall development level, a high employment especially of elder workers, general patterns of societal tolerance vis-à-vis outgroups, the Effective Democracy Index and other democracy measures, the comparative price level, the ESI Index Component Social and Institutional Capacity and other environment performance data, mean years of schooling and other education data, life satisfaction data (Gallup), a Protestant, Catholic or Buddhist heritage, well-functioning libraries with items on Jewish

---

[10] https://www.academia.edu/44295690/GULF_2020_Covid_19_PUBLIC_ACCESS_Data_base_for_the_book_Factors_affecting_the_Gulfs_future_Global_cycles_and_value_change_Springer_Publishing_Series_https_www_springer_com_series_16417_Gulf_Studies_edited_by_Prof_Mizanur_Rahman_Qatar_University.

history (the Shoah), social security data, environmental performance data, health expenditures, the World Bank Deregulation Index, happy life years, trust in other people, general government final consumption expenditures, and the share of Jews per total population are to be noted here (Table 7.2).

Our first multiple OLS regression is based on 92 countries and explains (adjusted) 72.7% of societal Antisemitism according to ADL, 2014. The equation is significant to the highest degree ($F = 41{,}420$; error $p = .000$). By comparison, Berggren and Nilsson worked with 87 countries, and their adjusted $R^2$ oscillates between 71.4 and 77.1%. Several of their predictors in their multiple OLS regression are not significant, and in our case, one predictor—the second part of the Kuznets curve—has an error probability of 10.5% and the employment rate of elder workers has an error probability of 6.5%. Our equation tells a simple and parsimonious story: Antisemitism is strongest at middle income levels, and closing the gender gap and creating employment, especially for older workers, reduces Antisemitism, while, *ceteris paribus*—an Orthodox and a Muslim heritage all increase Antisemitism.

Figure 7.7 portrays the standardized regression coefficients (Beta weights) of employment, the Kuznets curve, patriarchy, and religious adherence on Antisemitism.

Now I invite readers to compare the choropleth map in Fig. 7.8—which is based on the residuals from the regression of Table 7.3—with the choropleth map in Fig. 7.1. All of a sudden, the situation in Saudi Arabia—and also in Iran and Turkey—is in the "blue zone" of relatively lower Antisemitism, compared to the conditioning factors, while in several predominantly Catholic countries, like Colombia, Peru, Paraguay, but also France, Poland, and Lithuania, Antisemitism is much higher than one would had to have expected.

In the light of Table 7.3, the highest residuals and thus the most disturbing rates of Antisemitism then have to be diagnosed for Panama, United Arab Emirates, Qatar, Peru, and South Africa, and the lowest rates of Antisemitism, compared to the conditioning factors, mentioned in Table 7.3 are to be observed in Nigeria, Azerbaijan, Iran, Singapore, and the Czech Republic. Table 7.4 documents these residuals and their global ranks for the Gulf countries:

We now document the Pearson correlation coefficients of the residuals explaining Antisemitism by indicators of patriarchy, employment, religious adherence and the Kuznets curve with standard indicators of global development and human values. Again, the variables for this exercise are available to our readers.[11] The Pearson correlations speak in a way for themselves (Table 7.5).

Now, in Table 7.6, we extend the model used in our regression of Table 7.3 and now ask ourselves whether standard variables of international development accounting, gathered by the World Bank (2017) and the UNDP (2019, current issues), as well as Alesina's societal trust variable (Alesina & Guiliano, 2013, 205; Alesina & Ferrara,

[11] Solomon and Tausch (2020, 2021), Grinin et al. (2018), furthermore https://www.academia.edu/44295690/GULF_2020_Covid_19_PUBLIC_ACCESS_Data_base_for_the_book_Factors_affecting_the_Gulfs_future_Global_cycles_and_value_change_Springer_Publishing_Series_https_www_springer_com_series_16417_Gulf_Studies_edited_by_Prof_Mizanur_Rahman_Qatar_University.

**Table 7.2** Partial correlations with Antisemitism, based on the ADL 100 survey data and standard indicators of global development. The Kuznets curve based on income levels is kept constant

| | Partial correlation with Antisemitism (Kuznets curve constant) | Error probability | *df* |
|---|---|---|---|
| Share of Muslims per total population | 0.786 | 0.000 | 94 |
| Membership in the Islamic Conference OIC | 0.697 | 0.000 | 93 |
| ImmigrationShare of population 2005 (%) | 0.559 | 0.000 | 94 |
| Civil and Political Liberties violations | 0.536 | 0.000 | 94 |
| military personnel rate ln (MPR + 1) | 0.529 | 0.000 | 92 |
| Share of total immigration from OIC countries, 2013 in % | 0.524 | 0.000 | 94 |
| Share of total immigration OIC countries, 2017 in % | 0.516 | 0.000 | 94 |
| Annual population growth rate, 1975–2005 (%) | 0.515 | 0.000 | 92 |
| Net migration rate | 0.437 | 0.000 | 94 |
| Military expenditures per GDP | 0.434 | 0.000 | 80 |
| Carbon emissions per capita | 0.420 | 0.000 | 94 |
| Total unemployment rate of immigrants (both genders) | 0.414 | 0.000 | 94 |
| Economic growth IMF prediction growth rate in 2009 | 0.394 | 0.000 | 93 |
| Combined Failed States Index | 0.344 | 0.001 | 94 |
| Crisis Performance Factor 2008–2011 | 0.325 | 0.002 | 89 |
| Net international migration rate, 2005–2010 | 0.319 | 0.002 | 94 |
| Economic growth IMF prediction growth rate in 2010 | 0.292 | 0.004 | 93 |
| Unemployment rate | 0.287 | 0.007 | 86 |
| Openness-Index, 1990 (export-share per GDP + import-share per GDP) | 0.251 | 0.015 | 91 |
| Annual rate of growth international migration, 1960–2005 | 0.240 | 0.029 | 81 |
| Coefficient of human inequality 2013 | 0.233 | 0.032 | 83 |

(continued)

**Table 7.2** (continued)

| | Partial correlation with Antisemitism (Kuznets curve constant) | Error probability | *df* |
|---|---|---|---|
| FPZ (free production zones) employment as % of total population | 0.227 | 0.026 | 94 |
| Thousand asylum seekers per total population (in millions) | 0.216 | 0.037 | 92 |
| Asylum seekers as permille of total population | 0.216 | 0.037 | 92 |
| Share of Jews per total population | −0.209 | 0.041 | 94 |
| General government final consumption expenditure | −0.232 | 0.023 | 94 |
| Gallup poll about satisfaction: Health care quality | −0.234 | 0.022 | 94 |
| Share of adherents of other religions per total population | −0.236 | 0.021 | 94 |
| Members of EEA + EFTA | −0.239 | 0.019 | 94 |
| Share of Buddhists per total population | −0.242 | 0.018 | 94 |
| Gallup poll about satisfaction: Trust in other people | −0.247 | 0.020 | 86 |
| Share of Roman Catholics per total population | −0.248 | 0.015 | 94 |
| Happy life years | −0.255 | 0.015 | 88 |
| Gallup poll about satisfaction: Job | −0.263 | 0.010 | 94 |
| Pearson correlation 1992–2014 GDP pc growth | −0.275 | 0.016 | 75 |
| NL 4th root number of items with the word "Shoah" in the title | −0.275 | 0.031 | 60 |
| Life Satisfaction (0–10) | −0.283 | 0.007 | 88 |
| Gallup poll about satisfaction: Education quality | −0.285 | 0.005 | 92 |
| Gallup poll about satisfaction: Freedom of choice | −0.305 | 0.003 | 94 |
| Tertiary enrollment | −0.331 | 0.002 | 87 |
| Expenditure on education | −0.337 | 0.001 | 87 |
| Labour force participation rate of migrants (both sexes) | −0.339 | 0.001 | 94 |
| Share of other Christians per total population | −0.340 | 0.001 | 94 |

(continued)

**Table 7.2** (continued)

| | Partial correlation with Antisemitism (Kuznets curve constant) | Error probability | *df* |
|---|---|---|---|
| Share of people without religion per total population | −0.341 | 0.001 | 94 |
| Deregulation Index 2013, World Bank | −0.343 | 0.001 | 94 |
| Closing educational gender gap | −0.344 | 0.001 | 88 |
| Health expenditure as % of GDP | −0.351 | 0.000 | 94 |
| Environmental Performance Index (EPI) | −0.356 | 0.001 | 90 |
| Social protection (ILO) | −0.372 | 0.000 | 92 |
| Social security expenditure per GDP average 1990s (ILO) | −0.375 | 0.001 | 71 |
| UC Fourth root number of items with the word "Shoah" | −0.378 | 0.014 | 40 |
| Female share of seats in parliament | −0.382 | 0.000 | 94 |
| ESI-Index Environment Sustainability Index (Yale Columbia) | −0.386 | 0.000 | 90 |
| Per capita world class universities | −0.387 | 0.000 | 94 |
| UNDP education index | −0.387 | 0.000 | 94 |
| PROT00 share of Protestants per total population | −0.392 | 0.000 | 94 |
| Gallup poll about satisfaction: Overall life satisfaction index | −0.397 | 0.000 | 94 |
| Mean Years of Schooling 2013 | −0.424 | 0.000 | 94 |
| Female Population with at least some secondary education | −0.428 | 0.000 | 94 |
| % women in government, all levels | −0.446 | 0.000 | 90 |
| comparative price level 2013 | −0.464 | 0.000 | 92 |
| closing political gender gap | −0.487 | 0.000 | 88 |
| Democracy measure | −0.492 | 0.000 | 90 |
| ESI Index Component Social and Institutional Capacity | −0.520 | 0.000 | 90 |
| Effective Democracy Index | −0.524 | 0.000 | 93 |
| Muslim empowerment index, based on WVS | −0.542 | 0.011 | 19 |

(continued)

**Table 7.2** (continued)

| | Partial correlation with Antisemitism (Kuznets curve constant) | Error probability | *df* |
|---|---|---|---|
| MIPEX 2013 | −0.587 | 0.001 | 25 |
| Overall 35 variable development index | −0.611 | 0.000 | 94 |
| Overall 30 variable development index | −0.653 | 0.000 | 94 |
| LFPR 55–59 | −0.693 | 0.000 | 94 |
| Overall 35 variable development index, based on 7 dimensions | −0.696 | 0.000 | 86 |
| Closing economic gender gap | −0.731 | 0.000 | 88 |
| Closing of global gender gap overall score 2009 | −0.755 | 0.000 | 88 |
| Global tolerance index | −0.818 | 0.000 | 59 |
| WVS: dissent with the opinion: religious authorities should interpret the laws | −0.839 | 0.000 | 39 |
| Very favorable or somewhat favorable opinion of Israel (2013) | −0.886 | 0.000 | 16 |
| Gender empowerment index value | −0.968 | 0.000 | 58 |

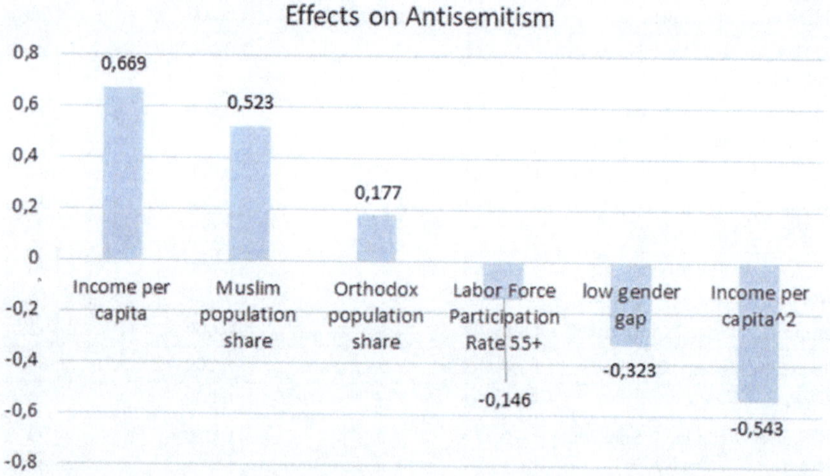


**Fig. 7.7** Effects of employment, the Kuznets curve, patriarchy, and religious adherence on Antisemitism: *Beta*-weights

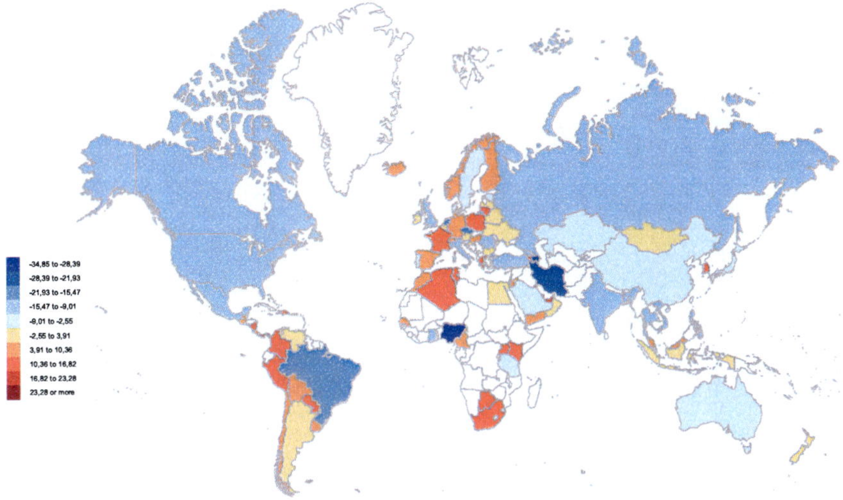


**Fig. 7.8** Residuals of Antisemitism explained by indicators of patriarchy, employment, religious adherence, and the Kuznets curve

**Table 7.3** Multiple OLS regression on societal Antisemitism (ADL data), explained by indicators of patriarchy, employment, religious adherence, and the Kuznets curve

| | Regression coefficient B | Standard error | Beta | *t*-value | Error *p* |
|---|---|---|---|---|---|
| Constant | 103.627 | 25.660 | | 4.039 | 0.000 |
| Closing of global gender gap overall score 2009 | −119.435 | 32.571 | −0.323 | −3.667 | 0.000 |
| LFPR 55–59 Labor Force Participation Rate of the 55- to 59-year-old population | −0.237 | 0.127 | −0.146 | −1.870 | 0.065 |
| Income per capita 2010 (n log EU =100) | 11.946 | 5.953 | 0.669 | 2.007 | 0.048 |
| Income per capita 2013 (n log EU =100)$^2$ | −1.488 | 0.909 | −0.543 | −1.636 | 0.105 |
| Share of Orthodox Christians per total population | 20.293 | 6.591 | 0.177 | 3.079 | 0.003 |
| Share of Muslims per total population | 33.827 | 5.599 | 0.523 | 6.041 | 0.000 |

$N = 92$; Adj $R^2 = 72.7\%$; $F = 41{,}420$; error $p = 0.000$

2000; Alesina et al., 2015) and Barro's religious adherence data (2003) sufficiently well explain the global ADL (2014) Antisemitism rates. We again expect that first there is an increase of Antisemitism with rising per capita incomes, levelling off at higher income levels. Mean years of education will be an important impediment against Antisemitism, and a climate of societal trust will diminish Antisemitism. Unfortunately, it is evident that with all these other factors being constant, adherence

**Table 7.4** Predicted values and residuals of societal Antisemitism (ADL 100 data), explained by indicators of patriarchy, employment, religious adherence, and the Kuznets curve

| Global rank (92 countries) | Country | Antisemitism predicted | Antisemitism residuals |
|---|---|---|---|
| 28 | Saudi Arabia | 81,035 | −7.035 |
| 44 | Oman | 76,492 | −0.492 |
| 76 | Bahrain | 69,411 | 11.589 |
| 83 | Kuwait | 68,043 | 13.957 |
| 90 | Qatar | 63,938 | 16.062 |
| 91 | United Arab Emirates | 61,831 | 18.169 |

**Table 7.5** Pearson correlation of the residuals explaining Antisemitism by indicators of patriarchy, employment, religious adherence, and the Kuznets curve with standard indicators of global development and human values

| Var label | Pearson correlation with the residuals | Error *p* | *N* |
|---|---|---|---|
| Female rejection of the market economy and democracy | 0.293 | 0.035 | 52 |
| Quintile share income difference between richest and poorest 20% | 0.272 | 0.012 | 84 |
| Annual population growth rate, 1975–2005 in % | 0.237 | 0.024 | 91 |
| Gallup poll about satisfaction: Safety | −0.229 | 0.030 | 90 |
| Pearson correlation 1992–2014 GDP pc growth with time axis | −0.238 | 0.039 | 76 |
| Share of people without religion per total population | −0.262 | 0.012 | 92 |
| Gallup poll about satisfaction: Trust in other people | −0.269 | 0.014 | 83 |
| Very favorable or somewhat favorable opinion of Israel (2013) | −0.537 | 0.018 | 19 |

to Catholicism, Orthodox Christianity, and Islam, all are still to be considered as significant drivers of the rate of societal Antisemitism, all reflecting the centuries of Antisemitism in the concrete, but different history of the religious environments in these cultures, contradicting the message of tolerance, which is at the roots of the "religions of the book" (Tausch et al., 2014a, 2014b). Our adjusted $R^2$ is now 73.9%; $N = 83$ countries; $F = 29.995$; and the error $p = 0.000$. The Labor Force Participation Rate (LFPR) of older workers, the second part of the Kuznets curve, and mean years of schooling, however, do not achieve an error probability of below 5%.

In Fig. 7.9, we again document for our readers the Beta-weights of this regression Table 7.6 above. With rising development, it is most difficult in countries with a

**Table 7.6** Global antisemitism (ADL, 2014) explained by indicators of patriarchy, employment, religious adherence, education, societal trust, and the Kuznets curve

| Variable | Regression coefficient B | Standard error | Beta | *t*-value | Error *p* |
|---|---|---|---|---|---|
| Constant | 98.999 | 27.796 | | 3.562 | 0.001 |
| closing of global gender gap overall score 2009 | −99.245 | 37.758 | −0.249 | −2.628 | 0.010 |
| LFPR 55–59 | −0.230 | 0.129 | −0.138 | −1.785 | 0.078 |
| income 2010 (EU =100) | 13.775 | 6.712 | 0.769 | 2.052 | 0.044 |
| *income 2013 (EU =100) ^2* | *−1.397* | *0.991* | *−0.498* | *−1.411* | *0.163* |
| share of Orthodox Christians per total population | 23.094 | 7.385 | 0.210 | 3.127 | 0.003 |
| share of Muslims per total population | 35.955 | 5.796 | 0.550 | 6.204 | 0.000 |
| Gallup poll about trust in other people | −0.319 | 0.123 | −0.176 | −2.600 | 0.011 |
| *Mean Years of Schooling 2013* | *−1.042* | *1.021* | *−0.115* | *−1.020* | *0.311* |

Data from the ADL (2014);World Bank, UNDP, Robert Barro (Harvard)
Adj. $R^2 = 73.9\%$; $N = 83$ countries; $F = 29.995$; error $p = 0.000$

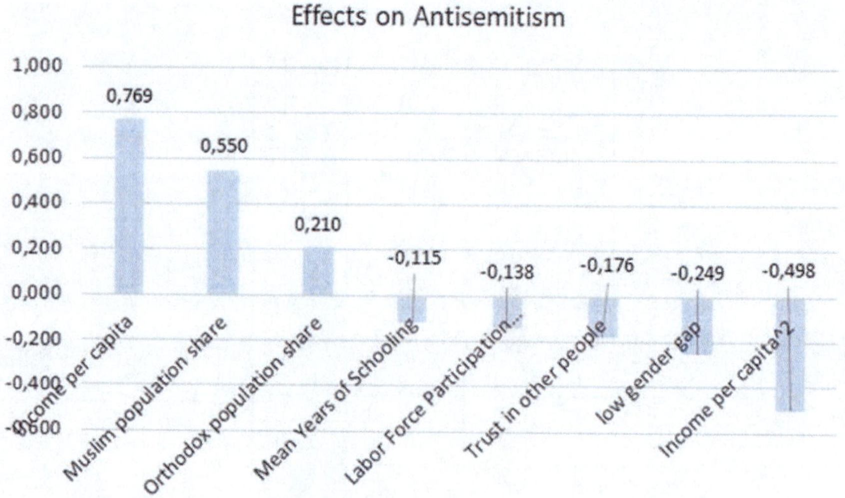


**Fig. 7.9** Effects of education, employment, the Kuznets curve, patriarchy, religious adherence, and societal trust on Antisemitism: *Beta*-weights

Muslim or an Orthodox heritage to avoid Antisemitism, while with rising development, gender empowerment, and the development of a climate of trust are the best receipts to create good conditions to avoid Antisemitism.

In Table 7.7, we document the residuals from equation Table 7.6 for the countries of the Gulf, and Fig. 7.10 projects the size of the residuals onto a choropleth map of the entire world. This time the diagnosis for the Gulf states is bleaker than in Fig. 7.8. The ten countries with the highest residual rate of Antisemitism are Panama, Korea, South, Morocco, Poland, Armenia, South Africa, Lithuania, Peru, United Arab Emirates, and Nicaragua, while the ten countries with the lowest residual degree of Antisemitism are Nigeria, Azerbaijan, Brazil, Singapore, Turkey, Bangladesh, Italy, Romania, Czech Republic, and the Philippines.

The factor trust, highlighted by Alesina, is an important stabilizing factor for an open society. In our analysis, this also holds true for the determination of the absence

**Table 7.7** Predicted values and residuals of global antisemitism (ADL, 2014) explained by indicators of patriarchy, employment, religious adherence, education, societal trust, and the Kuznets curve

| Global rank (of 83 countries) | Country | Antisemitism predicted | Antisemitism residuals |
|---|---|---|---|
| 26 | Saudi Arabia | 78,680 | −4.680 |
| 48 | Kuwait | 78,254 | 3.746 |
| 56 | Bahrain | 75,094 | 5.906 |
| 70 | Qatar | 69,494 | 10.506 |
| 75 | United Arab Emirates | 67,340 | 12.660 |

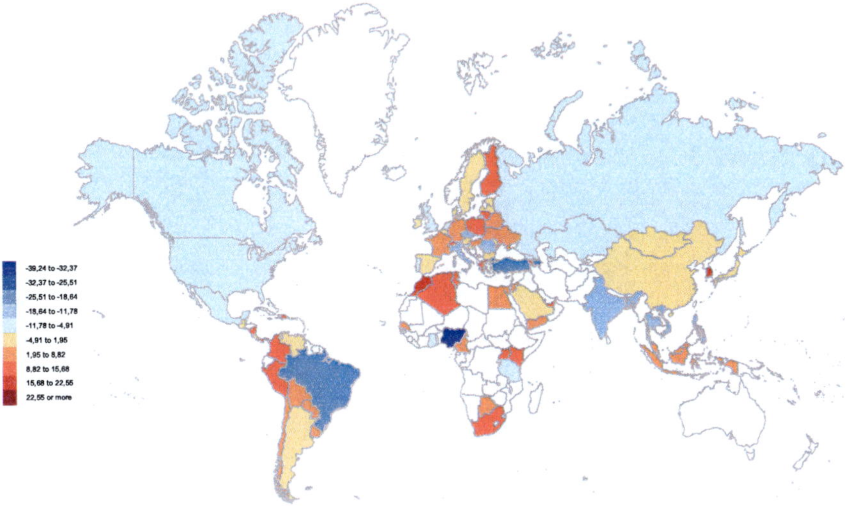


**Fig. 7.10** Residuals of global antisemitism (ADL, 2014) explained by indicators of patriarchy, employment, religious adherence, education, societal trust, and the Kuznets curve

of large-scale Antisemitism by the factor trust. There is again a so-called Kuznets curve (Kuznets, 1976) of Antisemitism, suggesting that Antisemitism coincides with the modernization crisis experienced by semi-industrial societies at middle stages of development (see also Tausch et al., 2014a, 2014b)

## 7.6 Results from the *World Values Survey*

In the 28 countries under comparison here, the following ranking of Antisemitic attitudes is to be observed (Table 7.8). In Iraq, Iran, Egypt, and India, more than half of the entire resident population rejected to have a Jewish neighbor, while in Canada, Argentina, Belarus, Uruguay, and the USA, this percentage was below 10%. It would be premature to interpret our results only along global denominational or world political conflict lines. Some Muslim majority societies, for example, Bangladesh and Albania, did even better than world political and military allies of the West, like South Korea, Slovakia, Japan, and Spain.

Table 7.9 contains the regression analytical results of our investigation showing the importance of the variables about a fundamentalist interpretation of the relationship between religion and state and the sexist inclination to reserve higher education only for male persons.

The dependent variable in the multiple standard OLS regression is Antisemitism (rejection of a Jewish neighbor; 0 = no rejection; 1 = rejection). The independent variables in our research endeavor were:

**Table 7.8** Percent of people saying that they reject to have a Jewish neighbor

| | % rejecting to have a Jewish neighbor (%) | *N* |
|---|---|---|
| **Iraq** | 90 | 4924 |
| **Iran** | 75 | 2657 |
| **Egypt** | 62 | 3000 |
| India | 53 | 2500 |
| South Korea | 41 | 1200 |
| Nigeria | 35 | 1001 |
| Slovakia | 34 | 466 |
| Japan | 28 | 1011 |
| Bosnia | 28 | 1200 |
| Venezuela | 26 | 1200 |
| Mexico | 26 | 3895 |
| Moldova | 25 | 1008 |
| South Africa | 24 | 3000 |
| Spain | 22 | 2719 |
| Uganda | 22 | 1002 |
| Bangladesh | 20 | 1500 |
| Kyrgyzstan | 20 | 1043 |
| Macedonia | 20 | 1055 |
| Zimbabwe | 19 | 1002 |
| Czech Republic | 18 | 924 |
| Albania | 17 | 1000 |
| Chile | 13 | 3700 |
| Russia | 11 | 4001 |
| United States | 9 | 1200 |
| Uruguay | 8 | 2000 |
| Belarus | 7 | 2092 |
| Argentina | 6 | 3361 |
| Canada | 4 | 1931 |

- Not important in life: Religion
- Important child qualities: religious faith
- Satisfaction with your life
- Reject opinion: Jobs scarce: Men should have more right to a job than women
- Satisfaction with financial situation of household
- Reject opinion: Men make better political leaders than women do
- Reject opinion: University is more important for a boy than for a girl
- Private versus state ownership of business
- Government responsibility

**Table 7.9** Explaining global antisemitism (rejection of Jewish neighbors) according to the data of the *World Values Survey* by multiple regression (standard OLS regression), based on indicators of patriarchy, socioeconomic status, gender, age, religious commitment, the role of government, and a secular worldview (reject the opinion that Politicians who do not believe in God are unfit for public office)

| Variable | Regression coefficient B | Standard error | Beta | *t*-value | Error *p* |
|---|---|---|---|---|---|
| Constant | 0,705 | 0,021 | | 33,605 | 0,000 |
| Not important in life: Religion | −0.044 | 0.004 | −0.082 | −10.956 | 0.000 |
| Important child qualities: religious faith | 0.100 | 0.007 | 0.102 | 14.684 | 0.000 |
| Satisfaction with your life | −0.013 | 0.001 | −0.070 | −10.095 | 0.000 |
| Reject opinion: Jobs scarce: Men should have more right to a job than women | −0.073 | 0.004 | −0.103 | −16.747 | 0.000 |
| Satisfaction with financial situation of household | 0.008 | 0.001 | 0.044 | 6.433 | 0.000 |
| Reject opinion: Men make better political leaders than women do | −0.073 | 0.003 | −0.153 | −22.369 | 0.000 |
| Reject opinion: University is more important for a boy than for a girl | −0.019 | 0.003 | −0.037 | −5.834 | 0.000 |
| Private vs state ownership of business | 0.013 | 0.001 | 0.081 | 13.705 | 0.000 |
| Government responsibility | 0.012 | 0.001 | 0.072 | 12.308 | 0.000 |
| How often do you attend religious services (never) (never) | 0.016 | 0.001 | 0.085 | 13.667 | 0.000 |
| Reject opinion: Politicians who don't believe in God are unfit for public office | −0.046 | 0.002 | −0.132 | −19.541 | 0.000 |
| Gender (female) | 0.001 | 0.006 | 0.001 | 0.161 | 0.872 |
| Age | −0.001 | 0.000 | −0.043 | −7.437 | 0.000 |
| Highest educational level attained | 0.005 | 0.001 | 0.022 | 3.768 | 0.000 |

$N = 24{,}858$; adj. $R^2 = 20.1\%$; $F = 448{,}668$; error $p = 0.000$

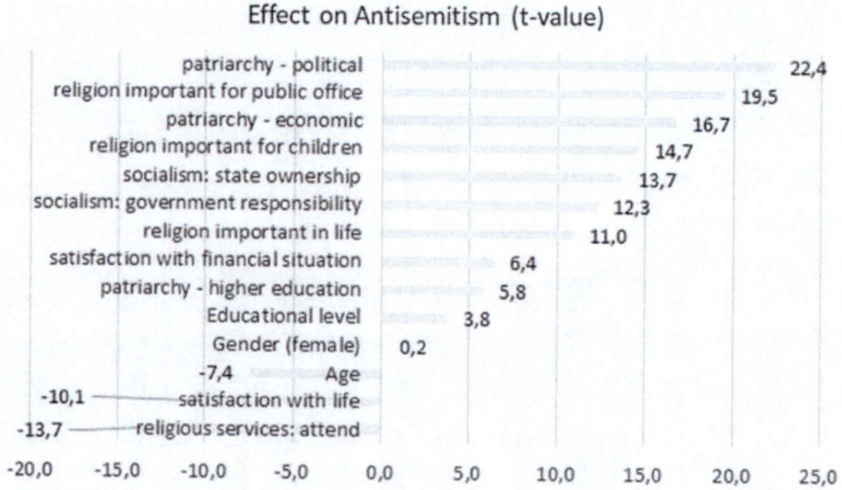


**Fig. 7.11** Drivers of antisemitism (*t*-test) as evidenced by a multiple regression based on *World Values Survey* data, ordered by the size of the effects

- How often do you attend religious services (never) (never)?
- Reject opinion: Politicians who don't believe in God are unfit for public office
- Gender (female)
- Age
- Highest educational level attained.

The model could be calculated for more than 27.000 *WVS* respondents, all indicators (except the higher education variable) wield a very significant effect on Antisemitism, the entire equation achieves an F-Ratio of more than 500 and is thus highly significant, and the explained $R^2$ is 14.8%. Table 7.9 shows the high weight—*ceteris paribus*—of fundamentalism and sexism in explaining Antisemitism.

Figures 7.11 and 7.12 further evaluate the results of Table 7.9 (*t*-tests) and groups these results by the size of the effects (Fig. 7.11) and by the group of variables (Fig. 7.12). Patriarchy and a fundamentalist reading of all—I repeat—all religious scriptures increase Antisemitism, and interestingly enough, also the quest for socialism does so as well. These results qualify Adorno's findings (2019) in an important way; at the time of writing his analysis, Adorno and associates believed that Antisemitism is a clear consequence of the left to right political spectrum, while today, patriarchy, religious fundamentalism of all sorts, and the quest for socialism and an active state are drivers of Antisemitism. We also highlight the fact that among the so-called structural factors, like age and life satisfaction, there are also noticeable exceptions from what one would have expected—the fact that people satisfied with the financial condition of their households, tend to be more antisemitic than others, and educational level is even positively associated with the level of Antisemitism.

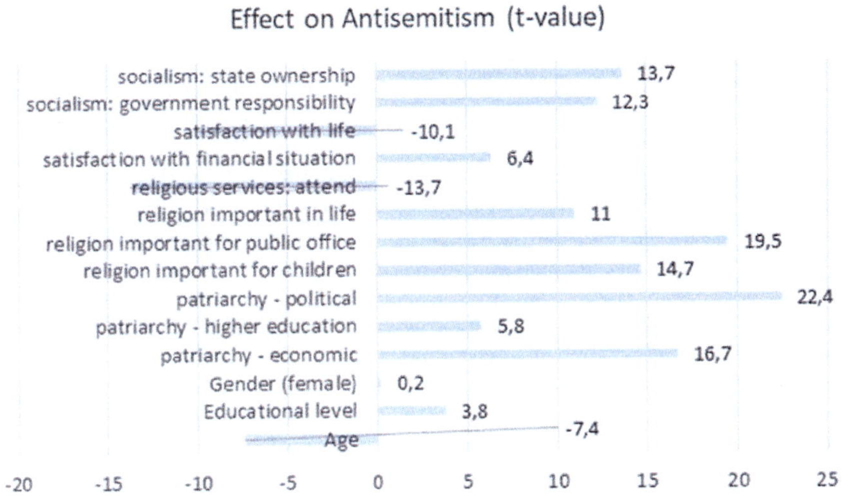


**Fig. 7.12** Drivers of antisemitism (*t*-test) as evidenced by a multiple regression based on *World Values Survey* data, ordered by the groups of variables, wielding an effect

## 7.7 Conclusions and Prospects for Future Changing Patterns of Relations of the Gulf States with Israel

Leading global think tanks, including from Israel, note the enormous variety of political patterns emerging from the Gulf, making a more profound study of societal values in the region all the more important. The Jerusalem Center for Public Affairs, in one of its recent analyses,[12] says:

> Interestingly, during this decade of turmoil, the process of normalization between Israel and some of its Arab neighbors has advanced a notch. It is particularly true in relations between Israel and the Gulf states. Delegations from Israel have been received in the Gulf states, and Israeli businessmen, as well as tourists and sports teams, are a common sight in Dubai, Bahrain, Qatar, and Morocco. […] High-ranking Saudi officials and businessmen met in the open with a Jerusalem Center for Public Affairs team headed by Jerusalem Center President Amb[assador] Dore Gold to discuss ways of containing the Iranian threat and developing bilateral relations. A trilateral meeting even took place between Indian, Israeli, and Saudi teams in New Delhi, far from the limelight of the press. Some Arab thinkers (writers, journalists) have expressed openly the need to normalize relations with Israel, and quite surprisingly, they were not reprimanded or threatened. Even the peace treaties with Egypt and Jordan and the agreement with the Palestinian Authority – while under severe criticism from the opposition – have survived the turmoil of the Middle East upheavals.[13]

James Dorsey, writing for the Begin/Sadat Center at Bar Ilan University in Israel, says[14]:

---

[12] https://jcpa.org/article/a-scorecard-on-the-first-decade-after-the-arab-spring/.

[13] See Footnote 12.

[14] https://besacenter.org/perspectives-papers/qatar-saudi-arabia-wahhabism/.

> Qatar is a traditional Gulf state and a Wahhabi state to boot, but it has never had a powerful religious establishment that could enforce ultra-conservative social norms. Nor did it implement absolute gender segregation. Non-Muslims could practice their faith in their own houses of worship and were exempt from bans on alcohol. Qatar became a sponsor of the arts, including a Doha version of the Tribeca Film Festival, and hosts the state-owned Al Jazeera television network, which revolutionized the region's-controlled media landscape and became one of the world's foremost global broadcasters. The UAE boasts many of the same traits minus the history of an ultra-conservative strand of Islam. Qatar's projection of a different approach to Wahhabism is rooted in the DNA of the Qatari state, which, from its founding, was determined not to emulate the Kingdom. Privately, Qataris distinguish between their "Wahhabism of the sea" as opposed to Saudi Arabia's "Wahhabism of the land.

Dorsey, in his analysis, speaks of a *"Qatari ambivalence toward Wahhabism"*. Generating a clerical class of its own would have enhanced the threat, because to do so, Qatar would have been dependent on Saudi scholars. That would have produced a religious establishment steeped in the Kingdom's austere theology and inspired by its history of political power-sharing, which it then would have demanded.

> As a result, Qatar lacks the institutions that have often held the Kingdom back. Similarly, Qatar does not have families known for producing religious scholars. Qatari religious schools are run by the ministry of education, not, as in Saudi Arabia, by the religious affairs authority.[15]

The Begin/Sadat Center study also highlights that Qatar in particular, but in many ways the UAE as well project a vision of a less restrictive and choking conservative Wahhabi society that grants individuals greater opportunities, irrespective of gender.[16] The Begin/Sadat Center study also underlines the fact that Qatar today is a long way from the mid-1990s, when Qatari women, as in Saudi Arabia until recently, were banned from driving, voting, or holding government jobs. Qatari women now occupy prominent positions in multiple sectors of society. With women accounting for 53% of the work force, Qatar outranks Middle Eastern and North African states by a large margin. Only Kuwait with 48% and the UAE with 42% come close.[17] It is a picture that long juxtaposed starkly with that of its Wahhabi big brother. Major problems identified by the Begin/Sadat study are that authoritarian modernizers cannot simply command a new attitude among their citizens. Opening cinemas and relaxing gender segregation may impress Saudi youth, but a new economy requires far more. Reformers in the UAE eventually realized—as Saudi rulers will discover, too—that they needed to adapt both the mindsets and the skill sets of the rising generation. In countries where people see a government job as a right, that means reshaping the very nature of citizenship. With a host of other observers, Dorsey believes that Qatari and Emirati promotion of knowledge, culture, and innovation in a bid to create globalized citizens has succeeded in developing civic attitudes including notions of tolerance and volunteerism, but have failed to alter economic perceptions of the rentier state rather than the private sector as the creator of jobs. Is there a proper

[15] See Footnote 14.

[16] See Footnote 14.

[17] See Footnote 14.

interest in entrepreneurship, or do people demand a citizen's right to a government job? Will top-down, government-engineered change and unilateral rewriting of social contracts produce results that fall short of what is required?

Yellinek,[18] in another current Israeli study on the Gulf states, emphasizes that recent developments suggest that there may be a future in which Arab states and Israel can engage in diplomatic and security cooperation and reduce their mutual animosity.[19]

The Qatarbased Doha Institute,[20] in a similar vein, also spoke about the normalization of relations between some Arab countries and Israel.[21] The study also highlights that in terms of intelligence and security, some Arab countries are believed to be major recipients of Israeli military and intelligence technology, as well as security services.[22]

> Due to the threat of Iran, Israel's relations with some Arab Gulf states have increased in strength. In August 2019, the Israeli Foreign Minister, Israel Katz, stated that Israel is part of an international coalition to protect cargo ships traveling through the Strait of Hormuz, which was established by the United States, and includes Saudi Arabia, the Emirates, Bahrain, Britain, Australia and Albania, and aims to enhance the security and safety of ships Commercial passing through the maritime corridors.[23]

The latest data from the Arab Opinion Index on preferred relations with Israel in the Arab World tell us the following[24]: A majority (88%) of Arabs still disapprove of recognition of Israel by their home countries, with only 6% accepting formal diplomatic recognition. When looking at Arab public opinion regarding this issue by country, the highest rate of refusal to recognize Israel was among respondents in Algeria at 99%, followed by those in Lebanon at 94%, and Tunisia and Jordan at 93% each. In the Gulf region; 88% of respondents in Qatar and Kuwait reject the idea of their countries recognizing Israel, while in Saudi Arabia, 65% expressed their rejection as contrasted with 6% who agreed to recognition and 29% refused to express their opinion. Only 13% of the Sudanese agreed that their country should recognize Israel, compared to 79% of them who rejected such a step. When examining the different age groups, there were no differences in refusal to recognize Israel: among all age groups, between 87 and 90% opposed diplomatic recognition of Israel by their country.

But in 2017/2018, only 7% of all Arabs surveyed by the ACRPS in Doha, Qatar, had a positive, and a further 14% a somewhat positive opinion of Iran's policy in the Arab region. 48% of all Arabs had a negative opinion of Iran's policy in the Arab

---

[18] https://besacenter.org/perspectives-papers/israel-gulf-states-relations/.

[19] See Footnote 18.

[20] https://www.dohainstitute.org/en/PoliticalStudies/Pages/Arab-Normalization-with-Israel.aspx.

[21] See Footnote 20.

[22] See Footnote 20.

[23] See Footnote 20.

[24] http://arabcenterdc.org/wp-content/uploads/2020/11/Arab-Opinion-Index-2019-2020-Summary-Final.pdf.

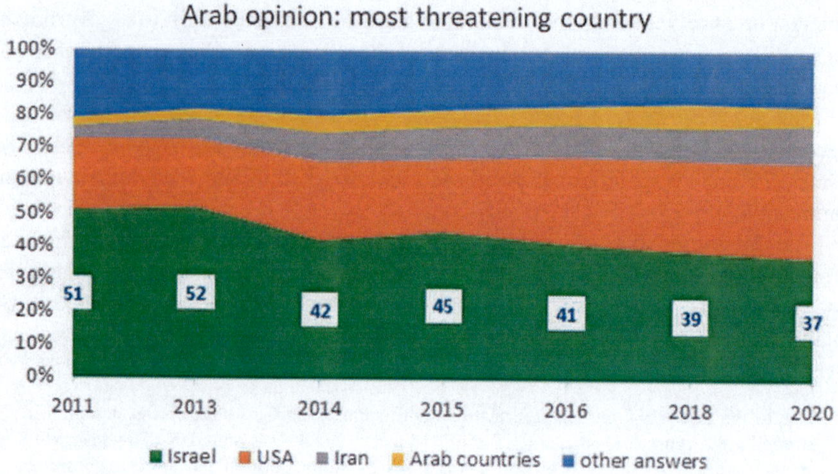


**Fig. 7.13** Arab center for research and policy studies, Doha: Who is the most threatening nation for the Arab countries?

region, and a further 16% held a somewhat negative opinion of Iran's policy in the Arab region (Fig. 7.13).

But with the ongoing value change in the region (see also Chaps. 6 and 9), we believe that the factors which will rise in future, most notably Feminism, will contribute to reducing Antisemitism in the Gulf region as well. Our data clearly supported the view that feminism is one of the most noteworthy blocks against societal Antisemitism, while immigration variables, human rights violations, a high share of military personnel, and military expenditure contribute to prolonging Antisemitism.

The most decisive blocks against Antisemitism appear to be, according to our patrial correlation analysis, gender equality, a secular worldview that rejects a too prominent role of religious authorities interpreting laws, a high overall development level, a high employment especially of elder workers, general patterns of societal tolerance vis-à-vis outgroups, the Effective Democracy Index and other democracy measures, the comparative price level, the ESI Index Component Social and Institutional Capacity and other environment performance data, mean years of schooling and other education data, life satisfaction data, well-functioning libraries with items on Jewish history (the Shoah), social security data, health expenditures, the World Bank Deregulation Index, happy life years, and trust in other people. In our reading of events, all these processes could be compatible with the dynamics, initiated by the ambitious Saudi Vision 2030 plan.

In our multivariate analysis of the ADL 100 data, we could show a simple and parsimonious story: Antisemitism is strongest at middle income levels, and closing the gender gap and creating employment, especially for older workers, reduces Antisemitism, while, *ceteris paribus*—an Orthodox and a Muslim heritage all increase

Antisemitism. Our multiple regressions on a country-to-country level led us to expect that first there was an increase of Antisemitism with rising per capita incomes, but that this will be levelling off at higher income levels. Mean years of education will be an important impediment against Antisemitism, and a climate of societal trust will diminish Antisemitism. Unfortunately, it is evident that with all these other factors being constant, adherence to Catholicism, Orthodox Christianity, and Islam all are still to be considered as significant drivers of the rate of societal Antisemitism, all reflecting the centuries of Antisemitism in the concrete, but different history of the religious environments in these cultures, contradicting the message of tolerance, which is at the roots of the "religions of the book".

In the 28 countries from the *World Values Survey* under comparison here, the following ranking of Antisemitic attitudes is to be observed. In Iraq, Iran, Egypt, and India, more than half of the entire resident population rejected to have a Jewish neighbor, while in Canada, Argentina, Belarus, Uruguay, and the USA, this percentage was below 10%. Some Muslim majority societies, for example, Bangladesh and Albania, did even better than world political and military allies of the West, like South Korea, Slovakia, Japan, and Spain.

Our regression analytical results were showing the importance of the variables about a fundamentalist interpretation of the relationship between religion and the state and the sexist inclination to reserve higher education only for male persons. The model could be calculated for more than 27.000 WVS respondents, all indicators (except the higher education variable) wield a very significant effect on Antisemitism, the entire equation achieves an F-ratio of more than 500 and is thus highly significant, and the explained $R^2$ is 14.8%.

Patriarchy and a fundamentalist reading of all—I repeat—all religious scriptures increase Antisemitism, and interestingly enough, also the quest for socialism does so as well. These results qualify Adorno's findings (2019) in an important way; at the time of writing his analysis, Adorno and associates believed that Antisemitism is a clear consequence of the left to right political spectrum, while today, patriarchy, religious fundamentalism of all sorts, and the quest for socialism and an active state are drivers of Antisemitism. We also highlight the fact that among the so-called structural factors, like age and life satisfaction, there are also noticeable exceptions from what one would have expected—the fact that people satisfied with the financial condition of their households, tend to be more antisemitic than others, and educational level is even positively associated with the level of Antisemitism.

For the Gulf countries and the State of Israel, which now opened diplomatic relations, our analysis might perhaps highlight some of the stumbling blocks to watch out which might lay ahead on the road toward the full realization of the enormous potential of these so vital Abraham accords.

## References


Acemoglu, D., & Robinson, J. A. (2002). The political economy of the Kuznets curve. *Review of Development Economics, 6*(2), 183–203.

Adams, M., Bell, L., & Griffin, P. (1997). *Teaching for diversity and social justice: A sourcebook.* Routledge.

ADL—Anti-Defamation League (2014). ADL 100 Index. http://global100.ADL.org/

Adorno, T., Frenkel-Brenswik, E., Levinson, D. J., & Sanford, R. N. (2019). *The authoritarian personality*. Verso Books.

Alesina, A., Algan, Y., Cahuc, P., & Giuliano, P. (2015). Family values and the regulation of labor. *Journal of the European Economic Association, 13*(4), 599–630.

Alesina, A., & Ferrara, E. L. (2000). *The determinants of trust* (No. w7621). National bureau of economic research.

Alesina, A., & La Ferrara, E. (2002). Who trusts others? *Journal of Public Economics, 85*(2), 207–234.

Alesina, A., & Guiliano, P. (2013). *Culture and institutions.* NBER Working Paper Series, Working Paper 19750, available at http://scholar.harvard.edu/alesina/publicationss

Alexander, S. (2018). *Does social and political instability increase intolerance? Predicting levels of anti-semitism across 101 countries.* Doctoral dissertation, University of Pittsburgh.

Alexander, A. C., Inglehart, R., & Welzel, C. (2012). Measuring effective democracy: A defense. *International Political Science Review, 33*(1), 41–62.

Auestad, L. (2015). *Respect, plurality, and prejudice: A psychoanalytical and philosophical enquiry into the dynamics of social exclusion and discrimination.* Karnac Books.

Barro, R. J. (2003). *Religion adherence data.* Harvard University, Department of Economics: https://scholar.harvard.edu/Barro/publications/religion-adherence-data

Barro, R. J., & Lee, J. W. (1996). International measures of schooling years and schooling quality. *The American Economic Review, 86*(2), 218–223.

Bauer, Y. (1993). Antisemitism as a European and world problem. *Patterns of Prejudice, 27*(1), 15–24.

Berggren, N., & Bjørnskov, C. (2019). Do voters dislike liberalizing reforms? New evidence using data on satisfaction with democracy. *Journal of Institutional Economics, 15*(4), 631–648.

Berggren, N., & Jordahl, H. (2006). Free to trust: Economic freedom and social capital. *Kyklos, 59*(2), 141–169.

Berggren, N., & Nilsson, T. (2013). Does economic freedom foster tolerance? *Kyklos, 66*(2), 177–207.

Berggren, N., & Nilsson, T. (2014). Market institutions bring tolerance, especially where there is social trust. *Applied Economics Letters, 21*(17), 1234–1237.

Berggren, N., & Nilsson, T. (2015). Globalization and the transmission of social values: The case of tolerance. *Journal of Comparative Economics, 43*(2), 371–389.

Berggren, N., & Nilsson, T. (2016). Tolerance in the United States: Does economic freedom transform racial, religious, political and sexual attitudes? *European Journal of Political Economy, 45*(December), 53–70.

Berggren, N., Ljunge, M., & Nilsson, T. (2019). Roots of tolerance among second-generation immigrants. *Journal of Institutional Economics, 15*(6), 999–1016.

Berggren, N., & Nilsson, T. (2020). *Economic freedom and antisemitism.* IFN Working Paper No. 1357. Available at SSRN: https://ssrn.com/abstract=3698723

Beyer, H. (2019). The globalization of resentment: antisemitism in an inter- and transnational context. *Social Science Quarterly, 100*(5), 1503–1522.

von Bieberstein, J. R. (1977). The story of the Jewish-Masonic conspiracy (1776–1945). *Patterns of Prejudice, 11*(6), 1–21.

Bobo, L., & Licari, F. C. (1989). Education and political tolerance: Testing the effects of cognitive sophistication and target group affect. *Public Opinion Quarterly, 53*(3), 285–308.

Brym, R. J., Shaffir, W., & Weinfeld, M. (Eds.). (1993). *The Jews in Canada.* Oxford University Press Canada.

Chzhen, Y., & Bruckauf, Z. (2017). Is university education more important for a boy than for a girl? Social approval of unequal educational opportunity across 21 countries. *UNICEF Innocenti Research Brief, 11.* Available at https://www.unicef-irc.org/publications/pdf/IRB_2017_11.pdf

Cohen, J. (2018a). From antisemitism to philosemitism? Trends in American attitudes toward Jews from 1964 to 2016. *Religions, 9*(4), 107.

Cohen, J. H. (2018b). Left, right, and antisemitism in European Public opinion. *Политикологија Религије, 12*(2), 341–371.

Cole, M. (2017). *Education, equality and human rights: Issues of gender, 'race', sexuality, disability and social class.* Routledge.

Curtis, M. (1997). *Antisemitism: Different perspectives. In sociological forum* (Vol. 12, No. 2, pp. 321–330). Springer Netherlands.

Dinda, S. (2004). Environmental Kuznets curve hypothesis: A survey. *Ecological Economics, 49*(4), 431–455.

Don, Y. (1986). The economic dimensions of antisemitism: Anti-Jewish legislation in Hungary 1938–1944. *East European Quarterly, 20*(4), 447.

Eisenstein, Z. (2014). *Hatreds: Racialized and sexualized conflicts in the 21st century.* Routledge.

Farnen, R. F., Dekker, H., De Landtsheer, C., Sünker, H., & German, D. B. (2005). *Democratization, Europeanization, and globalization trends: Cross-national analysis of authoritarianism, socialisation, communications, youth, and social policy.* Peter Lang.

Fox, J. (2007). Religious discrimination: A world survey. *Journal of International Affairs, 61*, 47–67.

Fox, J. (2011). Building composite measures of religion and state. *Interdisciplinary Journal of Research on Religion, 7*(8), 1–39.

Fox, J. (2013). Religious discrimination against religious minorities in Middle Eastern Muslim states. *Civil Wars, 15*(4), 454–470.

Fox, J. (2014). Is it really God's century? An evaluation of religious support and discrimination from 1990 to 2008. *Politics and Religion, 7*(1), 4–27.

Fox, J. (2015). *Political secularism, religion, and the state: A time Series analysis of worldwide data.* Cambridge University Press.

Fox, J. (2016). *The unfree exercise of religion: A world survey of discrimination against religious minorities.* Cambridge University Press.

Fox, J. (2019). Religious regulation: Discrimination against religious minorities. *Oxford Research Encyclopedia of Politics.* https://doi.org/10.1093/acrefore/9780190228637.013.839

Fox, J., & Flores, D. (2009). Religions, constitutions, and the state: A cross-national study. *The Journal of Politics, 71*(4), 1499–1513.

Fox, J., Finke, R., & Mataic, D. R. (2018). New data and measures on societal discrimination and religious minorities. *Interdisciplinary Journal of Research on Religion, 14*(14), 1–37.

Gendell, M. (2008). Older workers: Increasing their labor force participation and hours of work. *Monthly Labor Review, 131*, 41.

Glock, C. Y., & Stark, R. (1966). *Christian beliefs and anti-Semitism*, Vol. 1. New York: Harper & Row.

Gouda, M., & Gutmann, J. (2019). Islamic constitutions and religious minorities. *Public Choice.* https://doi.org/10.1007/s11127-019-00748-7. Available at https://link.springer.com/article/10.1007/s11127-019-00748-7.

Grinin, L., Korotayev, A., & Tausch, A. (2018). *Islamism, Arab spring, and the future of democracy.* Cham: Springer

Guillaumin, C. (2002). *Racism, sexism, power and ideology.* Routledge.

Hausmann, R., Tyson, L. D. A., & Zahidi, S. (2012). *The global gender gap report.* World Economic Forum.

Inglehart, R. F. (2018). *Cultural evolution: People's motivations are changing, and reshaping the world.* Cambridge University Press.

Inglehart, R. F., Haerpfer, C., Moreno, A., Welzel, C., Kizilova, K., Diez-Medrano, J., Lagos, M., Norris, P., Ponarin, E. & Puranen B. et al. (Eds.). (2014). World values survey: All rounds—Country-pooled datafile version. JD Systems Institute. https://www.worldvaluessurvey.org/WVSDocumentationWVL.jsp

Inglehart, R. F., & Norris, P. (2012). The four horsemen of the apocalypse: Understanding human security. *Scandinavian Political Studies, 35*(1), 71–95.

Jikeli, G. (2015). *European Muslim antisemitism: Why young urban males say they don't like Jews.* Indiana University Press.

Jikeli, G., & Allouche-Benayoun, J. (Eds.). (2012). *Perceptions of the holocaust in Europe and Muslim communities: Sources, comparisons and educational challenge* (Vol. 5). Springer Science, & Business Media.

Kaplan, E. (2006). Anti-Israel sentiment predicts anti-semitism in Europe. *The Journal of Conflict Resolution Journal of the Peace Science Society (international), 50*(4), 548–561.

Kertzer, D. I. (2007). *The Popes against the Jews: The Vatican's role in the rise of modern anti-semitism.* Vintage.

Klein, C., Lühr, M., & Streib, H. (2018). Extant empirical research on religiosity and prejudice. In *Xenosophia and religion. Biographical and statistical paths for a culture of welcome* (pp. 23–84). Springer.

Klein, C., & Streib, H. (2018). Design, methods, and sample characteristics of the Bielefeld study on xenosophia and religion. In *Xenosophia and religion. Biographical and statistical paths for a culture of welcome* (pp. 107–152). Springer.

Kurthen, H., Bergmann, W., & Erb, R. (Eds.). (1997). *Antisemitism and xenophobia in Germany after unification.* Oxford University Press.

Kuznets, S. (1955). Economic growth and income inequality. *American Economic Review, 45*(1), 1–28.

Kuznets, S. (1976). *Modern economic growth: Rate, structure and spread.* Yale University Press.

Legge, J. S., Jr. (1996). An economic theory of antisemitism? Exploring attitudes in the new German state. *Political Research Quarterly, 49*(3), 617–630.

Lutz, H., Phoenix, A., & Yuval-Davis, N. (1995). *Crossfires: Nationalism, racism, and gender in Europe.* Pluto Press.

Maccoby, H. (2006). *Antisemitism and modernity: Innovation and continuity.* Routledge.

Maritain, J. (1939). *Antisemitism.* Geoffrey Bles.

Michael, R. (2008). *A history of Catholic antisemitism: The dark side of the church.* Springer.

Modras, R. (2005). *The Catholic church and antisemitism.* Routledge.

Rosenfeld, A. H. (2013). *Resurgent antisemitism: Global perspectives.* Indiana University Press.

Rossman, V. J. (2002). *Russian intellectual antisemitism in the post-communist era.* University of Nebraska Press.

Smith, J. (Ed.). (1988). *Racism, sexism, and the world-system* (No. 84). Greenwood Press.

Solomon, H., & Tausch, A. (2020). *Islamism, crisis and democratization: Implications of the World Values Survey for the Muslim world [Perspectives on development in the Middle East and North Africa (Mena region)].* Springer.

Solomon, H., & Tausch, A. (2021). *Arab MENA countries: Vulnerabilities and constraints against democracy on the eve of the global Covid-19 crisis [Perspectives on development in the Middle East and North Africa (Mena region)].* Springer.

Tausch, A. (2014). The new global antisemitism: Implications from the recent ADL-100 data. *Middle East Review of International Affairs, 18* (3) (Fall 2014). Available at SSRN: https://ssrn.com/abstract=2549654 or https://doi.org/10.2139/ssrn.2549654

Tausch, A. (2016a, September). Islamism and antisemitism. Preliminary evidence on their relationship from cross-national opinion data. *Social Evolution & History, 15* (2), 50–99 (Uchitel Publishing House).

Tausch, A. (2016, April). The civic culture of the Arab world: A comparative analysis based on *World Values Survey* data. *Middle East Review of International Affairs* (Rubin Center, Research in International Affairs, IDC Herzliya). Available at: SSRN: https://ssrn.com/abstract=2827232 or https://doi.org/10.2139/ssrn.2827232

Tausch, A. (2017). Better look the other way? Islamism and gender issues in the Muslim world. *Telos, 2017*(178), 217–231. https://doi.org/10.3817/0317178217

Tausch, A. (2017). *Occidentalism, terrorism, and the Shari'a state: New multivariate perspectives on Islamism based on international survey data*. Available at SSRN 2731640. Available at SSRN: https://ssrn.com/abstract=2731640 or https://doi.org/10.2139/ssrn.2731640.

Tausch, A. (2019). Migration from the Muslim world to the west: Its most recent trends and effects. *Jewish Political Studies Review, 30* (1–2), 65–225, available at http://jcpa.org/article/migration-from-the-muslim-world-to-the-west-its-most-recent-trends-and-effects/ (with data definitions and sources). Free data download available from https://www.academia.edu/37568941/Migration_from_the_Muslim_World_to_the_West_Its_Most_Recent_Trends_and_Effects

Tausch, A., Heshmati, A., & Karoui, H. (2014a). *The political algebra of global value change. General models and implications for the Muslim world*. Nova Science Publishers.

Tausch, A., Heshmati, A., & Karoui, H. (2014b). *The political algebra of global value change: General models and implications for the Muslim world*. Nova Science Publishers. https://www.scopus.com/inward/record.uri?eid=2-s2.0-84952781328&partnerID=40&md5=7559153e0e895584b810889f47d26721

Tausch, A., & Obirek, S. (2019). *Global Catholicism, tolerance and the open society. An empirical study of the value systems of Roman Catholics*. Springer. https://doi.org/10.1007/978-3-030-23239-9

Tausch, A., & Obirek, S. (2020). The second Vatican council in the hearts and minds of Global Catholics: The open society, catholic antisemitism, and the effects of Nostra Aetate. *Global Catholicism, Tolerance and the Open Society*, pp. 33–98. Springer.

Tibi, B. (1993). Die Verschwörung: das Trauma arabischer Politik. Hoffmann u. Campe.

Tibi, B. (2007). The totalitarianism of jihadist Islamism and its challenge to Europe and to Islam. *Totalitarian Movements and Political Religions*, 8 (1), 35–54.

Tibi, B. (2009). Islamists approach Europe. Turkey's Islamist danger. *Middle East Quarterly, Winter, 2009*, 47–54.

Tibi, B. (2012). *Islamism and Islam*. Yale University Press.

Tibi, B. (2013a). The Islamist venture of the politicization of Islam to an ideology of Islamism: A critique of the dominating narrative in Western Islamic Studies. *Soundings: An Interdisciplinary Journal, 96* (4), 431–449.

Tibi, B. (2013b). *The Sharia State: Arab spring and democratization*. Routledge.

Tibi, B. (2015). *From Sayyid Qutb to Hamas: The Middle East Conflict & the Islamization of Antisemitism*. Online ISSN: 1940-610X ISBN: 978-0-9819058-8-4. Institute for the Study of Global Antisemitism and Policy (ISGAP), 10022, available at http://isgap.org/working-paper-bassam-tibi/

UNDP (2019, current issues). *UNDP Human Development Report*. Available at http://www.hdr.undp.org/

Verkhovsky, A. (2002). The role of the Russian Orthodox Church in nationalist, xenophobic and anti-western tendencies in Russia today: Not nationalism, but fundamentalism. *Religion, State & Society, 30*(4), 333–345.

Verkhovsky, A. (2004). Who is the enemy now? Islamophobia and antisemitism among Russian Orthodox nationalists before and after September 11. *Patterns of Prejudice, 38*(2), 127–143.

Whiting, E. (2005). The labour market participation of older people. *Labour Market Trends, 113*(7), 285–296.

Williams, F. (1989). *Social policy: A critical introduction: Issues of race, gender, and class*. Polity Press.

Wistrich, R. S. (2004). Anti-Zionism and anti-semitism. *Jewish Political Studies Review, 2004*, 27–31.

Wistrich, R. S. (2007). *Anti-semitism and multiculturalism: The uneasy connection*. The Hebrew University of Jerusalem.

Wistrich, R. S. (2010). *A lethal obsession: Anti-semitism from antiquity to the global Jihad*. Random House.

World Bank (2017). World Bank open data. World Bank: data.worldbank.org.

Worrell, M. P. (2019). The disintegration of fordism and the transformation of black anti-semitism in America, 1945–2005. *Fast Capitalism, 4* (1).

Young-Bruehl, E. (1996). *The anatomy of prejudices*. Harvard University Press.

# Chapter 8
# Saudi Arabia—Religion, Gender, and the Desire for Democracy

**Abstract** The social scientific analysis of Saudi Arabia is at a certain crossroad. The UNDP's Human Development Report, (2019) already put its emphasis on inequality and still existing restrictive gender norms as the most decisive issues of global development in the twenty-first century. The official Saudi government plan "Saudi Arabia 2030" focused equally reform and gender. With the availability of World Values Survey data, a more thorough probe into these realities of Saudi society is possible. Even a precursory view of the World Values Survey data from Saudi Arabia reveals highly interesting and contradictory perspectives and conclusions about a political system, which is officially an absolute monarchy governed by a system of Shari'a law. Our data reveal a considerable incipient support for less restrictive gender norms and democracy already being held by Saudi publics. Our analysis—the most comprehensive multivariate analysis of Saudi opinions ever undertaken in the literature—first of all presents an exploratory 54 variable promax factor analysis of the World Values Survey data on Saudi Arabia. We then proceeded to analyze gender norms in Saudi Arabia in a larger context of World Values Survey data from the Muslim world. A fairly good WVS coverage of these issues is available for Algeria, Bangladesh, Egypt, Indonesia, Iran, Jordan, Nigeria, Saudi Arabia, and Turkey. In a final round of analysis, we attempted a second-order factor analysis of the opinions of Saudi publics again based on 54 variables. We especially tested the correlations between these second-order promax factors and also analyzed the factor scores of those 23% of Saudi Arab respondents who are in favor of a radical change of society.This methodology allows us to assess the long-term perspectives of change in Saudi society. Saudi publics in favor of a radical change of society are in the league of the six nations of the world with the most widespread desire for a radical break with the past. Our data thus suggest a considerable potential for a "Saudi Spring" of opposition to the current system. The social values of this opposition are characterized by religious faith among people rejecting polygamy, feminism, and religious tolerance and trust in social movements. By and large, our research confirms the hypotheses that the undeniable economic advancement of women in Saudi society will bring about a change in the official normative system hitherto favoring very restrictive gender norms.



A. Tausch, *The Future of the Gulf Region*, Gulf Studies 2,
https://doi.org/10.1007/978-3-030-78299-3_8

**Keywords** Saudi Arabia · World values survey · Restrictive gender norms · Polygamy · Inequality · Desire for radical social change

## 8.1 Introduction to Saudi Arabia—Religion, Gender and the Desire for Democracy

In a pioneering study on the issues being dealt with in this chapter written by Moaddel, in 2006, the starting point was the fact that fifteen of the nineteen terrorists who attacked the World Trade Center and the Pentagon on 9/11, 2001 in America were Saudi citizens. Moaddel's question is as urgent today as it was in 2006: Which social conditions have produced such violent personalities capable of the mass taking of innocent lives and devastating an entire city, if not a nation? In his empirical study of Saudi values, Moaddel already came to the conclusion that on the other hand a lower percentage of Saudis described themselves as religious than Iranians, Jordanians, Egyptians, or even Americans. A clear majority expressed even favorable attitudes toward democracy.

With the UNDP Human Development Report (2019) and its emphasis on inequality and still existing restrictive gender norms in mind, one cannot escape but to note that even a precursory view of the *World Values Survey* data from Saudi Arabia reveals highly interesting and contradictory perspectives and conclusions about a political system, which is officially still an absolute monarchy governed by a system modeled completely by Shari'a law. While only a handful of Saudi men (1%) and Saudi women (9%) already openly state their disagreement with the proposition that a "*wife must obey*", 20% of Saudi men and 44% of Saudi women already were brave enough to have openly expressed their disagreement with polygamy, which is still a cornerstone of Saudi family law. Equally astonishing, 74.2% of the Saudi public consented with the view that democracy may have problems but it is a better form of government than other systems of government. Asked about the respect for individual human rights nowadays, 72% of the Saudi public now think that there is a lot or at least some respect for human rights nowadays. And while 68.9% of the Saudi public think that only laws of the *Shari'a* should govern the country, an almost equal number—69.8%—also thinks that laws should be based on people's wishes.

Questions from the *World Values Survey* typically revealing value change along gender role dimensions also indicate that Saudi Arabia is not (any more) a monolithic bloc, holding traditional views on marriage and the family. 66.6% of the Saudi public already think that husband and wife should both contribute to the household income, and 25.4% of the Saudi public now disagree or strongly disagree with the view that men make better political leaders than women do; and 38.1% now disagree or strongly disagree with the view that University is more important for a boy than for a girl.

Our re-analysis of all these questions in the context of the debates about the plan "Saudi Arabia 2030" first of all performs a 54 variable promax factor analysis of the *World Values Survey* data on Saudi Arabia. These data are from the dimensions of gender roles and family models, happiness, personal health, life satisfaction, important child qualities, religion, societal norms, socio-economic background variables, the ideal political system, the ideal society, trust, what is important in life for Saudi publics, and finally xenophobia and racism.

We then proceeded to analyze gender norms in Saudi Arabia in a larger context of *World Values Survey* data from the Muslim world. A fairly good WVS coverage of these issues is available from Algeria, Bangladesh, Egypt, Indonesia, Iran, Jordan, Nigeria, Saudi Arabia, and Turkey. Our choice of variables concerned the views of the publics of those countries on confidence in the media, democracy, importance assigned to equal access of women to University education, child qualities, polygamy, income equality, rejecting neighbors, religious life, satisfaction with life, traits in a woman, what is important in life, and must a wife obey?

In a final round of analysis, we attempted a second-order factor analysis of the opinions of Saudi publics. We especially tested the correlations between the second-order promax factors and also analyzed the first-order and second-order factor scores of Saudi Arab respondents who are in favor of a radical change of society.

An astonishing 23% of Saudi publics are now openly in favor of a radical change of society, thus placing the Kingdom in the league of the six nations of the world with a really widespread desire for a radical break with the past. Our data suggest a considerable, if not seismic potential for a "Saudi Spring" of opposition to the current system. The social values of this opposition are characterized by religious faith among people rejecting polygamy, feminism, and religious tolerance and trust in social movements. By and large, our research confirms the hypotheses of those who think that with the undeniable economic advancement of women in Saudi society, a change in the official normative system hitherto favoring very restrictive gender norms is now inevitable.

## 8.2 Background on Religion, Gender and the Desire for Democracy

In Chap. 1 of the present book, we already highlighted the path-breaking character of the *"Abraham accords"* between Israel and the Gulf countries Bahrain and United Arab Emirates and the recent establishment of diplomatic relations between Israel and a number of Arab States for the entire Middle East. In Chap. 6 of the present book, we highlighted values and value change in the Gulf region. In Chap. 7, we could establish the close relationship between Antisemitism and still existing restrictive gender norms. We agree with Beck (2019) that a rapprochement between Saudi Arabia and Israel would be a decisively new element on the political map of the Middle East:

> A strategic alliance between the Gulf States and Israel would be a first major step toward Israel's transformation from being a middle power located in the Middle East to being the leading Middle Eastern regional power, because such an alliance could facilitate the deployment of superior Israeli technology and know-how to the financially highly potent

Gulf States. Thus, Peres' vision of a "New Middle East" could come true in the end [...]. (Beck, , 2019)

But what do we know in the end, in terms of hard facts from political science data, about the sustainability of such a reshaping of the Middle East political maps? And will the "earthquake" of changing gender relations in Saudi Arabia make such a détente in the Middle East more viable? We definitely think yes, and we will produce large-scale evidence on the landscape of social and gender values in Saudi Arabia as one of the most important allies of the West in the Arab World.

The ***Arab Opinion Index*** of the Arab Center for Research and Policy Studies in Doha, **Qatar** (***Arab Opinion Index*** 2017/2018; see Chaps. 2 and 6, above) presented some data which contributed to laying the groundwork for future empirical social science research on Saudi Arabia.[1] The ***Arab Opinion Index*** is currently the largest public opinion survey in the Arab World (see also Chaps. 2 and 6 of this work).

- 19% of Saudis thought that democracy is defined by the preservation of civil and public liberties. 5% of Saudis identify democracy above all with equality and justice for all citizens and 10% with a democratic system of, while also 10% identify democracy above all with safety, security, and stability. 49% of Saudi respondents said they do not know or gave no answer.
- 17% of Saudi respondents strongly agreed that no religious authority is entitled to declare followers of other religions as infidels, and 26% agreed. 23% disagreed, and 11% strongly disagreed, while 23% of all respondents declined an answer.

Studies on civil society in Saudi Arabia and the other Gulf states, based on systematic use of international survey data, and being carried out in Saudi Arabia itself are only at a beginning (Ali & Hasan, 2019). Surveying the existing, predominantly Western literature on the political development of Saudi Arabia, one immediately finds that the systematic analysis of value development of the most populous Gulf State, Saudi Arabia, is hitherto not too frequent. Saudi Arabia, at the time of writing this chapter, just finished its chair of the G-20 and rather seems to present itself at first sight as an "enigma" to international scholarship. An enigma, which it always seemed to be for Western scholarship. From the scattered pieces of evidence, already published in the literature (Abalkhail, 2016; Afiouni, 2014; Ali & Hasan, 2018, 2019; Amoranto et al., 2010; Chapra, 2007; Gengler et al., 2013; Grinin et al., 2018; Heath & Li, 2015; Iyer, 2016; Kostenko et al., 2019; Li et al., 2018; Metcalfe, 2011; Norris & Inglehart, 2002; Rees, 2009; Ruck et al., 2018; Shehadeh & Lai, 2020; Solomon & Tausch, 2020, 2021; Tausch, 2014, 2015, 2016, 2019; Tausch & Moaddel, 2009; Tausch et al., 2014) we still cannot draw too many consistent conclusions on the subject.

In one of the few studies carried out in a truly cross-cultural setting, undertaken by two researchers, one with a Christian, the other two with a Muslim cultural background, Tausch et al. (2014) concluded that Saudi Arabia certainly belongs to the group of countries, experiencing the strongest "*clash of educational levels*"

[1] https://www.dohainstitute.org/en/Pages/index.aspx and http://arabcenterdc.org/wp-content/uploads/2018/07/Arab-Opinion-Index-2017-2018-1.pdf.

between the generations. Concerning feminism, Saudi Arabia, Iran and Bangladesh were real laggards compared to other Muslim and non-Muslim countries.

Now, add to this the potential of instability in the region. Not only since 9/11, with its fifteen of the nineteen terrorists with a Saudi background, such questions loom very large in all Western debates about the country. What if radical Islamists take over control of Saudi Arabia? Support for radical change in Saudi Arabia is relatively high by international standards. According to our current research in preparation of this Chapter, the *World Values Survey* data 1981–2014 demonstrate that more than 20% of the population in Viet Nam, Nigeria, Tanzania, Czech Republic, Azerbaijan, Saudi Arabia (23.40%), Pakistan, Philippines, and Turkey (in descending order) want a radical change of society, while in Moldova, South Africa, Brazil, Armenia, Poland, Kyrgyzstan, India, Mexico, Russia, Macedonia, Slovakia, Dominican Republic, Zimbabwe, Bosnia, South Korea, Bangladesh, Bosnia, Venezuela, Albania, Belarus, Latvia, and Serbia, this percentage is between 20 and 10%, and in Georgia, Lithuania, Jordan, Uganda, Ukraine, Uruguay, Montenegro, Romania, Algeria, Colombia, Slovenia, Argentina, Canada, Peru, Bulgaria, USA, Switzerland, Chile, El Salvador, Australia, Indonesia, China, Egypt, New Zealand, Morocco, Croatia, Spain, Puerto Rico, Sweden, Germany, Japan, Taiwan, Estonia, Norway, Hungary, and Finland, it is below 10%.

Surveying a number of high-profile Western studies, we find on the one hand the very pessimistic and stark assessment by Inglehart (2017) about the *"sexual clash of civilizations"* now separating the Islamic world—and especially the Arab countries, and the West. But a number of other high-profile studies have qualified the relationship between Shari'a law, religiosity in Islam, and the support for democracy (Berger, 2019a, 2019b; Fox et al., 2016; Glas & Spierings, 2019, 2020; Glas et al., 2019; Kilavuz & Sumaktoyo, 2020; Spierings, 2019, 2020). Our analysis is well aware of these reservations.

## 8.3 Research Design in the Context of Earlier Research on the Subject

Current research with a significance for our research question—religion, gender, and the desire for democracy in the eyes of the Saudi publics—has been put forward, among others, by Abbott (2017), Blaydes and Linzer (2008), MacCulloch and Pezzini (2002), Moaddel (2006), Rizzo et al., (2007), Seguino (2007), Yassine-Hamdan and Strate (2020). We will shortly review these important studies before we present our own research design.

In the pioneering study by Moaddel (2006), the image of Saudi Arabia in the US and international media is debated at some length: Most of the media saw Saudi youth brainwashed by the most extremist school in Islam, namely Wahhabism, and it was maintained that educational institutions are promoting a vehement Antisemitism, anti-Western attitudes, and intolerance of other religions. Saudi society was and still is also condemned in the media for having a corrupt and backward political system.

But Moaddel could show that:

> Among the Saudi respondents, despite the rejection of democracy by both the Saudi rulers and clerics, a clear majority expressed favorable attitudes toward democracy, as 58 percent espoused the view that democracy is the best form of government, compared with 23 percent who disagreed. These figures were 42 percent and 19 percent for Iran; 75 percent and 9 percent for Jordan; 91 percent and 2 percent for Egypt; and 85 percent and 12 percent for the United States. (Moaddel, 2006)

MacCulloch and Pezzini (2002) used micro-data based on the surveys of revolutionary preferences of 130,000 people living in 61 nations between 1980 and 1997, based on the mentioned *World Values Survey* data on support for radical societal change. What places Saudi publics among the six countries most inclined for radical change? Controlling for personal characteristics, country and year fixed effects, more freedom and economic growth both reduce revolutionary support. Losing one level of freedom, equivalent to a shift from the US to Turkey, increases support for revolt by 4 percentage points. To reduce support by the same amount requires adding 14 percentage points onto the GDP growth rate. Being Muslim in a free country has no effect on the probability of supporting revolt compared to a non-religious person. However, being Muslim in a country that is not free increases it by 13 percentage points. Being Christian in a free country decreases the chance of supporting revolt by 4 percentage points, compared to a non-religious person, and in a not-free country by 1 percentage point.

Abbott (2017) in her *World Values Survey* analysis found that there is that there is little support among either men or women in the MENA region for gender equality and the empowerment of women. Since her data are from the MENA region only, a generalization including the much higher developed Gulf States and Saudi Arabia in particular is uncertain. But some aspects of this study must nevertheless be considered in our own hypothesis testing. Women, according to the Abbott (2017) study, are much more supportive of gender equality and empowerment of women than men, although even among women support is low. The gap in support between men and women is noticeably larger in Morocco, Jordan and Iraq and lowest in Libya, with Tunisia and Egypt lying between. The more educated, the better off and those living in urban areas are more supportive and those who support all status law being based on Shari'a are less supportive. Age makes no difference, indicating that young people are no more supportive than older ones and confirming that there has been no generational shift to more liberal gender values. The differences between countries are statistically significant, with Iraq being the most supportive, closely followed by Morocco and Tunisia, and Libya the least supportive closely followed by Egypt. Jordan lies between the two groups.

Blaydes and Linzer (2008), using cross-national public opinion data from eighteen countries with significant Muslim populations, applied a latent class model to characterize respondents according to their views on gender norms, political Islam, and personal religiosity. Among women, lack of economic opportunity is a stronger predictor of fundamentalist belief systems than socio-economic class.

Cross-nationally, fundamentalism among women is most prevalent in poor countries and those with a large male–female wage gap.

Rizzo et al., (2007) debated Huntington's "*Clash of Civilization*" thesis using data from the *World Values Survey* fourth wave conducted between 1999 and 2003 to compare Arab Muslim countriès to non-Arab Muslim countries on the issues of gender equality and democratization. All of the Muslim majority countries then available in the WVS were included in this analysis. Rizzo et al. classified countries as Arab if they belong to the League of Arab States and had Arabic as the official language. They included Egypt, Algeria, Morocco, Saudi Arabia and Jordan. The non-Arab Muslim majority countries consisted of Turkey, Iran, Bangladesh, Pakistan, Nigeria and Indonesia. The average number of respondents surveyed in each country was about 1400. Rizzo, et al. vehemently questioned the extent to which a "*clash of civilizations*" between Islam and the West rested on issues of gender equality, and thus diametrically are in opposition to the writings of Richard F. Inglehart and Pippa Norris, quoted at length especially in Chap. 6 of the present book. Rizzo et al. maintain that in non-Arab Muslim societies, there were higher levels of support for women's rights, and those who supported gender equality were much more likely to support democracy. In Arab Muslim societies, the reverse was true. Rizzo et al. think that this manifests a concern that governance based on the *shar'ia* could follow a move to democratic governance. Often, in Arab Muslim societies, advocates for women's rights reluctantly ally themselves with secular yet authoritarian regimes. Doing so provides them protection against citizens who decry Western cultural invasion and favor democracy so that they can stem its tide. Rizzo et al. think that complexities in Arab Muslim majority countries over women's rights and democracy abound and, in fact, may dwarf the importance of divisions between the West and Muslim populations over gender. The clashes between groups within Muslim societies and even within individual nations are significant and not captured by a general theory addressing the place of gender in conflict between Islam and the West. As a political conclusion, Rizzo et al. maintain that gender inequality is a hurdle to overcome in order for democratization to be complete and unbiased. Therefore, Rizzo et al. propose that having the procedures of democracy such as elections, political parties, checks and balances, and other institutional forms is not enough. This can be clearly seen in places like Afghanistan and Iraq at one extreme but also in Egypt, Algeria, Morocco and Jordan.

> Unless citizenship rights, participation and inclusion into these institutions and processes are protected and encouraged by the rule of law then those groups in the Arab World that would, under different circumstances, be supporters of democracy – feminists, ethnic and religious minorities, intellectuals – become the allies of secular autocratic regimes. (Rizzo, et al., 2007)

In the last paper to be discussed here, Yassine-Hamdan and Strate (2020), just as the present book well in the framework of the debates initiated by the UNDP Human Development Report, (2019), empirically analyze the predictive power of such variables as polygamy for still existing restrictive gender norms (see Chap. 1 of this book). On average, gender inequality in Arab countries is much higher than in OECD countries. Gender inequality varies among Arab countries mainly due to differences in per capita incomes linked with economic development. In Arab countries, the sources of gender inequality according to Yassine-Hamdan and Strate (2020) are multiple. The distinctive sources include, especially, traditional patriarchal practices, such as civil law and customs that allow polygamy, and in all but a few Arab countries, very high levels of corruption. Yassine-Hamdan and Strate (2020) think that Arab countries can adopt policies to reduce gender inequality. Most important is the promotion of women's education that leads to women's greater employment and higher per capita income. Also, top-down strategies such as legal reform is very instrumental.

> Nothing diminishes patriarchy and subsequently gender inequality more than women's ability to acquire and control their own resources. (Yassine-Hamdan & Strate, 2020)

During the Arab Spring, many Arab citizens, including many women, went out onto the streets to show their contempt for regimes they viewed to be autocratic, corrupt, and illegitimate. They wanted democracy and a voice. A few Arab rulers were driven from power. Others used force to curb demonstrations, arrest opponents, and stay in power. There has been uneven progress with respect to women's rights. With increasing modernization, however, Arab rulers will need to listen to Arab women holding institutional authority and demanding their human rights and an end to abusive patriarchal practices.

Finally, Seguino (2007) concluded in her study combining *World Values Survey* data from different waves of interviews over time with econometric country level data that the evidence suggests that norms and stereotypes are shifting in a gender-equitable direction on a global scale. This is true for both men and women; although the gap between men's and women's attitudes continues to exist, it has closed on a few issues. Seguino (2007) also emphasizes that even though women internalize gender norms and stereotypes, they appear to do so to a lesser extent than men. The regression results further suggest that women's economic empowerment is one factor in this shift, whether that occurs by example or through persuasion. The results presented by Seguino (2007)also show that women's increased share of employment has an independent effect on norms and stereotypes, even after controlling for region, macro-economic conditions, and the structure of production, although each of these additional variables also explains some of the differences and trends in norms and stereotypes.

Our own research design starts with an exploratory promax factor analysis of the totality of available *World Values Survey* data with a sufficiently satisfactory coverage from Saudi Arabia, suitable for multivariate analysis. These data are from the following dimensions:

- Gender role and family models
- Happiness, personal health, life satisfaction
- Important child qualities
- Religion
- Societal norms

- Socio-economic background
- The ideal political system, the ideal society
- Trust
- What is important in life
- Xenophobia and racism.

The *World Values Survey* items, used in our analysis of Saudi society, are to be seen in our publicly available electronic data appendix.[2] We will analyze all these dimensions by promax factor analysis, allowing us to calculate the relationships (correlations) between these dimensions. We then proceed to analyze gender norms in Saudi Arabia in a larger context of *World Values Survey* data from the Muslim world. A fairly good WVS coverage of these issues is available from Algeria, Bangladesh, Egypt, Indonesia, Iran, Jordan, Nigeria, Saudi Arabia, and Turkey. Our choice of variables concerns the views of the publics of those countries on:

- Confidence in the media of the country (best cross-national coverage in the *World Values Survey*: Television)
- Democracy may have problems but is better
- Importance assigned to equal access of women to university education
- Important child qualities: hard work, tolerance and respect for other people, religious faith, obedience
- Opinions on polygamy
- Positions on income equality
- Rejecting neighbors: People of a different race, Immigrants/foreign workers
- Religious life (How often do you attend religious services? Believe in: God, hell, heaven, how important is God in your life?
- Satisfaction with life
- Traits in a woman (what a woman should be): woman wearing a weil, woman a good wife, woman religious, woman educated
- What is important in life: Family, Friends, Leisure time, Politics, Work, Religion
- Wife must obey?

The research question in light of the literature in this context is which relationships exist then in the Muslim world between.

- Authoritarian values, stressing obedience in education,
- Egalitarianism,
- Feminism,
- Ideas and values about family roles,
- Racism and xenophobia,
- Religious faith,
- Secularism,
- Support or rejection of democracy, and
- Support or rejection of polygamy,

[2] https://www.academia.edu/45321329/Factors_affecting_the_Gulfs_future_Global_cycles_and_value_change_Electronic_Appendix.

- Values in education,
- Views on a socially active or passive life, involving family and friends,
- Work ethics.

We will again use promax factor analysis to find out the relationships between these dimensions, and to test, what place the values of Saudi Arab publics really have among other major Muslim countries. Moaddel (2006), already concluded:

> On polygamy and obedience, the attitudes of the Saudi public were conservative and unfavorable to women. Only 28 percent of Saudis disagreed with polygamy in contrast with 80 percent of Egyptians, 71 percent of Jordanians, and 76 percent of Iranians. This does not mean that a majority of Saudis agreed with polygamy: 45 percent indicated that it was acceptable for a man to have more than one wife, while 26 percent neither agreed nor disagreed. These figures were 11 percent and 11 percent for Iran, 19 percent and 10 percent for Jordan, and 10 percent and 9 percent for Egypt. Likewise, on obedience, fully 81 percent of the Saudi respondents agreed that a wife must always obey her husband, compared with 52 percent of Iranians, 74 percent of Jordanians, and 78 percent of Egyptians. Only 5 percent of Saudi citizens disagreed with this statement, compared with 29 percent of Iranians, 19 percent of Jordanians, and 9 percent of Egyptians. (Moaddel, 2006)

In a final round of analysis, we will then attempt a secondary-order factor analysis of the analysis of Saudi publics based on 54 variables. Second-order factor analyses are a well-established technique, especially in psychology with all the experience that this social scientific discipline has with large-scale questionnaires, while in the literature, dealing with the *World Values Survey*, it is rarely applied (see Marsh & Hocevar, 1985; Rindskopf & Rose, 1988; Milfont & Duckitt, 2004). We especially test the correlations between these second-order promax factors, and also analyze the first-order and second-order factor scores of those Saudi Arab respondents who are in favor of a radical change of society.

## 8.4 Saudi Arabia on the Radar Screens of the *World Values Survey*

Running an exploratory promax factor analysis of the well available items from the *World Values Survey* data on Saudi Arabia revealed a number of destabilizing and stabilizing factors of Saudi opinion.

The resulting factors can be seen as the dimensions, most sharply polarizing public opinion in Saudi Arabia. The "yes" and "no" "electrifies" Saudi publics.

The following potentially **destabilizing dimensions** of Saudi opinion must be highlighted.

- Poverty
- The shadow economy
- Rejecting the religious role model for women
- Female rejection of polygamy
- Rejection of traditional gender and family norms,

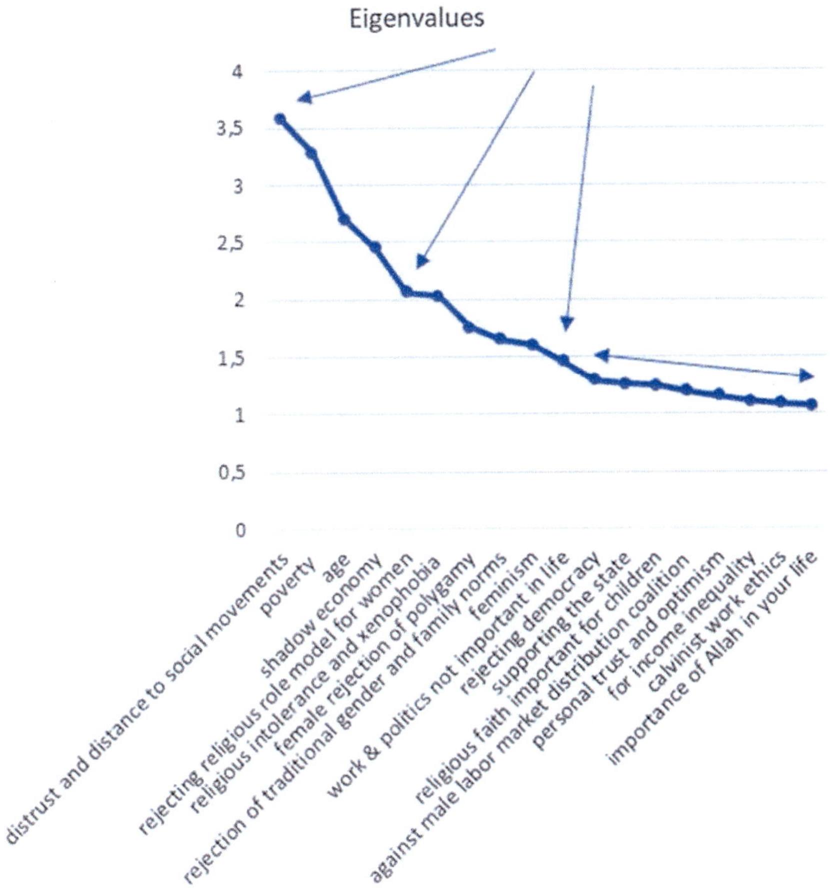


**Fig. 8.1** Eigenvalues for the factor analytical model of Saudi Arabian opinion

- Feminism
- Work and politics not important in life
- Against male labor market distribution coalition.

The factor analytical model, as revealed in our electronic appendix for the specialists,[3] explains more than 59% of the total variance of all the variables. In the following, we will concentrate on the factors whose so-called eigenvalues (Tausch, Heshmati, & Karoui) are exceptionally high (the so-called *scree-test*) (Fig. 8.1):

One by one, these factors, which really divide Saudi opinion, are:

[3] https://www.academia.edu/45321329/Factors_affecting_the_Gulfs_future_Global_cycles_and_value_change_Electronic_Appendix.

- **Factor 1: 58.2% of the Saudi public already have a (high) confidence in the environmental protection movement, and 43.5% have a (high) confidence in the women's movement. Thus, as correctly predicted by Alesina and associates, trustor distrustis a defining dimension of any economy** (Alesina & Ferrara, 2000, 2002; Alesina & Giuliano, 2015; Alesina et al., 2015.

Distrust and distance to social movements have the following factor loadings with more than 10% of variance in common with the following variables:

| | Distrust and distance to social movements |
|---|---|
| No confidence: The Environmental Protection Movement | 0.705 |
| No confidence: Major Companies | 0.696 |
| No confidence: The Women's Movement | 0.689 |
| No confidence: The Press | 0.673 |
| No confidence: Television | 0.667 |
| No confidence: The Civil Services | 0.645 |

In the following, we also list the factor correlations higher or equal to ± 0.100 with the other factors of our model:

| | Distrust and distance to social movements |
|---|---|
| Work and politics not important in life | 0.195 |
| Religious faith important for children | 0.152 |
| Female rejection of polygamy | –0.132 |
| Personal trust and optimism | 0.120 |
| Religious intolerance and xenophobia | –0.101 |

- **Factor 2: The next defining factor of Saudi life is well-being versus poverty. On a scale, ranging from 1 (low satisfaction)—10 (high satisfaction), only 22.7% of Saudi publics are not (very) satisfied with their lives, and only 23.9% of Saudi publics are not (very) satisfied with the financial situation of their household.**

Poverty has the following factor loadings (> 10% of variance in common) with the variables:

| | Poverty |
|---|---|
| Satisfaction with your life | –0.756 |
| Satisfaction with financial situation of household | –0.707 |
| Feeling of unhappiness | 0.646 |

(continued)

(continued)

| | Poverty |
|---|---|
| State of health bad (subjective) | 0.547 |
| Highest educational level attained | −0.388 |

We can observe the following more noticeable correlations with the other factors of our model:

| | Poverty |
|---|---|
| Female rejection of polygamy | 0.150 |
| Religious faith important for children | −0.148 |
| Rejecting religious role model for women | 0.144 |
| Shadow economy | 0.115 |
| Personal trust and optimism | 0.103 |

- **Factor 3: The next factor is the number of children and age**
  The more noticeable factor loadings of this factor are to be observed with the following variables:

| | Age |
|---|---|
| How many children do you have | 0.818 |
| Age | 0.815 |
| Not living with your parents | −0.755 |

This factor has the following noticeable correlations with the other factors of the model:

| | Age |
|---|---|
| Religious intolerance and xenophobia | −0.128 |
| Feminism | 0.123 |
| Supporting the state | −0.106 |

- **Factor 4: The next factor which polarizes Saudi opinion is the shadow economy and corruption.**

55% of the Saudi publics would never claim government benefits without a proper justification. Of the 99 countries and territories with available data in the *World Values Survey* data file "WVS_Longitudinal_1981_2014_spss_v2015_04_18.sav", Saudi Arabia ranks only 55 of the 99 countries and territories. Compared to the global leaders Netherlands, Bangladesh and Turkey, where cheating the state to get illegally government benefits is simply unthinkable, the performance of the Gulf States in the fight against abuse of social benefits is rather poor. Bahrain is rank 32; Qatar is rank 80, and Kuwait is rank 92 and thus among the global laggards Lebanon; Mexico; Albania; Hong Kong; Serbia; Montenegro; Slovakia; and the Czech Republic.

The noticeable factor loadings of this lamentable factor are:

| | Shadow economy |
|---|---|
| Justifiable: claiming government benefits | 0.752 |
| Justifiable: someone accepting a bribe | 0.725 |
| Justifiable: avoiding a fare on public transport | 0.612 |
| Cultural invasion by the west no threat | 0.548 |

All indications are that the shadow economy is also a *"survival strategy"* of the poorer strata in Saudi society. The correlations of this factor with the other factors are:

| | Shadow economy |
|---|---|
| Supporting the state | 0.219 |
| For income inequality | –0.184 |
| Rejecting religious role model for women | 0.180 |
| Rejection of traditional gender and family norms | 0.159 |
| Importance of Allah in your life | 0.139 |
| Poverty | 0.115 |
| Feminism | 0.112 |
| Against male labor market distribution coalition | 0.108 |

- **Factor 5: The next defining axis, which divides Saudi opinion, is consent or rejection of the idea, that an important trait in a female person should be that this person is religious.**

94.4% of the Saudi public think that this is the case. Consent to or rejection of this idea coincides with the importance, assigned to religion in one's life, the importance assigned to a woman's education, the importance assigned to wearing the weil, and the importance of the family in one's life. The "rejectionists" (hardly 5% of the Saudi public), who do not care whether a woman is religious or not, who see no importance of religion for their lives, who do not assign importance to a woman's education or a woman's weil, and who do not think that the family is important in their lives, also do not care about an interesting job. All indications are that this combination of opinions is held among poorer strata of the population, who are (as yet) marginalized by the attempts of reforming female employment and the situation of women in the Kingdom. Typically, in an article published in March, 2020,[4] the semi-official

[4] https://saudigazette.com.sa/article/590572.

English language daily "*Saudi Gazette*" referred to the concept of "*nation building*" developed in political science over the decades especially by Deutsch and Foltz (2010), and boasted:

> The number of employment opportunities for women across industries in Saudi Arabia is rising as the government shifts its focus on an approach that provides more opportunities for women. [...] Recognizing women's role in nation building and accelerating the country's progress in the modern world, Saudi Vision 2030 emphasizes the importance of increased female participation in the workforce.

The social realities of the country seem to imply that there is an *"optimist majority"* which cares whether a woman is religious or not, assigns a greater importance of religion for its lives, assigns importance to a woman's education and a woman's weil, and thinks that the family is important in lives, and also cares about an interesting job.

| | Rejecting religious role model for women |
|---|---|
| Reject opinion: traits in a woman: Woman religious | 0.723 |
| Not important in life: Religion | 0.572 |
| Reject opinion: traits in a woman: Woman educated | 0.540 |
| Reject opinion: traits in a woman: Woman wearing weil | 0.525 |
| Not important in life: Family | 0.455 |
| Important in a job: a job that is interesting | −0.355 |

All indications are that the factor reflects a protest potential combining the rejection a religious role model for women with feminism, and—interestingly and somewhat surprisingly—with the importance of Allah in one's life (on a ten-point scale). Rejecting a (purely) religious role model for women correlates also with the rejection of income inequality and positively correlates with the importance assigned to religion in the education of children. Rejecting a purely religious role model for women correlates positively with the shadow economy and with poverty, and also with supporting the state. Those segments of the Saudi public, rejecting a (purely) religious role model for women, however are not exhibiting religious intolerance and xenophobia, they do not support a Calvinist work ethics, and they also do not share the female rejection of polygamy, now becoming popular nowadays among some educated and high-income strata.

Saudi society, which is becoming very mobile in recent years, witnessed these debates. In the *"Saudi Gazette"*, we read for example on September 10, 2016[5]:

> Many religious scholars still adopt a rigid stance against renewed readings of Islamic texts. They remain in confrontation with modernity. However, in recent years some moderate Muslim scholars have gained popularity among many young Muslims who became disappointed at the hard-line position of ulemas (scholars) failing to provide guidance to the troubled Muslim world. [...] The vast popularity of these figures is an indication of how

[5] https://saudigazette.com.sa/article/163090/A-renewed-reading-of-Islamic-texts.

young Arabs today are looking for moderate religious leaders to help them become better Muslims in a troubled Arab World. Meanwhile, there are also prominent religious scholars who are advocating a moderate interpretation of the Islamic texts.

The factor correlations of rejecting the religious role model for women are:

| | Rejecting the religious role model for women |
|---|---|
| Feminism | 0.316 |
| Importance of Allah in your life | 0.245 |
| For income inequality | −0.218 |
| Religious faith important for children | 0.208 |
| Shadow economy | 0.180 |
| Poverty | 0.144 |
| Religious intolerance and xenophobia | −0.144 |
| Calvinist work ethics | −0.137 |
| Female rejection of polygamy | −0.108 |
| Personal trust and optimism | −0.108 |
| Supporting the state | 0.106 |

The factor correlations between all the factors are revealed in Table 8.1:

## 8.5 Potential Destabilization of the Saudi System?

In Fig. 8.2, we present the *"political landscape"* of the potential for continuity and change in Saudi politics. A high 23.4% of the total Saudi public are in favor of radical change, while 43.1% want gradual improvements by reform, and 33.6% want to valiantly defend the existing system.

Figure 8.3 analyzes which values Saudi parents transmit to the future generations. "Conservative values" such a good manners, religious faith and obedience are highly represented, while responsibility and tolerance and respect for other people are shared by less than 60% of respondents.

On the face of it, comparable data from the *World Values Survey* (rollout 2016) reveal a noticeable degree of satisfaction of Saudi publics with their political system. Rating the quality of the political system for governing the country on a 10-point scale, only 2.4% of the Saudi public scaled the Saudi political system as "very bad", and only 13.9% of the Saudi publics scaled their political system with the bottom values of the scale, 1–4, on the 10-point scale. More than 90,000 representative citizens from across the globe answered this *World Values Survey* item, and Table 8.2 lists the results for Saudi Arabia by international comparison. Provided that these data do not reflect interviewer pressure in semi-totalitarian or totalitarian societies, we would have to say that satisfaction with the political system in Vietnam is almost

**Table 8.1** Factor correlations of opinions of the Saudi public according to our multivariate analysis of *World Values Survey* data

| Correlation between components | Distrust and distance to social movements | Poverty | Age | Shadow economy | Rejecting religious role model for women | Religious intolerance and xenophobia | Female rejection of polygamy | Rejection of traditional gender and family norms | Feminism |
|---|---|---|---|---|---|---|---|---|---|
| Distrust and distance to social movements | 1.000 | 0.030 | 0.074 | 0.043 | 0.049 | −0.101 | −0.132 | 0.039 | 0.021 |
| Poverty | 0.030 | 1.000 | −0.007 | 0.115 | 0.144 | 0.031 | 0.150 | −0.027 | 0.043 |
| Age | 0.074 | −0.007 | 1.000 | −0.079 | 0.024 | −0.128 | −0.081 | −0.035 | 0.123 |
| Shadow economy | 0.043 | 0.115 | −0.079 | 1.000 | 0.180 | 0.062 | 0.013 | 0.159 | 0.112 |
| Rejecting religious role model for women | 0.049 | 0.144 | 0.024 | 0.180 | 1.000 | −0.144 | −0.108 | 0.059 | 0.316 |
| Religious intolerance and xenophobia | −0.101 | 0.031 | −0.128 | 0.062 | −0.144 | 1.000 | 0.100 | −0.014 | −0.087 |
| Female rejection of polygamy | −0.132 | 0.150 | −0.081 | 0.013 | −0.108 | 0.100 | 1.000 | −0.077 | −0.052 |
| Rejection of traditional gender and family norms | 0.039 | −0.027 | −0.035 | 0.159 | 0.059 | −0.014 | −0.077 | 1.000 | −0.005 |
| Feminism | 0.021 | 0.043 | 0.123 | 0.112 | 0.316 | −0.087 | −0.052 | −0.005 | 1.000 |

(continued)

**Table 8.1** (continued)

| Correlation between components | Distrust and distance to social movements | Poverty | Age | Shadow economy | Rejecting religious role model for women | Religious intolerance and xenophobia | Female rejection of polygamy | Rejection of traditional gender and family norms | Feminism |
|---|---|---|---|---|---|---|---|---|---|
| Work and politics not important in life | 0.195 | −0.014 | −0.026 | 0.052 | −0.092 | 0.071 | −0.082 | 0.095 | −0.005 |
| Rejecting democracy | 0.027 | −0.041 | −0.045 | −0.098 | −0.049 | −0.030 | −0.064 | 0.001 | −0.109 |
| Supporting the state | −0.059 | 0.007 | −0.106 | 0.219 | 0.106 | 0.041 | −0.042 | 0.031 | −0.035 |
| Religious faith important for children | 0.152 | −0.148 | 0.068 | 0.046 | 0.208 | −0.101 | −0.316 | 0.022 | −0.058 |
| Rejecting male labor market distribution coalition | −0.069 | 0.092 | −0.073 | 0.108 | 0.075 | 0.053 | 0.003 | 0.054 | −0.196 |
| Trust | 0.120 | 0.103 | 0.060 | 0.039 | −0.108 | 0.175 | 0.197 | −0.064 | −0.069 |
| For income inequality | −0.049 | −0.056 | 0.095 | −0.184 | −0.218 | 0.091 | 0.236 | −0.084 | 0.069 |
| Calvinist work ethics | −0.023 | 0.035 | 0.006 | −0.008 | −0.137 | 0.171 | −0.007 | 0.074 | 0.122 |
| Importance of Allah in your life | 0.008 | 0.025 | 0.061 | 0.139 | 0.245 | −0.027 | −0.150 | −0.128 | 0.180 |

(continued)

**Table 8.1** (continued)

| Correlation between components | Work and politics not important in life | Rejecting democracy | Supporting the state | Religious faith important for children | Rejecting male labor market distribution coalition | Trust | For income inequality | Calvinist work ethics | Importance of Allah in your life |
|---|---|---|---|---|---|---|---|---|---|
| Distrust and distance to social movements | 0.195 | 0.027 | −0.059 | 0.152 | −0.069 | 0.120 | −0.049 | −0.023 | 0.008 |
| Poverty | −0.014 | −0.041 | 0.007 | −0.148 | 0.092 | 0.103 | −0.056 | 0.035 | 0.025 |
| Age | −0.026 | −0.045 | −0.106 | 0.068 | −0.073 | 0.060 | 0.095 | 0.006 | 0.061 |
| Shadow economy | 0.052 | −0.098 | 0.219 | 0.046 | 0.108 | 0.039 | −0.184 | −0.008 | 0.139 |
| Rejecting religious role model for women | −0.092 | −0.049 | 0.106 | 0.208 | 0.075 | −0.108 | −0.218 | −0.137 | 0.245 |
| Religious intolerance and xenophobia | 0.071 | −0.030 | 0.041 | −0.101 | 0.053 | 0.175 | 0.091 | 0.171 | −0.027 |
| Female rejection of polygamy | −0.082 | −0.064 | −0.042 | −0.316 | 0.003 | 0.197 | 0.236 | −0.007 | −0.150 |
| Rejection of traditional gender and family norms | 0.095 | 0.001 | 0.031 | 0.022 | 0.054 | −0.064 | −0.084 | 0.074 | −0.128 |
| Feminism | −0.005 | −0.109 | −0.035 | −0.058 | −0.196 | −0.069 | 0.069 | 0.122 | 0.180 |

(continued)

**Table 8.1** (continued)

| Correlation between components | Work and politics not important in life | Rejecting democracy | Supporting the state | Religious faith important for children | Rejecting male labor market distribution coalition | Trust | For income inequality | Calvinist work ethics | Importance of Allah in your life |
|---|---|---|---|---|---|---|---|---|---|
| Work and politics not important in life | 1.000 | −0.184 | −0.094 | 0.053 | −0.022 | 0.157 | 0.092 | 0.292 | −0.162 |
| Rejecting democracy | −0.184 | 1.000 | 0.104 | 0.165 | 0.100 | −0.160 | −0.179 | −0.121 | 0.026 |
| Supporting the state | −0.094 | 0.104 | 1.000 | 0.138 | 0.214 | 0.018 | −0.223 | −0.233 | 0.345 |
| Religious faith important for children | 0.053 | 0.165 | 0.138 | 1.000 | 0.066 | 0.042 | −0.351 | −0.070 | 0.298 |
| Rejecting male labor market distribution coalition | −0.022 | 0.100 | 0.214 | 0.066 | 1.000 | −0.173 | −0.247 | −0.062 | −0.014 |
| Trust | 0.157 | −0.160 | 0.018 | 0.042 | −0.173 | 1.000 | 0.163 | 0.124 | 0.114 |
| For income inequality | 0.092 | −0.179 | −0.223 | −0.351 | −0.247 | 0.163 | 1.000 | 0.096 | −0.224 |
| Calvinist work ethics | 0.292 | −0.121 | −0.233 | −0.070 | −0.062 | 0.124 | 0.096 | 1.000 | −0.215 |
| Importance of Allah in your life | −0.162 | 0.026 | 0.345 | 0.298 | −0.014 | 0.114 | −0.224 | −0.215 | 1.000 |

**Fig. 8.2** 23.4% of the Saudi public are for radical change

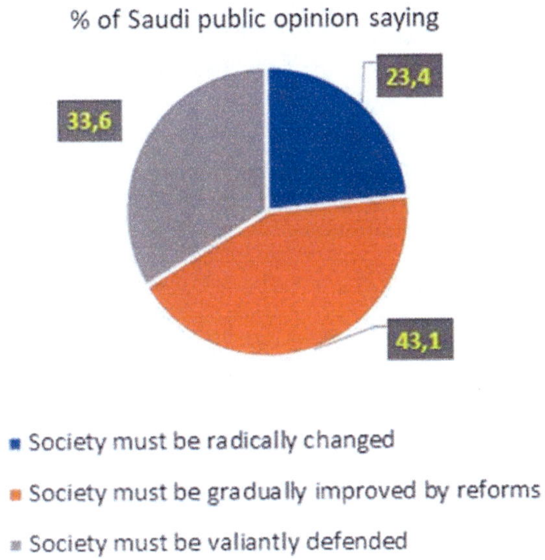


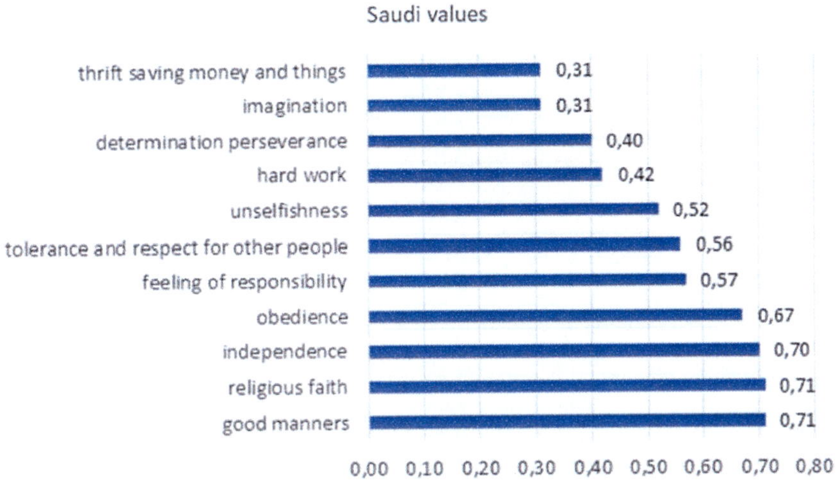


**Fig. 8.3** Saudi values for the education of children

double the size of the satisfaction registered in the USA, and that satisfaction with the political system in Saudi Arabia is higher than in many Western democracies.

And yet, in the Saudi political system, there is an enormous potential for reform and change. If we take trust in the environmental protection movement and in the women's movement as indicators of attitudes how Saudi publics would like their country to adapt to the major challenge s of the twenty-first century, then the data

**Table 8.2** How Saudi publics rate their political system—a 60 nation cross-country comparison

| Country/region | Rate political system for governing country (1, worst to 10, best) | *N* | Standard deviation |
|---|---|---|---|
| Vietnam | 8.74 | 950 | 1.495 |
| Japan | 6.92 | 929 | 1.752 |
| Saudi Arabia | 6.82 | 1443 | 2.295 |
| Azerbaijan | 6.73 | 1786 | 1.371 |
| Tanzania | 6.70 | 1132 | 3.430 |
| Norway | 6.29 | 1114 | 1.698 |
| Jordan | 6.06 | 1119 | 2.277 |
| Iran | 5.84 | 1953 | 2.200 |
| South Africa | 5.72 | 5786 | 2.520 |
| Morocco | 5.64 | 1158 | 2.476 |
| Peru | 5.63 | 2608 | 2.340 |
| Taiwan | 5.52 | 752 | 1.783 |
| Chile | 5.29 | 961 | 1.904 |
| Spain | 5.23 | 2255 | 2.139 |
| Croatia | 5.20 | 1178 | 2.067 |
| Puerto Rico | 5.20 | 1864 | 2.515 |
| Switzerland | 5.20 | 1131 | 1.814 |
| Philippines | 5.18 | 1189 | 2.224 |
| Indonesia | 5.01 | 975 | 2.171 |
| Albania | 4.98 | 1983 | 1.981 |
| Montenegro | 4.95 | 225 | 2.432 |
| Germany | 4.94 | 2000 | 2.007 |
| India | 4.86 | 3381 | 2.270 |
| Bulgaria | 4.82 | 1008 | 2.182 |
| Poland | 4.78 | 1070 | 2.051 |
| United States | 4.78 | 1503 | 2.087 |
| Argentina | 4.76 | 2257 | 2.377 |
| Finland | 4.71 | 930 | 1.856 |
| El Salvador | 4.68 | 1136 | 2.631 |
| Hungary | 4.65 | 616 | 2.010 |
| Estonia | 4.63 | 983 | 1.940 |
| Slovakia | 4.60 | 1058 | 2.170 |
| Slovenia | 4.58 | 964 | 1.881 |
| Czech Rep. | 4.50 | 1114 | 2.025 |
| Czech Rep. | 4.50 | 1114 | 2.025 |
| Sweden | 4.49 | 979 | 1.903 |
| Australia | 4.42 | 2015 | 2.105 |
| Algeria | 4.36 | 1178 | 2.529 |
| Lithuania | 4.30 | 965 | 2.144 |

(continued)

**Table 8.2** (continued)

| | | | |
|---|---|---|---|
| Brazil | 4.24 | 1117 | 2.600 |
| Latvia | 4.12 | 1169 | 1.961 |
| Georgia | 4.08 | 1947 | 2.277 |
| Moldova | 3.94 | 1828 | 2.264 |
| Armenia | 3.90 | 1947 | 2.184 |
| Uruguay | 3.89 | 965 | 2.313 |
| Bosnia Herzegovina | 3.86 | 1192 | 2.203 |
| Pakistan | 3.80 | 1691 | 1.863 |
| Dominican Rep. | 3.79 | 413 | 2.400 |
| Serbia | 3.73 | 1243 | 2.379 |
| Mexico | 3.69 | 1467 | 2.396 |
| Belarus | 3.57 | 1898 | 1.866 |
| Turkey | 3.39 | 5254 | 2.224 |
| New Zealand | 3.37 | 1133 | 1.915 |
| Ukraine | 3.32 | 2470 | 1.992 |
| Macedonia | 3.20 | 2005 | 2.273 |
| Romania | 3.02 | 1221 | 1.885 |
| Nigeria | 2.87 | 1936 | 2.034 |
| Russia | 2.77 | 1958 | 1.763 |
| Venezuela | 2.34 | 1183 | 2.167 |

from the *World Values Survey* hold a lot of surprise in store. A staggering 58.2% of the Saudi public have a great deal or quite a lot of confidence in the environment protection movement, and 43.5% of the Saudi public have a great deal or quite a lot of confidence in the women's movement. In the age of the Internet and the social media, *World Values Survey* also suggest that in their very own interests, the political elites of the country would do well in allowing press freedom and permit not only a feminization but also a "greening" of Saudi politics:

- 8.1% of the Saudi public completely distrust the press, another 28.7% holds not very much confidence in the press
- 6.4% of the Saudi public completely distrust television, another 26.2% has not very much confidence in television.

In Fig. 8.4, we permit our readers a close-up of the drivers of dissatisfaction with the political system, based on the factor loadings of the promax factors with the variable "dissatisfaction with the political system". People decidedly sharing a "Calvinist" work ethics, distrust and distance to social movements, and also the poor segments of society are dissatisfied with the current political system, while the female rejection of polygamy is even a stabilizing factor, indicating that segments of Saudi feminism are grateful to the political system for the relative amount of freedom they enjoy, especially compared to other countries in the region or the "alternative" of harsher Islamist currents taking power in the Kingdom.

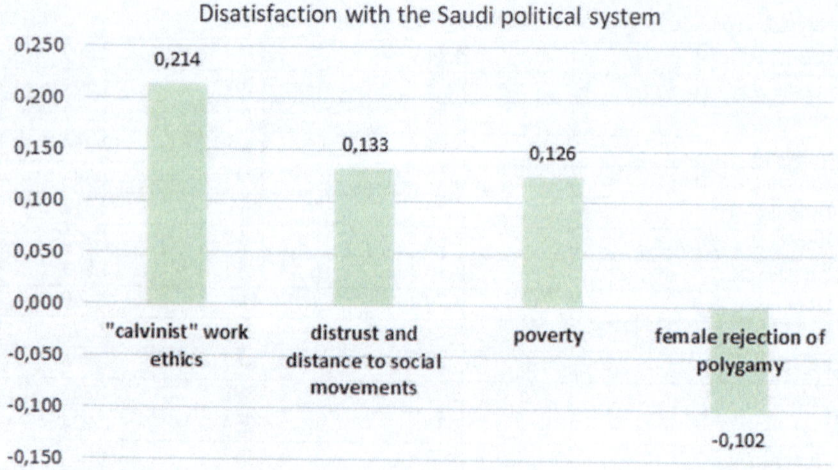


**Fig. 8.4** Drivers of dissatisfaction with the political system

## 8.6 Gender Norms, Inequality, and Civil Society in Saudi Arabia by International Comparison

In the following, we will analyze more closely the relationships between gender norms, inequality, and civil society in the framework of the analytical perspective of the UNDP Human Development Report, (2019) perspective. Dorsey (2021) correctly remarked in an illuminating article published recently that the major battle for Muslim religious soft power that pits Saudi Arabia, Iran, the United Arab Emirates, Qatar, Turkey, and Indonesia against one another is largely about enhancing countries' global and regional influence. Our comparison of societal values in Muslim majority countries comprises the publics in.

- Algeria
- Bangladesh
- Egypt
- Indonesia
- Iran
- Jordan
- Nigeria
- Saudi Arabia
- Turkey.

The values from the *World Values Survey* we analyzed by promax factor analysis were from a very wide range of societal phenomena and are to be seen in our publicly available electronic data appendix.[6] The extracted factors explain 55.95% of total variance, and we name these factors in accordance with the factor loadings, reproduced in our publicly available electronic appendix[7] as:

- Religious faith
- Egalitarianism, rejection of polygamy
- Feminism
- Rejecting traditional family roles,
- Nihilism
- Racism and xenophobia
- Rejection of Calvinist work ethics and secularism
- Obedience
- Skepticism about democracy
- Authoritarianism.

Table 8.3 lists the factor loadings, while Table 8.4 mentions the country factor scores.

The following factors would represent value orientations in the direction of what most comparative value researchers in the world would consider as an outlook, compatible with a free and prosperous society (see also Chap. 6 of this work):

- Egalitarianism, rejection of polygamy
- Feminism
- Rejecting traditional family roles,
- Absence of racism and xenophobia
- Calvinist work ethics and secularism
- Favoring democracy
- Rejection of authoritarianism.

For the general readership of our book, it might suffice to remember that the combined factor scores from the above-mentioned factors suggest that—in descending order—Turkey (most); Algeria; Iran; Jordan; Egypt; Indonesia; Bangladesh; Saudi Arabia; Nigeria (least) combine the seven value outlooks, we consider to be most compatible with the essence of the UNDP's policy recommendations regarding inequality and still existing restrictive gender norms.

In most respects, not only publics in Turkey, but also, in Iran are less inclined to accept the "logic" of still existing restrictive gender norms, analyzed here, while publics in Saudi Arabia are still more strongly favoring the kind of restrictive roles analyzed in the UNDP's Human Development Report (2019) and which also emerge from our factor analysis.

Analyzing the factor scores for the factor *"Egalitarianism, rejection of polygamy"* among Muslim populations in nine Muslim majority countries, it emerges that Turkey

[6] https://www.academia.edu/45321329/Factors_affecting_the_Gulfs_future_Global_cycles_and_value_change_Electronic_Appendix.

[7] https://www.academia.edu/45321329/Factors_affecting_the_Gulfs_future_Global_cycles_and_value_change_Electronic_Appendix.

**Table 8.3** Factor correlations of the nine-country comparative investigation about gender norms

| Correlations of components | Religious faith | Egalitarianism, rejection of polygamy | Feminism | Rejecting traditional family roles | Nihilism | Racism and xenophobia | Rejection of Calvinist work ethics and secularism | Obedience | Skepticism about democracy | Authoritarianism |
|---|---|---|---|---|---|---|---|---|---|---|
| Religious faith | 1.000 | −0.282 | −0.134 | −0.307 | 0.156 | 0.115 | −0.216 | 0.177 | 0.104 | −0.058 |
| Egalitarianism, rejection of polygamy | **−0.282** | 1.000 | 0.207 | 0.162 | −0.044 | −0.129 | −0.048 | −0.038 | 0.013 | 0.145 |
| Feminism | **−0.134** | **0.207** | 1.000 | 0.055 | −0.080 | −0.117 | 0.062 | 0.056 | −0.137 | −0.095 |
| Rejecting traditional family roles | **−0.307** | **0.162** | 0.055 | 1.000 | −0.060 | −0.130 | 0.259 | −0.195 | −0.104 | 0.082 |
| Nihilism | **0.156** | −0.044 | −0.080 | −0.060 | 1.000 | 0.029 | −0.185 | 0.193 | −0.012 | 0.033 |
| Racism and xenophobia | **0.115** | **−0.129** | **−0.117** | **−0.130** | 0.029 | 1.000 | −0.015 | −0.058 | 0.057 | −0.027 |
| Rejection of Calvinist work ethics and secularism | **−0.216** | −0.048 | 0.062 | **0.259** | **−0.185** | −0.015 | 1.000 | −0.235 | −0.083 | −0.086 |
| Obedience | **0.177** | −0.038 | 0.056 | **−0.195** | **0.193** | −0.058 | **−0.235** | 1.000 | −0.085 | −0.199 |
| Skepticism about democracy | **0.104** | 0.013 | **−0.137** | **−0.104** | −0.012 | 0.057 | −0.083 | −0.085 | 1.000 | 0.119 |
| Authoritarianism | −0.058 | **0.145** | −0.095 | 0.082 | 0.033 | −0.027 | −0.086 | **−0.199** | **0.119** | 1.000 |

**Table 8.4** Factor scores: gender norms, politics and religion among Muslim populations in nine countries

| | Religious faith | Egalitarianism, rejection of polygamy | Feminism | Rejecting traditional family roles | Nihilism | Racism and xenophobia | Rejection of Calvinist work ethics and secularism | Obedience | Skepticism about democracy | Authoritarianism | *N* |
|---|---|---|---|---|---|---|---|---|---|---|---|
| Algeria | 0.084 | −0.142 | −0.088 | 0.276 | 0.061 | −0.381 | −0.342 | 0.108 | −0.150 | 0.386 | 892 |
| Bangladesh | −0.017 | −0.549 | −0.064 | 0.071 | −0.235 | 0.907 | 0.159 | −0.842 | −0.237 | −0.196 | 1241 |
| Egypt | 0.193 | −0.660 | 0.066 | −0.168 | 0.368 | 0.129 | 0.055 | 0.312 | −0.528 | −0.229 | 2667 |
| Indonesia | 0.247 | −0.191 | 0.248 | −0.215 | 0.104 | 0.036 | −0.627 | 0.190 | 0.697 | 0.055 | 741 |
| Iran | −0.118 | 0.002 | 0.298 | 0.425 | 0.085 | −0.463 | 0.028 | −0.293 | 0.459 | 0.197 | 1006 |
| Jordan | 0.171 | −0.374 | 0.310 | −0.134 | 0.173 | −0.234 | 0.144 | 0.141 | −0.112 | −0.211 | 940 |
| Nigeria | 0.036 | −0.134 | −1.270 | −0.442 | −0.299 | −0.124 | −0.501 | 0.106 | 1.081 | 0.372 | 578 |
| Saudi Arabia | 0.124 | −0.440 | −0.680 | 0.013 | −0.339 | −0.076 | 0.478 | 0.184 | 0.509 | −0.006 | 1106 |
| Turkey | −0.348 | 1.325 | 0.253 | 0.078 | −0.191 | −0.116 | 0.008 | −0.046 | −0.082 | 0.095 | 2734 |

**Table 8.5** Noticeable factor loadings of egalitarianism and the rejection of polygamy with the variables of Muslim public opinion in nine Muslim majority countries

| | Egalitarianism, rejection of polygamy |
|---|---|
| Not important trait in a woman: Woman wearing weil | 0.756 |
| No confidence: Television | 0.504 |
| Not important trait in a woman: Woman religious | 0.489 |
| Not important in life: Religion | 0.336 |
| Important child qualities: hard work | 0.312 |
| Important child qualities: religious faith | −0.556 |
| we need larger income differences | −0.606 |

is the most progressive Muslim society in the sample, while Egypt is the least progressive Muslim majority society. Table 8.5 with the noticeable factor loadings of egalitarianism and the rejection of polygamy with the variables of Muslim public opinion in nine Muslim majority countries, and finally, Table 8.6, with the loadings of the variables corresponding to the UNDP (2019) gender indicators with the factors of our model for Muslim public opinion in our nine Muslim majority countries should be considered in addition. For the general readership of our book, it might suffice to say here that, in a nutshell, Saudi Arabia would be well advised to head the recommendations of the UNDP's (2019) Human Development Report and that its social structures are still far away from those of less gender restrictive societies in the

**Table 8.6** Loadings of the variables corresponding to the UNDP (2019) gender indicators with the factors of our model for Muslim public opinion in nine Muslim majority countries

| | GNI1 | GNI2 | INEQ |
|---|---|---|---|
| | University equally important for a boy and a girl | Reject opinion: wife must obey | We need larger income differences |
| Religious faith | −0.081 | **−0.158** | **0.129** |
| Feminism | ***0.686*** | ***0.623*** | 0.072 |
| Racism and xenophobia | **−0.126** | **−0.149** | 0.059 |
| Nihilism | 0.051 | **−0.146** | 0,036 |
| Obedience | **0.187** | **−0.155** | 0.034 |
| Authoritarianism | 0.011 | 0.056 | −0.051 |
| Rejecting traditional family roles | −0.050 | **0.260** | −0.059 |
| Rejection of Calvinist work ethics and secularism | −0.080 | **0.194** | −0.098 |
| Skepticism about democracy | **−0.100** | 0.086 | **−0.116** |
| Egalitarianism, rejection of polygamy | **0.212** | **0.143** | ***−0.606*** |

Muslim world. The specialists, consulting our Tables and Images, will note perhaps that the factor *"Egalitarianism, and the rejection of polygamy"*, especially present among less privileged strata of society, especially the working class, is still connected with what Inglehart throughout his vast work has called *"materialism"*, in our case favoring such values as the obedience of children in education Inglehart and Baker (2000). Table 8.6, in methodological terms, is the direct link of our factor analytical results to the categories used in the UNDP's Human Development Report (2019).

In the following, we will analyze that whatever strength restrictive gender norms still might have in Saudi society, social change will contribute to work in the directions of change.

## 8.7 Value Change Will Fundamentally Transform Saudi Society

We start our analysis by zooming in on which factor loadings are to be observed for those forces in Saudi society openly associating themselves in the *World Values Survey* interviews with people favoring democracy and feminism. Table 8.7 mentions these factor loadings. All indications are that both currents are rather presented among the poorer and less privileged strata of the Saudi population. Interestingly enough, pro-democracy currents rather hold the opinion that *"the cultural invasion by the*

**Table 8.7** Factor loadings of the pro-democracy currents and feminist currents in Saudi society

| | Saudi Pro-democracy current (factor loadings) | Saudi feminism (factor loadings) |
|---|---|---|
| State of health bad (subjective) | 0.246 | 0.240 |
| Reject opinion: traits in a woman: Woman wearing weil | 0.123 | 0.271 |
| Feeling of unhappiness | 0.116 | 0.115 |
| Rejecting neighbors: People of a different religion | −0.102 | −0.130 |
| Highest educational level attained | −0.109 | −0.117 |
| Cultural invasion by the west no threat | −0.129 | 0.247 |
| Important child qualities: obedience | −0.143 | −0.248 |
| Reject opinion: wife must obey | −0.151 | 0.353 |
| No confidence: The Press | −0.160 | −0.212 |
| No confidence: Television | −0.173 | −0.127 |
| Rejecting neighbors: Immigrants/foreign workers | −0.205 | −0.174 |

*West*" is a threat, while feminist currents tend to view *the cultural invasion by the West* as not constituting a threat:

Table 8.8 analyzes the factor scores of those Saudi respondents who think that Saudi society must be radically changed. Their attitudes combine the female rejection of polygamy, feminism, poverty and the rejection of a religious role model for women. They will be not supporting the state, they will not think that religious faith is important for children, they will be distant from religious intolerance and xenophobia, and they will be accepting democracy.

In the following, we also attempted a second-order factor analysis of our data on Saudi Arabia. Five second-order factors explain 53,985% of the total variance. These factors are to be interpreted as:

- Egalitarianism
- Calvinist work ethics
- Importance of religious faith among people rejecting polygamy
- Feminism
- Trust and importance of Allah in one's life
- Religious tolerance and trust in social movements.

In the following Tables 8.9 and 8.10, we briefly list the results of this operation.

Figure 8.5 reveals the clash of values among the Saudi generations. Egalitarianism is stronger among the 15–29-year-old and is weakest among the population aged 50 or more years, and religious tolerance and trust in social movements is strongest among the population aged 50 or more and decreases among the younger people.

Figure 8.6 reveals another important contradiction and destabilizing element of Saudi society: the differences in the support for radical change versus the defense of society by marital status of male persons. The system of polygamy, which diminishes the chances poorer young males ever to be married to women of their age group, considerably increases the potential of support for radical social and political change among these male persons. Figure 8.6 reveals that while only 16.40% of married

**Table 8.8** Critical factor scores for Saudi respondents who think that society must be radically changed

| | Society must be radically changed (factor scores) |
|---|---|
| Female rejection of polygamy | 0.289 |
| Feminism | 0.215 |
| Poverty | 0.154 |
| Rejecting religious role model for women | 0.148 |
| Rejecting democracy | −0.101 |
| Religious intolerance and xenophobia | −0.109 |
| Religious faith important for children | −0.115 |
| Supporting the state | −0.167 |

**Table 8.9** Results from the promax second-order factor analysis of the value systems of Saudi publics

| | Egalitarianism | Calvinist work ethics | Importance of religious faith among people rejecting polygamy | Feminism | Trust and importance of Allah in one's life | Religious tolerance and trust in social movements |
|---|---|---|---|---|---|---|
| Distrust and distance to social movements | 0.062 | 0.116 | −0.124 | 0.003 | 0.120 | **0.701** |
| Poverty | 0.307 | −0.023 | **0.507** | 0.220 | 0.147 | 0.342 |
| Age | −0.312 | −0.098 | −0.077 | 0.159 | −0.009 | 0.352 |
| Shadow economy | **0.537** | 0.095 | 0.011 | 0.439 | 0.290 | −0.026 |
| Rejecting religious role model for women | 0.332 | −0.250 | −0.153 | **0.691** | 0.015 | 0.254 |
| Religious intolerance and xenophobia | 0.081 | 0.271 | 0.106 | −0.168 | 0.386 | **−0.514** |
| Female rejection of polygamy | −0.077 | −0.001 | **0.667** | −0.132 | 0.151 | −0.158 |
| Rejection of traditional gender and family norms | 0.279 | 0.356 | −0.025 | 0.066 | −0.243 | 0.061 |
| Feminism | −0.176 | 0.035 | −0.020 | **0.728** | −0.013 | −0.014 |
| Work and politics not important in life | −0.015 | **0.653** | −0.014 | −0.086 | 0.137 | 0.186 |
| Rejecting democracy | 0.170 | −0.354 | −0.259 | −0.301 | −0.197 | 0.076 |
| Supporting the state | **0.543** | −0.401 | −0.293 | 0.188 | 0.414 | −0.225 |
| Religious faith important for children | 0.286 | −0.226 | **−0.727** | 0.104 | 0.227 | 0.275 |
| Rejecting male labor market distribution coalition | **0.596** | −0.102 | 0.020 | −0.141 | −0.107 | 0.019 |
| Trust | −0.095 | 0.148 | 0.125 | −0.027 | **0.725** | 0.070 |
| For income inequality | **−0.630** | 0.270 | 0.444 | −0.151 | −0.003 | −0.193 |
| Calvinist work ethics | −0.184 | **0.684** | 0.130 | −0.031 | −0.019 | −0.120 |
| Importance of Allah in your life | 0.251 | **−0.536** | −0.476 | 0.495 | **0.553** | −0.045 |

**Table 8.10** Second-order factor correlations of the value systems of Saudi publics

| Correlations | Egalitarianism | Calvinist work ethics | Importance of religious faith among people rejecting polygamy | Feminism | Trust and importance of Allah in one's life |
|---|---|---|---|---|---|
| Calvinist work ethics | **−0.209** | | | | |
| Importance of religious faith among people rejecting polygamy | **−0.195** | **0.313** | | | |
| Feminism | **0.209** | **−0.173** | **−0.134** | | |
| Trust and importance of Allah in one's life | **0.215** | **−0.115** | **−0.123** | **0.233** | |
| Religious tolerance and trust in social movements | **0.100** | −0.041 | −0.050 | **0.112** | **−0.109** |

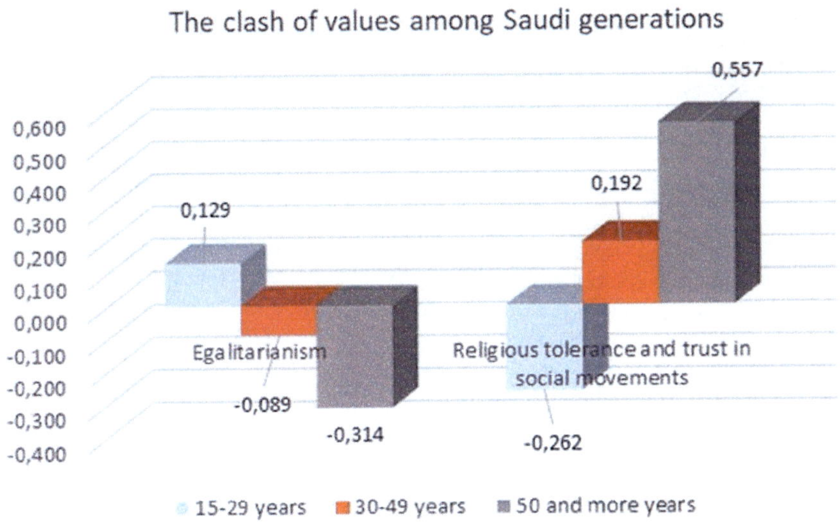


**Fig. 8.5** Results from the second-order factor analysis: the clash of values among the Saudi generations

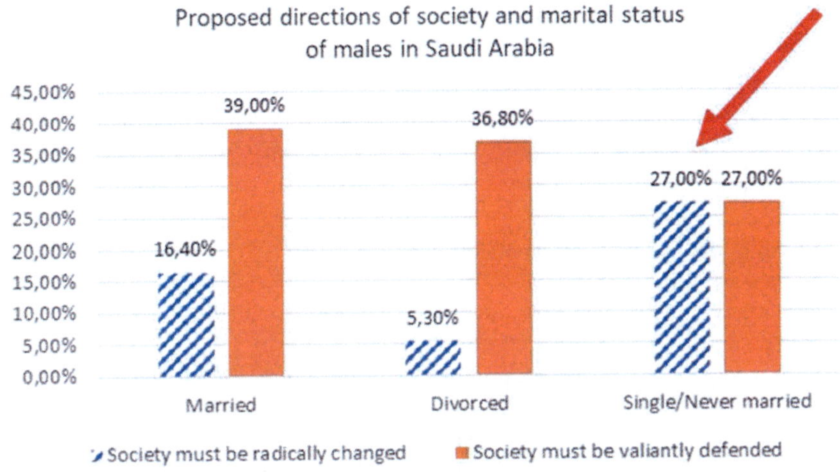


**Fig. 8.6** Proposed directions of society and marital status of males in Saudi Arabia

males favor radical change, 27% of single male persons are of this opinion, incidentally the same percentage as those who think that society must be valiantly defended. Among married men, the percentage of those wanting to valiantly defend society are about double the size (39%) as those wishing a radical change.

Table 8.11 reveals the quantitative size of the single male population stratum, marginalized by the system of polygamy, favoring the chances of the rich to get

**Table 8.11** Percentage of males and females being married in societies with a high incidence of polygamy based on *World Values Survey* data

| | % of males married (%) | % of women married (%) | % difference (%) |
|---|---|---|---|
| Nigeria | 45.80 | 58.00 | 12.20 |
| Pakistan | 64.70 | 75.80 | 11.10 |
| Saudi Arabia | 52.90 | 63.10 | 10.20 |

married. In Nigeria, Pakistan and Saudi Arabia, the percentage of males being married is ten percent or even more considerable.

## 8.8 Conclusions: Gender Roles and Inequality in the Kingdom of Saudi Arabia

Our in-depth multivariate opinion survey analysis of Saudi opinion revealed that the still existing restrictive gender norms dimension will be decisive for the future of the Kingdom. Our analysis shows that the percentage of people openly favoring a radical political break with the past is among the highest in the world (currently 23% in Saudi Arabia), and we analyze from where these frustrations and dissatisfactions stem from: dissatisfactions along gender lines, dissatisfactions among the poorer strata of the population with the system of polygamy, which is mainly supported by the rich, and the incipient but nevertheless substantial quest for democracy.

The following potentially destabilizing dimensions of Saudi opinion were highlighted in our analysis.

- Poverty
- The shadow economy
- Rejecting the religious role model for women
- Female rejection of polygamy
- Rejection of traditional gender and family norms
- Feminism
- Work and politics not important in life
- Against male labor market distribution coalition.

The factor analytical model explained more than 59% of the total variance of all the variables.

All indications are that there is a considerable protest potential combining the rejection a religious role model for women with feminism, and—interestingly and somewhat surprisingly—with the importance of Allah in one's life (on a ten-point scale).

As already stated, a high 23.4% of the total Saudi public are in favor of radical change, while 43.1% want gradual improvements by reform, and 33.6% want to valiantly defend the existing system.

On the face of it, comparable data from the *World Values Survey* (rollout 2016) reveal a noticeable degree of satisfaction of Saudi publics with their political system. Rating the quality of the political system for governing the country on a 10-point scale, only 2.4% of the Saudi public scaled the Saudi political system as "very bad", and only 13.9% of the Saudi publics scaled their political system with the bottom values of the scale, 1–4, on the 10-point scale. Provided that these data do not reflect interviewer pressure in semi-totalitarian or totalitarian societies, we would have to say that satisfaction with the political system in Vietnam is almost double the size of the satisfaction registered in the USA, and that satisfaction with the political system in Saudi Arabia is higher than in many Western democracies.

And yet, in the Saudi political system, there is an enormous potential for reform and change. If we take trust in the environmental protection movement and in the women's movement as indicators of attitudes how Saudi publics would like their country to adapt to the major challenges of the twenty-first century, then the data from the *World Values Survey* hold a lot of surprise in store. A staggering 58.2% of the Saudi public have a great deal or quite a lot of confidence in the environment protection movement, and 43.5% of the Saudi public have a great deal or quite a lot of confidence in the women's movement. In the age of the Internet and the social media, the *World Values Survey* also suggests that in their very own interests, the political elites of the country would do well in allowing press freedom and permit not only a feminization but also a "greening" of Saudi politics. People decidedly sharing a "Calvinist" work ethics, distrust and distance to social movements, and also the poor segments of society are dissatisfied with the current political system, while the female rejection of polygamy is even a stabilizing factor, indicating that segments of Saudi feminism are grateful to the political system for the relative amount of freedom they enjoy, especially compared to other countries in the wider Arab region or the "alternative" of harsher Islamist currents taking power in the Kingdom.

Our comparison of societal values in nine Muslim majority countries suggest that—in descending order—Turkey (most); Algeria; Iran; Jordan; Egypt; Indonesia; Bangladesh; Saudi Arabia; Nigeria (least) combine the seven value outlooks, we consider to be most compatible with the essence of the UNDP's policy recommendations regarding inequality and still existing restrictive gender norms:

- Egalitarianism, rejection of polygamy
- Feminism
- Rejecting traditional family roles
- Absence of racism and xenophobia
- Calvinist work ethics and secularism
- Favoring democracy
- Rejection of authoritarianism

In most respects, not only publics in Turkey, but also, in Iran are less inclined to accept the "logic" of still existing restrictive gender norms, analyzed here, while publics in Saudi Arabia are still more strongly favoring the kind of restrictive roles analyzed in the UNDP's Human Development Report (2019) and which also emerge from our factor analysis.

All indications are that both currents are rather presented among the poorer and less privileged strata of the Saudi population. Interestingly enough, pro-democracy currents rather hold the opinion that "the cultural invasion by the West" is a threat, while feminist currents tend to view the cultural invasion by the West as not constituting a threat. Attitudes of Saudi publics wishing a radical change combine the female rejection of polygamy, feminism, poverty and the rejection of a religious role model for women. They will be not supporting the state, they will not think that religious faith is important for children, they will be distant from religious intolerance and xenophobia, and they will be accepting democracy. We also attempted a second-order factor analysis of our data on Saudi Arabia. Five second-order factors explain 53,985% of the total variance. These factors are to be interpreted as:

- Egalitarianism
- Calvinist work ethics
- Importance of religious faith among people rejecting polygamy
- Feminism
- Trust and importance of Allah in one's life
- Religious tolerance and trust in social movements.

Egalitarianism is stronger among the 15–29-year-old and is weakest among the population aged 50 or more years, and religious tolerance and trust in social movements is strongest among the population aged 50 or more and decreases among the younger people.

There is another important contradiction and destabilizing element of Saudi society: the differences in the support for radical change versus the defense of society by marital status of male persons. The system of polygamy, which diminishes the chances poorer young males ever to be married to women of their age group, considerably increases the potential of support for radical social and political change among these male persons. While only 16.40% of married males favor radical change, 27% of single male persons are of this opinion, incidentally the same percentage as those who think that society must be valiantly defended. Among married men, the percentage of those wanting to valiantly defend society are about double the size (39%) as those wishing a radical change.

## References

Abalkhail, J., et al. (2016). "Wasta" and women's careers in the Arab Gulf States. *Gender in Management an International Journal, 31*(3), 162–180.

Abbott, P. (2017). *Gender equality and MENA women's Empowerment in the aftermath of the 2011 uprisings*. Arab Transformations Working Paper 10, Available at SSRN: https://ssrn.com/abstract=2902890 or https://doi.org/10.2139/ssrn.2902890

Afiouni, F., et al. (2014). HRM in the middle east: Toward a greater understanding. *International Journal of Human Resource Management, 25*(2), 133–143.

Alesina, A., & Ferrara, E. L. (2000). *The determinants of trust* (No. w7621). National bureau of economic research.

Alesina, A., & Giuliano, P. (2015). Culture and institutions. *Journal of Economic Literature, 53*(4), 898–944.

Alesina, A., & La Ferrara, E. (2002). Who trusts others? *Journal of Public Economics, 85*(2), 207–234.

Alesina, A., Algan, Y., Cahuc, P., & Giuliano, P. (2015). Family values and the regulation of labor. *Journal of the European Economic Association, 13*(4), 599–630.

Ali, S. S., & Hasan, H. (2018). Measuring deprivation from Maqasid al-Shariah dimensions in OIC countries: Ranking and policy focus. *Journal of King Abdul-Aziz University: Islamic Economics, 31*(1), 3–24.

Ali, S. S., & Hasan, H. (2019). Does religiosity affect multidimensional poverty? Evidence from *World Values Survey* (2010–2014). *Journal of King Abdul-Aziz University: Islamic Economics, 32* (2), available at https://www.kau.edu.sa/Files/320/Researches/72833_45992.pdf.

Almutairi, S., Heller, M., & Yen, D. (2020). Reclaiming the heterogeneity of the Arab states. *Cross Cultural & Strategic Management*. Available at: https://bura.brunel.ac.uk/bitstream/2438/21344/2/FullText.pdf

Amoranto, G., Chun, N., & Deolalikar, A. B. (2010). Who are the middle class and what values do they hold? Evidence from the *World Values Survey*. Evidence from the *World Values Survey* (October 1, 2010). Asian Development Bank Economics Working Paper Series, (229).

Beck, M. (2019). *Israel and the Arab gulf: An Israeli-Saudi alliance in the making (News Analysis of SDU's Resource Center on Middle East Studies).* University of Southern Denmark.

Berger, L. (2019) Sharia, Islamism and Arab support for democracy. *Democratization, 26* (2), 309–326. ISSN 1351-0347. https://doi.org/10.1080/13510347.2018.1527316; orcid.org/0000-0001-7846-804X

Berger, L. (2019). Democratic values and the Micro-foundations of Arab support for peace with Israel. *Conflict Management and Peace Science, 36*(3), 270–290.

Blaydes, L., & Linzer, D. A. (2008). The political economy of women's support for fundamentalist Islam. *World Politics, 60*(4), 576–609.

Chapra, M. U. (2007). *The Islamic vision of development in the light of Maqasid Al-Shari'ah.* Islamic Research and Training Institute, Islamic Development Bank.

Deutsch, K. W., & Foltz, W. J. (2010). *Nation building in comparative contexts.* Aldine Transaction.

Dorsey, J. (2021). *The battle for the Soul of Islam. The begin Sadat Center for strategic studies.* Mideast Security and Policy Studies No. 186, available at https://besacenter.org/wp-content/uploads/2020/12/186WEB.pdf

Fox, A. M., Abdelkarim Alzwawi, S., & Refki, D. (2016). Islamism, secularism and the woman question in the aftermath of the Arab spring: Evidence from the Arab barometer. *Politics and Governance, 4*(4), 40–57.

Gengler, J., Tessler, M., Al-Emadi, D., & Diop, A. (2013). Civic life and democratic citizenship in Qatar findings from the First Qatar *World Values Survey. Middle East Law and Governance, 5*(3), 258–279.

Glas, S., & Spierings, N. (2019). Support for feminism among highly religious Muslim citizens in the Arab region. *European Journal of Politics and Gender, 2*(2), 283–310.

Glas, S., & Spierings, N. (2020). Connecting contextual and individual drivers of Anti-Americanism in Arab countries. *Political Studies*, 0032321720923261.

Glas, S., Spierings, N., Lubbers, M., & Scheepers, P. (2019). How polities shape support for gender equality and religiosity's impact in Arab countries. *European Sociological Review, 35*(3), 299–315.

Grinin, L., Korotayev, & Tausch, A. (2016). *Economic cycles, crises, and the global periphery.* Springer.

Grinin, L., Korotayev, A. V., & Tausch, A. (2018). Islamism, Arab Spring, and the future of democracy: World system and world values perspectives Ser. Perspectives on development in the Middle East and North Africa (Mena region). Springer.

Heath, A., & Li, Y. (2015). *Review of the relationship between religion and poverty: An analysis for the Joseph Rowntree Foundation* (Centre for Social Investigation Working Paper No. 2015-01). http://csi.nuff.ox.ac.uk/wpcontent/uploads/2015/03/religion-and-povertyworking-paper.pdf

Inglehart, R. F. (2017). Changing values in the Islamic world and the West. *Values, Political Action, and Change in the Middle East and the Arab Spring*, 3–24.

Inglehart, R. F., & Baker, W. E. (2000). Modernization, cultural change, and the persistence of traditional values. *American Sociological Review, 65*(1), 19–51.

Inglehart, R. F., Haerpfer, C., Moreno, A., Welzel, C., Kizilova, K., Diez-Medrano, J., Lagos, M., Norris, P., Ponarin, E., & Puranen, B. et al. (Eds.). (2014). World values survey: Round six—Country-pooled datafile version. JD Systems Institute. https://www.worldvaluessurvey.org/WVSDocumentationWV6.jsp

Inglehart, R. F., Haerpfer, C., Moreno, A., Welzel, C., Kizilova, K., Diez-Medrano, J., Lagos, M., Norris, P., Ponarin, E., Puranen, B. et al. (Eds.). (2014). World values survey: All rounds—Country-pooled datafile version. JD Systems Institute. https://www.worldvaluessurvey.org/WVSDocumentationWVL.jsp

Iyer, S. (2016). The new economics of religion. *Journal of Economic Literature, 54*(2), 395–441.

Kilavuz, M. T., & Sumaktoyo, N. G. (2020). Hopes and disappointments: Regime change and support for democracy after the Arab Uprisings. *Democratization*, 1–20.

Kingdom of Saudi Arabia. (2020). *Vision 2030*. https://www.vision2030.gov.sa/sites/default/files/report/Saudi_Vision2030_EN_2017.pdf

Kostenko, V. V., Ponarin, E. D., Strebkova, O. N. (2019). *Historical legacies and gender attitudes in the Middle East*. Vostok (Oriens) (2), 126–146. https://www.scopus.com/inward/record.uri?eid=2-s2.0-85078602468&doi=10.31857%2fS086919080004472-6&partnerID=40&md5=a02c86affb47be799ebcfb6d7a71b5e2. https://doi.org/10.31857/S086919080004472-6

Li, Q., Wang, B., Deng, H., & Yu, C. (2018). A quantitative analysis of global environmental protection values based on the *World Values Survey* data from 1994 to 2014. *Environmental Monitoring and Assessment, 190*(10), 593.

MacCulloch, R., & Pezzini, S. (2002). The role of freedom, growth and religion in the taste for revolution (September 2002). The Suntory Centre, Suntory and Toyota International Centres for Economics and Related Disciplines, London School of Economics and Political Science, available at http://eprints.lse.ac.uk/6646/1/The_Role_of_Freedom,_Growth_and_Religion_in_the_Taste_for_Revolution.pdf

Marsh, H. W., & Hocevar, D. (1985). Application of confirmatory factor analysis to the study of self-concept: First-and higher order factor models and their invariance across groups. *Psychological Bulletin, 97*(3), 562.

Metcalfe, B. (2011). Women, empowerment and development in Arab Gulf states: A critical appraisal of governance, culture and national human resource development (hrd) frameworks. *Human Resource Development International, 14*(2), 131–148.

Milfont, T. L., & Duckitt, J. (2004). The structure of environmental attitudes: A first-and second-order confirmatory factor analysis. *Journal of Environmental Psychology, 24*(3), 289–303.

Moaddel, M. (2006). The Saudi public speaks: Religion, gender, and politics. *International Journal of Middle East Studies, 38*(1), 79–108.

Norris, P., & Inglehart, R. (2002). Islamic culture and democracy: Testing the 'clash of civilizations' thesis. *Comparative Sociology, 1*(3–4), 235–263.

Rees, T. J. (2009). Is personal insecurity a cause of cross-national differences in the intensity of religious belief? *Journal of Religion and Society, 11*, 1–24.

Rindskopf, D., & Rose, T. (1988). Some theory and applications of confirmatory second-order factor analysis. *Multivariate Behavioral Research, 23*(1), 51–67.

Rizzo, H., Abdel-Latif, A. H., & Meyer, K. (2007). The relationship between gender equality and democracy: A comparison of Arab versus non-Arab Muslim societies. *Sociology, 41*(6), 1151–1170.

Ruck, D. J., Bentley, R. A., & Lawson, D. J. (2018). Religious change preceded economic change in the 20th century. *Science Advances, 4* (7). http://advances.sciencemag.org/content/advances/4/7/eaar8680.full.pdf

Seguino, S. (2007). PlusÇa change? Evidence on global trends in gender norms and stereotypes. *Feminist Economics, 13*(2), 1–28.

Shehadeh, L., Lai, Q. (2020). Political economy of exclusion: State-based incitement of intolerance as a distraction from a downturn in the economy. *Politics and Religion.* https://www.scopus.com/inward/record.uri?eid=2-s2.0-85084813239&doi=10.1017%2fS1755048320000206&partnerID=40&md5=6b1c15925091e6906c033744617285d0. https://doi.org/10.1017/S1755048320000206

Solomon, H., & Tausch, A. (2020). *Islamism, crisis and democratization: Implications of the World Values Survey for the Muslim world [Perspectives on development in the Middle East and North Africa (Mena region)].* Springer.

Solomon, H., & Tausch, A. (2021). Arab MENA countries: Vulnerabilities and constraints against democracy on the eve of the global Covid-19 crisis [Perspectives on development in the Middle East and North Africa (Mena region)]. Springer.

Spierings, N. (2019). The multidimensional impact of Islamic religiosity on ethno-religious social tolerance in the Middle East and North Africa. *Social Forces, 97*(4), 1693–1730.

Spierings, N. (2020). Democratic disillusionment? Desire for democracy after the Arab uprisings. *International Political Science Review, 41*(4), 522–537.

Tausch, A., & Moaddel, M. (2009). What 1.3 billion Muslims really think. An answer to a recent Gallup study, based on the '*World Values Survey*'. Nova Science Publishers.

Tausch, A., Heshmati, A., & Karoui, H. (2014). *The political algebra of global value change. General models and implications for the Muslim world.* Nova Science Publishers.

Tausch, A. (2014). The new global antisemitism: Implications from the recent ADL-100 data. *Middle East Review of International Affairs, 18* (3) (Fall 2014).

Tausch, A. (2015). The fertile grounds for ISIL terrorism. *Telos* (Critical Theory of the Contemporary) (171), Summer, 54–75.

Tausch, A. (2016, September). Islamism and antisemitism. Preliminary evidence on their relationship from cross-national opinion data. *Social Evolution & History, 15* (2), 50–99 (Uchitel Publishing House).

Tausch, A. (2019). Migration from the Muslim world to the West: Its most recent trends and effects. *Jewish Political Studies Review, 30* (1–2), 65–225, available at http://jcpa.org/article/migration-from-the-muslim-world-to-the-west-its-most-recent-trends-and-effects/ (with data definitions and sources). Free data download available from https://www.academia.edu/37568941/Migration_from_the_Muslim_World_to_the_West_Its_Most_Recent_Trends_and_Effects

Tausch, A., & Heshmati, A. (2013). *Globalization, the human condition, and sustainable development in the twenty-first century: Cross-national perspectives and European implications.* Anthem Press.

Tausch, A., & Prager, F. (1993). *Towards a socio-liberal theory of world development.* Palgrave Macmillan/St. Martins Press.

United Nations Development Programme (UNDP) (2019). *Human Development Report.* Oxford and New York: Oxford University Press. Available at http://www.hdr.undp.org/.

World Economic Forum. (2010). *The kingdom of Saudi Arabia and the world: Scenarios to 2025.* http://www3.weforum.org/docs/WEF_Scenario_SoudiaArabiaWorld2025_Report_2010.pdf

Yassine-Hamdan, N., & Strate, J. (2020). Gender inequality in the Arab World: A comparative perspective. *Contemporary Arab Affairs, 13*(3), 25–50.

# Chapter 9
# Executive Summary, Conclusions, and Perspectives

**Abstract** The Abraham Accords between the Gulf states of Bahrain and the United Arab Emirates and the State of Israel are a role model for the peaceful coexistence and cooperation between Israel and its Arab neighbors and for the entire Middle East. The present book analyzes the main factors which will affect the trajectory of the Gulf countries in the future. Based on a rigorous quantitative methodology of the analysis of international relations, using cross-national and time series analyses, including time series reaching very far back in human history, global opinion surveys and macro-quantitative national data, our book identifies inequality and restrictive gender norms as the basic problems facing the region.

Chapter 1 deals with this background, and Chap. 2 explains the methodology. Chapters 3–5 analyze the world public health, economic and political environment which the Gulf will face in these uncertain times.

With COVID-19 disastrously affecting the Free West, global output falling in unprecedented proportions, and inequalities certainly rising as a result of every major pandemic and depression, the world economic outlook also for the Gulf region is rather bleak. Our analysis focuses on the long cycles of global politics and economics, based on time series analyses from 1495 and 1740 onwards, suggesting that tensions between the major powers will increase. All indications are that the world economy faces a Great Depression in the proportions of the crisis of the 1930s which led up to the Second World War.

These negative factors, which the leaderships in the Middle East, also of those moderate Arab countries which are friends of the West will face, are compounded by an acute leadership crisis at a global level. Merkel, Macron, successive US Presidents, let alone Putin or Xi Xinping, no global leadership figure today commands the support of more than 2/3 of the global population, and our analyses, based on PEW and World Values Survey data, also reveals an acute value crisis in many Western democracies, facing social decay and secularization at the same time.

The remaining chapters of the book (Chaps. 6–8) analyze the values held by the publics in the Gulf countries by international comparison.

A UNDP-Human Development-type Index of Civil Society (nonparametric Index) combining



A. Tausch, *The Future of the Gulf Region*, Gulf Studies 2,
https://doi.org/10.1007/978-3-030-78299-3_9

- Important child qualities: tolerance and respect for other people.
- Accepting neighbors: people of a different religion.
- Men do not make better political leaders than women do.
- University is equally important for a boy and a girl.
- Democracy: religious authorities should not interpret the laws.
- Not justifiable: for a man to beat his wife.

places Qatar on rank 50 of 77 countries and Bahrain on rank 76.

A nonparametric, i.e., UNDP Human Development type of Index of Global Tolerance, based on

1. Disagree or strongly disagree: The only acceptable religion is my religion (mean),
2. Agree or strongly agree: All religions should be taught in public schools (mean),
3. Agree or strongly agree: People who belong to different religions are probably just as moral as those who belong to mine (mean),
4. Trust completely or somewhat: People of another religion (mean),
5. Meaning of religion: Do good to other people (%—percentages).

places the Gulf countries as follows among the 59 countries of the world:

Bahrain 34,
Kuwait 50,
Qatar 55.

Combing World Values Survey data in a factor analysis (promax factor analysis), which resulted in the following factors

1. The non-violent and law-abiding society,
2. Democracy movement,
3. Climate of personal non-violence,
4. Trust in institutions,
5. Happiness, good health,
6. No redistributive religious fundamentalism,
7. Accepting the market economy,
8. Feminism,
9. Involvement in politics,
10. Optimism and engagement,
11. No welfare mentality, acceptancy of the Calvinist work ethics.

assigns the (parametric) rank of the Gulf countries among 59 nations of the world is as follows:

Qatar rank number 9,
Kuwait rank number 41,
Bahrain rank number 54.

With the growing "thaw" and "détente" under way between the State of Israel and the countries of the Gulf region, the book also analyzes the drivers of the still existing and lamentably high Antisemitism in the region, based on a multivariate analysis of global and Gulf ADL-100 Index data on Antisemitism (Chap. 7). Developing analyses, already presented in Solomon and Tausch (Islamism, crisis and democratization: implications of the world values survey for the muslim world (Perspectives on development in the Middle East and North Africa (Mena region). Springer, 2020), the present analysis now specifically takes the all-important gender policy variables (again UNDP, United Nations Development Programme (UNDP) (Human development report. Oxford University Press, 2019) on still existing restrictive gender norms) and employment variables properly into consideration. The quantitative, empirical proof of these cross-national aspects of the rates of Antisemitism, based on multiple regression analysis, is one of the methodological innovations of this study. If it is really true that growing gender parity really reduces the propensity of a social system for Antisemitism, we can be at least partially hopeful. Although current data indicate that Antisemitism in the Gulf, as defined by the ADL-100 Index data, is still one of the highest in the entire world, there is some reason for hope that the Gulf will re-discover its own Jewish heritage and that growing female employment and gender parity will contribute to a social climate less favorable for the rise of Antisemitism.

In Chap. 8, the book offers several multivariate analyses of opinions in Saudi Arabia, the biggest country of the region, on religion, gender and the desire for democracy, based on World Values Survey data, underlining the general and also comparative approach of the book. Our debate should be considered as a contribution in the framework of the Saudi Vision 2030, which the present author considers to be a step in the right direction of focusing on a development perspective of human development, social inclusion and gender justice, inherent for many years in the analyses of the United Nations Human Development Reports.

What is the exact relationship between opinions on polygamy, gender inequality, inequality, and trust in the political system of this key Gulf country? Chapter 8 is the first in-depth multivariate opinion survey analysis of Saudi opinion undertaken in the literature. In fact, this analysis is the most comprehensive survey ever undertaken of what Saudi publics think about the world. And, it attempts comparisons of Saudi society with other countries in the region and internationally as well. Without question, overcoming the restrictive gender norms dimension will be decisive for the future of the Kingdom. Our analysis shows that the percentage of people openly favoring a radical political break with the past is among the highest in the world (currently 23% in Saudi Arabia), and we also analyze from where these frustrations and dissatisfactions stem from: again dissatisfactions along gender lines, dissatisfactions among the poorer strata of the population with the system of polygamy, which is still mainly supported by the rich, and the incipient but nevertheless substantial quest for democracy among Saudi publics. Our data about Saudi public support must be considered as one of the main surprises of this research project.

For the Gulf countries and the State of Israel, which now opened diplomatic relations, our analysis might perhaps highlight some of the stumbling blocks to watch out which might lay ahead on the road toward the full realization of the enormous potential of the Abraham accords, for the good of the people of the Middle East and for the good of humanity.

**Keywords** World values survey · Arab barometer · Gulf countries · G-20 · OECD · COVID-19 · Economic cycles · Global leadership · Abraham Accords

## 9.1 Overall Results of Our Study

These days of the global health emergency due to COVID-19 and the profound crisis of democracy in America, which culminated so far in the 01/06 attack on Capitol Hill, present an urgent necessity for social scientists, especially political scientists, to look anew at the countries of the Gulf region. The establishment of diplomatic relations between Bahrain, Morocco, Sudan, and the United Arab Emirates, and the State of Israel are a world historic event which calls for a thorough re-thinking of established social scientific knowledge on the Gulf.

Our world view is the world view, expressed in the yearly report by the Institute for National Security Studies, in Tel Aviv, Israel (INSS, 2020), called the "*Strategic Survey for Israel 2019–2020.*" So, with the INSS, we see the following political forces at work in the Gulf region and the greater Middle East:

- **The radical Shiite axis**: This cluster is led by Iran and includes Bashar al-Assad's Syria, Hezbollah, the Houthis in Yemen, the Shiite militias operating in various arenas throughout the Middle East, and Palestinian Islamic Jihad (despite its Sunni identity). Our estimates of the strength of support for this camp among Arab publics was concluded in our study from the latest version of the *Arab Opinion Index* (2020). Twenty-five percentage of all interviewed inhabitants in thirteen Arab countries and territories have a (very) positive view of Iranian foreign policy. Opposition to this camp among Arab publics today is truly considerable; 58% hold a negative view of Iranian foreign policy, and 12% think that today Iran poses the greatest security threat to the Arab World.
- **The pragmatic Sunni states**: This bloc includes Egypt, Jordan, Morocco, Saudi Arabia, the United Arab Emirates, and other Arab Gulf states. These actors are advancing a pro-Western, anti-Iranian, anti-Islamist, and nationalist vision. In our book, we analyze data from the *Arab Barometer* and suggest that 41% of all Arabs now hold a negative view of the Muslim Brotherhood, which is the strongest and most cohesive force of the Sunni Islamists (see below).

- **The Sunni Islamists**: This group includes supporters of Muslim Brotherhood-style "*Political Islam*": Turkey, and surely Hamas and remnants of the Brotherhood and its derivative movements throughout the region, such as *Ennahda*, the dominant political party in Tunisia, belong to this camp (see also Solomon & Tausch, 2020, 2021). Our data from the *Arab Barometer* suggest that 49% of all Arabs hold positive views of the leadership figure of contemporary Sunni Islamism, Turkish President Recep Tayyip Erdoğan.
- **The jihadists**: This camp includes the Islamic State (ISIS) and al-Qaeda and the terrorist organizations associated with them. According to our calculation from the latest *Arab Opinion Index*, 3% of all Arabs now openly and strongly support IIL (Daesh), and 2% support it to some extent. Support for ISIL (Daesh) declined since 2014, when 4% of all inhabitants of the Arab World had a very positive view of ISIL (Daesh) and a further 7% had a to some extent positive view of ISIL (Daesh).

The recent UNDP (2019) highlighted what in our opinion are the real issues at stake, and which a good part of academic writing on the Middle East tended to overlook in recent years: the dire issues of global values on inequality and gender norms (Tausch, 2019). The UNDP (2019) explicitly applied comparative global value research based on the World Values Survey in an exemplary way to these two most pressing issues of global development: income and wealth inequality and the way people think about these inequalities, and still existing restrictive gender norms, as they change along development and with the advancement of human security. Our book faces up to these challenges, and features on the multivariate relationships between *World Values Survey* data on opinions on inequality and gender norms, and how these opinions and norms interrelate with more general value perspectives including the role democracies should perform.

In Chap. 1 of our study, we present this world view in a larger context. In Chap. 2, we present the rigorous quantitative methodology of our approach to analyze the problems of the region in the global context: cross-national opinion survey research, comparative economics, and time series analysis involving long-term data.

The current COVID-19 pandemic will further shift the weight of world economic leadership away from the old and in many ways decaying West to the benefit of China and India (see Chaps. 3–5 of this book); and the Gulf countries, maybe rather unexpectedly, will be forced to assume a much larger role in global leadership than they did before.

With the World Bank, we expect COVID-19 mutations and delays in vaccine deployment which will disappoint global financial markets and trigger global inflation. Cumulative output losses over 2020–25 could amount to 68% of 2019 output globally (World Bank, 2021). One tool used in Chap. 3, dedicated to the analysis of the global COVID-19 pandemic, was the comparison of the effects of earlier pandemics on the long cycles of economic development over the last two thousand years. Especially, the role of the Spanish Flu, 1918–1920 as a "rare economic disaster" of enormous magnitude has been utterly underestimated in previous research on the Great Depression and the end of democracy in many countries around the world in the 1920s and 1930s.

The world ahead of us can be guessed by the data and perspectives contained in Barro and Ursúa (2011), Grinin et al. (2016) and Solomon and Tausch (2021). A lesson of the 1918–1920 pandemic is that the severity of the then pandemic, as it impacted the different countries of the world, contributes to rising income inequality over time. We also attempted to estimate the effects of the severity of the 1918/20 pandemic on international income convergence/divergence over time, as the limited time series data available allow us to do. The result was that the severity of an epidemic negatively influences international income convergence over time.

The rise of China to a global position of dominance is now irreversible and seems to revert the world to the geographical and economic order which existed before the age of industrialization. Arrighi (1995, 2007), as well as Frank (1998) were already very clear on this, and in fact, also the Maddison database confirms this perspective.

The public health fallout from the current crisis is already beyond our imaginations. Of the 211 countries and territories with complete data, the rates of COVID-19 deaths per 1 million population indicate that the Gulf countries were not at the forefront of the most seriously affected countries up to now. The Gulf countries are ranked the following way on the indicator "COVID-19 deaths per 1 million population":

43 Oman,
52 Bahrain,
53 Kuwait,
55 Saudi Arabia,
75 Qatar,
100 United Arab Emirates.

It clearly emerges that the West is at the receiving end of the current global crisis. So, what COVID-19 does is nothing more and nothing less than undermining the fabric of the Free World, and benefitting totalitarian regimes. COVID-19 deaths per 1 million population show that the Americas, Latin Europe, Sweden and the UK are at the forefront of those countries, suffering most severely from the pandemic. In Chap. 3, we also ask ourselves which countries—given the global infection rates—still handled to protect their populations in the best possible manner from dying from the disease. The Gulf states did remarkably well according to this scale.

Our research also presents calculations about the already observable economic fallout of the COVID-19 pandemic on the basis of the IMF World Economic Outlook data base. Inexorably, the center of gravitation of the world economy is further propelled by COVID-19 toward the two giants which will dominate the twenty-first century, China and India. Unemployment is currently highest in the larger Mediterranean and MENA area, in Brazil and South Africa. The implosion of economic growth, due to COVID-19, is scattered around the following group of countries:

- some, but not all Western democracies,
- most of the Mediterranean and MENA countries,
- Southern Africa,
- some former Communist countries.

To ascertain the multivariate effects on COVID-19 death rates per 1 million population and COVID-19 fatality rates in a partial correlation analysis on a country-to-country basis, we kept the following variables constant:

- Share of Muslims per total population,
- Membership in the Arab League,
- Human Development Index,
- Human Development Index.

On the one hand, it is clear that countries with a high share of Roman Catholics per total population, years of membership in the EU, 2010, a high degree of democracy, a former high standard of living, a well-developed system of tertiary enrollment, and an equally well-developed system of social protection as well as a high Happy Planet Index, HPI were the infection centers in the current pandemic. However, we must also consider that countries with a high social inequality, as well as a high unemployment and a high rate of outward investment of their transnational enterprises were among the centers of the spread of the disease. Now, which conditions contributed to a relatively lower spread of the disease? Trust in institutions seems to be a decisive factor against a rapid spread of the disease. A relatively well-developed health system, measured by the availability of hospital beds and a great number of nurses and midwives per total population, a high share of the poorest in total incomes, and a high gender empowerment index all contributed toward the ability of countries to control the spread of the COVID-19 virus. Contrary to the widely debated opinion that economic globalization is to blame for the spread of the virus, the opposite is true, because globalized systems could react more rapidly to the emergency. Contrary to the opinion voiced by President Trump, debated above ("China virus"), the presence of Chinese immigrants per total immigrants did not create conditions for a rapid spread of the virus, and also the share of Buddhists and also the share of the adherents of (other) Eastern religions per total population was not a "driver" of high infection rates.

Analyzing COVID-19 fatality rates, the results are at times even more surprising. It emerges again as confirmed by the facts that countries with a relatively high standard of living and a developed democracy were the countries not only where there were many infections but also where those infected had a higher death rate.

With the world now facing the starkest economic depression since the Great Depression of the 1930s in the wake of the current COVID-19 crisis (see Chaps. 3 and 4), we further analyze the global world political and economic environment the Gulf will most probably face in the near future. The current depression, like in the 1930s, can it lead to global war? What are the lessons we have to draw from the past long cycles of global politics and economics? Our book presents time series analyses of the long cycles of global politics and economics on a global level in the framework of the debates started by the American political scientist Joshua Goldstein in the late 1980s. Prophetically, Goldstein (1985, 1988) highlighted the strong statistical relationships between global warfare (since 1495) and economic depressions (since 1740, see also Chap. 4 of the present book).

Our work must be regarded as a confirmation of the most dismal world systems literature with the advanced method of Spectral Analysis. Our analysis clearly confirms the existence of the Kuznets cycles (22 years); Barro cycles (36 years); Kondratieff cycles (54 years) and Wallerstein cycles (140 years). Our readers should be reminded of the fact that these results were achieved without any prior transformation of the data and without any omission of time periods of measurement between 1740 and 2020. Also, the war cycle literature, initiated by Goldstein (1985, 1988) is again confirmed, with a shorter 30-year war cycle and a longer 130-year war cycle. Cross-correlation analysis also confirms another hypothesis well compatible with the world system literature of the 1980s and 1990s, mainly a time-lag of about 4 years between major global depressions and global warfare. For the Gulf countries, this chapter implies a grave warning of an impeding world political tsunami.

Our updated data series suggest that the current COVID-19 depression casts indeed a very long shadow over the world and brings along with it the rising risk of global confrontations, fitting in with these global cycle theories [see also the more recent volume by Grinin et al. (2016)]. Our rigorous time series analysis of the available data of major battle fatalities from major power wars (current permanent members of the UN Security Council plus Germany), 1495 to the present, and world industrial production growth, 1740 to the present, based on the sophisticated methods of spectral density analysis and autocorrelation analysis clearly confirms the original assumptions made by Goldstein (1985, 1988): economic cycles of around 50 years duration (Kondratieff cycles) besides shorter and even longer fluctuations, and shorter (30 years) and longer (130 years) war cycles. Our analyses also show that as in the Great Depression, the shorter and the longer waves of economic depressions coincide in a spring tide with the presumable devastating effect of a tsunami. The sophisticated time series analysis of cross-correlations also seems to suggest that there is a lag of only half a decade between the trough of a Great Depression and the onset of major power wars.

With Chinese power now sweeping across the globe, with which visions of global leadership the political and economic elites of the countries of the Gulf will team up? The author analyses the emergent dangerous vacuum of global leadership reflected in current PEW opinion surveys as the second factor shaping the Gulf's future. Do the visions offered by world political leaders such as Emmanuel Macron, Angela Merkel, Vladimir Putin, the President of the USA (in the PEW Spring Survey 2018 Donald Trump, now removed from of office by the US presidential election of November 3, 2020), or Xi Xinping command the sufficient support of a significant portion of humanity, and what are the multivariate relationships of these support rates? Or is there an emergent real vacuum in global leadership? This emerging vacuum in global leadership is the second factor under scrutiny here, and it will substantially affect the long-term options for the Gulf leadership in the twenty-first century. No global leadership—Xi, Trump, Putin, Macron, Merkel,—commanded the sufficient backing of more than 2/3 of the global population, and none seemed to have offered a credible guidance in the current crisis. In the respective chapter of our book (Chap. 5), we pinpoint the underlying trends and contradictions of what the people of our globe really think about these leading global actors and global constellations. For the countries of the Gulf, this global leadership vacuum, exacerbated by an acute value crisis and the loss of trust in the West, analyzed in Chap. 5 of the present book with the latest data, 2017–2020 from the World Values Survey from 79 countries, implies to have second thoughts about "playing the China card," since only the free West guarantees the Gulf states a "constitution of liberty" (Hayek) with the sufficient respect for property values and family values. These values are, following Hayek, at the basis of all successful global religions. It also implies that the Gulf countries themselves must assume a positive and strategic role in such institutions as the United Nations, the IMF and the World Bank, the G-20, and hopefully also the OECD. Our analysis of World Values Survey data, 2017–2017 indicate that large areas of the world today are affected by a combination of

- No confidence in societal institutions,
- Distrust,
- Corruption and shadow economy,
- Marginalization and hopelessness of the poorest strata.

Truly vast areas of Latin America, Latin Europe, Eastern, and Southeastern Europe, the mainly Orthodox regions of the former USSR, and the Arab World are affected by this strong social decay, while only a few fortunate European countries (Scandinavia, Estonia, Belarus, UK, Belgium, Portugal, Macedonia, Montenegro, the Slovak Republic, Turkey, and the mainly German-speaking countries Austria, Germany, and Switzerland) seem to be able to escape from social decay. Inexorably, the center of gravitation not only of the world economy, but also of value development already has shifted toward East and Southeast Asia and the Pacific region, with its dynamism affecting also already the Indian Ocean region and the countries in sub-Saharan Africa. Value decay is highest in Peru, Mexico, Bolivia, Tunisia, Guatemala, Nicaragua, Iraq, Lebanon, Egypt, and Colombia. The countries with the lowest value decay are Denmark, Norway, Tajikistan, China, Netherlands, Vietnam, Bangladesh, Iceland, Burma/Myanmar, and Finland.

But value change in a positive direction, brought about by decades of modernization and development will affect the governance structures of the Gulf states, as new generations, which benefitted from a decent tertiary education at the Gulf's world class universities and social globalization will enter into the arena of public affairs and will have a decisive voice in the future of their countries (see Chaps. 6–8 of this book). These changes will not be free of contradictions, and comparative social science can at least estimate the seismographic fault lines of these processes as they emerge from the empirical analysis of the opinion configurations of the populations of the region.

It is clear that the Gulf monarchies are ruled by extended families that control most of the centers of power in their countries. As long as these families succeeded in cultivating consensus among their different branches, this centralized power has, over the years, contributed to the relative stability of the monarchies (Guzansky & Segal, 2020).

Our book focused on the multivariate analysis of civil society in the countries of Gulf. This is the third factor under consideration here, and we devoted the remaining portion of our book to the analysis of this factor (Chaps. 6–8). These analyses and value rankings apply global comparisons, as data about the values of the citizens of our planet and from the Arab region emerge from the representative opinion surveys of the *World Values Survey* and the *Arab Barometer*.

We looked at the differential development patterns of global society and the Arab countries along rising human development and the advancement of world class tertiary education. Our book, developing approaches already tested in Grinin et al. (2018), Solomon and Tausch (2020, 2021), analyzed not only general value change and value decay, but also such factors as the population's trust in the institutions of the state. And while the publics of the Gulf countries increasingly think that—with all its deficiencies—democracies are the best form of government, and other values typical for countries with high development levels and human security increasingly take hold in the Gulf region as well, there are as yet no viable structures of democracy and constitutional monarchy implemented in the countries of the region. The data of this book could help decision makers and civil societies of the region toward an orderly and smooth transition to such constitutional monarchies.

A UNDP-Human Development-type Index of Civil Society (nonparametric Index) combining.

- Important child qualities: tolerance and respect for other people.
- Accepting neighbors: people of a different religion.
- Men do not make better political leaders than women do.
- University is equally important for a boy and a girl.
- Democracy: religious authorities should not interpret the laws.
- Not justifiable: for a man to beat his wife.

places Qatar on rank 50 of 77 countries, and Bahrain on rank 76.

A nonparametric, i.e., UNDP Human Development type of Index of Global Tolerance, based on

1. Disagree or strongly disagree: The only acceptable religion is my religion (mean).
2. Agree or strongly agree: All religions should be taught in public schools (mean).

3. Agree or strongly agree: People who belong to different religions are probably just as moral as those who belong to mine (mean).
4. Trust completely or somewhat: people of another religion (mean).
5. meaning of religion: do good to other people (%—percentages).

Places the Gulf countries as follows among the 59 countries of the world:

Bahrain 34,
Kuwait 50,
Qatar 55.

Combing World Values Survey data in a factor analysis (promax factor analysis),

1. The non-violent and law-abiding society,
2. Democracy movement,
3. Climate of personal non-violence,
4. Trust in institutions,
5. Happiness, good health,
6. No redistributive religious fundamentalism,
7. Accepting the market economy,
8. Feminism,
9. Involvement in politics
10. Optimism and engagement,
11. No welfare mentality, acceptancy of the Calvinist work ethics.

The (parametric) rank (achieved by multivariate analysis) of the Gulf countries among 59 nations of the world is as follows:

Qatar 9,
Kuwait 41,
Bahrain 54.

Comparative global value research in the tradition of Inglehart and Norris and associates has maintained for a long time that manifestations of xenophobia and racism have a strong tendency to be related to the lack of human security, and that with rising human security, which can be measured among others by the UNDP's Human Development Index (UNDP, 2019), these manifestations will diminish. With a growing "thaw" and "détente" under way between the State of Israel and the countries of the Gulf region, the book also analyses the drivers of the still existing and lamentably high Antisemitism in the region, based on a multivariate analysis of global and Gulf ADL-100 Index data on Antisemitism (Chap. 7). Developing analyses, already presented in Solomon and Tausch (2020), the present analysis now specifically takes the all-important gender policy variables (again UNDP, 2019 on still existing restrictive gender norms) and employment variables properly into consideration. *Ceteris paribus*, what is their effect on the global rates of Antisemitism and on Antisemitism in Gulf? And could we expect that with changing gender policy variables and employment variables Antisemitism in the region will diminish? Although current data indicate that Antisemitism in the Gulf, as defined by the ADL-100 Index data, which are the best source of comparative data on Antisemitism today, is still one of the highest in the entire world, there is some reason for hope that the Gulf will re-discover its own Jewish heritage.

In Chap. 8, the book offers several multivariate analyses of opinions in Saudi Arabia, the biggest country of the region, on religion, gender and the desire for democracy, based on *World Values Survey* data, underlining the general and also comparative approach of the book. Our debate should be considered as a contribution in the framework of the Saudi Vision 2030, which the present author considers to be a step in the right direction of focusing on a development perspective of human development, social inclusion and gender justice, inherent for many years in the analyses of the United Nations Human Development Reports.

What is the exact relationship between opinions on polygamy, gender inequality, inequality, and trust in the political system of this key Gulf country? Our Chap. 8 is the first in-depth multivariate opinion survey analysis of Saudi opinion undertaken in the literature. In fact, this analysis is the most comprehensive survey ever undertaken of what Saudi publics think about the world. And it attempts comparisons of Saudi society with other countries in the region and internationally as well. Without question, the still existing restrictive gender norms dimension will be decisive for the future of the Kingdom. Our analysis shows that the percentage of people openly favoring a radical political break with the past is among the highest in the world (currently 23% in Saudi Arabia), and we analyze from where these frustrations and dissatisfactions stem from: dissatisfactions along gender lines, dissatisfactions among the poorer strata of the population with the system of polygamy, which is mainly supported by the rich, and the incipient but nevertheless substantial quest for democracy.

For the Gulf countries and the State of Israel, which now opened diplomatic relations, our analysis might perhaps highlight some of the stumbling blocks to watch out which might lay ahead on the road toward the full realization of the enormous potential of the Abraham accords.

## 9.2 The Quest for the Constitution of Liberty (Hayek) in the Gulf Region and Peace with Israel

Beyond of what we have said already in this book, it is now time to reflect somewhat on the future. Two leading Israeli specialists in the field of Gulf studies, Guzansky and Lindenstrauss (2020), put the current situation like this:

> The recent renewed signals regarding thawing the tension among the Gulf states, first and foremost between the United Arab Emirates and Saudi Arabia on the one hand and Qatar on the other hand, also raise questions regarding Turkey's regional standing. Compromise between the Gulf states is liable to do away with at least some of Qatar's financial support for Turkey and for the Muslim Brotherhood movement in the region, and may distance Ankara from the Gulf arena. Israel must consider whether it is possible and worthwhile for it to play a mediating role among the Gulf states, in light of its good relations with Qatar, the United Arab Emirates, and Saudi Arabia. At the same time, given its regional aspirations, Ankara may continue posing a threat to these states, as well as to Israel. (Guzansky & Lindenstrauss, 2020)

Available opinion data clearly demonstrate the deep interest which Arab publics nowadays express in democracy, and throughout this book, we have dealt with this dimension in its manifold aspects (Table 9.1).

Our book is a straightforward and clear argument in favor of safeguarding the "*Constitution of Liberty*" and to be at the forefront of such developments, bravely initiated by Bahrain and the United Arab Emirates in establishing diplomatic relations with the state of Israel. Throughout our book, we are critical of "*Political Islam*," which today is centered around Iran and Turkey, Hamas, and the remnants of the Muslim Brotherhood and its derivative movements throughout the region (see also Solomon & Tausch, 2020, 2021).[1]

For Qatar, in many ways the most special of the special cases of the Gulf countries, the choices are on the table. After all, the followers of "*Political Islam*" (see Solomon & Tausch, 2020, 2021) will never condone that Qatar is anyway the home to the US Air Force Al Udeid Air Base, the largest United States Air Base in the region, which includes even the forward headquarters of US Air Forces Central Command (AFCENT) as well as the 379th Air Expeditionary Wing, which supports US operations throughout the US Central Command area of responsibility and supports the Joint Task Force-Horn of Africa.

At the end of the day, we should reiterate what Winter andGuzansky already highlighted recently: that for many years, and particularly over the last decade, the United Arab Emirates (UAE) has striven to disseminate a religious-political doctrine that defines peace as an Islamic value and a fundamental element of national identity. This is the path, which Qatar, with all its potential of the factor social trust which Qatari publics hold in the institutions of the state, its human capital endowment and its world class Universities and world class media as well should follow in the years to come. The United Arab Emirates (UAE) poses this stance as an ideological alternative to the radical concepts of "*Political Islam*," advocated by the Muslim

---

[1] https://www.state.gov/the-abraham-accords/.

**Table 9.1** Arab Barometer: to what extent do you agree or disagree with the following statement?

| | Algeria | Egypt | Iraq | Jordan | Kuwait | Lebanon | Libya | Morocco | West Bank and Gaza | Sudan | Tunisia | Yemen |
|---|---|---|---|---|---|---|---|---|---|---|---|---|
| Strongly agree (%) | 18.9 | 28.0 | 20.8 | 26.2 | 8.7 | 56.7 | 23.3 | 23.4 | 12.8 | 34.9 | 34.4 | 22.1 |
| I somewhat agree (%) | 50.6 | 42.9 | 50.5 | 48.2 | 64.1 | 27.6 | 45.5 | 46.0 | 65.0 | 38.2 | 35.7 | 44.1 |
| I somewhat disagree (%) | 12.7 | 7.6 | 18.8 | 14.7 | 17.3 | 8.5 | 15.6 | 14.4 | 15.6 | 12.3 | 8.3 | 17.0 |
| I strongly disagree (%) | 2.4 | 5.5 | 4.0 | 2.8 | 7.0 | 6.3 | 2.4 | 8.4 | 2.7 | 8.9 | 6.4 | 7.4 |
| Don't know (%) | 15.0 | 15.3 | 4.5 | 7.9 | 2.6 | 0.7 | 11.6 | 6.5 | 3.8 | 4.9 | 14.5 | 8.5 |
| Refuse (%) | 0.5 | 0.7 | 1.3 | 0.2 | 0.3 | 0.3 | 1.5 | 1.3 | 0.1 | 0.8 | 0.6 | 0.9 |
| N | 1220 | 1196 | 1215 | 1795 | 1021 | 1200 | 1247 | 1116 | 1200 | 1200 | 1199 | 1200 |
| Total (%) | 100.0 | 100.0 | 100.0 | 100.0 | 100.0 | 100.0 | 100.0 | 100.0 | 100.0 | 100.0 | 100.0 | 100.0 |

A democratic system may have problems, yet it is better than other systems?

Brotherhood and the Salafi-jihadist forces in the region (see also Solomon & Tausch, 2020, 2021). Such efforts by the UAE help to foster a national ethos in the country that supports the anti-Islamist vision it seeks to promote in the Middle East. These efforts also contribute to the UAE's international image as a stronghold of religious freedom, pluralism, and multiculturalism (Guzansky & Winter, 2020).

A better and more profound understanding of the issues at stake in the debates about the Middle East was also provided by Professor Lars Berger from the German Federal University of Applied Administrative Sciences, Departmental Branch of the Intelligence Services, Federal Intelligence Service Division in Berlin. In a series of high-profile articles published in leading journals of social science, he challenges many of the assumptions of much of the current Western academia. Israeli scholars recently even spoke about a "red-green" alliance between the academic left and Islamist currents. Behr and Berger (2009), in focusing on the challenge of talking about terrorism in the European Union and in the Arab countries came to the conclusion that while there is an obvious divide between the Arab and EU debates regarding the assessment of past and present forms of Western politics, both merge in the statement and concession that past forms of western imperialism and enforced modernization are responsible for contemporary radicalization in the Arab World. The underlying analysis, (Behr & Berger, 2009) maintain, fits well with the EU's focus on multilateral means of conflict resolution and trade promotion as counterterrorism strategies. From the perspective of political reform in the Middle East, the problem according to Behr and Berger (2009) arises that it ends up negotiating the debate into rhetorical territory easily amendable to the interests of the authoritarian Arab governments. While there are certainly external factors at play, both debates manage, or at least play a certain role in, producing and sustaining what Behr and Berger (2009) call "*a red herring*" that legitimizes, or, as in the case of the EU, does at least not strictly delegitimize the emerging Islamist terrorist violence which claims to arise from conflicts such as the one between Israel and its Arab neighbors. Behr and Berger (2009) speak about marginalized liberal and self-critical voices in the Arab World which coincide with cautious assessments from within the EU that are also prepared to focus on totalitarian aspects of radical and violent Islamist movements. Both debates, following (Behr & Berger, 2009) can be characterized by a critical inward focus on their own societies in order to make out causes for terrorism instead of looking for such causes in foreign countries/regions and thereby distracting (intentionally, or not) from problems at home.

> The distinctions between radical, totalitarian aspects of Islamism and moderate, peaceful understandings of Islam are undoubtedly based on appropriate political judgement. However, the EU faces the problem that certain interpretations of Islam have lent themselves to the fortification of authoritarian regimes in the Arab World in exchange for a (relatively) free hand in the excommunication of divergent liberal religious views. This challenge became even more virulent with the post-Madrid and post-London recognition that what may be considered as the appropriate means of 'defending Islam' in some quarters of the Arab World can pose a problem through the conversion and transnational radicalization of migrant communities for the EU itself. (Behr & Berger, 2009)

In this context, Berger (2019), analyzing large-scale survey data from the *Arab Barometer* (see Solomon & Tausch, 2021) came to the conclusion that there is robust evidence for the impact of democratic norms on Arab perceptions of peace with Israel. The finding that support for secularism and gender equality are firmly linked with greater support for peace with Israel among Arab publics corroborates the insistence that the democratic peace thesis is better described as a "*joint freedom proposition*" and the conceptualization of a Democratic Peace as resting on liberalism's respect for the moral equality of individuals. By demonstrating the explanatory power of attitudes on gender equality, Berger maintains, his data analysis offers an example of how the perceived gap between rationalist security studies and "emancipative" gender studies can be overcome:

> This study's results only reinforce [… the] conclusion that those interested in peace would do well to invest more into organizations supporting the equal status of women. These results also further undermine Huntington's […] primordialist view on Islam. Contrary to the Orientalists' essentializing view of Islam as a never-changing, one-dimensional causal factor in the political behavior of Muslims, this study shows that as cultural change in the Muslim world occurs along the path set in Western societies, albeit at a slower pace […] the chances of the acceptance of Israel in the region increase. […] The positive impact of greater socio-economic security as well as support for political gender equality and the separation of religion and politics would suggest that further socio-economic and socio-cultural development across the Arab World will strengthen public support for peace with Israel. (Berger, 2019)

Berger also argues as follows:

> This study shows that, if the United States and partners like the EU can bring Israelis and Palestinians to agree to a compromise whose main contours have been very clear for more than a decade now […] then the parties involved can count on the existence of a clearly identifiable set of Arab men and women whose economic considerations as well as normative political preferences lead them to support such a compromise. Contrary to the claims of Israeli skeptics, […] the social-cultural transformation of the Arab World toward the greater acceptance of a reduced role of religion in politics as well as greater support for gender equality is likely to increase the chances of a 'warm peace' to emerge between Israel and its Arab neighbors. The expectation of early liberal Zionists that modernization and Westernization of the Arab World would make the region more likely to accept Israel's existence […] therefore does find empirical support. (Berger, 2019)

Our study shares with Berger the hope that humanity is adopting more peaceful, feminine, and humanistic orientations as opportunities arise that make these orientations more useful in mastering life, providing an increasingly solid mass basis for interstate peace. Berger is correct in maintaining in this context that while approval of democracy in itself is not linked to greater support for peace, it would still be short-sighted to treat this finding as an argument in favor of authoritarian stability. This is because robust democratic structures would allow rationalist and normative dimensions of Arab support for peace with Israel to influence government policies:

> If Arab governments would truly listen to their populations, then the economic considerations of Arab men and women as well as the normative preferences in favor of a reduced role of religion in politics and in support of gender equality would make a considerable number of them look very favorably at post-conflict normalization between Israel and its Arab neighbors. (Berger, 2019)

## 9.3 Measuring Some Final Critical Parameters in the Gulf Region and the Societal Support for Radical Change in the Region

Data-driven as our analysis was projected, we would like to visit finally two analytical tables which reflect the entire ambiguity of the situation, in which we are currently in in the Gulf and in the world at large. The specialists will find more analytical tables on a country-to-country basis in freely available electronic statistical appendix.[2] This appendix offers to our global readership country files for each country of the Gulf, with econometric time series analyses, and global economic, social, and value development performance scales.

To make these long stories short, we can state here the following: the most positive performance of the Gulf region, on average, was to be found in such areas as

- Gross national income (GNI) per capita,
- Involvement in politics,
- Social globalization, de facto index,
- Trust in institutions,
- Overcoming the fatigue of democracy 2 (it is good to have a democratic political system),
- Economic Globalization, de facto index,
- Human Development Index (HDI),
- Happiness, good health,
- Economic Globalization, overall index.

where the Gulf countries were among the top global 25%, while undoubtedly, still existing restrictive gender norms and inequality were the most problematic aspects of the Gulf development model (Table 9.2).

Table 9.3 summarizes the *World Values Survey*-based evidence about support for radical societal change.

With this, we have ended our journey to the Gulf.

[2] https://www.academia.edu/45321329/Factors_affecting_the_Gulfs_future_Global_cycles_and_value_change_Electronic_Appendix.

**Table 9.2** Gulf countries—percentile development performances measured on a global scale

| Indicator direction | Indicator | Percentile performance Bahrain | Percentile performance Kuwait | Percentile performance Oman | Percentile performance Qatar | Percentile performance Saudi Arabia | Percentile performance United Arab Emirates | Average Gulf countries |
|---|---|---|---|---|---|---|---|---|
| Positive | Gross national income (GNI) per capita | 14.4 | 2.7 | 15.4 | 0.5 | 7.4 | 3.7 | 7.4 |
| Positive | Involvement in politics | 5.1 | 8.5 | xx | 11.9 | xx | xx | 8.5 |
| Positive | Social Globalization, de facto index | 14.6 | 13.6 | 25.1 | 8.0 | 15.1 | 3.0 | 13.2 |
| Positive | Trust in institutions | 10.2 | 37.3 | xx | 3.4 | xx | xx | 16.9 |
| Positive | Overcoming the fatigue of democracy 2 (it is good to have a democratic political system) | 7.3 | 36.5 | xx | xx | xx | xx | 21.9 |
| Positive | Economic Globalization, de facto index | 5.3 | 7.9 | 52.6 | 13.7 | 53.7 | 2.6 | 22.6 |
| Positive | Human development index (HDI) | 23.9 | 30.3 | 25.0 | 21.8 | 19.1 | 18.6 | 23.1 |
| Positive | HDI 2018 | 23.9 | 30.3 | 25.0 | 21.8 | 19.1 | 18.6 | 23.1 |
| Positive | Happiness, good health | 62.7 | 8.5 | xx | 1.7 | xx | xx | 24.3 |
| Positive | Economic Globalization, overall index | 10.1 | 21.8 | 38.8 | 18.1 | 52.7 | 4.8 | 24.4 |
| Positive | DYN GEN Happiness, good health | 10.2 | 49.2 | xx | 18.6 | xx | xx | 26.0 |
| Positive | Life expectancy at birth | 28.2 | 40.4 | 26.1 | 17.0 | 42.6 | 25.0 | 29.9 |
| Positive | DYN GEN Accepting the market economy | 15.3 | 72.9 | xx | 1.7 | xx | xx | 29.9 |
| Positive | Economic Globalization, de jure index | 24.0 | 42.5 | 30.2 | 25.7 | 48.6 | 11.2 | 30.4 |
| Positive | DYN GEN Involvement in politics | 5.1 | 35.6 | xx | 50.8 | xx | xx | 30.5 |
| Positive | No welfare mentality, acceptancy of the Calvinist work ethics | 71.2 | 23.7 | xx | 6.8 | xx | xx | 33.9 |
| Positive | DYN GEN The law-abiding society | 5.1 | 89.8 | xx | 13.6 | xx | xx | 36.2 |
| Positive | Social Globalization, overall index | 35.6 | 35.1 | 39.6 | 32.2 | 41.6 | 36.1 | 36.7 |
| Positive | Expected years of schooling | 25.0 | 44.1 | 34.0 | 66.0 | 9.0 | 45.2 | 37.2 |
| Positive | DYN GEN No welfare mentality, acceptancy of the Calvinist work ethics | 18.6 | 94.9 | xx | 6.8 | xx | xx | 40.1 |
| Positive | Optimism and engagement | 100.0 | 22.0 | xx | 1.7 | xx | xx | 41.2 |
| Positive | DYN GEN Climate of personal non-violence | 5.1 | 79.7 | xx | 44.1 | xx | xx | 42.9 |
| Positive | The law-abiding society | 39.0 | 72.9 | xx | 18.6 | xx | xx | 43.5 |
| Negative | Low Global terrorism index | 82.1 | 24.5 | 23.8 | 23.8 | 74.2 | 37.1 | 44.3 |
| Positive | Mean years of schooling | 45.2 | 65.4 | 41.5 | 44.1 | 43.6 | 28.2 | 44.7 |
| Positive | DYN GEN Overall Civil Society Index | 16.9 | 100.0 | xx | 23.7 | xx | xx | 46.9 |
| Positive | DYN GEN Democracy movement | 10.2 | 52.5 | xx | 83.1 | xx | xx | 48.6 |
| Positive | DYN GEN Trust in institutions | 52.5 | 76.3 | xx | 27.1 | xx | xx | 52.0 |
| Positive | DYN GEN Feminism | 83.1 | 32.2 | xx | 49.2 | xx | xx | 54.8 |
| Positive | DYN HDI 2010-2018 | 46.5 | 83.4 | 47.6 | 85.6 | 38.5 | 41.2 | 57.1 |
| Positive | Overall Civil Society Index | 91.5 | 69.5 | xx | 15.3 | xx | xx | 58.8 |
| Positive | Social Globalization, de jure index | 58.8 | 59.8 | 56.8 | 63.8 | 65.8 | 66.8 | 62.0 |
| Positive | Accepting the market economy | 100.0 | 40.7 | xx | 50.8 | xx | xx | 63.8 |
| Positive | Effective Democracy Index | 61.2 | 44.8 | 67.3 | 72.7 | 95.8 | 66.7 | 68.1 |
| Positive | No redistributive religious fundamentalism | 33.9 | 78.0 | xx | 93.2 | xx | xx | 68.4 |
| Positive | Climate of personal non-violence | 69.5 | 71.2 | xx | 67.8 | xx | xx | 69.5 |
| Positive | DYN HDI 1990-2018 | 78.3 | 81.1 | xx | 89.5 | 43.4 | 55.2 | 69.5 |
| Positive | DYN GEN Optimism and engagement | 96.6 | 81.4 | xx | 33.9 | xx | xx | 70.6 |
| Negative | Low GSN Physical integrity | xx | 68.0 | xx | 80.0 | xx | xx | 74.0 |
| Negative | Low GSN Educational | xx | 82.7 | xx | 66.7 | xx | xx | 74.7 |
| Negative | Low Fatigue of democracy (democracy may have problems but is not better) | xx | xx | xx | xx | 81.8 | xx | 81.8 |
| Negative | Low GSN Economic | xx | 77.3 | xx | 88.0 | xx | xx | 82.7 |
| Negative | Low richest 1% | 74.3 | 81.4 | 78.6 | 98.6 | 80.0 | 91.4 | 84.0 |
| Negative | Low GSNI | xx | 72.0 | xx | 97.3 | xx | xx | 84.7 |
| Positive | Share of people with no bias | xx | 73.3 | xx | 97.3 | xx | xx | 85.3 |
| Positive | DYN GEN No redistributive religious fundamentalism | 89.8 | 72.9 | xx | 100.0 | xx | xx | 87.6 |
| Negative | Low GSNI2 | xx | 82.7 | xx | 94.7 | xx | xx | 88.7 |
| Positive | Feminism | 96.6 | 100.0 | xx | 76.3 | xx | xx | 91.0 |
| Positive | Democracy movement | 100.0 | 86.4 | xx | 93.2 | xx | xx | 93.2 |
| Negative | Low GSN Political | xx | 93.3 | xx | 98.7 | xx | xx | 96.0 |

**Table 9.3** Potential for radical change in the Arab world according to the *World Values Survey*

| | Society must be radically changed (%) | *N* |
|---|---|---|
| Viet Nam | 53.80 | 909 |
| Nigeria | 29.90 | 4758 |
| Tanzania | 27.00 | 1111 |
| Czech Rep. | 23.90 | 1957 |
| Azerbaijan | 23.50 | 1891 |
| **Saudi Arabia** | **23.40** | **1374** |
| Pakistan | 21.80 | 2488 |
| Philippines | 21.70 | 2360 |
| Turkey | 20.80 | 5732 |
| Moldova | 18.40 | 1773 |
| South Africa | 17.00 | 9086 |
| Brazil | 16.30 | 1713 |
| Armenia | 16.00 | 1874 |
| Poland | 15.30 | 1880 |
| Kyrgyzstan | 15.10 | 1028 |
| India | 14.40 | 5619 |
| Mexico | 14.40 | 6298 |
| Russia | 13.90 | 3526 |
| Macedonia | 13.80 | 1881 |
| Slovakia | 13.80 | 1483 |
| Dominican Rep. | 12.60 | 406 |
| Zimbabwe | 12.60 | 934 |
| Bosnia | 12.20 | 1052 |
| South Korea | 12.10 | 4453 |
| Bangladesh | 12.00 | 2803 |
| Bosnia | 12.00 | 769 |
| Venezuela | 12.00 | 2291 |
| Albania | 11.10 | 1688 |
| Belarus | 11.00 | 2893 |
| Latvia | 10.90 | 1157 |
| Serbia | 10.80 | 2242 |
| Georgia | 9.80 | 1921 |
| Lithuania | 9.70 | 877 |
| Jordan | 9.50 | 1106 |
| Uganda | 9.40 | 980 |
| Ukraine | 8.70 | 2447 |
| Uruguay | 8.50 | 926 |
| Montenegro | 8.30 | 300 |
| Romania | 7.80 | 1102 |
| Algeria | 7.30 | 1219 |
| Colombia | 7.10 | 2941 |
| Slovenia | 7.10 | 916 |
| Argentina | 7.00 | 4031 |

(continued)

**Table 9.3** (continued)

| | | |
|---|---|---|
| Canada | 7.00 | 1782 |
| Peru | 6.90 | 2549 |
| Bulgaria | 6.60 | 879 |
| United States | 6.60 | 2546 |
| Switzerland | 6.50 | 1110 |
| Chile | 6.40 | 3412 |
| El Salvador | 6.20 | 1077 |
| Australia | 5.40 | 3121 |
| Indonesia | 5.30 | 885 |
| China | 5.20 | 958 |
| Egypt | 5.20 | 2919 |
| New Zealand | 5.10 | 953 |
| Morocco | 5.00 | 1148 |
| Croatia | 4.80 | 1107 |
| Spain | 4.50 | 3550 |
| Puerto Rico | 4.40 | 1798 |
| Sweden | 4.40 | 909 |
| Germany | 4.20 | 1976 |
| Japan | 3.50 | 3247 |
| Taiwan | 3.00 | 722 |
| Estonia | 2.90 | 968 |
| Norway | 2.70 | 1105 |
| Hungary | 2.00 | 594 |
| Finland | 1.80 | 938 |

## References


Arab Opinion Index. (2020). Available at https://www.dohainstitute.org/en/Lists/ACRPS-PDFDocumentLibrary/Arab-Opinion-Index-2019-2020-Inbreef-English-Version.pdf

Arrighi, G. (1995). *The long 20th century. Money, power, and the origins of our times*. Verso.

Arrighi, G. (2007). *Adam Smith in Beijing: Lineages of the twenty-first century*. Verso.

Barro, R. J., & Ursúa, J. F. (2011). *Barro-Ursúa macroeconomic data*. Available at: https://scholar.harvard.edu/barro/publications/barro-ursua-macroeconomic-data

Behr, H., & Berger, L. (2009). The challenge of talking about terrorism: The EU and the Arab debate on the causes of Islamist terrorism. *Terrorism and Political Violence, 21*(4), 539–557.

Berger, L. (2019). Democratic values and the micro foundations of Arab support for peace with Israel. *Conflict Management and Peace Science, 36*(3), 270–290.

Frank, A. G. (1998). *ReOrient: Global economy in the Asian age*. University of California Press.

Goldstein, J. S. (1985). Kondratieff waves as war cycles. *International Studies Quarterly, 29*(4), 411–444.

Goldstein, J. S. (1988). *Long cycles: Prosperity and war in the modern age*. Yale University Press.

Grinin, L., Korotayev, & Tausch, A. (2016). *Economic cycles, crises, and the global periphery*. Springer.

Grinin, L., Korotayev, A. V., & Tausch, A. (2018). Islamism, Arab Spring, and the future of democracy: World system and world values perspectives Ser. In *Perspectives on development in the Middle East and North Africa* (Mena region). Springer.

Guzansky, Y., & Lindenstrauss, G. (2020). *A dysfunctional family: Disagreements and clashes in the sunni world*. INSS Insight No. 1330, June 4, 2020. Available at: https://www.jstor.org/stable/pdf/resrep25521.pdf

Guzansky, Y., & Segal, E. (2020). All in the family: Leadership changes in the Gulf. INSS Insight No. 1378, August 30, 2020. Available at: https://www.INSS.org.il/publication/gulf-royal-families/?offset=1&posts=201&outher=Yoel%20Guzansky

Guzansky, Y., & Winter, O. (2020). *Apolitical normalization: A new approach to jews in Arab States*. INSS Insight No. 1332, June 8, 2020. Available at: https://www.INSS.org.il/publication/judaism-in-the-arab-world/?offset=5&posts=201&outher=Yoel%20Guzansky

INSS (Institute for National Security Studies), Tel Aviv=. (2020). *Strategic survey for Israel 2019–2020*. Available at https://www.INSS.org.il/publication/strategic-survey-for-israel-2019-2020/

Solomon, H., & Tausch, A. (2020). *Islamism, crisis and democratization: Implications of the World Values Survey for the Muslim world (Perspectives on development in the Middle East and North Africa (Mena region)*. Springer.

Solomon, H., & Tausch, A. (2021). *Arab MENA countries: Vulnerabilities and constraints against democracy on the eve of the global Covid-19 crisis. Perspectives on development in the Middle East and North Africa (Mena region)*. Springer.

Tausch, A. (2019). Migration from the Muslim world to the West: Its most recent trends and effects. *Jewish Political Studies Review, 30*(1–2), 65–225. Available at http://jcpa.org/article/migration-from-the-muslim-world-to-the-west-its-most-recent-trends-and-effects/ (with data definitions and sources. Free data download available from https://www.academia.edu/37568941/Migration_from_the_Muslim_World_to_the_West_Its_Most_Recent_Trends_and_Effects

United Nations Development Programme (UNDP). (2019). *Human development report*. Oxford University Press. Available at http://www.hdr.undp.org/

World Bank. (2021). *Global economic prospects*. A World Bank Flagship Report. World Bank. Available at https://www.worldbank.org/en/publication/global-economic-prospects

# Appendix A
# Country Profiles

## Sources


Cordesman A. H. (2020). *The Greater Middle East: From the "Arab Spring" to the "Axis of failed States."* Center for Strategic and International Studies. Available at https://www.csis.org/analysis/greater-middle-east-arab-spring-axis-failed-states

European Commission, European Union External Action. (2020). *EU in the World. Countries*. Available at https://eeas.europa.eu/headquarters/headquarters-homepage/area/geo_en

Federal Foreign Office, Federal Republic of Germany. (2020). *Aussenpolitik. Laender*. Available at https://www.auswaertiges-amt.de/de/aussenpolitik/laender

Freedom House. (2020). *Countries and territories*. Available at https://freedomhouse.org/countries/freedom-world/scores

IMF. (2020). *World economic outlook database*. Available at https://www.imf.org/external/pubs/ft/weo/2020/01/weodata/index.aspx.

SCIMAGO SIR. (2020). *Scimago institutions rankings*. Available at https://www.scimagoir.com/rankings.php

UNDP. (2020). *Global human development indicators*. Available at http://hdr.undp.org/en/countries

United States Central Intelligence Agency. (2021). *The world factbook*. Available at https://www.cia.gov/the-world-factbook/

United States Central Intelligence Agency (2021). *World leaders and cabinet members of Foreign Governments*. Available at https://www.cia.gov/resources/world-leaders/





A. Tausch, *The Future of the Gulf Region*, Gulf Studies 2,
https://doi.org/10.1007/978-3-030-78299-3

United States Department of State. (2020). *Country reports on human rights practices*. Available at https://www.state.gov/reports/2019-country-reports-on-human-rights-practices/

World Bank. (2020). *MENA economic update*. Available at https://www.worldbank.org/en/region/mena/publication/mena-economic-update-april-2020-how-transparency-can-help-the-middle-east-and-north-africa

## Bahrain

### *Political Background—Federal Foreign Office, Federal Republic of Germany, United States Department of State, and European Commission*

Country name: Kingdom of Bahrain, Mamlakat Al-Bahrayn

Form of government: since 2002 constitutional monarchy

Head of State: King Hamad bin Isa Al-Khalifa, Head of State since 1999, King since 2002

Head of Government: Prime Minister: Salman bin Hamad Al Khalifa, since November 11, 2020

Foreign Minister: Abdul Latif bin Rashid Al Zayani, since 2020

Human Development Index: Rank 45

The Kingdom of Bahrain is an archipelago consisting of 33 islands off the coast of Saudi Arabia. With a size of 760 $km^2$ and 1.5 million inhabitants (slightly over 50% foreigners), it is the smallest member state in the Gulf Cooperation Council (GCC). Relations with Saudi Arabia and the other GCC members are close. In June 2017, Bahrain (with Saudi Arabia, United Arab Emirates, and Egypt) broke off diplomatic relations with Qatar due to regional political differences.

The free exercise of religion is very important in Bahrain: Up to 70% of the local population are Shiites, and there is a small Jewish community and 18 Christian communities. The Sunni Al-Khalifa family has ruled Bahrain since 1783. Since 2002, Bahrain has had a two-chamber system with an elected House of Representatives and an advisory assembly appointed by the king. The Shiite majority population is only represented by 14 out of 40 MPs. The main opposition associations (Al-Wefaq and Al-Waad) were dissolved in 2016/2017. Freedom of the press and freedom of expression as well as public demonstrations are severely restricted.

King Hamad bin Isa Al-Khalifa, the head of state, appoints the cabinet, consisting of 24 ministers; 12 of the ministers were members of the Al-Khalifa ruling family. The king, who holds ultimate authority over most government decisions, also appoints the prime minister–the head of government–who does not have to be a member of parliament. There is also an appointed Upper House of Parliament, the Shura (Consultative) Council, with 40 seats. The country holds parliamentary elections every four years, and according to the government, 67% of eligible voters participated in the most recent elections, held in November 2018. The government did not permit international election monitors. Domestic monitors generally concluded that the authorities administered the elections without significant procedural irregularities.

The Ministry of Interior is responsible for internal security and controls the public security force and the specialized security units responsible for maintaining internal order. The Coast Guard is also under its jurisdiction. The Bahrain Defense Force is primarily responsible for defending against external threats, while the Bahrain National Guard is responsible for both external and internal threats. Civilian authorities maintained effective control over the security forces.

## *United States Department of State (2020) Country Reports on Human Rights Practices*

Significant human rights issues included: allegations of torture; arbitrary detention; political prisoners; arbitrary or unlawful interference with privacy; restrictions on freedom of expression, the press, and the Internet, including censorship, site blocking, and criminal libel; substantial interference with the rights of peaceful assembly and freedom of association, including restrictions on independent non-governmental organizations (NGOs) keeping them from freely operating in the country; restrictions on freedom of movement, including revocation of citizenship; and restrictions on political participation, including banning former members of the al-Wifaq and Wa'ad political societies from running as candidates in elections.

The government prosecuted low-level security force members accused of human rights abuses, following investigations by government or quasi-governmental institutions. Human rights organizations claimed investigations were slow and lacked transparency.

## *UNDP Data*

Human Development Index

0.838

Global Rank 45

Life expectancy at birth (years)

77.2

Expected years of schooling (years)

15.3

Gross national income (GNI) per capita (2011 PPP $)

40,399

Gender Development Index (GDI)

0.937

Employment to population ratio (% ages 15 and older)

72.1

Homicide rate (per 100,000 people)

0.5

Exports and imports (% of GDP)

159.2

Internet users, total (% of population)

98.6

Carbon dioxide emissions, per capita (tonnes)

20.8

Total population (millions)

1.6

Skilled labor force (% of labor force)

19.3.

## *Overall Assessment for Bahrain with the Data from this Study*

| Indicator direction | Indicator | Global percentile performance Bahrain |
|---|---|---|
| Positive | Involvement in politics | 5.1 |
| Positive | DYN GEN involvement in politics | 5.1 |
| Positive | DYN GEN the law-abiding society | 5.1 |
| Positive | DYN GEN climate of personal non-violence | 5.1 |
| Positive | Economic globalisation, de facto index | 5.3 |
| Positive | Overcoming the fatigue of democracy 2 (it is good to have a democratic political system) | 7.3 |
| Positive | Economic globalisation, overall index | 10.1 |
| Positive | Trust in institutions | 10.2 |
| Positive | DYN GEN happiness, good health | 10.2 |
| Positive | DYN GEN democracy movement | 10.2 |
| Positive | Gross national income (GNI) per capita | 14.4 |
| Positive | Social globalisation, de facto index | 14.6 |
| Positive | DYN GEN accepting the market economy | 15.3 |
| Positive | DYN GEN overall civil society index | 16.9 |
| Positive | DYN GEN no welfare mentality, acceptancy of the Calvinist work ethics | 18.6 |
| Positive | Human development index (HDI) | 23.9 |
| Positive | HDI 2018 | 23.9 |
| Positive | Economic globalisation, de jure index | 24.0 |
| Positive | Expected years of schooling | 25.0 |
| Positive | Life expectancy at birth | 28.2 |
| Positive | No redistributive religious fundamentalism | 33.9 |
| Positive | Social globalisation, overall index | 35.6 |
| Positive | The law-abiding society | 39.0 |
| Positive | Mean years of schooling | 45.2 |
| Positive | DYN HDI 2010–2018 | 46.5 |
| Positive | DYN GEN trust in institutions | 52.5 |
| Positive | Social globalisation, de jure index | 58.8 |
| Positive | Effective democracy Index | 61.2 |
| Positive | Happiness, good health | 62.7 |
| Positive | Climate of personal non-violence | 69.5 |
| Positive | No welfare mentality, acceptancy of the Calvinist work ethics | 71.2 |
| Negative | Low richest 1% | 74.3 |
| Positive | DYN HDI 1990–2018 | 78.3 |
| Negative | Low global terrorism index | 82.1 |
| Positive | DYN GEN feminism | 83.1 |
| Positive | DYN GEN No redistributive religious fundamentalism | 89.8 |
| Positive | Overall civil society index | 91.5 |

(continued)

(continued)

| | | |
|---|---|---|
| Positive | DYN GEN optimism and engagement | 96.6 |
| Positive | Feminism | 96.6 |
| Positive | Optimism and engagement | 100.0 |
| Positive | Accepting the market economy | 100.0 |
| Positive | Democracy movement | 100.0 |

## *World Bank Overall Assessment*

Bahrain overall economic growth is expected to decelerate this year as economic disruption associated with COVID-19 will weaken oil demand and weigh heavily on non-oil activity. The fiscal and external deficits are expected to reverse the narrowing path observed in 2019. Then the budget deficit is projected to only gradually narrow given lower oil revenues and the large off-budget spending. A key source of stress is large interest payments on the external debt, intensifying pressure on reserves. Downside risks arise from the twin crises for the Gulf of continued plunge in oil prices and COVID-19 outbreak.

## *SCIMAGO—SIR: World Class Universities and Research Centers*

| Institution | Sector | Global Rank |
|---|---|---|
| Arabian Gulf University | Universities | 744 |
| University of Bahrain | Universities | 794 |

## *Assessment by the Center for Strategic and International Studies*

Bahrain has experienced progress in many areas, but its Sunni minority and ruling family are increasingly repressive in dealing with the native Shi'ite majority. Bahrain is also heavily dependent on Saudi aid and foreign workers at expensive of own citizens.

Five Year Prognosis: Bahrain has high potential if it can reduce repression and heal divisions between its Sunni and Shi'ite population.

### *Assessment by Freedom House*

Bahrain

NOT FREE

11

100

Bahrain was once viewed as a promising model for political reform and democratic transition, but it has become one of the Middle East's more repressive states. Since violently crushing a popular prodemocracy protest movement in 2011, the Sunni-led monarchy has systematically eliminated a broad range of political rights and civil liberties, dismantled the political opposition, and cracked down harshly on persistent dissent concentrated among the Shiite population.

## Kuwait

### *Political Background—Federal Foreign Office, Federal Republic of Germany, United States Department of State, and European Commission*

Country name: State of Kuwait, "Daulat Al-Kuwait"

Form of government: hereditary principality (Emirate), monarchy with parliamentary participation (National Assembly ("Majlis Al-Umma") which consists of a chamber with 50 freely elected members

Head of State: S.H. Sheikh Nawaf Al-Ahmad Al-Jaber Al-Sabah, Emir of the State of Kuwait, took office: September 30, 2020

Representative: S.H. Sheikh Mishal Al-Ahmad Al-Jaber Al-Sabah, Deputy Emir and Crown Prince, took office: October 8, 2020

Head of Government: H.E. Sheikh Sabah Al-Khaled Al-Hamad Al-Sabah, Prime Minister, took office on November 19, 2019

Foreign Minister: H.E. Sheikh Dr. Ahmad Nasser Al-Mohammad Al-Sabah, took office: December 16, 2019

National Day: February 25 (immediately followed by Liberation Day: February 26)

Independence: June 19, 1961

Kuwait is a hereditary principality. The system of government is based on the separation of powers and combines elements of a traditional monarchy with the parliamentary form of government.

Kuwait, which is financially strong due to its oil wealth, has crude oil reserves of around 8.5% of the world's deposits, which at the current level of production will last for over 90 years. An extensive welfare state is financed with the income from the oil business. Since 2010, the government has tried to diversify and reduce spending in multiannual economic plans ("New Kuwait").

Kuwait avoids exposing itself in foreign policy and seeks both cooperation with the members of the Gulf Cooperation Council, in the framework of which it often acts as a mediator, and liaison with major powers outside the region (especially the USA).

Freedom of expression and freedom of the press are guaranteed in the constitution. The coverage of the private press is also lively on controversial topics but is subject to restrictions. Criticizing the person of the Emir, the ruling family and insulting religions are prohibited.

While there is also a democratically elected parliament, the Emir holds ultimate authority over most government decisions. The last parliamentary general election was held in 2016 and was generally free and fair with members of the opposition winning seats.

Police have sole responsibility for the enforcement of laws not related to national security, and the Kuwait State Security (KSS) oversees national security matters; both report to the Ministry of Interior, as does the Kuwait Coast Guard. The armed forces are responsible for external security and report to the Ministry of Defense. The Kuwait National Guard is a separate entity responsible for critical infrastructure protection, support for the Ministries of Interior and Defense, and the maintenance of national readiness. Civilian authorities maintained effective control over the security forces.

## *United States Department of State (2020) Country Reports on Human Rights Practices*

Significant human rights issues included: reports of torture; arbitrary detention; political prisoners; arbitrary or unlawful interference with privacy; restrictions on free expression, the press, and the Internet, including criminalization of libel, censorship, and Internet site blocking; interference with the rights of peaceful assembly and freedom of association; restrictions on freedom of movement; trafficking in persons; criminalization of consensual adult male same-sex sexual conduct; and reports of forced labor, principally among foreign workers.

The government took steps in some cases to prosecute and punish officials who committed abuses, whether in the security services or elsewhere in the government. Impunity was a problem in corruption cases.

## *UNDP Data*

Human Development Index

0.808

Global Rank

57

Life expectancy at birth (years)

75.4

Expected years of schooling (years)

13.8

Gross national income (GNI) per capita (2011 PPP $)

71,164

Employment to population ratio (% ages 15 and older).

72.4

Homicide rate (per 100,000 people)

1.8

Exports and imports (% of GDP)

98.4

Internet users, total (% of population)

99.6

Carbon dioxide emissions, per capita (tonnes)

22.8

Total population (millions)

4.1

Skilled labor force (% of labor force)

n.a.

## *Overall Assessment for Kuwait with the Data from this Study*

| Indicator direction | Indicator | Global percentile performance Kuwait |
|---|---|---|
| Positive | Gross national income (GNI) per capita | 2.7 |
| Positive | Economic globalisation, de facto index | 7.9 |
| Positive | Involvement in politics | 8.5 |
| Positive | Happiness, good health | 8.5 |
| Positive | Social globalisation, de facto index | 13.6 |
| Positive | Economic globalisation, overall index | 21.8 |
| Positive | Optimism and engagement | 22.0 |
| Positive | No welfare mentality, acceptancy of the Calvinist work ethics | 23.7 |
| Negative | Low global terrorism index | 24.5 |
| Positive | Human development index (HDI) | 30.3 |
| Positive | HDI 2018 | 30.3 |
| Positive | DYN GEN feminism | 32.2 |
| Positive | Social Globalisation, overall index | 35.1 |
| Positive | DYN GEN involvement in politics | 35.6 |
| Positive | Overcoming the fatigue of democracy 2 (it is good to have a democratic political system) | 36.5 |
| Positive | Trust in institutions | 37.3 |
| Positive | Life expectancy at birth | 40.4 |
| Positive | Accepting the market economy | 40.7 |
| Positive | Economic globalisation, de jure index | 42.5 |
| Positive | Expected years of schooling | 44.1 |
| Positive | Effective democracy Index | 44.8 |
| Positive | DYN GEN happiness, good health | 49.2 |
| Positive | DYN GEN democracy movement | 52.5 |
| Positive | Social globalisation, de jure index | 59.8 |
| Positive | Mean years of schooling | 65.4 |
| Negative | Low GSN physical integrity | 68.0 |
| Positive | Overall civil Society Index | 69.5 |
| Positive | Climate of personal non-violence | 71.2 |
| Negative | Low GSNI | 72.0 |
| Positive | DYN GEN accepting the market economy | 72.9 |
| Positive | The law-abiding society | 72.9 |
| Positive | DYN GEN no redistributive religious fundamentalism | 72.9 |
| Positive | Share of people with no bias | 73.3 |
| Positive | DYN GEN trust in institutions | 76.3 |
| Negative | Low GSN economic | 77.3 |
| Positive | No redistributive religious fundamentalism | 78.0 |
| Positive | DYN GEN climate of personal non-violence | 79.7 |

(continued)

(continued)

| | | |
|---|---|---|
| Positive | DYN HDI 1990–2018 | 81.1 |
| Positive | DYN GEN optimism and engagement | 81.4 |
| Negative | Low richest 1% | 81.4 |
| Negative | Low GSN educational | 82.7 |
| Negative | Low GSNI2 | 82.7 |
| Positive | DYN HDI 2010–2018 | 83.4 |
| Positive | Democracy movement | 86.4 |
| Positive | DYN GEN the law-abiding society | 89.8 |
| Negative | Low GSN political | 93.3 |
| Positive | DYN GEN no welfare mentality, acceptancy of the Calvinist work ethics | 94.9 |
| Positive | DYN GEN overall civil society index | 100.0 |
| Positive | Feminism | 100.0 |

## *World Bank Overall Assessment*

Subdued oil prices and lower oil production led to slower overall economic growth in 2019 in Kuwait, but robust public spending and credit growth are expected to underpin non-oil growth through the medium term. The steep downturn in oil prices since March and slower global growth spurred by the coronavirus will be absorbed by fiscal and financial buffers, at the expense of sustainability and diversification. This underscores the need for implementing fiscal and structural reforms to diversify away from hydrocarbons, support private sector activity, and lay the foundations for a more sustainable growth model.

## *SCIMAGOSIR: World Class Universities and Research Centers*

| Institution | Sector | Global Rank |
|---|---|---|
| Kuwait University Health Science Center | Health | 701 |
| Kuwait University | Universities | 717 |
| Kuwait Institute for Scientific Research | Government | 800 |
| College of Technological Studies | Universities | 821 |

(continued)

(continued)

| Institution | Sector | Global Rank |
|---|---|---|
| American University of the Middle East | Universities | 824 |
| Kuwait Oil Company | Government | 885 |

## *Assessment by the Center for Strategic and International Studies*

The feuding between Qatar and Saudi Arabia, the UAE, and Bahrain has seriously undermined the already limited capabilities of the Gulf Cooperation Council, and left Kuwait more exposed to Iran. Kuwait has a Popular Assembly, but its performance has been unstable, the opposition is divided, and popular governance has limited impact. Some divisions in the royal family, such as illness and the age of ruler, raise some succession issues. There is little repression of the native population, but there is poor treatment of foreign workers.

Five Year Prognosis: Kuwait's mid-term potential is good if its royal family does not divide and the global petroleum market has at least a modest recovery.

## *Assessment by Freedom House*

Kuwait

PARTLY FREE

36

100

Kuwait is a constitutional emirate ruled by the Sabah family. While the monarchy holds executive power and dominates most state institutions, the elected parliament plays an influential role, often challenging the government. Partly due to friction between lawmakers and the executive, government reshuffles and snap parliamentary elections have been frequent since 2011. State authorities have constrained freedoms of speech and assembly in recent years.

## Oman

### *Political Background—Federal Foreign Office, Federal Republic of Germany, United States Department of State, and European Commission*

Country name: Sultanate of Oman, Saltanat 'Uman

Form of government: monarchy

Head of State: S.M. Haitham bin Tariq bin Taimur Al Said, since January 11, 2020

Head of Government: S.M. Sultan Haitham; two representatives: S.H. Sayyid Fahd bin Mahmoud Al Said and S.H. Sayyid Asa'ad bin Tariq bin Taimur Al Said

Foreign Minister: formally S.M. Sultan Haitham; Minister responsible: H.E. Yousuf bin Alawi bin Abdullah

Oman was strongly influenced by the almost 50-year rule of Sultan Qabus bin Said Al Said, who died on January 10th, 2020. The new Sultan Haitham bin Tariq Al Said has announced that he will maintain the political lines of his predecessor. Under Sultan Qaboos, the state and society managed the balancing act between tradition and modernity in domestic politics. In foreign policy, he pursued a policy of moderation, balance, and dialogue. Oman is not militarily involved in any of the regional conflicts and is a member of the Gulf Cooperation Council and the Arab League.

The constitutional order ("Basic Statute of the State") issued in 1996 affirms the position of the Sultan and defines the rights and obligations of citizens. The Sultan rules as head of state, head of government, foreign, finance and defense minister over the 4.66 million inhabitants (56% Omanis and 44% foreigners). He appoints the Council of Ministers, enacts laws, and approves agreements.

There is a parliament-like assembly with two chambers, which acts in an advisory capacity according to the "shura" principle without popular sovereignty. All citizens over the age of 21 are entitled to vote. There are no political parties.

The bicameral Majlis Oman (parliament) can draft laws on nonsecurity-related matters, and citizens may provide input through their elected representatives. The Majlis Oman is composed of the Majlis al-Dawla (upper house or State Council), whose 85 members are appointed by the Sultan, and the elected 86-member Majlis al-Shura (lower house or Consultative Assembly). In October 2019, nearly 350,000 citizens participated in the Majlis al-Shura elections for the Consultative Assembly; there were no notable claims of improper government interference.

The Royal Office controls internal and external security and coordinates all intelligence and security policies. Under the Royal Office, the Internal Security Service investigates all matters related to domestic security. The Royal Oman Police (ROP), including the ROP Coast Guard, is also subordinate to the Royal Office and performs regular police duties. The Royal Office and Royal Diwan–the sultan's personal offices–maintained effective control over the security forces.

## *United States Department of State (2020) Country Reports on Human Rights Practices*

Significant human rights issues included: allegations of torture of prisoners and detainees in government custody; restrictions on free expression, the press, and the Internet, including censorship, site blocking, and criminal libel; substantial interference with the rights of peaceful assembly and freedom of association; required exit permits for foreign workers; restrictions on political participation; and criminalization of lesbian, gay, bisexual, transgender, and intersex (LGBTI) conduct.

Authorities generally held security personnel and other government officials accountable for their actions. The government acted against corruption during the year, with cases proceeding through the court system.

## *UNDP Data*

Human Development Index

0.808

Global Rank

57

Life expectancy at birth (years)

75.4

Expected years of schooling (years)

13.8

Gross national income (GNI) per capita (2011 PPP $)

71,164

Employment to population ratio (% ages 15 and older)

72.4

Homicide rate (per 100,000 people)

1.8

Exports and imports (% of GDP)

98.4

Internet users, total (% of population)

99.6

Carbon dioxide emissions, per capita (tonnes)

22.8

Total population (millions)

4.1

Skilled labor force (% of labor force)

n.a.

### *Overall Assessment for Oman with the Data from this Study*

| Indicator direction | Indicator | Global percentile performance Oman |
|---|---|---|
| Positive | Gross national income (GNI) per capita | 15.4 |
| Negative | Low global terrorism index | 23.8 |
| Positive | Human development index (HDI) | 25.0 |
| Positive | HDI 2018 | 25.0 |
| Positive | Social globalisation, de facto index | 25.1 |
| Positive | Life expectancy at birth | 26.1 |
| Positive | Economic globalisation, de jure index | 30.2 |
| Positive | Expected years of schooling | 34.0 |
| Positive | Economic globalisation, overall index | 38.8 |
| Positive | Social globalisation, overall index | 39.6 |
| Positive | Mean years of schooling | 41.5 |
| Positive | DYN HDI 2010–2018 | 47.6 |
| Positive | Economic globalisation, de facto index | 52.6 |
| Positive | Social globalisation, de jure index | 56.8 |
| Positive | Effective democracy index | 67.3 |
| Negative | Low richest 1% | 78.6 |

### *World Bank Overall Assessment*

Oman's economy is expected to show a marked slowdown in 2020 due to the oil price slide and concerns related to the coronavirus outbreak. An increase in gas output and infrastructure spending plans in the non-oil sector will help growth to recover over 2021–2022. Fiscal and external deficits will remain under immense strain due to low oil and gas prices. Rigid recurrent spending will keep public debt high at an estimated 60% of GDP in 2020 and to increase further in the years to come. Key risks to the outlook arise if the oil price crash is long-lived, which would include higher domestic and external borrowing and result in further sovereign rating downgrades and higher financing costs.

## *Assessment by the Center for Strategic and International Studies*

Summary: Oman has experienced slow but steady progress in development, political reform, and meeting popular needs over time. It is less repressive than most Gulf governments and less corrupt. Oman is overdependent on oil and gas resources, which can generate between and 68 and 85% of government revenue, and its diversification efforts have so far had limited impact. Oman has distanced itself from Saudi Arabia and UAE in the Gulf Cooperation, over the boycott of Qatar and in support of the war in Yemen. The crisis in world petroleum prices has affected Oman's national budget, and its military spending consumes an exceptionally high percentage of GNP. Oman does actively seek to avoid confrontation with Iran.

Five Year Prognosis: Oman's potential is uncertain, and the new Sultan needs time to show his ability to lead the government

## *Assessment by Freedom House*

Oman

NOT FREE

23

100

Oman is a hereditary monarchy, and power is concentrated in the hands of the sultan. The regime restricts virtually all political rights and civil liberties, imposing criminal penalties for criticism and dissent.

# Qatar

## *Political Background—Federal Foreign Office, Federal Republic of Germany, United States Department of State, and European Commission*

Country name: State of Qatar, State of Qatar, Dawlat Qatar (Arabic)

Form of government: Absolute monarchy with Consultative Assembly (Majlis al-Shura, members are appointed by the Emir)

Head of State and Prime Minister: Emir Sheikh Tamim bin Hamad Al-Thani (since June 26, 2013)

Prime Minister and Interior Minister: Sheikh Abdullah bin Nasser bin Khalifa Al-Thani (since June 26, 2013).

Deputy Prime Minister (since 11.11.2017) and Foreign Minister (since 27.01.2016): Sheikh Mohammed bin Abdulrahman bin Jassim Al-Thani

Deputy Prime Minister (since 11/11/2017) and Minister of State for Defense: Dr. Khaled bin Mohamed al-Attiyah (since January 27, 2016)

Qatar is a hereditary monarchy. The ruling Emir is striving to build a modern Islamic state of a liberal character with a cautious social modernization course. As the state religion, Islam shapes social life.

The constitution, which came into force in 2005, provides for a Consultative Assembly, two-thirds of which are to be elected and one-third to be appointed by the Emir. So far, however, all members of this body have been appointed. So far, elections have only taken place at the local level.

There are no political parties and union activities are subject to strict regulations. Freedom of expression and freedom of the press are guaranteed in the constitution, but Qatari media practice self-censorship.

Emir Sheikh Tamim bin Hamad Al Thani exercises full executive power. The constitution provides for hereditary rule by men in the Emir's branch of the Al Thani family. The most recent elections were in 2019 for the Central Municipal Council, an advisory and consultative body. Observers considered these elections free and fair. All cabinet members are appointed by the Emir, including the prime minister.

The national police and Ministry of Interior forces maintain internal security that addresses, among other matters, terrorism, cyberattacks, and espionage. The national police oversee general law enforcement. The army is responsible for external security. Civilian authorities maintained effective control over security forces.

## *United States Department of State (2020) Country Reports on Human Rights Practices*

Significant human rights issues included: criminalization of libel; restrictions on peaceful assembly and freedom of association, including prohibitions on political parties and labor unions; restrictions on the freedom of movement for migrant workers' travel abroad; refusal to grant asylum despite risk of arrest and torture; limits on the ability of citizens to choose their government in free and fair elections; criminalization of consensual same-sex sexual activity; and reports of forced labor.

The government took limited steps to prosecute those suspected of committing human rights abuses. The government took steps to address forced labor.

## *UNDP Data*

Human Development Index

0.848

Global Rank

41

Life expectancy at birth (years)

80.1

Expected years of schooling (years)

12.2

Gross national income (GNI) per capita (2011 PPP $)

110,489

Employment to population ratio (% ages 15 and older)

86.8

Homicide rate (per 100,000 people)

0.4

Exports and imports (% of GDP)

88.3

Internet users, total (% of population)

99.7

Carbon dioxide emissions, per capita (tonnes)

29.8

Total population (millions)

2.8

Skilled labor force (% of labor force)

43.9.

## *Overall Assessment for Qatar with the Data from this Study*

| Indicator direction | Indicator | Global percentile performance Qatar |
|---|---|---|
| Positive | Gross national income (GNI) per capita | 0.5 |
| Positive | Happiness, good health | 1.7 |
| Positive | DYN GEN accepting the market economy | 1.7 |
| Positive | Optimism and engagement | 1.7 |
| Positive | Trust in institutions | 3.4 |
| Positive | No welfare mentality, acceptancy of the Calvinist work ethics | 6.8 |
| Positive | DYN GEN No welfare mentality, acceptancy of the Calvinist work ethics | 6.8 |
| Positive | Social globalisation, de facto index | 8.0 |
| Positive | Involvement in politics | 11.9 |
| Positive | DYN GEN the law-abiding society | 13.6 |
| Positive | Economic globalisation, de facto index | 13.7 |
| Positive | Overall civil society index | 15.3 |
| Positive | Life expectancy at birth | 17.0 |
| Positive | Economic globalisation, overall index | 18.1 |
| Positive | DYN GEN happiness, good health | 18.6 |
| Positive | The law-abiding society | 18.6 |
| Positive | Human development index (HDI) | 21.8 |
| Positive | HDI 2018 | 21.8 |
| Positive | DYN GEN overall civil society index | 23.7 |
| Negative | Low global terrorism index | 23.8 |
| Positive | Economic globalisation, de jure index | 25.7 |
| Positive | DYN GEN trust in institutions | 27.1 |
| Positive | Social globalisation, overall index | 32.2 |
| Positive | DYN GEN optimism and engagement | 33.9 |
| Positive | DYN GEN climate of personal non-violence | 44.1 |
| Positive | Mean years of schooling | 44.1 |
| Positive | DYN GEN feminism | 49.2 |
| Positive | DYN GEN involvement in politics | 50.8 |
| Positive | Accepting the market economy | 50.8 |
| Positive | Social globalisation, de jure index | 63.8 |
| Positive | Expected years of schooling | 66.0 |
| Negative | Low GSN educational | 66.7 |
| Positive | Climate of personal non-violence | 67.8 |
| Positive | Effective democracy index | 72.7 |
| Positive | Feminism | 76.3 |
| Negative | Low GSN physical integrity | 80.0 |
| Positive | DYN GEN democracy movement | 83.1 |
| Positive | DYN HDI 2010–2018 | 85.6 |

(continued)

(continued)

| | | |
|---|---|---|
| Negative | Low GSN economic | 88.0 |
| Positive | DYN HDI 1990–2018 | 89.5 |
| Positive | No redistributive religious fundamentalism | 93.2 |
| Positive | Democracy movement | 93.2 |
| Negative | Low GSNI2 | 94.7 |
| Negative | Low GSNI | 97.3 |
| Positive | Share of people with no bias | 97.3 |
| Negative | Low richest 1% | 98.6 |
| Negative | Low GSN political | 98.7 |
| Positive | DYN GEN no redistributive religious fundamentalism | 100.0 |

## *World Bank Overall Assessment*

Growth had been rebalancing as FIFA-related infrastructure investment tapered while other large projects in the gas and non-hydrocarbon sectors are underway. Fiscal and external balances had returned to surplus but remain highly vulnerable to volatility in hydrocarbon prices. Expanding liquified natural gas (LNG) capacity is expected to underpin growth in the medium term. Drags on growth include energy price volatility, COVID-19 disruptions, and the continued diplomatic rift with some Gulf neighbors. The diplomatic rift vindicated the infrastructure-focused diversification strategy, but COVID-19 has worsened under-utilization of capacity.

## *SCIMAGO—SIR: World Class Universities and Research Centers*

| Institution | Sector | Global Rank |
|---|---|---|
| Weill Cornell Medical College in Qatar | Universities | 586 |
| University of Qatar | Universities | 629 |
| Hamad Medical Corp | Health | 663 |
| Texas A&M University at Qatar | Universities | 686 |
| Hamad Bin Khalifa University | Universities | 711 |

### *Assessment by the Center for Strategic and International Studies*

Summary: Qatar has experienced significant progress in many areas, such as the rule of law, low levels of repression and security measures for native population, and some elements of political and governance reform, particularly in the treatment of foreign workers. Qatar has, however, overspent on infrastructure and showpiece projects, but its gas and oil exports protect the economy and deficits are not critical. Tensions with Saudi Arabia, UAE, and other neighbors. Qatari tolerance of media attacks on Saudi Arabia as well as its covert role in conflicts and civil tensions of other states has produced little success and has rather been largely disruptive.

Five Year Prognosis: Qatar's potential is good if it can reconcile with neighbors and avoid clashes with Iran.

### *Assessment by Freedom House*

Qatar

NOT FREE

25

100

Qatar's hereditary Emir holds all executive and legislative authority, and ultimately controls the judiciary as well. Political parties are not permitted, and the only elections are for an advisory municipal council. While Qatari citizens are among the wealthiest in the world, the vast majority of the population consists of noncitizens with no political rights, few civil liberties, and limited access to economic opportunity.

## Saudi Arabia

### *Political Background—Federal Foreign Office, Federal Republic of Germany, United States Department of State, and European Commission*

Country name: Kingdom of Saudi Arabia; al-Mamlaka al-Arabiyya al-Saudiyya

Climate: predominantly arid desert climate, high humidity near the sea, little mountain forest in the extreme southwest

Location: Arabian Peninsula

Size of the country: 2.15 million km (around 80% of the Arabian Peninsula)

Capital: Riyadh (approx. 7 million inhabitants)

Population: around 33 million inhabitants, of which around 11 million are foreigners, especially from Arab countries (especially Egypt, Jordan, Sudan) and from Asia (especially Pakistan, India, Philippines, Indonesia)

National language: Arabic, English is also used as a business language

Religion: Islam (Sunni Islam state religion); Shiite minority mainly in the Eastern Province

National Day: September 23

Independence: September 23, 1932 (Proclamation of the Kingdom)

Head of State and Prime Minister: King Salman bin Abdulaziz Al Saud (Title: "Guardian of the two Holy Places and King of Saudi Arabia")

Crown Prince, Deputy Chairman of the Council of Ministers and Minister of Defense: Muhammad bin Salman bin Abdulaziz Al Saud

Foreign Minister: Dr. Ibrahim bin Abdulaziz bin Abdullah Al-Assaf

Parliament: no elected parliament, but a Consultative Assembly ("Majlis al-Shura") with 150 members appointed by the king, fixed quota of women of at least 20%

Governing party: none (there are no parties)

Unions: None

Administrative structure: 13 provinces, divided into a total of 118 administrative districts. At the head of the regional administration are Emirs with the rank of a minister, who are appointed by the king and all belong to the royal family.

Membership in international organizations: United Nations and special organizations, WTO (World Trade Organization), OPEC, OAPEC (Organization of Arab Petroleum Exporting Countries), Arab League, Organization for Islamic Cooperation (OIC; seat in Jeddah), Gulf Cooperation Council (GCC; General Secretariat in Riyadh)). The influential non-governmental organization "Muslim World League" (MWL) is based in Mecca.

Main media: Al-Arabiya (from Dubai, in Saudi ownership), private entertainment channels of the Saudi media group "MBC," state radio and television (news channel "Al-Ikhbariya"); Daily newspapers: "Al-Hayat," "Ash-sharq Al Awsat" (both with headquarters abroad, but in Saudi ownership), "Okaz," "Al-Riyadh," "Al-Watan," "Arab News," and "Saudi Gazette"(in English).

## Constitutional State Structure

Saudi Arabia is an absolute monarchy. Islam is the state religion. The Koran and the "Sunna" form the basis of the state order. The "Basic Law" of 1992 determines the essential characteristics of the state and society. The head of state is the king. He bears the title "Guardian of the two holy places," Mecca and Medina. Until recently, the succession to the throne was limited to the sons of the state's founder, King Abdulaziz (d. 1953), and was based on the principle of seniority and suitability.

After the death of his half-brother Abdallah on January 23, 2015, the current King Salman bin Abdulaziz Al Saud ascended to the throne. On June 21, 2017, King Salman appointed his son Muhammad bin Salman as the new Crown Prince. The appointment continues the ascent of Muhammad bin Salman in the rulership of Saudi Arabia, who also de facto now determines the guidelines of politics.

The government consists of the Council of Ministers, chaired by the king, who also holds the post of prime minister. Deputy Chairman of the Council of Ministers and Minister of Defense is the Crown Prince. Since December 2018, Dr. Ibrahim Al-Assaf is Foreign Minister.

## Houses of Parliament

There is no elected parliament, but the Consultative Assembly ("Majlis al-Shura" / "Shura Council") has existed since 1992. The king appoints the 150 members for four years each. The Shura Council advises the government, gives its opinion on legislative proposals, and can introduce its own legislative proposals. It does not have budget rights. The Shura Council has had female members since January 2013, and the (mandatory) proportion of women is at least one-fifth of the MPs.

## *United States Department of State (2020) Country Reports on Human Rights Practices*

Significant human rights issues included: unlawful killings; executions for nonviolent offenses; forced disappearances; torture of prisoners and detainees by government agents; arbitrary arrest and detention; political prisoners; arbitrary interference with privacy; criminalization of libel, censorship, and site blocking; restrictions on freedoms of peaceful assembly, association, and movement; severe restrictions of religious freedom; citizens' lack of ability and legal means to choose their government through free and fair elections; trafficking in persons; violence and official discrimination against women, although new women's rights initiatives were implemented; criminalization of consensual same-sex sexual activity; and prohibition of trade unions.

In several cases, the government did not punish officials accused of committing human rights abuses, contributing to an environment of impunity. Following the high-profile October 2018 killing of journalist Jamal Khashoggi in Turkey, a court sentenced five officials to death and three officials to prison on December 23, 2019. The court ruled that guilt could not be established in the case of three other defendants.

In September 2019, state-owned oil processing facilities in Abqaiq and Khurais were attacked by drones and missiles. Houthi militants in Yemen claimed responsibility, but the Saudi government concluded Iran was responsible for the attack. Houthi militants were also responsible for numerous other attacks on civilian infrastructure inside Saudi Arabia, including airports, schools, hospitals, and oil facilities. Saudi Arabia continued air operations in Yemen throughout the year as leader of a coalition formed in 2015 to counter the 2014 Houthi takeover of government institutions and facilities. Saudi-led coalition airstrikes in Yemen reportedly resulted in civilian casualties and damage to infrastructure on multiple occasions. The pace of airstrikes declined in the fall, as the warring parties pursued a negotiated settlement to the conflict. The coalition's Joint Incident Assessment Team (JIAT) investigated allegations of civilian casualties, but the Saudi government did not prosecute any cases based on JIAT findings.

## *UNDP Data*

Human Development Index

0.857

Global Rank

36

Life expectancy at birth (years)

75.0

Expected years of schooling (years)

17.0

Gross national income (GNI) per capita (2011 PPP $)

49,338

Employment to population ratio (% ages 15 and older)

53.2

Homicide rate (per 100,000 people)

1.3

Exports and imports (% of GDP)

66.5

Internet users, total (% of population)

93.3

Carbon dioxide emissions, per capita (tonnes)

16.3

Total population (millions)

33.7

Skilled labor force (% of labor force)

58.6.

### *Overall Assessment for Saudi Arabia with the Data from this Study*

| Indicator direction | Indicator | Global percentile performance Saudi Arabia |
|---|---|---|
| Positive | Gross national income (GNI) per capita | 7.4 |
| Positive | Expected years of schooling | 9.0 |
| Positive | Social globalisation, de facto index | 15.1 |
| Positive | Human development index (HDI) | 19.1 |
| Positive | HDI 2018 | 19.1 |
| Positive | DYN HDI 2010–2018 | 38.5 |
| Positive | Social globalisation, overall index | 41.6 |
| Positive | Life expectancy at birth | 42.6 |
| Positive | DYN HDI 1990-2018 | 43.4 |
| Positive | Mean years of schooling | 43.6 |
| Positive | Economic globalisation, de jure index | 48.6 |
| Positive | Economic globalisation, overall index | 52.7 |
| Positive | Economic globalisation, de facto index | 53.7 |
| Positive | Social globalisation, de jure index | 65.8 |
| Negative | Low global terrorism index | 74.2 |
| Negative | Low richest 1% | 80.0 |
| Negative | Low fatigue of democracy (democracy may have problems but is not better) | 81.8 |
| Positive | Effective democracy index | 95.8 |

## World Bank Overall Assessment

In Saudi Arabia, the contracting oil sector led to sluggish growth in 2019, despite strong performance of non-oil sectors. The outlook for 2020 remains very weak in the wake of COVID-19 and oil supply shocks. Medium-term fiscal balances are estimated to continue in deficit—risking the ability in realizing Vision 2030 fiscal targets. Vision 2030 related reforms are critical for diversification and progress was made on business environment reforms. The ambitious reforms agenda will pose implementation challenges for the public sector.

## SCIMAGO—SIR: World Class Universities and Research Centers

| Institution | Sector | Global Rank |
|---|---|---|
| King Abdullah University of Science and Technology | Universities | 234 |
| King Abdulaziz University | Universities | 287 |
| King Saud University | Universities | 371 |
| King Fahd University of Petroleum and Minerals | Universities | 572 |
| King Abdulaziz City for Science and Technology | Government | 636 |
| King Faisal Specialist Hospital and Research Center | Health | 638 |
| King Abdullah International Medical Research Center | Health | 662 |
| Alfaisal University | Universities | 669 |
| Saudi Arabian Oil Company | Companies | 669 |
| King Abdulaziz Medical City, Riyadh | Health | 682 |
| King Saud bin Abdulaziz University for Health Sciences | Universities | 702 |
| Tabuk University | Universities | 717 |
| Umm Al-Qura University | Universities | 720 |
| Taibah University | Universities | 721 |
| King Fahad Medical City | Health | 733 |
| Prince Sattam Bin Abdulaziz University | Universities | 739 |
| Majmaah University | Universities | 740 |
| King Faisal University | Universities | 742 |
| Najran University | Universities | 742 |
| King Khalid University Hospital | Health | 745 |
| Prince Sultan University | Universities | 747 |
| Imam Abdulrahman Bin Faisal University | Universities | 751 |
| King Khalid University | Universities | 759 |
| Prince Mohammad Bin Fahd University | Universities | 763 |

(continued)

(continued)

| Institution | Sector | Global Rank |
|---|---|---|
| Jazan University | Universities | 763 |
| Shaqra University | Universities | 763 |
| Qassim University | Universities | 773 |
| Effat University | Universities | 774 |
| Taif University | Universities | 775 |
| Princess Nora bint Abdul Rahman University | Universities | 784 |
| University of Jeddah | Universities | 785 |
| University of Hail | Universities | 785 |
| Albaha University | Universities | 788 |
| Aljouf University | Universities | 802 |
| Imam Muhammad ibn Saud Islamic University | Universities | 807 |
| Northern Border University | Universities | 823 |

## *Assessment by the Center for Strategic and International Studies*

Saudi Arabia has made important progress in social reform but not in political reform. It has become more repressive and experienced a decline in the rule of law. Its broad reform goals in its 2030 and other plans set good goals, but its execution is poor, wasteful, and often directed toward grandiose goals and projects of uncertain practical benefit.

Saudi Arabia overspends on military security forces with mixed actual security benefits, but it does face a continuing threat from Iran, terrorism and extremism. Military and security spending is a high percent of its GDP, which puts a serious strain on the economy. Efforts to create new Arab alliances have so far had largely a cosmetic impact, and its role in the boycotts against Qatar has further weakened the Gulf Cooperation Council. Leadership has been compromised by its role in repression, treatment of other members of the royal family, and possible ties to arrests and deaths.

Its role in the Yemeni civil war has so far been an expensive failure.

Five Year Prognosis: Saudi Arabia will face serious problems if it does not limit repression and sectarian divisions, stop feuding with Qatar, and take a more realistic and well managed approach to it military spending, 2030 projections, and other economic reforms.gen.

### *Assessment by Freedom House*

Saudi Arabia

NOT FREE

7

100

Saudi Arabia's absolute monarchy restricts almost all political rights and civil liberties. No officials at the national level are elected. The regime relies on pervasive surveillance, the criminalization of dissent, appeals to sectarianism and ethnicity, and public spending supported by oil revenues to maintain power. Women and religious minorities face extensive discrimination in law and in practice. Working conditions for the large expatriate labor force are often exploitative.

## United Arab Emirates

### *Political Background—Federal Foreign Office, Federal Republic of Germany, United States Department of State, and European Commission*

Country name: United Arab Emirates (UAE), Daulat al-Imaraat al-Arabiyya al-Muttahida

Form of government and state: federal constitutional monarchy (seven Emirates)

Head of State: S.H. President Sheikh Khalifa bin Zayed Al Nahyan (since November 3rd, 2004), elected by the rulers of the seven Emirates and also ruler of the Emirate of Abu Dhabi. Since 2014 his half-brother, S.H. Crown Prince Sheikh Mohammed bin Zayed Al Nahyan, in fact the official business.

Representative of the head of state and head of government: H.H. Sheikh Mohammed bin Rashid Al Maktoum, ruler of the Emirate of Dubai and Minister of Defense (since January 5, 2006).

Foreign Minister: H.H. Sheikh Abdullah bin Zayed Al Nahyan (since February 9, 2006).

The UAE has a federal system of rule in which determining forces continue to be family and tribal ties. At the top are the rulers of the seven semi-autonomous Emirates, who form the Supreme Council as the highest constitutional body and elect the president from among their number.

The federal government's design options are limited. The individual Emirates, especially Abu Dhabi, continue to exert great influence. In the economic field, their autonomy is not affected.

The 40-member National Federal Council is an advisory body. Half of the MPs are appointed by the seven Emirs, the other half are determined in national elections, most recently on October 5, 2019.

On the one hand, public life is characterized by far-reaching liberality for the region; on the other hand, civil rights such as freedom of the press, freedom of expression and assembly or the rights of millions of foreign workers are subject to significant restrictions. Political parties and trade unions are prohibited. There have been no domestic political shocks to date, not least thanks to the generally high level of prosperity.

The Emirates are under patriarchal rule with political allegiance defined by loyalty to tribal leaders, leaders of the individual Emirates, and leaders of the federation. A limited, appointed electorate participates in periodic elections for the partially elected Federal National Council (FNC). As already said above, the consultative body examines, reviews, and recommends changes to legislation and may discuss topics for legislation. Citizens may express their concerns directly to their leaders through traditional consultative mechanisms such as the open *Majlis (forum)*.

Civilian authorities maintained effective control over the security forces. Each Emirate maintained a local police force called a general directorate, which was officially a branch of the federal Ministry of Interior. All Emirate-level general directorates of police enforced their respective Emirate's laws autonomously. They also enforced federal laws within their Emirate in coordination with each other under the federal ministry. The federal government maintained federal armed forces under the Ministry of Defense for external security.

## *United States Department of State (2020) Country Reports on Human Rights Practices*

Significant human rights issues included allegations of torture in detention; arbitrary arrest and detention, including incommunicado detention, by government agents; political prisoners; government interference with privacy rights; undue restrictions on free expression and the press, including criminalization of libel, censorship, and Internet site blocking; substantial interference with the rights of peaceful assembly and freedom of association; the inability of citizens to choose their government in free and fair elections; and criminalization of same-sex sexual activity, although no cases were publicly reported during the year. The government did not permit workers to join independent unions and did not effectively prevent physical and sexual abuse of foreign domestic servants and other migrant workers.

The government investigated, prosecuted, and punished officials who committed abuses. There were no public reports of impunity involving officials, but there was also no publicly available information on whether authorities investigated complaints of police abuses, including prison conditions and mistreatment.

The United Nations, human rights groups, and others alleged UAE military operations as part of the Saudi-led coalition in Yemen killed civilians, damaged civilian infrastructure, and obstructed delivery of humanitarian aid. Human rights groups alleged UAE-backed security forces in Yemen committed torture, sexual assault, and mistreatment against detainees. The government rejected allegations that members of its security forces serving in Yemen had committed human rights abuses, and there was no publicly available information on whether the government carried out any investigations into these reported incidents.

## *UNDP Data*

Human Development Index

0.866

Global Rank

35

Life expectancy at birth (years)

77.8

Expected years of schooling (years)

13.6

Gross national income (GNI) per capita (2011 PPP $)

66,912

Employment to population ratio (% ages 15 and older)

80.9

Homicide rate (per 100,000 people)

0.5

Exports and imports (% of GDP)

161.8

Internet users, total (% of population)

98.5

Carbon dioxide emissions, per capita (tonnes)

20.5

Total population (millions)

9.6

Skilled labor force (% of labor force)

52.8.

## *Overall Assessment for the United Arab Emirates with the Data from this Study*

| Indicator direction | Indicator | Global percentile performance United Arab Emirates |
|---|---|---|
| Positive | Economic globalisation, de facto index | 2.6 |
| Positive | Social globalisation, de facto index | 3.0 |
| Positive | Gross national income (GNI) per capita | 3.7 |
| Positive | Economic globalisation, overall index | 4.8 |
| Positive | Economic globalisation, de jure index | 11.2 |
| Positive | Human development index (HDI) | 18.6 |
| Positive | HDI 2018 | 18.6 |
| Positive | Life expectancy at birth | 25.0 |
| Positive | Mean years of schooling | 28.2 |
| Positive | Social globalisation, overall index | 36.1 |
| Negative | Low global terrorism index | 37.1 |
| Positive | DYN HDI 2010–2018 | 41.2 |
| Positive | Expected years of schooling | 45.2 |
| Positive | DYN HDI 1990–2018 | 55.2 |
| Positive | Effective democracy Index | 66.7 |
| Positive | Social globalisation, de jure index | 66.8 |
| Negative | Low richest 1% | 91.4 |

## *World Bank Overall Assessment*

Emirates overall economic growth stabilized in 2019. In that year, there was a recovery in the hydrocarbon (HC) sector. Fiscal stimulus programs targeting business costs were launched to facilitate an acceleration. In 2020, plunging oil prices and efforts to contain the coronavirus will weigh heavily on the non-HC sector which was already facing challenging debt burdens and intensified traded services competition.

The medium-term outlook depends on a rebound in trade/travel and on structural reforms that can reinvigorate productivity and innovation.

## *SCIMAGO—SIR: World Class Universities and Research Centers*

| Institution | Sector | Global Rank |
|---|---|---|
| Khalifa University | Universities | 606 |
| United Arab Emirates University | Universities | 659 |
| New York University Abu Dhabi | Universities | 687 |
| University of Sharjah | Universities | 780 |
| American University of Sharjah | Universities | 793 |
| Al Ain University | Universities | 808 |
| Zayed University | Universities | 809 |
| American University of Ras Al Khaimah | Universities | 810 |
| Abu Dhabi University | Universities | 816 |
| Higher Colleges of Technology | Universities | 836 |

## *Assessment by the Center for Strategic and International Studies*

The United Arab Emirates has not made major recent progress in political reform and has become more repressive. It has diversified more than most Gulf exporting counties and invested in its poorer northern Emirates, but it has overspent on security and military forces while Dubai's economy has become steadily more speculative. The UAE overspends on military forces with limited security benefits and at levels of GNP that put a strain on its economy. The leadership has been compromised by possible ties to arrests and deaths. Its role in the Yemeni civil war has so far been an expensive failure. The boycott of Qatar has also been counterproductive.

Five Year Prognosis: The UAE will decline in status if it does not limit repression, stop feuding with Qatar, and take a more realistic and well managed approach to military spending and economic reforms.

## *Assessment by Freedom House*

United Arab Emirates

NOT FREE

17

100

The United Arab Emirates (UAE) is a federation of seven emirates led in practice by Abu Dhabi, the largest by area and richest in natural resources. Limited elections are held for a federal advisory body, but political parties are banned, and all executive, legislative, and judicial authority ultimately rests with the seven hereditary rulers. The civil liberties of both citizens and noncitizens, who make up an overwhelming majority of the population, are subject to significant restrictions.

# Appendix B
# Combined Results of Our Study for the Gulf Countries


A. Tausch, *The Future of the Gulf Region*, Gulf Studies 2,
https://doi.org/10.1007/978-3-030-78299-3

| Indicator direction | Indicator | Average Gulf countries—global percentile performance |
|---|---|---|
| Positive | Gross national income (GNI) per capita | 7.4 |
| Positive | Involvement in politics | 8.5 |
| Positive | Social globalisation, de facto index | 13.2 |
| Positive | Trust in institutions | 16.9 |
| Positive | Overcoming the fatigue of democracy 2 (it is good to have a democratic political system) | 21.9 |
| Positive | Economic globalisation, de facto index | 22.6 |
| Positive | Human development index (HDI) | 23.1 |
| Positive | HDI 2018 | 23.1 |
| Positive | Happiness, good health | 24.3 |
| Positive | Economic globalisation, overall index | 24.4 |
| Positive | DYN GEN happiness, good health | 26.0 |
| Positive | Life expectancy at birth | 29.9 |
| Positive | DYN GEN accepting the market economy | 29.9 |
| Positive | Economic globalisation, de jure index | 30.4 |
| Positive | DYN GEN involvement in politics | 30.5 |
| Positive | No welfare mentality, acceptancy of the Calvinist work ethics | 33.9 |
| Positive | DYN GEN the law-abiding society | 36.2 |
| Positive | Social globalisation, overall index | 36.7 |
| Positive | Expected years of schooling | 37.2 |
| Positive | DYN GEN no welfare mentality, acceptancy of the Calvinist work ethics | 40.1 |
| Positive | Optimism and engagement | 41.2 |
| Positive | DYN GEN climate of personal non-violence | 42.9 |
| Positive | The law-abiding society | 43.5 |
| Negative | Low global terrorism index | 44.3 |
| Positive | Mean years of schooling | 44.7 |
| Positive | DYN GEN overall civil society index | 46.9 |
| Positive | DYN GEN democracy movement | 48.6 |
| Positive | DYN GEN trust in institutions | 52.0 |
| Positive | DYN GEN feminism | 54.8 |
| Positive | DYN HDI 2010–2018 | 57.1 |
| Positive | Overall civil society index | 58.8 |
| Positive | Social globalisation, de jure index | 62.0 |
| Positive | Accepting the market economy | 63.8 |
| Positive | Effective democracy index | 68.1 |

(continued)

(continued)

| | | |
|---|---|---|
| Positive | No redistributive religious fundamentalism | 68.4 |
| Positive | Climate of personal non-violence | 69.5 |
| Positive | DYN HDI 1990–2018 | 69.5 |
| Positive | DYN GEN optimism and engagement | 70.6 |
| Negative | Low GSN physical integrity | 74.0 |
| Negative | Low GSN educational | 74.7 |
| Negative | Low Fatigue of democracy (democracy may have problems but is not better) | 81.8 |
| Negative | Low GSN economic | 82.7 |
| Negative | Low richest 1% | 84.0 |
| Negative | Low GSNI | 84.7 |
| Positive | Share of people with no bias | 85.3 |
| Positive | DYN GEN No redistributive religious fundamentalism | 87.6 |
| Negative | Low GSNI2 | 88.7 |
| Positive | Feminism | 91.0 |
| Positive | Democracy movement | 93.2 |
| Negative | Low GSN political | 96.0 |

# Author Index





A. Tausch, *The Future of the Gulf Region*, Gulf Studies 2,
https://doi.org/10.1007/978-3-030-78299-3

**C**



**D**



**E**



**F**

# Subject Index




A. Tausch, *The Future of the Gulf Region*, Gulf Studies 2,
https://doi.org/10.1007/978-3-030-78299-3

**D**



**E**

**P**



**Q**



**R**

## U



## V



## W